Optimization and Dynamical Systems

Communications and Control Engineering Series

Editors: B.W. Dickinson · A. Fettweis · J.L. Massey · J.W. Modestino
E.D. Sontag · M. Thoma

CCES published titles include:

Sampled-Data Control Systems
J. Ackermann

Interactive System Identification
T. Bohlin

The Riccatti Equation
S. Bittanti, A.J. Laub and J.C. Willems (Eds.)

Nonlinear Control Systems
A. Isidori

Analysis and Design of Stream Ciphers
R.A. Rueppel

Sliding Modes in Control Optimization
V.I. Utkin

Fundamentals of Robotics
M. Vukobratović

*Parametrizations in Control, Estimation and Filtering Problems:
Accuracy Aspects*
M. Gevers and G. Li

*Parallel Algorithms for Optimal Control of Large Scale
Linear Systems*
Zoran Gajić and Xuemin Shen

Loop Transfer Recovery: Analysis and Design
A. Saberi, B.M. Chen and P. Sannuti

Markov Chains and Stochastic Stability
S.P. Meyn and R.L. Tweedie

Robust Control: Systems with Uncertain Physical Parameters
J. Ackermann in co-operation with A. Bartlett, D. Kaesbauer, W. Sienel and
R. Steinhauser

Uwe Helmke and John B. Moore

with a foreword by R. Brockett

Optimization and Dynamical Systems

With 33 Figures

Springer-Verlag
London Berlin Heidelberg New York
Paris Tokyo Hong Kong
Barcelona Budapest

Uwe Helmke, PhD
Department of Mathematics, University of Regensburg,
Universitätsstrasse 31, 93053 Regensburg, Germany

John B. Moore, PhD
Department of Systems Engineering, Australian National University,
Canberra ACT 0200, Australia

ISBN 3-540-19857-1 Springer-Verlag Berlin Heidelberg New York
ISBN 0-387-19857-1 Springer-Verlag New York Berlin Heidelberg

British Library Cataloguing in Publication Data
Helmke, Uwe
  Optimization and Dynamical Systems. -
(Communications & Control Engineering
Series)
  I. Title II. Moore, John B. III. Series
515.352
ISBN 3-540-19857-1

Library of Congress Cataloging-in-Publication Data
A catalog record for this book is available from the Library of Congress

Apart from any fair dealing for the purposes of research or private study, or criticism or review, as permitted under the Copyright, Designs and Patents Act 1988, this publication may only be reproduced, stored or transmitted, in any form or by any means, with the prior permission in writing of the publishers, or in the case of reprographic reproduction in accordance with the terms of licences issued by the Copyright Licensing Agency. Enquiries concerning reproduction outside those terms should be sent to the publishers.

© Springer-Verlag London Limited 1994
Printed in Great Britain

The publisher makes no representation, express or implied, with regard to the accuracy of the information contained in this book and cannot accept any legal responsibility or liability for any errors or omissions that may be made.

Typesetting: Camera ready by authors
Printed by Antony Rowe Ltd., Chippenham, Wiltshire
69/3830-543210 Printed on acid-free paper

Foreword

Differential equations have provided the language for expressing many of the important ideas of automatic control. Questions involving optimization, dynamic compensation and stability are effectively expressed in these terms. However, as technological advances in VLSI reduce the cost of computing, we are seeing a greater emphasis on control problems that involve real-time computing. In some cases this means using a digital implementation of a conventional linear controller, but in other cases the additional flexibility that digital implementation allows is being used to create systems that are of a completely different character. These developments have resulted in a need for effective ways to think about systems which use both ordinary dynamical compensation and logical rules to generate feedback signals. Until recently the dynamic response of systems that incorporate if- then rules, branching, etc. has not been studied in a very effective way.

From this latter point of view it is useful to know that there exist families of ordinary differential equations whose flows are such as to generate a sorting of the numerical values of the various components of the initial conditions, solve a linear programming problem, etc. A few years ago, I observed that a natural formulation of a steepest descent algorithm for solving a least-squares matching problem leads to a differential equation for carrying out such operations. During the course of some conversation's with Anthony Bloch it emerged that there is a simplified version of the matching equations, obtained by recasting them as flows on the Lie algebra and then restricting them to a subspace, and that this simplified version can be used to sort lists as well. Bloch observed that the restricted equations are identical to the Toda lattice equations in the form introduced by Hermann Flaschka nearly twenty years ago. In fact, one can see in Jürgen Moser's early paper on the solution of the Toda Lattice, a discussion of sorting couched in the language of scattering theory and presumably one could have developed the subject from this, rather different, starting point. The fact that a sorting flow can be viewed as a gradient flow and that each simple Lie algebra defines a slightly different version of this gradient flow was the subject of a systematic analysis in a paper that involved collaboration with Bloch and Tudor Ratiu.

The differential equations for matching referred to above are actually formulated on a compact matrix Lie group and then rewritten in terms of matrices that evolve in the Lie algebra associated with the group. It is only with respect to a particular metric on a smooth submanifold of this Lie algebra (actually the manifold of all matrices having a given set of eigenvalues) that the final equations appear to be in gradient form. However, as equations on this manifold, the descent equations can be set up so as to find the eigenvalues of the initial condition matrix. This then makes contact with the subject of numerical linear algebra and a whole line of interesting work going back at least to Rutishauser in the mid 1950s and continuing to the present day. In this book Uwe Helmke and John Moore emphasize problems, such as computation of eigenvalues, computation of singular values, construction of balanced realizations, etc. involving more structure than just sorting or solving linear programming problems. This focus gives them a natural vehicle to introduce and interpret the mathematical aspects of the subject. A recent Harvard thesis by Steven Smith contains a detailed discussion of the numerical performance of some algorithms evolving from this point of view.

The circle of ideas discussed in this book have been developed in some other directions as well. Leonid Faybusovich has taken a double bracket equation as the starting point in a general approach to interior point methods for linear programming. Wing Wong and I have applied this type of thinking to the assignment problem, attempting to find compact ways to formulate a gradient flow leading to the solution. Wong has also examined similar methods for nonconvex problems and Jeffrey Kosowsky has compared these methods with other flow methods inspired by statistical physics. In a joint paper with Bloch we have investigated certain partial differential equation models for sorting continuous functions, i.e. generating the monotone equimeasurable rearrangements of functions, and Saveliev has examined a family of partial differential equations of the double bracket type, giving them a cosmological interpretation.

It has been interesting to see how rapidly the literature in this area has grown. The present book comes at a good time, both because it provides a well reasoned introduction to the basic ideas for those who are curious and because it provides a self-contained account of the work on balanced realizations worked out over the last few years by the authors. Although I would not feel comfortable attempting to identify the most promising direction for future work, all indications are that this will continue to be a fruitful area for research.

Roger Brockett

Preface

This work is aimed at mathematics and engineering graduate students and researchers in the areas of optimization, dynamical systems, control systems, signal processing, and linear algebra. The motivation for the results developed here arises from advanced engineering applications and the emergence of highly parallel computing machines for tackling such applications. The problems solved are those of linear algebra and linear systems theory, and include such topics as diagonalizing a symmetric matrix, singular value decomposition, balanced realizations, linear programming, sensitivity minimization, and eigenvalue assignment by feedback control.

The tools are those, not only of linear algebra and systems theory, but also of differential geometry. The problems are solved via *dynamical systems* implementation, either in continuous time or discrete time, which is ideally suited to distributed parallel processing. The problems tackled are indirectly or directly concerned with dynamical systems themselves, so there is feedback in that dynamical systems are used to understand and optimize dynamical systems. One key to the new research results has been the recent discovery of rather deep existence and uniqueness results for the solution of certain matrix least squares *optimization* problems in geometric invariant theory. These problems, as well as many other optimization problems arising in linear algebra and systems theory, do not always admit solutions which can be found by algebraic methods. Even for such problems that do admit solutions via algebraic methods, as for example the classical task of singular value decomposition, there is merit in viewing the task as a certain matrix optimization problem, so as to shift the focus from algebraic methods to geometric methods. It is in this context that gradient flows on manifolds appear as a natural approach to achieve construction methods that complement the existence and uniqueness results of geometric invariant theory.

There has been an attempt to bridge the disciplines of engineering and mathematics in such a way that the work is lucid for engineers and yet suitably rigorous for mathematicians. There is also an attempt to teach, to provide insight, and to make connections, rather than to present the results as a *fait accompli*.

The research for this work has been carried out by the authors while visiting each other's institutions. Some of the work has been written in conjunction with the writing of research papers in collaboration with PhD students Jane Perkins and Robert Mahony, and post doctoral fellow Weiyong Yan. Indeed, the papers benefited from the book and vice versa, and consequently many of the paragraphs are common to both. Uwe Helmke has a background in global analysis with a strong interest in systems theory, and John Moore has a background in control systems and signal processing.

Acknowledgements

This work was partially supported by grants from Boeing Commercial Airplane Company, the Cooperative Research Centre for Robust and Adaptive Systems, and the German-Israeli Foundation. The LaTeX setting of this manuscript has been expertly achieved by Ms Keiko Numata. Assistance came also from Mrs Janine Carney, Ms Lisa Crisafulli, Mrs Heli Jackson, Mrs Rosi Bonn, and Ms Marita Rendina. James Ashton, Iain Collings, and Jeremy Matson lent LaTeX programming support. The simulations have been carried out by our PhD students Jane Perkins and Robert Mahony, and post-doctoral fellow Weiyong Yan. Diagrams have been prepared by Anton Madievski. Bob Mahony was also a great help in proof reading and polishing the manuscript.

Finally, it is a pleasure to thank Anthony Bloch, Roger Brockett and Leonid Faybusovich for helpful discussions. Their ideas and insights had a crucial influence on our way of thinking.

Contents

1 Matrix Eigenvalue Methods   **1**
   1.1 Introduction 1
   1.2 Power Method for Diagonalization 5
   1.3 The Rayleigh Quotient Gradient Flow 14
   1.4 The QR Algorithm 29
   1.5 Singular Value Decomposition (SVD) 34
   1.6 Standard Least Squares Gradient Flows 36

2 Double Bracket Isospectral Flows   **45**
   2.1 Double Bracket Flows for Diagonalization 45
   2.2 Toda Flows and the Riccati Equation 59
   2.3 Recursive Lie-Bracket Based Diagonalization 69

3 Singular Value Decomposition   **85**
   3.1 SVD via Double Bracket Flows 85
   3.2 A Gradient Flow Approach to SVD 88

4 Linear Programming   **107**
   4.1 The Role of Double Bracket Flows 107
   4.2 Interior Point Flows on a Polytope 116
   4.3 Recursive Linear Programming/Sorting 122

5 Approximation and Control — 133
- 5.1 Approximations by Lower Rank Matrices 133
- 5.2 The Polar Decomposition 150
- 5.3 Output Feedback Control 152

6 Balanced Matrix Factorizations — 169
- 6.1 Introduction 169
- 6.2 Kempf-Ness Theorem 171
- 6.3 Global Analysis of Cost Functions 173
- 6.4 Flows for Balancing Transformations 177
- 6.5 Flows on the Factors X and Y 190
- 6.6 Recursive Balancing Matrix Factorizations 196

7 Invariant Theory and System Balancing — 205
- 7.1 Introduction 205
- 7.2 Plurisubharmonic Functions 208
- 7.3 The Azad-Loeb Theorem 210
- 7.4 Application to Balancing 212
- 7.5 Euclidean Norm Balancing 223

8 Balancing via Gradient Flows — 231
- 8.1 Introduction 231
- 8.2 Flows on Positive Definite Matrices 233
- 8.3 Flows for Balancing Transformations 246
- 8.4 Balancing via Isodynamical Flows 249
- 8.5 Euclidean Norm Optimal Realizations 257

9 Sensitivity Optimization — 267
- 9.1 A Sensitivity Minimizing Gradient Flow 267

9.2	Related L^2-Sensitivity Minimization Flows	279
9.3	Recursive L^2-Sensitivity Balancing	287
9.4	L^2-Sensitivity Model Reduction	291
9.5	Sensitivity Minimization with Constraints	293

A Linear Algebra   307

A.1	Matrices and Vectors	307
A.2	Addition and Multiplication of Matrices	308
A.3	Determinant and Rank of a Matrix	308
A.4	Range Space, Kernel and Inverses	309
A.5	Powers, Polynomials, Exponentials and Logarithms	310
A.6	Eigenvalues, Eigenvectors and Trace	310
A.7	Similar Matrices	311
A.8	Positive Definite Matrices and Matrix Decompositions	312
A.9	Norms of Vectors and Matrices	313
A.10	Kronecker Product and Vec	314
A.11	Differentiation and Integration	314
A.12	Lemma of Lyapunov	315
A.13	Vector Spaces and Subspaces	315
A.14	Basis and Dimension	316
A.15	Mappings and Linear Mappings	316
A.16	Inner Products	317

B Dynamical Systems   319

B.1	Linear Dynamical Systems	319
B.2	Linear Dynamical System Matrix Equations	320
B.3	Controllability and Stabilizability	321
B.4	Observability and Detectability	322

- B.5 Minimality 323
- B.6 Markov Parameters and Hankel Matrix 323
- B.7 Balanced Realizations 323
- B.8 Vector Fields and Flows 324
- B.9 Stability Concepts 326
- B.10 Lyapunov Stability 327

C Global Analysis — 331

- C.1 Point Set Topology 331
- C.2 Advanced Calculus 334
- C.3 Smooth Manifolds 337
- C.4 Spheres, Projective Spaces and Grassmannians 339
- C.5 Tangent Spaces and Tangent Maps 342
- C.6 Submanifolds 345
- C.7 Groups, Lie Groups and Lie Algebras 347
- C.8 Homogeneous Spaces 350
- C.9 Tangent Bundle 353
- C.10 Riemannian Metrics and Gradient Flows 355
- C.11 Stable Manifolds 357
- C.12 Convergence of Gradient Flows 360

Bibliography — 363

Author Index — 381

Subject Index — 385

Chapter 1

Matrix Eigenvalue Methods

1.1 Introduction

Optimization techniques are ubiquitous in mathematics, science, engineering, and economics. *Least Squares* methods date back to Gauss who developed them to calculate planet orbits from astronomical measurements. So we might ask: "What is new and of current interest in *Least Squares Optimization*?" Our curiosity to investigate this question along the lines of this work was first aroused by the conjunction of two "events".

The first "event" was the realization by one of us that recent results from geometric invariant theory by Kempf and Ness had important contributions to make in systems theory, and in particular to questions concerning the existence of unique optimum least squares balancing solutions. The second "event" was the realization by the other author that constructive procedures for such optimum solutions could be achieved by *dynamical systems*. These dynamical systems are in fact ordinary matrix differential equations (ODE's), being also gradient flows on manifolds which are best formulated and studied using the language of differential geometry.

Indeed, the beginnings of what might be possible had been enthusiastically expounded by Roger Brockett in a 1988 Conference on Decision and Control. He showed, quite surprisingly to us at the time, that certain problems in linear algebra could be solved by calculus. Of course, his results had their origins in earlier work dating back to that of Fischer (1905), Courant (1922) and von Neumann (1937), as well as that of Rutishauser in the 1950s. Also there were parallel efforts in numerical analysis by Chu (1988). We also mention the influence of the ideas of Hermann (1979) on the development of applications of differential geometry in systems theory and linear algebra. Brockett showed that the tasks of diagonalizing a matrix, linear programming, and sorting, could all be solved by dynamical systems, and in particular by finding the limiting solution of certain well behaved ordinary matrix differential equations. Moreover, these construction procedures were actually mildly disguised solutions to matrix least squares minimization problems.

Of course, we are not used to solving problems in linear algebra by calculus, nor does it seem that matrix differential equations are attractive for replacing linear algebra computer packages. So why proceed along such lines? Here we must look at the cutting edge of current applications which result in matrix formulations involving quite high dimensional matrices, and the emergent computer technologies with distributed and parallel processing such as in the connection machine, the hypercube, array processors, systolic arrays, and artificial neural networks. For such "neural" network architectures, the solutions of high order nonlinear matrix differential (or difference) equations is not a formidable task, but rather a natural one. We should not exclude the possibility that new technologies, such as charge coupled devices, will allow N digital additions to be performed simultaneously rather than in N operations. This could bring about a new era of numerical methods, perhaps permitting the dynamical systems approach to optimization explored here to be very competitive.

The subject of this book is currently in an intensive state of development, with inputs coming from very different directions. Starting from the seminal work of Khachian and Karmarkar, there has been a lot of progress in developing interior point algorithms for linear programming and nonlinear programming, due to Bayer, Lagarias and Faybusovich, to mention a few. In numerical analysis there is the work of Kostant, Symes, Deift, Chu, Tomei and others on the Toda flow and its connection to completely integrable Hamiltonian systems. This subject also has deep connections with torus actions and symplectic geometry. Starting from the work of Brockett, there is now an emerging theory of completely integrable gradient flows on manifolds which is developed by Bloch, Brockett, Flaschka and Ratiu. We also mention the work of Bloch on least squares estimation with relation to completely integrable Hamiltonian systems. In our own work we have tried to develop the applications of gradient flows on manifolds to systems theory, signal processing and control theory. In the future we expect more applications to optimal control theory. We also mention the obvious connections to artificial neural networks and nonlinear approximation theory. In all these research directions, the development is far from being complete and a definite picture has not yet appeared.

It has not been our intention, nor have we been able to cover thoroughly all these recent developments. Instead, we have tried to draw the emerging interconnections between these different lines of research and to raise the reader's interests in these fascinating developments. Our window on these developments, to which we invite the reader to share, is of course our own research of recent years.

We see then that a dynamical systems approach to optimization is rather timely. Where better to start than with least squares optimization? The first step for the approach we take is to formulate a cost function which when minimized over the constraint set would give the desired result. The next step is to formulate a Riemannian metric on the tangent space of the constraint set, viewed as a manifold, such that a gradient flow on the manifold can be readily implemented, and such that the flow converges to the desired algebraic solution. We do not offer a systematic approach for achieving any "best" selections of the metric, but rather demonstrate the approach by examples. In the first chapters of the monograph, these examples will be associated with fundamental and classical tasks in linear algebra and in linear system theory, therefore representing more subtle, rather than dramatic, advances. In the later chapters new problems, not previously addressed by any complete theory, are tackled.

Of course, the introduction of methods from the theory of dynamical systems to optimization is well established, as in modern analysis of the classical steepest descent gradient techniques and the Newton method. More recently, feedback control techniques are being applied to select the step size in numerical integration algorithms. There are interesting applications of optimization theory to dynamical systems in the now well established field of optimal control and estimation theory. This book seeks to catalize further interactions between optimization and dynamical systems.

We are familiar with the notion that Riccati equations are often the dynamical systems behind many least squares optimization tasks, and engineers are now comfortable with implementing Riccati equations for estimation and control. Dynamical systems for other important matrix least squares optimization tasks in linear algebra, systems theory, sensitivity optimization, and inverse eigenvalue problems are studied here. At first encounter these may appear quite formidable and provoke caution. On closer inspection, we find that these dynamical systems are actually Riccati-like in behaviour and are often induced from linear flows. Also, it is very comforting that they are exponentially convergent, and converge to the set of global optimal solutions to the various optimization tasks. It is our prediction that engineers will become familiar with such equations in the decades to come.

To us the dynamical systems arising in the various optimization tasks studied in this book have their own intrinsic interest and appeal. Although we have not tried to create the whole zoo of such ordinary differential equations (ODE's), we believe that an introductory study is useful and worthy of such optimizing flows. At this stage it is too much to expect that each flow be competitive with the traditional numerical methods where these apply. However, we do expect applications in circumstances not amenable to standard numerical methods, such as in adaptive, stochastic, and time-varying environments, and to optimization tasks as explored in the later parts of this book. Perhaps the empirical tricks of modern numerical methods, such as the use of the shift strategies in the QR algorithm, can accelerate our gradient flows in discrete time so that they will be more competitive than they are now. We already have preliminary evidence for this in our own research not fully documented here.

In this monograph, we first study the classical methods of diagonalizing a symmetric matrix, giving a dynamical systems interpretation to the methods. The power method, the Rayleigh quotient method, the QR algorithm, and standard least squares algorithms which may include constraints on the variables, are studied in turn. This work leads to the gradient flow equations on symmetric matrices proposed by Brockett involving Lie brackets. We term these equations double bracket flows, although perhaps the term Lie-Brockett could have some appeal. These are developed for the primal task of diagonalizing a real symmetric matrix. Next, the related exercise of singular value decomposition (SVD) of possibly complex nonsymmetric matrices is explored by formulating the task so as to apply the double bracket equation. Also a first principles derivation is presented. Various alternative and related flows are investigated, such as those on the factors of the decompositions. A mild generalization of SVD is a matrix factorization where the factors are balanced. Such factorizations are studied by similar gradient flow techniques, leading into a later topic of balanced realizations. Double bracket flows are then applied to linear programming. Recursive discrete-time versions of these flows are proposed and rapprochement with earlier linear algebra techniques is made.

Moving on from purely linear algebra applications for gradient flows, we tackle balanced realizations and diagonal balancing as topics in linear system theory. This part of the work can be viewed as a generalization of the results for singular value decompositions.

The remaining parts of the monograph concerns certain optimization tasks arising in signal processing and control. In particular, the emphasis is on quadratic index minimization. For signal processing the parameter sensitivity costs relevant for finite-word-length implementations are minimized. Also constrained parameter sensitivity minimization is covered so as to cope with scaling constraints in a digital filter design. For feedback control, the quadratic indices are those usually used to achieve trade-offs between control energy and regulation or tracking performance. Quadratic indices are also used for eigenvalue assignment.

For all these various tasks, the geometry of the constraint manifolds is important and gradient flows on these manifolds are developed. Digressions are included in the early chapters to cover such topics as Projective Spaces, Riemannian Metrics, Gradient Flows, and Lie groups. Appendices are included to cover the relevant basic definitions and results in linear algebra, dynamical systems theory, and global analysis including aspects of differential geometry.

1.2 Power Method for Diagonalization

In this chapter we review some of the standard tools in numerical linear algebra for solving matrix eigenvalue problems. Our interest in such methods is not to give a concise and complete analysis of the algorithms, but rather to demonstrate that some of these algorithms arise as discretizations of certain continuous-time dynamical systems. This work then leads into the more recent matrix eigenvalue methods of Chapter 2, termed double bracket flows, which also have application to linear programming and topics of later chapters.

There are excellent textbooks available where the following standard methods are analyzed in detail; one choice is Golub and Van Loan (1989).

The Power Method The power method is a particularly simple iterative procedure to determine a dominant eigenvector of a linear operator. Its beauty lies in its simplicity rather than its computational efficiency. Appendix A gives background material in matrix results and in linear algebra.

Let $A : \mathbb{C}^n \to \mathbb{C}^n$ be a diagonalizable linear map with eigenvalues $\lambda_1, \cdots, \lambda_n$ and eigenvectors $v_1, \cdots, v_n$. For simplicity let us assume that A is nonsingular and $\lambda_1, \cdots, \lambda_n$ satisfy $|\lambda_1| > |\lambda_2| \geq \cdots \geq |\lambda_n|$. We then say that λ_1 is a *dominant eigenvalue* and v_1 a *dominant eigenvector*. Let

$$||x|| = (\sum_{i=1}^{n} |x_i|^2)^{1/2} \tag{2.1}$$

denote the standard Euclidean norm of $\mathbb{C}^n$. For any initial vector x_0 of $\mathbb{C}^n$ with $||x_0|| = 1$, we consider the infinite normalized Krylov-sequence (x_k) of unit vectors of $\mathbb{C}^n$ defined by the discrete-time dynamical system

$$\boxed{x_k = \frac{Ax_{k-1}}{||Ax_{k-1}||} = \frac{A^k x_0}{||A^k x_0||}, \quad k \in \mathbb{N}.} \tag{2.2}$$

Appendix B gives some background results for dynamical systems.

Since the growth rate of the component of $A^k x_0$ corresponding to the eigenvector v_1 dominates the growth rates of the other components we would expect that (x_k) converges to the dominant eigenvector of A:

$$\lim_{k \to \infty} x_k = \lambda \frac{v_1}{||v_1||} \quad \text{for some } \lambda \in \mathbb{C} \text{ with } |\lambda| = 1 . \tag{2.3}$$

Of course, if x_0 is an eigenvector of A so is x_k for all $k \in \mathbb{N}$. Therefore we would expect (2.3) to hold only for generic initial conditions, that is for almost all $x_0 \in \mathbb{C}^n$. This is indeed quite true; see Golub and Van Loan (1989), Parlett and Poole (1973).

Examples   **1.** Let

$$A = \begin{bmatrix} 1 & 0 \\ 0 & 2 \end{bmatrix} \quad \text{with } x_0 = \frac{1}{\sqrt{2}} \begin{bmatrix} 1 \\ 1 \end{bmatrix} .$$

Then

$$x_k = \frac{1}{\sqrt{1 + 2^{2k}}} \begin{bmatrix} 1 \\ 2^k \end{bmatrix} , \quad k \geq 1,$$

which converges to $\begin{bmatrix} 0 \\ 1 \end{bmatrix}$ for $k \to \infty$.

2. Let

$$A = \begin{bmatrix} 1 & 0 \\ 0 & -2 \end{bmatrix} , \quad x_0 = \frac{1}{\sqrt{2}} \begin{bmatrix} 1 \\ 1 \end{bmatrix} .$$

Then

$$x_k = \frac{1}{\sqrt{1 + 2^{2k}}} \begin{bmatrix} 1 \\ (-2)^k \end{bmatrix} , \quad k \in \mathbb{N},$$

and the sequence $(x_k \mid k \in \mathbb{N})$ has

$$\left\{ \begin{bmatrix} 0 \\ 1 \end{bmatrix} , \begin{bmatrix} 0 \\ -1 \end{bmatrix} \right\}$$

as a limit set, see Figure 1.2.1.

In both of the above examples, the power method defines a discrete-dynamical system on the unit circle $\{(a, b) \in \mathbb{R}^2 \mid a^2 + b^2 = 1\}$

$$\begin{bmatrix} a_{k+1} \\ b_{k+1} \end{bmatrix} = \frac{1}{\sqrt{a_k^2 + 4 b_k^2}} \begin{bmatrix} a_k \\ 2\varepsilon b_k \end{bmatrix} \tag{2.4}$$

1.2. POWER METHOD FOR DIAGONALIZATION 7

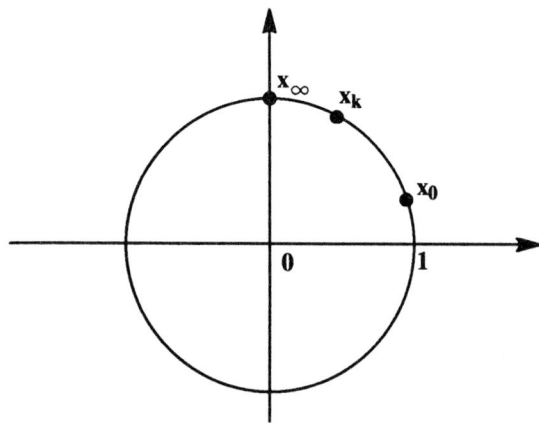

Figure 1.2.1: Power estimates of $[0\ 1]'$

with initial conditions $a_0^2 + b_0^2 = 1$, and $\varepsilon = \pm 1$. In the first case, $\varepsilon = 1$, all solutions (except for those where $\begin{bmatrix} a_0 \\ b_0 \end{bmatrix} = \pm \begin{bmatrix} 1 \\ 0 \end{bmatrix}$) of (2.4) converge to either $\begin{bmatrix} 0 \\ 1 \end{bmatrix}$ or to $-\begin{bmatrix} 0 \\ 1 \end{bmatrix}$ depending on whether $b_0 > 0$ or $b_0 < 0$. In the second case, $\varepsilon = -1$, no solution with $b_0 \neq 0$ converges; in fact the solutions oscillate infinitely often between the vicinity of $\begin{bmatrix} 0 \\ 1 \end{bmatrix}$ and that of $\begin{bmatrix} 0 \\ -1 \end{bmatrix}$, approaching $\left\{ \begin{bmatrix} 0 \\ 1 \end{bmatrix}, \begin{bmatrix} 0 \\ -1 \end{bmatrix} \right\}$ closer and closer. ∎

The dynamics of the power iterations (2.2) become particularly transparent if eigenspaces instead of eigenvectors are considered. This leads to the interpretation of the power method as a discrete-time dynamical system on the complex projective space; see Ammar and Martin (1986), Parlett and Poole (1973). We start by recalling some elementary terminology and facts about projective spaces, see also Appendix C on differential geometry.

Digression: Projective Spaces and Grassmann Manifolds The n-dimensional *complex projective space* $\mathbb{CP}^n$ is defined as the set of all one-dimensional complex linear subspaces of $\mathbb{C}^{n+1}$. Likewise, the n-dimensional real projective space $\mathbb{RP}^n$ is the set of all one-dimensional real linear subspaces of $\mathbb{R}^{n+1}$. Thus, using stereographic projection, $\mathbb{RP}^1$ can be depicted as the unit circle in $\mathbb{R}^2$. In a similar way, $\mathbb{CP}^1$ coincides with the familiar Riemann sphere $\mathbb{C} \cup \{\infty\}$ consisting of the complex plane and the point at infinity, as illustrated in Figure 1.2.2.

Since any one-dimensional complex subspace of $\mathbb{C}^{n+1}$ is generated by a unit vec-

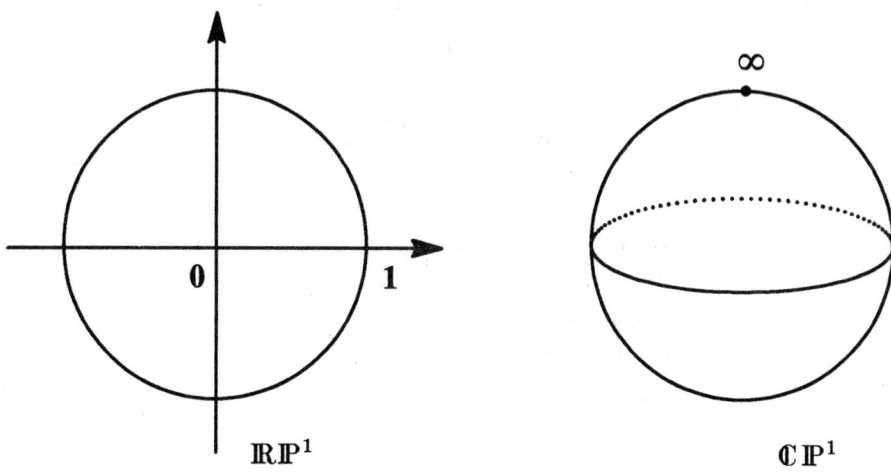

Figure 1.2.2: Circle $\mathbb{RP}^1$ and Riemann Sphere $\mathbb{CP}^1$

tor in $\mathbb{C}^{n+1}$, and since any two unit row vectors $z = (z_0, \cdots, z_n)$, $w = (w_0, \cdots, w_n)$ of $\mathbb{C}^{n+1}$ generate the same complex line if and only if

$$(w_0, \cdots, w_n) = (\lambda z_0, \cdots, \lambda z_n)$$

for some $\lambda \in \mathbb{C}, |\lambda| = 1$, one can identify $\mathbb{CP}^n$ with the set of equivalence classes $[z_0 : \cdots : z_n] = \{(\lambda z_0, \cdots, \lambda z_n) \mid \lambda \in \mathbb{C}, |\lambda| = 1\}$ for unit vectors $(z_0, \cdots, z_n)$ of $\mathbb{C}^{n+1}$. Here $z_0, \cdots, z_n$ are called the *homogeneous coordinates* for the complex line $[z_0 : \cdots : z_n]$. Similarly, we denote by $[z_0 : \cdots : z_n]$ the complex line which is, generated by an arbitrary nonzero vector $(z_0, \cdots, z_n) \in \mathbb{C}^{n+1}$.

Now, let $\mathcal{H}_1(n+1)$ denote the set of all one-dimensional Hermitian projection operators on $\mathbb{C}^{n+1}$. Thus $H \in \mathcal{H}_1(n+1)$ if and only if

$$H = H^*, \quad H^2 = H, \quad \text{rank } H = 1. \tag{2.5}$$

By the spectral theorem every $H \in \mathcal{H}_1(n+1)$ is of the form $H = x \cdot x^*$ for a unit column vector $x = (x_0, \cdots, x_n)' \in \mathbb{C}^{n+1}$. The map

$$f : \mathcal{H}_1(n+1) \to \mathbb{CP}^n$$

$$H \mapsto \text{Image of } H = [x_0 : \cdots : x_n] \tag{2.6}$$

is a bijection and we can therefore identify the set of rank one Hermitian projection operators on $\mathbb{C}^{n+1}$ with the complex projective space $\mathbb{CP}^n$. This is what we refer to as the isospectral picture of the projective space. Note that this parametrization of the projective space is not given as a collection of local coordinate charts but rather as a global algebraic representation. For our purposes such global descriptions are of more interest than the local coordinate chart descriptions.

Similarly, the complex *Grassmann manifold* $\text{Grass}_{\mathbb{C}}(k, n+k)$ is defined as the set of all k-dimensional complex linear subspaces of $\mathbb{C}^{n+k}$. If $\mathcal{H}_k(n+k)$ denotes

1.2. POWER METHOD FOR DIAGONALIZATION

the set of all Hermitian projection operators H of $\mathbb{C}^{n+k}$ with rank k

$$H = H^*, \quad H^2 = H, \quad \text{rank } H = k,$$

then again, by the spectral theorem, every $H \in \mathcal{H}_k(n+k)$ is of the form $H = X \cdot X^*$ for a complex $(n+k) \times k$-matrix X satisfying $X^*X = I_k$. The map

$$f : \mathcal{H}_k(n+k) \to \text{Grass}_{\mathbb{C}}(k, n+k) \tag{2.7}$$

defined by

$$f(H) = \text{image}(H) = \text{column space of } X \tag{2.8}$$

is a bijection. The spaces $\text{Grass}_{\mathbb{C}}(k, n+k)$ and $\mathbb{C}\mathbb{P}^n = \text{Grass}_{\mathbb{C}}(1, n+1)$ are compact complex manifolds of (complex) dimension kn and n, respectively. Again, we refer to this as the isospectral picture of Grassmann manifolds.

In the same way the real Grassmann manifolds $\text{Grass}_{\mathbb{R}}(1, n+1) = \mathbb{R}\mathbb{P}^n$ and $\text{Grass}_{\mathbb{R}}(k, n+k)$ are defined; i.e. $\text{Grass}_{\mathbb{R}}(k, n+k)$ is the set of all k-dimensional real linear subspaces of $\mathbb{R}^{n+k}$ of dimension k. ∎

Power Method as a Dynamical System We can now describe the power method as a dynamical system on the complex projective space $\mathbb{C}\mathbb{P}^{n-1}$. Given a complex linear operator $A : \mathbb{C}^n \to \mathbb{C}^n$ with $\det A \neq 0$, it induces a map on the complex projective space $\mathbb{C}\mathbb{P}^{n-1}$ denoted also by A:

$$A : \mathbb{C}\mathbb{P}^{n-1} \to \mathbb{C}\mathbb{P}^{n-1},$$

$$\ell \mapsto A \cdot \ell, \tag{2.9}$$

which maps every one-dimensional complex vector space $\ell \subset \mathbb{C}^n$ to the image $A \cdot \ell$ of ℓ under A. The main convergence result on the power method can now be stated as follows:

Theorem 2.1 *Let A be diagonalizable with eigenvalues $\lambda_1, \ldots, \lambda_n$ satisfying $|\lambda_1| > |\lambda_2| \geq \ldots \geq |\lambda_n|$. For almost all complex lines $\ell_0 \in \mathbb{C}\mathbb{P}^{n-1}$ the sequence $(A^k \cdot \ell_0 \mid k \in \mathbb{N})$ of power estimates converges to the dominant eigenspace of A:*

$$\lim_{k \to \infty} A^k \cdot \ell_0 = \text{dominant eigenspace of } A.$$

Proof Without loss of generality we can assume that $A = \text{diag}(\lambda_1, \cdots, \lambda_n)$. Let $\ell_0 \in \mathbb{C}\mathbb{P}^{n-1}$ be any complex line of $\mathbb{C}^n$, with homogeneous coordinates of the form $\ell_0 = [1 : x_2 : \cdots : x_n]$. Then

$$\begin{aligned} A^k \cdot \ell_0 &= [\lambda_1^k : \lambda_2^k x_2 : \cdots : \lambda_n^k x_n] \\ &= [1 : (\frac{\lambda_2}{\lambda_1})^k x_2 : \cdots : (\frac{\lambda_n}{\lambda_1})^k x_n], \end{aligned}$$

which converges to the dominant eigenspace $\ell_\infty = [1 : 0 : \cdots : 0]$, since $\left|\frac{\lambda_i}{\lambda_1}\right| < 1$ for $i = 2, \cdots, n$. ∎

An important insight into the power method is that it is closely related to the Riccati equation. Let $F \in \mathbb{C}^{n \times n}$ be partitioned by

$$F = \begin{bmatrix} F_{11} & F_{12} \\ F_{21} & F_{22} \end{bmatrix}$$

where F_{11}, F_{12} are 1×1 and $1 \times (n-1)$ matrices and F_{21}, F_{22} are $(n-1) \times 1$ and $(n-1) \times (n-1)$. The *matrix Riccati differential equation* on $\mathbb{C}^{(n-1) \times 1}$ is then for $K \in \mathbb{C}^{(n-1) \times 1}$:

$$\boxed{\dot{K} = F_{21} + F_{22}K - KF_{11} - KF_{12}K, \quad K(0) \in \mathbb{C}^{(n-1) \times 1}.} \quad (2.10)$$

Given any solution $K(t) = (K_1(t), \cdots, K_{n-1}(t))'$ of (2.10), it defines a corresponding curve

$$\ell(t) = [1 : K_1(t) : \cdots : K_{n-1}(t)] \in \mathbb{CP}^{n-1}$$

in the complex projective space $\mathbb{CP}^{n-1}$. Let e^{tF} denote the matrix exponential of tF and let

$$e^{tF} : \mathbb{CP}^{n-1} \to \mathbb{CP}^{n-1}$$
$$\ell \mapsto e^{tF} \cdot \ell \quad (2.11)$$

denote the associated flow on $\mathbb{CP}^{n-1}$. Then it is easy to show and in fact well known that

$$\ell(t) = e^{tF} \cdot \ell_0 \quad (2.12)$$

for any solution of the Riccati equation. Conversely, if $\ell(t) = e^{tF} \cdot \ell_0$, $\ell(t) = [\ell_0(t) : \ell_1(t) : \cdots : \ell_{n-1}(t)]$, then

$$K(t) = (\ell_1(t)/\ell_0(t), \cdots, \ell_{n-1}(t)/\ell_0(t))'$$

is a solution of the Riccati equation, as long as $\ell_0(t) \neq 0$.

There is therefore a one-to-one correspondence between solutions of the Riccati equation and curves $t \mapsto e^{tF} \cdot \ell$ in the complex projective space.

The above correspondence between matrix Riccati differential equations and systems of linear differential equations is of course well known and is the basis for explicit solution methods for the Riccati equation. In the scalar case the idea is particularly transparent.

Example Let $x(t)$ be a solution of the scalar Riccati equation

$$\dot{x} = ax^2 + bx + c, \quad a \neq 0.$$

1.2. POWER METHOD FOR DIAGONALIZATION

Set
$$y(t) = e^{-a \int_{t_0}^{t} x(\tau)d\tau}$$

so that we have $\dot{y} = -axy$. Then a simple computation shows that $y(t)$ satisfies the second-order equation

$$\ddot{y} = b\dot{y} - acy.$$

Conversely if $y(t)$ satisfies $\ddot{y} = \alpha\dot{y} + \beta y$ then $x(t) = \frac{\dot{y}(t)}{y(t)}, y(t) \neq 0$, satisfies the Riccati equation

$$\dot{x} = -x^2 + \alpha x + \beta.$$

■

With this in mind, we can now easily relate the power method to the Riccati equation.

Let $F \in \mathbb{C}^{n \times n}$ be a matrix logarithm of a nonsingular matrix A

$$e^F = A \;,\; \log A = F\,. \tag{2.13}$$

Then for any integer $k \in \mathbb{N}, A^k = e^{kF}$ and hence $(A^k \cdot \ell_0 \mid k \in \mathbb{N}) = (e^{kF} \cdot \ell_0 \mid k \in \mathbb{N})$. Thus the power method for A is identical with constant time sampling of the flow $t \mapsto \ell(t) = e^{tF} \cdot \ell_0$ on $\mathbb{CP}^{n-1}$ at times $t = 0, 1, 2, \cdots$. Thus,

Corollary 2.2 *The power method for nonsingular matrices A corresponds to an integer time sampling of the Riccati equation with $F = \log A$.*

The above results and remarks have been for an arbitrary invertible complex $n \times n$ matrix A. In that case a logarithm F of A exists and we can relate the power method to the Riccati equation. If A is an invertible Hermitian matrix, a Hermitian logarithm does not exist in general, and matters become a bit more complex. In fact, by the identity

$$\det e^F = e^{\operatorname{tr} F}, \tag{2.14}$$

F being Hermitian implies that $\det e^F > 0$. We have the following lemma.

Lemma 2.3 *Let $A \in \mathbb{C}^{n \times n}$ be an invertible Hermitian matrix. Then there exists a pair of commuting Hermitian matrices $S, F \in \mathbb{C}^{n \times n}$ with $S^2 = I$ and $A = Se^F$. Also, A and S have the same signatures.*

Proof Without loss of generality we can assume A is a real diagonal matrix $\operatorname{diag}(\lambda_1, \cdots, \lambda_n)$. Define $F := \operatorname{diag}(\log |\lambda_1|, \cdots, \log |\lambda_n|)$ and $S := \operatorname{diag}(\frac{\lambda_1}{|\lambda_1|}, \cdots, \frac{\lambda_n}{|\lambda_n|})$. Then $S^2 = I$ and $A = Se^F$. ■

Applying the lemma to the Hermitian matrix A yields

$$A^k = (Se^F)^k = S^k e^{kF}$$
$$= \begin{cases} e^{kF} & \text{for } k \text{ even} \\ Se^{kF} & \text{for } k \text{ odd}. \end{cases} \quad (2.15)$$

We see that, for k even, the k-th power estimate $A^k \cdot l_0$ is identical with the solution of the Riccati equation for the Hermitian logarithm $F = \log A$ at time k. A straightforward extension of the above approach is in studying power iterations to determine the dominant k-dimensional eigenspace of $A \in \mathbb{C}^{n \times n}$. The natural geometric setting is that as a dynamical system on the Grassmann manifold $\text{Grass}_{\mathbb{C}}(k,n)$. This parallels our previous development. Thus let $A \in \mathbb{C}^{n \times n}$ denote an invertible matrix, it induces a map on the Grassmannian $\text{Grass}_{\mathbb{C}}(k,n)$ denoted also by A:

$$A : \text{Grass}_{\mathbb{C}}(k,n) \to \text{Grass}_{\mathbb{C}}(k,n)$$
$$V \mapsto A \cdot V$$

which maps any k-dimensional complex subspace $V \subset \mathbb{C}^n$ onto its image $A \cdot V = A(V)$ under A. We have the following convergence result.

Theorem 2.4 *Assume $|\lambda_1| \geq \cdots \geq |\lambda_k| > |\lambda_{k+1}| \geq \cdots \geq |\lambda_n|$. For almost all $V \in \text{Grass}_{\mathbb{C}}(k,n)$ the sequence $(A^\nu \cdot V | \nu \in \mathbb{N})$ of power estimates converges to the dominant k-dimensional eigenspace of A:*

$$\lim_{\nu \to \infty} A^\nu \cdot V = \text{the } k - \text{dimensional dominant eigenspace of } A.$$

Proof Without loss of generality we can assume that $A = \text{diag}(\Lambda_1, \Lambda_2)$ with $\Lambda_1 = \text{diag}(\lambda_1, \cdots, \lambda_k), \Lambda_2 = \text{diag}(\lambda_{k+1}, \cdots, \lambda_n)$. Given a full rank matrix $X \in \mathbb{C}^{n \times k}$, let $[X] \subset \mathbb{C}^n$ denote the k-dimensional subspace of $\mathbb{C}^n$ which is spanned by the columns of X. Thus $[X] \in \text{Grass}_{\mathbb{C}}(k,n)$. Let $V = \left[\begin{pmatrix} X_1 \\ X_2 \end{pmatrix} \right] \in \text{Grass}_{\mathbb{C}}(k,n)$ satisfy the genericity condition $\det(X_1) \neq 0$, with $X_1 \in \mathbb{C}^{k \times k}, X_2 \in \mathbb{C}^{(n-k) \times k}$. Thus

$$A^\nu \cdot V = \left[\begin{pmatrix} \Lambda_1^\nu \cdot X_1 \\ \Lambda_2^\nu \cdot X_2 \end{pmatrix} \right]$$
$$= \left[\begin{pmatrix} I_k \\ \Lambda_2^\nu X_2 X_1^{-1} \Lambda_1^{-\nu} \end{pmatrix} \right].$$

Estimating the 2-norm of $\Lambda_2^\nu X_2 X_1^{-1} \Lambda_1^{-\nu}$ gives

$$\|\Lambda_2^\nu X_2 X_1^{-1} \Lambda_1^{-\nu}\| \leq \|\Lambda_2^\nu\| \|\Lambda_1^{-\nu}\| \|X_2 X_1^{-1}\|$$
$$\leq \left(\frac{|\lambda_{k+1}|}{|\lambda_k|} \right)^\nu \|X_2 X_1^{-1}\|$$

1.2. POWER METHOD FOR DIAGONALIZATION

since $\|\Lambda_2^\nu\| = |\lambda_{k+1}|^\nu$, and $\|\Lambda_1^{-\nu}\| = |\lambda_k|^{-\nu}$ for diagonal matrices. Thus $\Lambda_1^\nu X_2 X_1^{-1} \Lambda_1^{-\nu}$ converges to the zero matrix and hence $A^\nu \cdot V$ converges to $\left[\begin{pmatrix} I_k \\ 0 \end{pmatrix}\right]$, that is to the k-dimensional dominant eigenspace of A. ∎

Proceeding in the same way as before, we can relate the power iterations for the k-dimensional dominant eigenspace of A to the Riccati equation with $F = \log A$. Briefly, let $F \in \mathbb{C}^{n \times n}$ be partitioned as

$$F = \begin{bmatrix} F_{11} & F_{12} \\ F_{21} & F_{22} \end{bmatrix}$$

where F_{11}, F_{12} are $k \times k$ and $k \times (n-k)$ matrices and F_{21}, F_{22} are $(n-k) \times k$ and $(n-k) \times (n-k)$. The matrix Riccati equation on $\mathbb{C}^{(n-k) \times k}$ is then given by (2.10) with $K(0)$ and $K(t)$ complex $(n-k) \times k$ matrices. Any $(n-k) \times k$ matrix solution $K(t)$ of (2.10) defines a corresponding curve

$$V(t) = \begin{bmatrix} I_k \\ K(t) \end{bmatrix} \in \text{Grass}_{\mathbb{C}}(k, n+k)$$

in the Grassmannian and

$$V(t) = e^{tF} \cdot V(0)$$

holds for all t. Conversely, if $V(t) = e^{tF} \cdot V(0)$ with $V(t) = \begin{bmatrix} X_1(t) \\ X_2(t) \end{bmatrix}, X_1 \in \mathbb{C}^{k \times k}, X_2 \in \mathbb{C}^{(n-k) \times k}$ then $K(t) = X_2(t) X_1^{-1}(t)$ is a solution of the matrix Riccati equation, as long as $\det X_1(t) \neq 0$.

Again the power method for finding the k-dimensional dominant eigenspace of A corresponds to integer time sampling of the $(n-k) \times k$ matrix Riccati equation defined for $F = \log A$.

Problem A differential equation $\dot{x} = f(x)$ is said to have finite escape time if there exists a solution $x(t)$ for some initial condition $x(t_0)$ such that $x(t)$ is not defined for all $t \in \mathbb{R}$. Prove

(a) The scalar Riccati equation $\dot{x} = x^2$ has finite escape time.

(b) $\dot{x} = ax^2 + bx + c, a \neq 0$, has finite escape time.

(c) $\dot{x} = \sum_{i=0}^n a_i x^i$, $a_n \neq 0$, has finite escape time if and only if n is even.

Main Points of Section The power method for determining a dominant eigenvector of a linear nonsingular operator $A : \mathbb{C}^n \to \mathbb{C}^n$ has a dynamical system interpretation. It is in fact a discrete-time dynamical system on a complex projective space $\mathbb{C}\mathbb{P}^{n-1}$, where the n-dimensional complex projective space $\mathbb{C}\mathbb{P}^n$ is the set of all one dimensional complex linear subspaces of $\mathbb{C}^{n+1}$.

The power method is closely related to a quadratic continuous-time matrix Riccati equation associated with a matrix F satisfying $e^F = A$. It is in fact a constant period sampling of this equation.

Grassmann manifolds are an extension of the projective space concept and provide the natural geometric setting for studying power iterations to determine the dominant k-dimensional eigenspace of A.

1.3 The Rayleigh Quotient Gradient Flow

Let $A \in \mathbb{R}^{n \times n}$ be real symmetric with eigenvalues $\lambda_1 \geq \cdots \geq \lambda_n$ and corresponding eigenvectors $v_1, \cdots, v_n$. The *Rayleigh quotient* of A is the smooth function $r_A : \mathbb{R}^n - \{0\} \to \mathbb{R}$ defined by

$$r_A(x) = \frac{x'Ax}{||x||^2}. \tag{3.1}$$

Let $S^{n-1} = \{x \in \mathbb{R}^n \mid ||x|| = 1\}$ denote the unit sphere in $\mathbb{R}^n$. Since $r_A(tx) = r_A(x)$ for all positive real numbers $t > 0$, it suffices to consider the Rayleigh quotient on the sphere where $||x|| = 1$. The following theorem is a special case of the celebrated Courant-Fischer minimax theorem characterizing the eigenvalues of a real symmetric $n \times n$ matrix, see Golub and Van Loan (1989).

Theorem 3.1 *Let λ_1 and λ_n denote the largest and smallest eigenvalue of a real symmetric $n \times n$-matrix A respectively. Then*

$$\lambda_1 = \max_{||x||=1} r_A(x), \tag{3.2}$$

$$\lambda_n = \min_{||x||=1} r_A(x). \tag{3.3}$$

More generally, the critical points and critical values of r_A are the eigenvectors and eigenvalues for A.

Proof Let $r_A : S^{n-1} \to \mathbb{R}$ denote the restriction of the Rayleigh quotient on the $(n-1)$-sphere. For any unit vector $x \in S^{n-1}$ the Fréchet-derivative of r_A is the linear functional $Dr_A|_x : T_x S^{n-1} \to \mathbb{R}$ defined on the tangent space $T_x S^{n-1}$ of S^{n-1} at x as follows. The tangent space is

$$T_x S^{n-1} = \{\xi \in \mathbb{R}^n \mid x'\xi = 0\}, \tag{3.4}$$

1.3. THE RAYLEIGH QUOTIENT GRADIENT FLOW

and
$$Dr_A|_x(\xi) = 2\langle Ax, \xi \rangle = 2x'A\xi. \quad (3.5)$$

Hence x is a critical point of r_A if and only if $Dr_A|_x(\xi) = 0$, or equivalently,

$$x'\xi = 0 \implies x'A\xi = 0,$$

or equivalently,

$$Ax = \lambda x$$

for some $\lambda \in \mathbb{R}$. Left multiplication by x' implies $\lambda = \langle Ax, x \rangle$. Thus the critical points and critical values of r_A are the eigenvectors and eigenvalues of A respectively. The result follows. ∎

This proof uses concepts from calculus, see Appendix C. The approach taken here is used to derive many subsequent results in the book. The arguments used are closely related to those involving explicit use of Lagrange multipliers, but here the use of such is implicit.

From the variational characterization of eigenvalues of A as the critical values of the Rayleigh quotient, it seems natural to apply gradient flow techniques in order to search for the dominant eigenvector. We now include a digression on these techniques, see also Appendix C.

Digression: Riemannian Metrics and Gradient Flows Let M be a smooth manifold and let TM and T^*M denote its tangent and cotangent bundle, respectively. Thus $TM = \bigcup_{x \in M} T_x M$ is the set theoretic disjoint union of all tangent spaces $T_x M$ of M while $T^*M = \bigcup_{x \in M} T_x^* M$ denotes the disjoint union of all cotangent spaces $T_x^* M = \text{Hom}(T_x M, \mathbb{R})$ (i.e. the dual vector spaces of $T_x M$) of M.

A *Riemannian metric* on M then is a family of nondegenerate inner products $\langle \, , \, \rangle_x$, defined on each tangent space $T_x M$, such that $\langle \, , \, \rangle_x$ depends smoothly on $x \in M$. Once a Riemannian metric is specified, M is called a *Riemannian manifold*. Thus a Riemannian metric on $\mathbb{R}^n$ is just a smooth map $Q: \mathbb{R}^n \to \mathbb{R}^{n \times n}$ such that for each $x \in \mathbb{R}^n$, $Q(x)$ is a real symmetric positive definite $n \times n$ matrix. In particular every nondegenerate inner product on $\mathbb{R}^n$ defines a Riemannian metric on $\mathbb{R}^n$ (but not conversely) and also induces by restriction a Riemannian metric on every submanifold M of $\mathbb{R}^n$. We refer to this as the *induced Riemannian metric* on M.

Let $\Phi: M \to \mathbb{R}$ be a smooth function defined on a manifold M and let $D\Phi: M \to T^*M$ denote the differential, i.e. the section of the cotangent bundle T^*M defined by

$$D\Phi(x): T_x M \to \mathbb{R}, \quad \xi \mapsto D\Phi(x) \cdot \xi,$$

where $D\Phi(x)$ is the derivative of Φ at x. We also often use the notation

$$D\Phi|_x(\xi) = D\Phi(x) \cdot \xi$$

to denote the derivative of Φ at x. To be able to define the gradient vector field of Φ we have to specify a Riemannian metric $\langle \, , \, \rangle$ on M. The gradient gradΦ of Φ

relative to this choice of a Riemannian metric on M is then uniquely characterized by the following two properties.

Tangency Condition
(i) $\operatorname{grad}\Phi(x) \in T_x M$ for all $x \in M$.

Compatibility Condition
(ii) $D\Phi(x) \cdot \xi = \langle \operatorname{grad}\Phi(x), \xi \rangle$ for all $\xi \in T_x M$.

There exists a uniquely determined vector field $\operatorname{grad}\Phi : M \to TM$ on M such that (i) and (ii) hold. $\operatorname{grad}\Phi$ is called the *gradient vector field* of Φ.

Note that the Riemannian metric enters into condition (ii) and therefore the gradient vector field $\operatorname{grad}\Phi$ will depend on the choice of the Riemannian metric; changing the metric will also change the gradient.

If $M = \mathbb{R}^n$ is endowed with its standard Riemannian metric defined by

$$\langle \xi, \eta \rangle = \xi' \eta \text{ for } \xi, \eta \in \mathbb{R}^n$$

then the associated gradient vector field is just the column vector

$$\nabla \Phi(x) = \left(\frac{\partial \Phi}{\partial x_1}(x), \cdots, \frac{\partial \Phi}{\partial x_n}(x) \right)'.$$

If $Q : \mathbb{R}^n \to \mathbb{R}^{n \times n}$ denotes a smooth map with $Q(x) = Q(x)' > 0$ for all x then the gradient vector field of $\Phi : \mathbb{R}^n \to \mathbb{R}$ relative to the Riemannian metric $\langle \ , \ \rangle$ on $\mathbb{R}^n$ defined by

$$\langle \xi, \eta \rangle_x := \xi' Q(x) \eta, \qquad \xi, \eta \in T_x(\mathbb{R}^n) = \mathbb{R}^n,$$

is

$$\operatorname{grad}\Phi(x) = Q(x)^{-1} \nabla \Phi(x).$$

This clearly shows the dependence of $\operatorname{grad}\Phi$ on the metric $Q(x)$.

The following remark is useful in order to compute gradient vectors. Let $\Phi : M \to \mathbb{R}$ be a smooth function on a Riemannian manifold and let $V \subset M$ be a submanifold. If $x \in V$ then $\operatorname{grad}(\Phi|V)(x)$ is the image of $\operatorname{grad}\Phi(x)$ under the orthogonal projection $T_x M \to T_x V$. Thus let M be a submanifold of $\mathbb{R}^n$ which is endowed with the induced Riemannian metric from $\mathbb{R}^n$ ($\mathbb{R}^n$ carries the Riemannian metric given by the standard Euclidean inner product), and let $\Phi : \mathbb{R}^n \to \mathbb{R}$ be a smooth function. Then the gradient vector $\operatorname{grad}(\Phi|_M)(x)$ of the restriction $\Phi : M \to \mathbb{R}$ is the image of $\nabla \Phi(x) \in \mathbb{R}^n$ under the orthogonal projection $\mathbb{R}^n \to T_x M$ onto $T_x M$. ∎

Returning to the task of achieving gradient flows for the Rayleigh quotient, in order to define the gradient of $r_A : S^{n-1} \to \mathbb{R}$, we first have to specify a Riemannian metric on S^{n-1}, i.e. an inner product structure on each tangent space $T_x S^{n-1}$ of S^{n-1} as in the digression above; see also Appendix C.

1.3. THE RAYLEIGH QUOTIENT GRADIENT FLOW

An obviously natural choice for such a Riemannian metric on S^{n-1} is that of taking the standard Euclidean inner product on $T_x S^{n-1}$, i.e. the Riemannian metric on S^{n-1} induced from the imbedding $S^{n-1} \subset \mathbb{R}^n$. That is we define for each tangent vectors $\xi, \eta \in T_x S^{n-1}$

$$\langle \xi, \eta \rangle := 2\xi' \eta . \tag{3.6}$$

where the constant factor 2 is inserted for convenience. The gradient ∇r_A of the Rayleigh quotient is then the uniquely determined vector field on S^{n-1} which satisfies the two conditions

$$\begin{align}
(i) \quad & \nabla r_A(x) \in T_x S^{n-1} \text{ for all } x \in S^{n-1}, \tag{3.7}\\
(ii) \quad & Dr_A|_x(\xi) = \langle \nabla r_A(x), \xi \rangle \\
& = 2 \nabla r_A(x)' \xi \text{ for all } \xi \in T_x S^{n-1} . \tag{3.8}
\end{align}$$

Since $Dr_A|_x(\xi) = 2x' A \xi$ we obtain

$$[\nabla r_A(x) - Ax]' \xi = 0 \quad \forall \xi \in T_x S^{n-1} . \tag{3.9}$$

By (3.4) this is equivalent to

$$\nabla r_A(x) = Ax + \lambda x$$

with $\lambda = -x' A x$, so that $x' \nabla r_A(x) = 0$ to satisfy (3.7). Thus the gradient flow for the Rayleigh quotient on the unit sphere S^{n-1} is

$$\boxed{\dot{x} = (A - r_A(x) I_n) x .} \tag{3.10}$$

It is easy to see directly that flow (3.10) leaves the sphere S^{n-1} invariant:

$$\begin{align}
\frac{d}{dt}(x'x) &= \dot{x}'x + x'\dot{x} = 2x'(A - r_A(x)I)x \\
&= 2[r_A(x) - r_A(x)] \|x\|^2 = 0.
\end{align}$$

We now move on to study the convergence properties of the Rayleigh quotient flow. First, recall that the *exponential rate of convergence* $\rho > 0$ of a solution $x(t)$ of a differential equation to an equilibrium point $\bar{x}$ refers to the maximal possible α occurring in the following lemma (see Appendix B).

Lemma 3.2 *Consider the differential equation $\dot{x} = f(x)$ with equilibrium point $\bar{x}$. Suppose that the linearization $Df|_{\bar{x}}$ of f has only eigenvalues with real part less than $-\alpha, \alpha > 0$. Then there exists a neighborhood N_α of $\bar{x}$ and a constant $C > 0$ such that*

$$\|x(t) - \bar{x}\| \leq C e^{-\alpha(t-t_0)} \|x(t_0) - \bar{x}\|$$

for all $x(t_0) \in N_\alpha, t \geq t_0$.

In the numerical analysis literature, often convergence is measured on a logarithmic scale so that exponential convergence here is referred to as *linear convergence*.

Theorem 3.3 *Let $A \in \mathbb{R}^{n \times n}$ be symmetric with eigenvalues $\lambda_1 \geq \cdots \geq \lambda_n$. The gradient flow of the Rayleigh quotient on S^{n-1} with respect to the (induced) standard Riemannian metric is given by (3.10). The solutions $x(t)$ of (3.10) exist for all $t \in \mathbb{R}$ and converge to an eigenvector of A. Suppose $\lambda_1 > \lambda_2$. For almost all initial conditions $x_0, \|x_0\| = 1, x(t)$ converges to a dominant eigenvector v_1 or $-v_1$ of A with an exponential rate $\rho = \lambda_1 - \lambda_2$ of convergence.*

Proof We have already shown that (3.10) is the gradient flow of $r_A : S^{n-1} \to \mathbb{R}$. By compactness of S^{n-1} the solutions of any gradient flow of $r_A : S^{n-1} \to \mathbb{R}$ exist for all $t \in \mathbb{R}$. It is a special case of a more general result in Duistermaat, et al. (1983) that the Rayleigh quotient r_A is a Morse-Bott function on S^{n-1}. Such functions are defined in the subsequent digression on Convergence of Gradient Flows. Moreover, using results in the same digression, the solutions of (3.10) converge to the critical points of r_A, i.e., by Theorem 3.1, to the eigenvectors of A. Let $v_i \in S^{n-1}$ be an eigenvector of A with associated eigenvalue λ_i. The linearization of (3.10) at v_i is then readily seen to be

$$\dot{\xi} = (A - \lambda_i I)\xi, \quad v_i'\xi = 0. \qquad (3.11)$$

Hence the eigenvalues of the linearization (3.11) are

$$(\lambda_1 - \lambda_i), \cdots, (\lambda_{i-1} - \lambda_i), (\lambda_{i+1} - \lambda_i), \cdots, (\lambda_n - \lambda_i).$$

Thus the eigenvalues of (3.11) are all negative if and only if $i = 1$, and hence only the dominant eigenvectors $\pm v_1$ of A can be attractors for (3.10). Since the union of eigenspaces of A for the eigenvalues $\lambda_2, \cdots, \lambda_n$ defines a nowhere dense subset of S^{n-1}, every solution of (3.10) starting in the complement of that set will converge to either v_1 or to $-v_1$. This completes the proof. ∎

Remark It should be noted that what is usually referred to as the *Rayleigh quotient method* is somewhat different to what is done here; see Golub and Van Loan (1989). The Rayleigh quotient method uses the recursive system of iterations

$$\boxed{x_{k+1} = \frac{(A - r_A(x_k)I)^{-1} x_k}{\|(A - r_A(x_k)I)^{-1} x_k\|}}$$

to determine the dominant eigenvector at A and is thus not simply a discretization of the continuous-time Rayleigh quotient gradient flow (3.10). The dynamics of the Rayleigh quotient iteration can be quite complicated. See Batterson and Smilie (1989) for analytical results as well as Beattie and

1.3. THE RAYLEIGH QUOTIENT GRADIENT FLOW

Fox (1989). For related results concerning generalised eigenvalue problems, see Auchmuty (1991). ∎

We now digress to study gradient flows, see also Appendices B and C.

Digression: Convergence of Gradient Flows Let M be a Riemannian manifold and let $\Phi : M \to \mathbb{R}$ be a smooth function. It is an immediate consequence of the definition of the gradient vector field gradΦ on M that the equilibria of the differential equation

$$\dot{x}(t) = -\text{grad}\Phi(x(t)) \tag{3.12}$$

are precisely the critical points of $\Phi : M \to \mathbb{R}$. For any solution $x(t)$ of (3.12)

$$\begin{aligned} \frac{d}{dt}\Phi(x(t)) &= \langle \text{grad}\Phi(x(t)), \dot{x}(t) \rangle \\ &= -\|\text{grad}\Phi(x(t))\|^2 \leq 0 \end{aligned} \tag{3.13}$$

and therefore $\Phi(x(t))$ is a monotonically decreasing function of t. The following standard result is often used in this book.

Proposition 3.4 *Let $\Phi : M \to \mathbb{R}$ be a smooth function on a Riemannian manifold with compact sublevel sets, i.e. for all $c \in \mathbb{R}$ the sublevel set $\{x \in M \mid \Phi(x) \leq c\}$ is a compact subset of M. Then every solution $x(t) \in M$ of the gradient flow (3.12) on M exists for all $t \geq 0$. Furthermore, $x(t)$ converges to a connected component of the set of critical points of Φ as $t \to +\infty$.*

Note that the condition of the proposition is automatically satisfied if M is compact. Moreover, in suitable local coordinates of M, the linearization of the gradient flow (3.12) around each equilibrium point is given by the Hessian H$_\Phi$ of Φ. Thus by symmetry, H$_\Phi$ has only real eigenvalues. The linearization is not necessarily given by the Hessian if gradΦ is expressed using an arbitrary system of local coordinate charts of M. However, the numbers of positive and negative eigenvalues of the Hessian H$_\Phi$ and of the linearization of gradΦ at an equilibriuim point are always the same. In particular, an equilibrium point $x_0 \in M$ of the gradient flow (3.12) is a locally stable attractor if the Hessian of Φ at x_0 is positive definite.

It follows from Proposition 3.4 that the solutions of a gradient flow have a particularly simple convergence behaviour. There are no periodic solutions, strange attractors or any chaotic behaviours. Every solution converges to a connected component of the set of equilibria points. Thus, if $\Phi : M \to \mathbb{R}$ has only finitely many critical points, then the solutions of (3.12) will converge to a single equilibrium point rather than to a set of equilibria. We state this observation as Proposition 3.5.

Recall that the ω-limit set $L_\omega(x)$ of a point $x \in M$ for a vector field X on M is the set of points of the form $\lim_{n \to \infty} \phi_{t_n}(x)$, where (ϕ_t) is the flow of X and $t_n \to +\infty$. Similarly, the α-limit set $L_\alpha(x)$ is defined by letting $t_n \to -\infty$ instead of $+\infty$.

Proposition 3.5 Let $\Phi : M \to \mathbb{R}$ be a smooth function on a Riemannian manifold M with compact sublevel sets. Then

(a) The ω-limit set $L_\omega(x)$, $x \in M$, of the gradient vector field $\mathrm{grad}\Phi$ is a nonempty, compact and connected subset of the set of critical points of $\Phi : M \to \mathbb{R}$.

(b) Suppose $\Phi : M \to \mathbb{R}$ has isolated critical points. Then $L_\omega(x), x \in M$, consists of a single critical point. Therefore every solution of the gradient flow (3.12) converges for $t \to +\infty$ to a critical point of Φ. ∎

In particular, the convergence of a gradient flow to a set of equilibria rather than to single equilibrium points occurs only in nongeneric situations. We now focus on the generic situation. The conditions in Proposition 3.6 below are satisfied in all cases studied in this book.

Let M be a smooth manifold and let $\Phi : M \to \mathbb{R}$ be a smooth function. Let $C(\Phi) \subset M$ denote the set of all critical points of Φ. We say Φ is a *Morse-Bott* function provided the following three conditions (i), (ii), (iii) are satisfied.

(i) $\Phi : M \to \mathbb{R}$ has compact sublevel sets.

(ii) $C(\Phi) = \bigcup_{j=1}^{k} N_j$ with N_j disjoint, closed and connected submanifolds of M such that Φ is constant on $N_j, j = 1, \cdots, k$.

(iii) $\mathrm{Ker}(\mathrm{H}_\Phi(x)) = T_x N_j \ \forall x \in N_j, j = 1, \cdots, k$.

Actually, the original definition of a Morse-Bott function also includes a global topological condition on the negative eigenspace bundle defined by the Hessian, but this condition is not relevant to us.

Here $\mathrm{Ker}(\mathrm{H}_\Phi(x))$ denotes the kernel of the Hessian of Φ at x, that is the set of all tangent vectors $\xi \in T_x M$ where the Hessian of Φ is degenerate. Of course, provided (ii) holds, the tangent space $T_x(N_j)$ is always contained in $\mathrm{Ker}(\mathrm{H}_\Phi(x))$ for all $x \in N_j$. Thus condition (iii) asserts that the Hessian of Φ is full rank in the directions normal to N_j at x.

Proposition 3.6 Let $\Phi : M \to \mathbb{R}$ be a Morse-Bott function on a Riemannian manifold M. Then the ω-limit set $L_\omega(x), x \in M$, for the gradient flow (3.12) is a single critical point of Φ. Every solution of the gradient flow (3.12) converges as $t \to +\infty$ to an equilibrium point.

Proof Since a detailed proof would take us a bit too far away from our objectives here, we only give a sketch of the proof. The experienced reader should have no difficulties in filling in the missing details.

By Proposition 3.5, the ω-limit set $L_\omega(x)$ of any element $x \in M$ is contained in a connected component of $C(\Phi)$. Thus $L_\omega(x) \subset N_j$ for some $1 \leq j \leq k$. Let $a \in L_\omega(x)$ and, without loss of generality, $\Phi(N_j) = 0$. Using the hypotheses on Φ and a straightforward generalization of the Morse lemma (see, e.g., Hirsch (1976) or Milnor (1963)), there exists an open neighborhood U_a of a in M and a diffeomorphism $f : U_a \to \mathbb{R}^n$, $n = \dim M$, such that

1.3. THE RAYLEIGH QUOTIENT GRADIENT FLOW

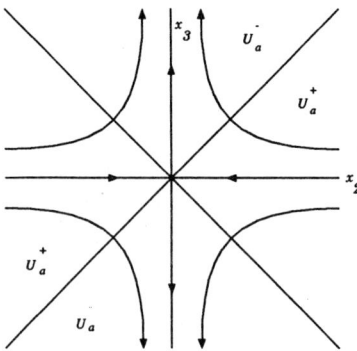

Figure 1.3.1: Flow around a saddle point

(i) $f(U_a \cap N_j) = \mathbb{R}^{n_0} \times \{0\}$

(ii) $\Phi \circ f^{-1}(x_1, x_2, x_3) = \frac{1}{2}(\|x_2\|^2 - \|x_3\|^2)$,
$x_1 \in \mathbb{R}^{n_0}, x_2 \in \mathbb{R}^{n_-}, x_3 \in \mathbb{R}^{n_+}, n_0 + n_- + n_+ = n$.

With this new system of coordinates on U_a, the gradient flow (3.12) of Φ on U_a becomes equivalent to the linear gradient flow

$$\dot{x}_1 = 0, \quad \dot{x}_2 = -x_2, \quad \dot{x}_3 = x_3 \tag{3.14}$$

of $\Phi \circ f^{-1}$ on $\mathbb{R}^n$, depicted in Figure 1.3.1. Note that the equilibrium point is a saddle point in x_2, x_3 space. Let $U_a^+ := \{f^{-1}(x_1, x_2, x_3) \mid \|x_2\| \geq \|x_3\|\}$ and $U_a^- = \{f^{-1}(x_1, x_2, x_3) \mid \|x_2\| \leq \|x_3\|\}$. Using the convergence properties of (3.14) it follows that every solution of (3.12) starting in $U_a^+ - \{f^{-1}(x_1, x_2, x_3) \mid x_3 = 0\}$ will enter the region U_a^-. On the other hand, every solution starting in $\{f^{-1}(x_1, x_2, x_3) \mid x_3 = 0\}$ will converge to a point $f^{-1}(x, 0, 0) \in N_j, x \in \mathbb{R}^{n_0}$ fixed. As Φ is strictly negative on U_a^-, all solutions starting in $U_a^- \cup U_a^+ - \{f^{-1}(x_1, x_2, x_3) \mid x_3 = 0\}$ will leave this set eventually and converge to some $N_i \neq N_j$. By repeating the analysis for N_i and noting that all other solutions in U_a converge to a single equilibrium point in N_j, the proof is completed. ∎

Returning to the Rayleigh quotient flow, (3.10) gives the gradient flow restricted to the sphere S^{n-1}. Actually (3.10) extends to a flow on $\mathbb{R}^n - \{0\}$. Is this extended flow again a gradient flow with respect to some Riemannian metric on $\mathbb{R}^n - \{0\}$? The answer is yes, as can be seen as follows.

A straightforward computation shows that the directional (Fréchet) derivative of $r_A : \mathbb{R}^n - \{0\} \to \mathbb{R}$ is given by

$$Dr_A|_x(\xi) = \frac{2}{\|x\|^2}(Ax - r_A(x)x)'\xi .$$

Consider the Riemannian metric defined on each tangent space $T_x(\mathbb{R}^n - \{0\})$

of $\mathbb{R}^n - \{0\}$ by

$$\langle\langle \xi, \eta \rangle\rangle := \frac{2}{\|x\|^2} \xi' \eta .$$

for $\xi, \eta \in T_x(\mathbb{R}^n - \{0\})$. The gradient of $r_A : \mathbb{R}^n - \{0\} \to \mathbb{R}$ with respect to this Riemannian metric is then characterized by

$$\begin{align} (i) \quad & \operatorname{grad} r_A(x) \in T_x(\mathbb{R}^n - \{0\}) \\ (ii) \quad & Dr_A|_x(\xi) = \langle\langle \operatorname{grad} r_A(x), \xi \rangle\rangle \end{align}$$

for all $\xi \in T_x(\mathbb{R}^n - \{0\})$. Note that (i) imposes no constraint on $\operatorname{grad} r_A(x)$ and (ii) is easily seen to be equivalent to

$$\operatorname{grad} r_A(x) = Ax - r_A(x)x$$

Thus (3.10) also gives the gradient of $r_A : \mathbb{R}^n - \{0\} \to \mathbb{R}$. Similarly, a stability analysis of (3.10) on $\mathbb{R}^n - \{0\}$ can be carried out completely.

Actually, the modified flow, termed here the *Oja flow*

$$\boxed{\dot{x} = (A - x'AxI_n)x} \tag{3.15}$$

is defined on $\mathbb{R}^n$ and can also be used as a means to determine the dominant eigenvector of A. This property is important in neural network theories for pattern classification, see Oja (1982). The flow seems to have certain additional attractive properties, such as being structurally stable for generic matrices A.

Examples 1. Again consider $A = \begin{bmatrix} 1 & 0 \\ 0 & 2 \end{bmatrix}$. The phase portrait of the gradient flow (3.10) is depicted in Figure 1.3.2. Note that the flow preserves $x'(t)x(t) = x'_0 x_0$.

2. The phase portrait of the Oja flow (3.15) on $\mathbb{R}^2$ is depicted in Figure 1.3.3. Note the apparent structural stability property of the flow. ∎

The Rayleigh quotient gradient flow (3.10) on the $(n-1)$ sphere S^{n-1} has the following interpretation. Consider the linear differential equation on $\mathbb{R}^n$

$$\dot{x} = Ax$$

for an arbitrary real $n \times n$ matrix A. Then

$$L_A(x) = Ax - (x'Ax)x$$

is equal to the orthogonal projection of $Ax, x \in S^{n-1}$, along x onto the tangent space $T_x S^{n-1}$ of S^{n-1} at x. See Figure 1.3.4.

1.3. THE RAYLEIGH QUOTIENT GRADIENT FLOW

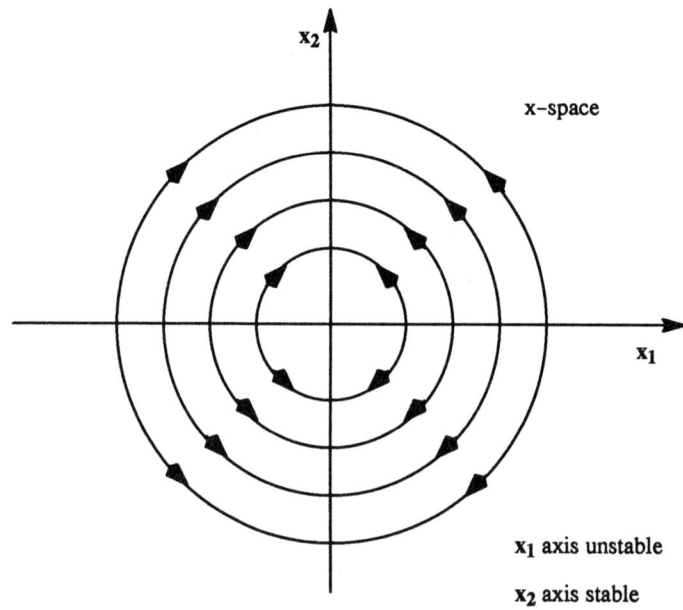

Figure 1.3.2: Phase portrait of Rayleigh quotient gradient flow

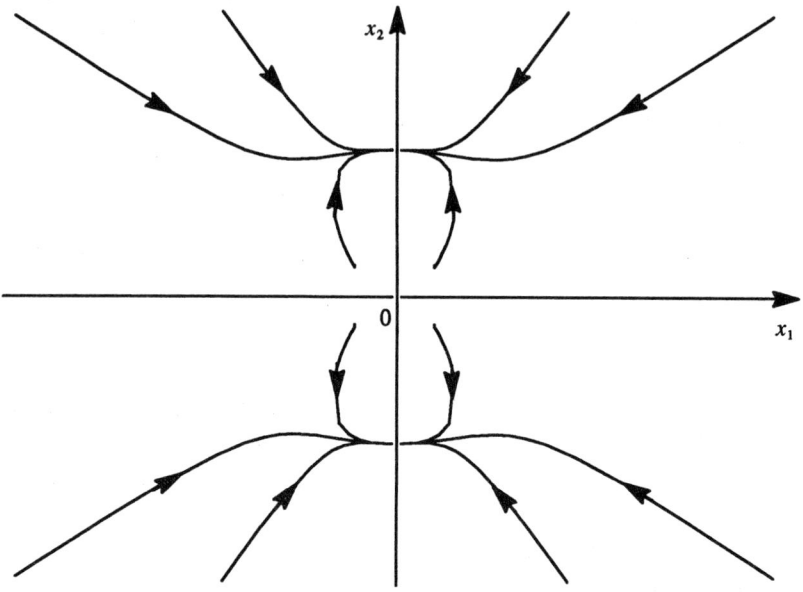

Figure 1.3.3: Phase portrait of the Oja flow

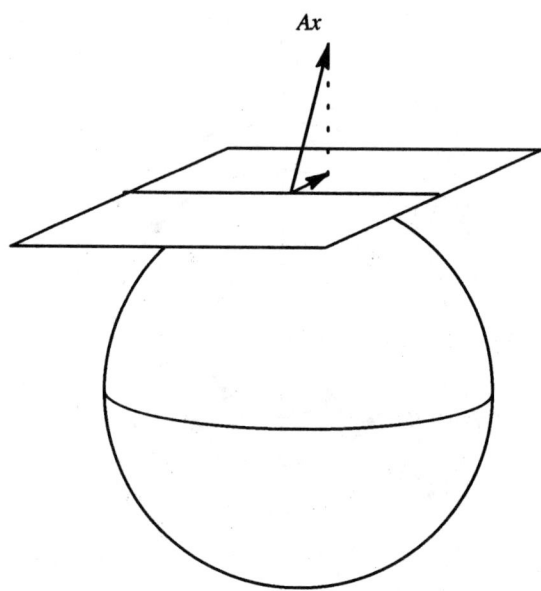

Figure 1.3.4: Orthogonal projection of Ax

Thus (3.10) can be thought of as the dynamics for the *angular part* of the differential equation $\dot{x} = Ax$. The analysis of this angular vector field (3.10) has proved to be important in the qualitative theory of linear stochastic differential equations.

An important property of dynamical systems is that of *structural stability* or robustness. In heuristic terms, structural stability refers to the property that the qualitative behaviour of a dynamical system is not changed by small perturbations in its parameters. Structural stability properties of the Rayleigh quotient gradient flow (3.10) have been investigated by de la Rocque Palis (1978). It is shown there that for $A \in \mathbb{R}^{n \times n}$ symmetric, the flow (3.10) on S^{n-1} is structurally stable if and only if the eigenvalues of A are all distinct.

Now let us consider the task of finding, for any $1 \leq k \leq n$, the k-dimensional subspace of $\mathbb{R}^n$ formed by the eigenspaces of the eigenvalues $\lambda_1, \cdots, \lambda_k$. We refer to this as the *dominant k-dimensional eigenspace of A*. Let

$$St(k,n) = \{X \in \mathbb{R}^{n \times k} \mid X'X = I_k\}$$

denote the *Stiefel manifold* of real orthogonal $n \times k$ matrices. The *generalized Rayleigh quotient* is the smooth function

$$R_A : St(k,n) \to \mathbb{R},$$

1.3. THE RAYLEIGH QUOTIENT GRADIENT FLOW

defined by
$$R_A(X) = \operatorname{tr}(X'AX) . \qquad (3.16)$$

Note that this function can be extended in several ways on larger subsets of $\mathbb{R}^{n \times k}$. A naive way of extending R_A to a function on $\mathbb{R}^{n \times k} - \{0\}$ would be as $R_A(X) = \frac{\operatorname{tr}(X'AX)}{\operatorname{tr}(X'X)}$. A more sensible choice which is studied later is as $R_A(X) = \operatorname{tr}(X'AX(X'X)^{-1})$.

The following lemma summarizes the basic geometric properties of the Stiefel manifold.

Lemma 3.7 $St(k,n)$ is a smooth, compact manifold of dimension $kn - \frac{1}{2}k(k+1)$. The tangent space at $X \in St(k,n)$ is

$$T_X St(k,n) = \{\xi \in \mathbb{R}^{n \times k} \mid \xi'X + X'\xi = 0\} . \qquad (3.17)$$

Proof Consider the smooth map $F : \mathbb{R}^{n \times k} \to \mathbb{R}^{\frac{1}{2}k(k+1)}$, $F(X) = X'X - I_k$, where we identified $\mathbb{R}^{\frac{1}{2}k(k+1)}$ with the vector space of $k \times k$ real symmetric matrices. The derivative of F at X, $X'X = I_k$, is the linear map defined by

$$DF|_X(\xi) = \xi'X + X'\xi . \qquad (3.18)$$

Suppose there exists $\eta \in \mathbb{R}^{k \times k}$ real symmetric with $\operatorname{tr}(\eta(\xi'X + X'\xi)) = 2\operatorname{tr}(\eta X'\xi) = 0$ for all $\xi \in \mathbb{R}^{n \times k}$. Then $\eta X' = 0$ and hence $\eta = 0$. Thus the derivative (3.18) is a surjective linear map for all $X \in St(k,n)$ and the kernel is given by (3.17). The result now follows from the fiber theorem; see Appendix C. ∎

The standard Euclidean inner product on the matrix space $\mathbb{R}^{n \times k}$ induces an inner product on each tangent space $T_X St(k,n)$ by

$$\langle \xi, \eta \rangle := 2\operatorname{tr}(\xi'\eta) .$$

This defines a Riemannian metric on the Stiefel manifold $St(k,n)$, which is called the induced Riemannian metric.

Lemma 3.8 The normal space $T_X St(k,n)^\perp$, that is the set of $n \times k$ matrices η which are perpendicular to $T_X St(k,n)$, is

$$\begin{aligned} T_X St(k,n)^\perp &= \{\eta \in \mathbb{R}^{n \times k} \mid \operatorname{tr}(\xi'\eta) = 0 \text{ for all } \xi \in T_X St(k,n)\} \\ &= \{X\Lambda \in \mathbb{R}^{n \times k} \mid \Lambda = \Lambda' \in \mathbb{R}^{k \times k}\} . \end{aligned} \qquad (3.19)$$

Proof For any symmetric $k \times k$ matrix Λ we have $2\operatorname{tr}((X\Lambda)'\xi) = \operatorname{tr}(\Lambda(X'\xi + \xi'X)) = 0$ for all $\xi \in T_X St(k,n)$ and therefore $\{X\Lambda \mid \Lambda = \Lambda' \in \mathbb{R}^{k \times k}\} \subset$

$T_X St(k,n)^\perp$. Both $T_X St(k,n)^\perp$ and $\{X\Lambda \in \mathbb{R}^{n\times k} \mid \Lambda = \Lambda' \in \mathbb{R}^{k\times k}\}$ are vector spaces of the same dimension and thus must be equal. ∎

The proofs of the following results are similar to those of Theorem 3.3, although technically more complicated, and are omitted.

Theorem 3.9 *Let $A \in \mathbb{R}^{n\times n}$ be symmetric and let $X = (v_1, \cdots, v_n) \in \mathbb{R}^{n\times n}$ be orthogonal with $X^{-1}AX = \mathrm{diag}(\lambda_1, \cdots, \lambda_n), \lambda_1 \geq \cdots \geq \lambda_n$. The generalized Rayleigh quotient $R_A : St(k,n) \to \mathbb{R}$ has at least $\binom{n}{k} = \frac{n!}{k!(n-k)!}$ critical points $X_I = (v_{i_1}, \cdots, v_{i_k})$ for $1 \leq i_1 < \cdots < i_k \leq n$. All other critical points are of the form $\Theta X_I \Psi$, where Θ, Ψ are arbitrary $n \times n$ and $k \times k$ orthogonal matrices with $\Theta' A \Theta = A$. The critical value of R_A at X_I is $\lambda_{i_1} + \cdots + \lambda_{i_k}$. The minimum and maximum value of R_A are*

$$\max_{X'X=I_k} R_A(X) = \lambda_1 + \cdots + \lambda_k , \qquad (3.20)$$

$$\min_{X'X=I_k} R_A(X) = \lambda_{n-k+1} + \cdots + \lambda_n . \qquad (3.21)$$

Theorem 3.10 *The gradient flow of the generalized Rayleigh quotient with respect to the induced Riemannian metric on $St(k,n)$ is*

$$\boxed{\dot{X} = (I - XX')AX.} \qquad (3.22)$$

The solutions of (3.22) exist for all $t \in \mathbb{R}$ and converge to some generalized k-dimensional eigenbasis $\Theta X_I \Psi$ of A. Suppose $\lambda_k > \lambda_{k+1}$. The generalized Rayleigh quotient $R_A : St(k,n) \to \mathbb{R}$ is a Morse-Bott function. For almost all initial conditions $X_0 \in St(k,n)$, $X(t)$ converges to a generalized k-dimensional dominant eigenbasis and $X(t)$ approaches the manifold of all k-dimensional dominant eigenbasis exponentially fast.

The equalities (3.20), (3.21) are due to Ky Fan (1949). As a simple consequence of Theorem 3.9 we obtain the following eigenvalue inequalities between the eigenvalues of symmetric matrices A, B and $A + B$.

Corollary 3.11 *Let $A, B \in \mathbb{R}^{n\times n}$ be symmetric matrices and let $\lambda_1(X) \geq \cdots \geq \lambda_n(X)$ denote the ordered eigenvalues of a symmetric $n \times n$ matrix X. Then for $1 \leq k \leq n$:*

$$\sum_{i=1}^{k} \lambda_i(A+B) \leq \sum_{i=1}^{k} \lambda_i(A) + \sum_{i=1}^{k} \lambda_i(B) .$$

1.3. THE RAYLEIGH QUOTIENT GRADIENT FLOW

Proof We have $R_{A+B}(X) = R_A(X) + R_B(X)$ for any $X \in St(k,n)$ and thus

$$\max_{X'X=I_k} R_{A+B}(X) \leq \max_{X'X=I_k} R_A(X) + \max_{X'X=I_k} R_B(X)$$

Thus the result follows from (3.20) ∎

It has been shown above that the Rayleigh quotient gradient flow (3.10) extends to a gradient flow on $\mathbb{R}^n - \{0\}$. This also happens to be true for the gradient flow (3.22) for the generalized Rayleigh quotient. Consider the extended Rayleigh quotient.

$$R_A(X) = \operatorname{tr}(X'AX(X'X)^{-1})$$

defined on the *noncompact Stiefel manifold* $ST(k,n) = \{X \in \mathbb{R}^{n \times k} \mid \operatorname{rk} X = k\}$ of all full rank $n \times k$ matrices. The gradient flow of $R_A(X)$ on $ST(k,n)$ with respect to the Riemannian metric on $ST(k,n)$ defined by

$$\langle\langle \xi, \eta \rangle\rangle := 2\operatorname{tr}(\xi'(X'X)^{-1}\eta)$$

on $T_X ST(k,n)$ is easily seen to be

$$\dot{X} = (I_n - X(X'X)^{-1}X')AX . \qquad (3.23)$$

For $X'X = I_k$ this coincides with (3.22). Note that every solution $X(t)$ of (3.23) satisfies $\frac{d}{dt}(X'(t)X(t)) = 0$ and hence $X'(t)X(t) = X'(0)X(0)$ for all t.

Instead of considering (3.23) one might also consider (3.22) with arbitrary initial conditions $X(0) \in ST(k,n)$. This leads to a more structurally stable flow on $ST(k,n)$ than (3.22) for which a complete phase portrait analysis is obtained in Yan, Helmke, and Moore (1993).

Remarks 1. The flows (3.10) and (3.22) are identical with those appearing in Oja's work on principal component analysis for neural network applications, see Oja (1982), Oja (1990). Oja shows that a modification of a Hebbian learning rule for pattern recognition problems leads naturally to a class of nonlinear stochastic differential equations. The Rayleigh quotient gradient flows (3.10), (3.22) are obtained using averaging techniques.

2. There is a close connection between the Rayleigh quotient gradient flows and the Riccati equation. In fact let $X(t)$ be a solution of the Rayleigh flow (3.22). Partition $X(t)$ as

$$X(t) = \begin{bmatrix} X_1(t) \\ X_2(t) \end{bmatrix},$$

where $X_1(t)$ is $k \times k$, $X_2(t)$ is $(n-k) \times k$ and $\det X_1(t) \neq 0$. Set

$$K(t) = X_2(t) X_1(t)^{-1}.$$

Then

$$\boxed{\dot{K} = \dot{X}_2 X_1^{-1} - X_2 X_1^{-1} \dot{X}_1 X_1^{-1}.}$$

A straightforward but somewhat lengthy computation, see subsequent Problem , shows that

$$\dot{X}_2 X_1^{-1} - X_2 X_1^{-1} \dot{X}_1 X_1^{-1} = A_{21} + A_{22} K - K A_{11} - K A_{12} K,$$

where

$$A = \begin{bmatrix} A_{11} & A_{12} \\ A_{21} & A_{22} \end{bmatrix}$$

∎

Problems 1. Show that every solution $x(t) \in S^{n-1}$ of the Oja flow (3.15) has the form

$$x(t) = \frac{e^{At} x_0}{\|e^{At} x_0\|}$$

2. Verify that for every solution $X(t) \in St(k,n)$ of (3.22), $H(t) = X(t) X'(t)$ satisfies the double bracket equation (Here, let $[A,B] = AB - BA$ denote the Lie bracket)

$$\boxed{\begin{aligned} \dot{H}(t) &= [H(t), [H(t), A]] \\ &= H^2(t) A + A H^2(t) - 2 H(t) A H(t) \,. \end{aligned}}$$

3. Verify the computations in Remark 2.

Main Points of Section The Rayleigh quotient of a symmetric matrix $A \in \mathbb{R}^{n \times n}$ is a smooth function $r_A : \mathbb{R}^n - \{0\} \to \mathbb{R}$. Its gradient flow on the unit sphere S^{n-1} in $\mathbb{R}^n$, with respect to the standard (induced) Riemannian metric, converges to an eigenvector of A. Moreover, the convergence is exponential. Generalized Rayleigh quotients are defined on Stiefel manifolds and their gradient flows converge to the k-dimensional dominant eigenspace, being a locally stable attractor for the flow. The notions of tangent spaces, normal spaces, as well as a precise understanding of the geometry of the constraint set are important for the theory of such gradient flows. There are important connections with Riccati equations and neural network learning schemes.

1.4 The QR Algorithm

In the previous sections we deal with the questions of how to determine a dominant eigenspace for a symmetric real $n \times n$ matrix. Given a real symmetric $n \times n$ matrix A, the QR algorithm produces an infinite sequence of real symmetric matrices $A_1, A_2, \cdots$ such that for each i the matrix A_i has the same eigenvalues as A. Also, A_i converges to a diagonal matrix as i approaches infinity.

The QR algorithm is described in more precise terms as follows. Let $A_0 = A \in \mathbb{R}^{n \times n}$ be symmetric with the QR-decomposition

$$A_0 = \Theta_0 R_0 , \tag{4.1}$$

where $\Theta_0 \in \mathbb{R}^{n \times n}$ is orthogonal and

$$R_0 = \begin{bmatrix} * & \cdots & * \\ & \ddots & \vdots \\ 0 & & * \end{bmatrix} \tag{4.2}$$

is upper triangular with nonnegative entries on the diagonal. Such a decomposition can be obtained by applying the Gram-Schmidt orthogonalization procedure to the columns of A_0.

Define
$$A_1 := R_0 \Theta_0 = \Theta_0' A_0 \Theta_0 . \tag{4.3}$$

Repeating the QR factorization for A_1,

$$A_1 = \Theta_1 R_1$$

with Θ_1 orthogonal and R_1 as in (4.2). Define

$$A_2 = R_1 \Theta_1 = \Theta_1' A_1 \Theta_1 = (\Theta_0 \Theta_1)' A_0 \Theta_0 \Theta_1 . \tag{4.4}$$

Continuing with the procedure a sequence $(A_i \mid i \in \mathbb{N})$ of real symmetric matrices is obtained with QR decomposition

$$\begin{align} A_i &= \Theta_i R_i , \tag{4.5} \\ A_{i+1} &= R_i \Theta_i = \Theta_i' A_i \Theta_i , \quad i \in \mathbb{N} . \tag{4.6} \end{align}$$

Thus for all $i \in \mathbb{N}$

$$A_i = (\Theta_0 \cdots \Theta_i)' A_0 (\Theta_0 \cdots \Theta_i) , \tag{4.7}$$

and A_i has the same eigenvalues as A_0. Now there is a somewhat surprising result; under suitable genericity assumptions on A_0 the sequence (A_i) always converges to a diagonal matrix, whose diagonal entries coincide with the eigenvalues of A_0. Moreover, the product $\Theta_0 \cdots \Theta_i$ of orthogonal matrices approximates, for i large enough, the eigenvector decomposition of A_0. This result is proved in the literature on the QR-algorithm, see Golub and Van Loan (1989), Chapter 7, and references.

The QR algorithm for diagonalizing symmetric matrices as described above, is not a particularly efficient method. More effective versions of the QR algorithm are available which make use of so-called variable shift strategies; see e.g. Golub and Van Loan (1989). Continuous-time analogues of the QR algorithm have appeared in recent years. One such analogue of the QR algorithm for tridiagonal matrices is the Toda flow, which is an integrable Hamiltonian system. Examples of recent studies are Symes (1982), Deift, Nanda and Tomei (1983), Nanda (1985), Chu (1984), Watkins (1984), Watkins and Elsner(1989). As pointed out by Watkins and Elsner, early versions of such continuous-time methods for calculating eigenvalues of matrices were already described in the early work of Rutishauser (1954, 1958). Here we describe only one such flow which interpolates the QR algorithm.

A differential equation

$$\dot{A}(t) = f(t, A(t)) \tag{4.8}$$

defined on the vector space of real symmetric matrices $A \in \mathbb{R}^{n \times n}$ is called *isospectral* if every solution $A(t)$ of (4.8) is of the form

$$A(t) = \Theta(t)' A(0) \Theta(t) \tag{4.9}$$

with orthogonal matrices $\Theta(t) \in \mathbb{R}^{n \times n}$, $\Theta(0) = I_n$. The following simple characterization is well-known, see the above mentioned literature.

Lemma 4.1 *Let $I \subset \mathbb{R}$ be an interval and let $B(t) \in \mathbb{R}^{n \times n}$, $t \in I$, be a continuous, time-varying family of skew-symmetric matrices. Then*

$$\boxed{\dot{A}(t) = A(t)B(t) - B(t)A(t)} \tag{4.10}$$

is isospectral. Conversely, every isospectral differential equation on the vector space of real symmetric $n \times n$ matrices is of the form (4.10) with $B(t)$ skew-symmetric.

Proof Let $\Theta(t)$ denote the unique solution of

$$\dot{\Theta}(t) = \Theta(t) B(t), \quad \Theta'(0) \Theta(0) = I_n . \tag{4.11}$$

Since $B(t)$ is skew-symmetric we have

$$\begin{aligned}
\frac{d}{dt} \Theta(t)' \Theta(t) &= \dot{\Theta}(t)' \Theta(t) + \Theta(t)' \dot{\Theta}(t) \\
&= B(t)' \Theta(t)' \Theta(t) + \Theta(t)' \Theta(t) B(t) \\
&= \Theta(t)' \Theta(t) B(t) - B(t) \Theta(t)' \Theta(t) ,
\end{aligned} \tag{4.12}$$

1.4. THE QR ALGORITHM

and $\Theta(0)'\Theta(0) = I_n$. By the uniqueness of the solutions of (4.12) we have $\Theta(t)'\Theta(t) = I_n$, $t \in I$, and thus $\Theta(t)$ is orthogonal. Let $\widehat{A}(t) = \Theta(t)'A(0)\Theta(t)$. Then $\widehat{A}(0) = A(0)$ and

$$\begin{aligned}\frac{d}{dt}\widehat{A}(t) &= \dot\Theta(t)'A(0)\Theta(t) + \Theta(t)'A(0)\dot\Theta(t) \\ &= \widehat{A}(t)B(t) - B(t)\widehat{A}(t) \, .\end{aligned} \quad (4.13)$$

Again, by the uniqueness of the solutions of (4.13), $\widehat{A}(t) = A(t), t \in I$, and the result follows. ∎

In the sequel we will frequently make use of the *Lie bracket* notation

$$[A, B] = AB - BA \quad (4.14)$$

for $A, B \in \mathbb{R}^{n \times n}$. Given an arbitrary $n \times n$ matrix $A \in \mathbb{R}^{n \times n}$ there is a unique decomposition

$$A = A_- + A_+ \quad (4.15)$$

where A_- is skew-symmetric and A_+ is upper triangular. Thus for $A = (a_{ij}) \in \mathbb{R}^{n \times n}$ symmetric then

$$A_- = \begin{bmatrix} 0 & -a_{21} & \cdots & -a_{n1} \\ a_{21} & 0 & & \vdots \\ \vdots & & \ddots & -a_{n,n-1} \\ a_{n1} & \cdots & a_{n,n-1} & 0 \end{bmatrix},$$

$$A_+ = \begin{bmatrix} a_{11} & 2a_{21} & \cdots & 2a_{n1} \\ 0 & a_{22} & \ddots & \vdots \\ \vdots & & \ddots & 2a_{n,n-1} \\ 0 & \cdots & 0 & a_{nn} \end{bmatrix}. \quad (4.16)$$

For any positive definite real symmetric $n \times n$ matrix A there exists a unique real symmetric logarithm of A

$$B = \log(A)$$

with $e^B = A$. The restriction to positive definite matrices avoids any possible ambiguities in defining the logarithm. With this notation we then have the following result, see Chu (1984) for a more complete theory.

Theorem 4.2 *The differential equation*

$$\boxed{\dot{A} = [A, (\log A)_-] = A(\log A)_- - (\log A)_- A, \quad A(0) = A_0} \quad (4.17)$$

is isospectral and the solution $A(t)$ exists for all $A_0 = A_0' > 0$ and $t \in \mathbb{R}$. If $(A_i \mid i \in \mathbb{N})$ is the sequence produced by the QR-algorithm for $A_0 = A(0)$, then

$$A_i = A(i) \text{ for all } i \in \mathbb{N}. \tag{4.18}$$

Proof The proof follows that of Chu (1984). Let

$$e^{t \cdot \log A(0)} = \Theta(t) R(t) \tag{4.19}$$

be the QR-decomposition of $e^{t \cdot \log A(0)}$. By differentiating both sides with respect to t, we obtain

$$\begin{aligned}\dot{\Theta}(t) R(t) + \Theta(t) \dot{R}(t) &= \log A(0) e^{t \cdot \log A(0)} \\ &= (\log A(0)) \cdot \Theta(t) R(t) \,.\end{aligned}$$

Therefore

$$\Theta(t)' \dot{\Theta}(t) + \dot{R}(t) R(t)^{-1} = \log(\Theta(t)' A(0) \Theta(t))$$

Since $\Theta(t)' \dot{\Theta}(t)$ is skew-symmetric and $\dot{R}(t) R(t)^{-1}$ is upper triangular, this shows

$$\dot{\Theta}(t) = \Theta(t)(\log(\Theta(t)' A(0) \Theta(t))_- \,. \tag{4.20}$$

Consider $A(t) := \Theta(t)' A(0) \Theta(t)$. By (4.20)

$$\begin{aligned}\dot{A} &= \dot{\Theta}' A(0) \Theta + \Theta' A(0) \dot{\Theta} \\ &= -(\log A)_- \Theta' A(0) \Theta + \Theta' A(0) \Theta (\log A)_- \\ &= [A, (\log A)_-] \,.\end{aligned}$$

Thus $A(t)$ is a solution of (4.17). Conversely, since $A(0)$ is arbitrary and by the uniqueness of solutions of (4.17), any solution $A(t)$ of (4.17) is of the form $A(t) = \Theta(t)' A(0) \Theta(t)$ with $\Theta(t)$ defined by (4.19). Thus

$$\begin{aligned}e^{t \cdot \log A(t)} &= e^{t \cdot \log(\Theta(t)' A(0) \Theta(t))} = \Theta(t)' e^{t \cdot \log A(0)} \Theta(t) \\ &= R(t) \Theta(t) \,.\end{aligned}$$

Therefore for $t = 1$, and using (4.19)

$$\begin{aligned}A(1) &= e^{\log A(1)} = R(1) \Theta(1) \\ &= \Theta(1)' \Theta(1) R(1) \Theta(1) \\ &= \Theta(1)' A(0) \Theta(1) \,.\end{aligned}$$

Proceeding by induction this shows that for all $k \in \mathbb{N}$

$$A(k) = \Theta(k)' \cdots \Theta(1)' A(0) \Theta(1) \cdots \Theta(k)$$

1.4. THE QR ALGORITHM

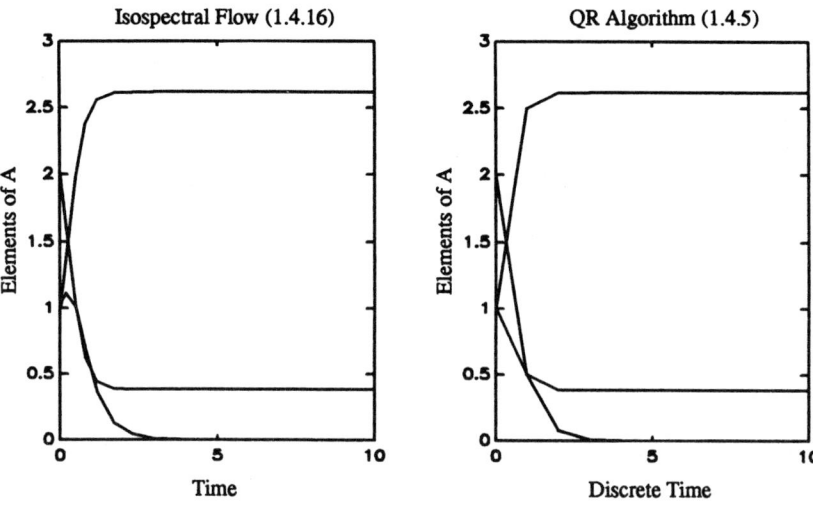

Figure 1.4.1: Isospectral flow and QR algorithm

is the k-th iterate produced by the QR algorithm. This completes the proof. ∎

As has been mentioned above, the QR algorithm is not a numerically efficient method for diagonalizing a symmetric matrix. Neither would we expect that integration of the continuous time flow (4.17), by e.g. a Runge-Kutta method, leads to a numerically efficient algorithm for diagonalization. Numerically efficient versions of the QR algorithms are based on variable shift strategies. These are not described here and we refer to e.g. Golub and Van Loan (1989) for a description.

Example For the case $A_0 = \begin{bmatrix} 1 & 1 \\ 1 & 2 \end{bmatrix}$, the isospectral flow (4.17) is plotted in Figure 1.4.1. The time instants $t = 1, 2, \cdots$ are indicated. ∎

Main Points of Section The QR algorithm produces a sequences of real symmetric $n \times n$ matrices A_i, each with the same eigenvalues, and which converges to a diagonal matrix A_∞ consisting of the eigenvalues of A. Isospectral flows are continuous-time analogs of the QR algorithm flows. One such is the Lie-bracket equation $\dot{A} = [A, (\log A)_-]$ where X_- denotes the skew-symmetric component of a matrix X. The QR algorithm is the integer time sampling of this isospectral flow. Efficient versions of the QR algorithm require shift strategies not studied here.

1.5 Singular Value Decomposition (SVD)

The *singular value decomposition* is an invaluable tool in numerical linear algebra, statistics, signal processing and control theory. It is a cornerstone of many reliable numerical algorithms, such as those for solving linear equations, least squares estimation problems, controller optimization, and model reduction. A variant of the QR algorithm allows the singular value decomposition of an arbitrary matrix $A \in \mathbb{R}^{m \times n}$ with $m \geq n$. The singular value decomposition (SVD) of A is

$$A = V\Sigma U' \quad \text{or} \quad V'AU = \Sigma, \quad \Sigma = \begin{bmatrix} \text{diag}(\sigma_1, \cdots, \sigma_n) \\ 0_{(m-n) \times n} \end{bmatrix}. \quad (5.1)$$

Here U, V are orthogonal matrices with $U'U = UU' = I_n$, $V'V = VV' = I_m$, and the $\sigma_i \in \mathbb{R}$ are the nonnegative singular values of A, usually ordered so that $\sigma_1 \geq \sigma_2 \geq \cdots \geq \sigma_n \geq 0$. The *singular values* $\sigma_i(A)$ are the nonnegative square roots of the eigenvalues of $A'A$. If A is Hermitian, then the singular values of A are identical to the absolute values of the eigenvalues of A. However, this is not true in general. The columns of V and U' are called the left and the right *singular vectors* of A, respectively.

Of course from (5.1) we have that

$$U'(A'A)U = \Sigma^2 = \text{diag}(\sigma_1^2, \cdots, \sigma_n^2),$$

$$V'(AA')V = \begin{bmatrix} \Sigma^2 & 0 \\ 0 & 0 \end{bmatrix} = \text{diag}(\sigma_1^2, \cdots, \sigma_n^2, \underbrace{0, \cdots, 0}_{m-n}). \quad (5.2)$$

Clearly, since $A'A, AA'$ are symmetric the application of the QR algorithm to $A'A$ or AA' would achieve the decompositions (5.1), (5.2). However, an unattractive aspect is that matrix squares are introduced with associated squaring of condition numbers and attendant loss of numerical accuracy.

An alternative reformulation of the SVD task which is suitable for application of the QR algorithm is as follows: Define

$$\widehat{A} = \begin{bmatrix} 0 & A \\ A' & 0 \end{bmatrix} = \widehat{A}', \quad (5.3)$$

then for orthogonal $\widehat{V}$,

$$\widehat{V}'\widehat{A}\widehat{V} = \begin{bmatrix} \Sigma & 0 & 0 \\ 0 & -\Sigma & 0 \\ 0 & 0 & 0 \end{bmatrix} = \text{diag}(\sigma_1, \cdots, \sigma_n, -\sigma_1, \cdots, -\sigma_n, \underbrace{0, \cdots, 0}_{m-n}). \quad (5.4)$$

1.5. SINGULAR VALUE DECOMPOSITION (SVD)

Simple manipulations show that

$$\widehat{V} = \frac{1}{\sqrt{2}} \begin{bmatrix} V_1 & V_1 & V_2 \\ U & -U & 0 \end{bmatrix},$$
$$V = \begin{bmatrix} V_1 & V_2 \\ n & m-n \end{bmatrix}. \quad (5.5)$$

with orthogonal matrices U, V satisfying (5.1). Actually, the most common way of calculating the SVD of $A \in \mathbb{R}^{m \times n}$ is to apply an implicit QR algorithm based on the work of Golub and Kahan (1965). In this text, we seek implementations via dynamical systems, and in particular seek gradient flows, or more precisely *self-equivalent* flows which preserve singular values. A key result on such flows is as follows.

Lemma 5.1 *Let $C(t) \in \mathbb{R}^{n \times n}$ and $D(t) \in \mathbb{R}^{m \times m}$ be a time-varying family of skew-symmetric matrices on a time interval. Then the flow on $\mathbb{R}^{m \times n}$,*

$$\boxed{\dot{H}(t) = H(t)C(t) + D(t)H(t)} \quad (5.6)$$

is self equivalent on this interval. Conversely, every self equivalent differential equation on $\mathbb{R}^{m \times n}$ is of the form (5.6) with $C(t), D(t)$ skew-symmetric.

Proof Let $(U(t), V(t))$ denote the unique solutions of

$$\begin{aligned} \dot{U}(t) &= U(t)C(t), U(0) = I_n, \\ \dot{V}(t) &= -V(t)D(t), V(0) = I_m. \end{aligned} \quad (5.7)$$

Since C, D are skew-symmetric, then U, V are orthogonal matrices. Let $A(t) = V'(t)H(0)U(t)$. Then $A(0) = H(0)$ and

$$\dot{A}(t) = V'(t)H(0)\dot{U}(t) + \dot{V}'(t)H(0)U(t) = A(t)C(t) + D(t)A(t).$$

Thus $A(t) = H(t)$ by uniqueness of solutions. ∎

Remarks 1. In the discrete-time case, any self equivalent flow takes the form

$$\boxed{H_{k+1} = S_k e^{C_k} H_k e^{D_k} T_k,} \quad (5.8)$$

where C_k, D_k are skew-symmetric with S_k, T_k appropriate sign matrices of the form $\mathrm{diag}(\pm 1, 1, \cdots, 1)$ and consequently e^{C_k}, e^{D_k} are orthogonal matrices. The proof is straightforward.

2. In Chu (1986) and Watkins and Elsner (1989), self-equivalent flows are developed which are interpolations of an explicit QR based SVD method. In particular, for square and nonsingular H_0 such a flow is

$$\boxed{\dot{H}(t) = [H(t), (\log(H'(t)H(t)))_-], \quad H(0) = H_0.} \quad (5.9)$$

Here, the notation A_- denotes the skew-symmetric part of A, see (4.15) - (4.16). This self-equivalent flow then corresponds to the isospectral QR flow of Chu (1986) and Watkins and Elsner (1989). Also so called shifted versions of the algorithm for SVD are considered by Watkins and Elsner (1989). ∎

Main Points of Section The singular value decomposition of $A \in \mathbb{R}^{m \times n}$ can be achieved using the QR algorithm on the symmetric matrix $\widehat{A} = \begin{bmatrix} 0 & A \\ A' & 0 \end{bmatrix}$. The associated flows on rectangular matrices A preserve the singular values and are called self-equivalent flows. General self-equivalent flows on $\mathbb{R}^{m \times n}$ are characterized by pairs of possibly time-variant skew-symmetric matrices. They can be considered as isospectral flows on symmetric $(m+n) \times (m+n)$ matrices.

1.6 Standard Least Squares Gradient Flows

So far in this chapter, we have explored dynamical systems evolving on rather simple manifolds such as spheres, projective spaces or Grassmannians for eigenvalue methods. The dynamical systems have not always been gradient flows nor were they always related to minimizing the familiar least squares measures. In the next chapters we explore more systematically eigenvalue methods based on matrix least squares. As a prelude to this, let us briefly develop standard vector least squares gradient flows. In the first instance there will be no side constraints and the flows will be on $\mathbb{R}^n$.

The standard (vector) least squares minimization task is to minimize over $x \in \mathbb{R}^n$ and for $A \in \mathbb{R}^{m \times n}$, $b \in \mathbb{R}^m$, the 2-norm index

$$\begin{aligned} \Phi(x) &= \frac{1}{2}\|Ax - b\|_2^2 \\ &= \frac{1}{2}(Ax-b)'(Ax-b) \end{aligned} \quad (6.1)$$

The directional derivative $D\Phi|_x(\xi)$ is given in terms of the gradient $\nabla \Phi(x) = (\frac{\partial \Phi}{\partial x_1}, \ldots, \frac{\partial \Phi}{\partial x_n})'$ as

$$D\Phi|_x(\xi) = (Ax-b)'A\xi =: \nabla \Phi(x)'\xi \quad (6.2)$$

Consequently, the gradient flow for least squares minimizing (6.1) in $\mathbb{R}^n$ is

$$\boxed{\dot{x} = -\nabla \Phi(x), \quad \nabla \Phi(x) = A'(Ax-b)} \quad (6.3)$$

1.6. STANDARD LEAST SQUARES GRADIENT FLOWS

Here $\mathbb{R}^n$ is considered in the trivial way as a smooth manifold for the flow, and $x(0) \in \mathbb{R}^n$. The equilibria satisfy

$$\nabla \Phi(x) = A'(Ax - b) = 0 \tag{6.4}$$

If A is injective, then $A'A > 0$ and there is a unique equilibrium $x_\infty = (A'A)^{-1}A'b$ which is exponentially stable. The exponential convergence rate is given by $\rho = \lambda_{\min}(A'A)$, that is by the square of the smallest singular value of A.

Constrained Least Squares Of more interest to the theme of the book are constrained least squares gradient flows. Let us impose for example an additional *affine equality constraint*, for a full row rank matrix $L \in \mathbb{R}^{p \times n}$, $c \in \mathbb{R}^p$

$$Lx = c \tag{6.5}$$

This affine constraint defines a smooth constraint manifold M in $\mathbb{R}^n$ with tangent spaces given by

$$T_x M = \{\xi | L\xi = 0\} \, , \, x \in M. \tag{6.6}$$

The gradient flow on M requires an orthogonal projection of the directional derivative into the tangent space, so we need that the gradient $\nabla \Phi(x)$ satisfy both (6.2) and $L \nabla \Phi(x) = 0$. Note that $P := I - (L')^{\#}L := I - L'(LL')^{-1}L$ is the linear projection operator from $\mathbb{R}^n$ onto $T_x M$. Thus

$$\dot{x} = -\nabla \Phi(x) \, , \, \nabla \Phi(x) = (I - (L')^{\#}L)A'(Ax - b) \tag{6.7}$$

is the gradient flow on M. Of course, the initial condition $x(0)$ satisfies $Lx(0) = c$. Here $\#$ denotes the pseudo-inverse. Again, there is exponential convergence to the equilibrium points satisfying $\nabla \Phi(x) = 0$. There is a unique equilibrium when A is full rank. The exponential convergence rate is given by $\rho = \lambda_{min}[(I - (L')^{\#}L)A'A]$.

As an example of a *nonlinear constraint*, let us consider *least squares estimation on a sphere*. Thus, consider the constraint $\|x\|_2 = 1$, defining the $(n-1)$-dimensional sphere S^{n-1}. Here the underlying geometry of the problem is the same as that developed for the Rayleigh quotient gradient flow of Section 1.3. Thus the gradient vector

$$\nabla \Phi(x) = A'(Ax - b) - (x'A'(Ax - b))x \tag{6.8}$$

satisfies both (6.2) on the tangent space $T_x S^{n-1} = \{\xi \in \mathbb{R}^n \mid x'\xi = 0\}$ and is itself in the tangent space for all $x \in S^{n-1}$, since $x' \nabla \Phi(x) = 0$ for all $\|x\|_2 = 1$. The gradient flow for least squares estimation on the sphere S^{n-1} is $\dot{x} = -\nabla \Phi(x)$, that is

$$\boxed{\dot{x} = (x'A'(Ax - b))x - A'(Ax - b)} \tag{6.9}$$

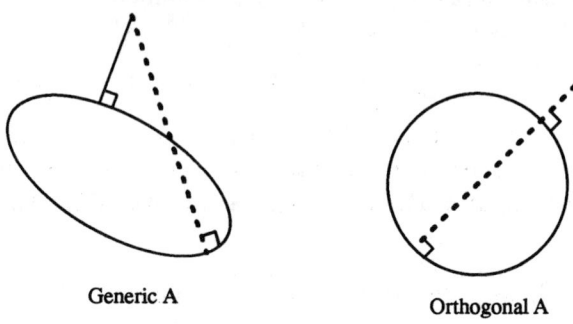

Figure 1.6.1: Least squares minimization

and the equilibria are characterized by

$$(x'A'(Ax - b))x = A'(Ax - b).$$

In general, there is no simple analytical formula for the equilibria x. However, for $m = n$, it can be shown that there are precisely two equilibria. There is a unique minimum (and a unique maximum), being equivalent to a minimization of the distance of $b \in \mathbb{R}^n$ to the surface of the ellipsoid $\{y \in \mathbb{R}^n \mid \|A^{-1}y\| = 1\}$ in $\mathbb{R}^n$, as depicted in Figure 1.6.1. If A is an orthogonal matrix, this ellipsoid is the sphere S^{n-1}, and the equilibria are $x^* = \pm\|A^{-1}b\|^{-1}A^{-1}b$. A linearization of the gradient flow equation at the equilibrium point with a positive sign yields an exponentially stable linear differential equation and thus ensures that this equilibrium is exponentially stable. The maximum is unstable, so that for all initial conditions different from the maximum point the gradient flow converges to the desired optimum.

Main Points of Section Certain constrained least squares optimization problems can be solved via gradient flows on the constraint manifolds. The cases discussed are the smooth manifolds $\mathbb{R}^n$, an affine space in $\mathbb{R}^n$, and the sphere S^{n-1}. The estimation problem on the sphere does not always admit a closed- form solution, thus numerical techniques are required to solve this problem.

The somewhat straightforward optimization procedures of this section and chapter form a prototype for some of the more sophisticated matrix least squares optimization problems in later chapters.

Notes for Chapter 1

For a general reference on smooth dynamical systems we refer to Palis and de Melo (1982). Structural stability properties of two-dimensional vector fields were investigated by Andronov and Pontryagin (1937), Andronov, Vitt and Khaikin (1987) and Peixoto (1962). A general investigation of structural stability properties of gradient vector fields on manifolds is due to Smale (1961). Smale shows that a gradient flow $\dot{x} = \text{grad } \Phi(x)$ on a smooth compact manifold is structurally stable if the Hessian of Φ at each critical point is nondegenerate, and the stable and unstable manifolds of the equilibria points intersect each other transversally. Morse has considered smooth functions $\Phi: M \to \mathbb{R}\mathbb{R}$ such that the Hessian at each critical point is nondegenerate. Such functions are called Morse functions and they form a dense subset of the class of all smooth functions on M. See Milnor (1962) and Bott (1982) for nice expositions on Morse theory. Other recent references are Henniart (1983) and Witten (1982). A generalization of Morse theory is due to Bott (1954) who considers smooth functions with nondegenerate manifolds of critical points. An extension to Morse theory on infinite dimensional spaces is due to Palais (1962).

Thom (1949) has shown that the stable and unstable manifolds of an equilibrium point of a gradient flow $\dot{x} = \text{grad } \Phi(x)$, for a smooth Morse function $\Phi: M \to \mathbb{R}$, are diffeomorphic to Euclidean spaces $\mathbb{R}^k$, i.e. they are cells. The decomposition of M by the stable manifolds of a gradient flow for a Morse function $\Phi: M \to \mathbb{R}$ is a cell decomposition. Refined topological properties of such cell decompositions were investigated by Smale (1960).

Convergence and stability properties of gradient flows may depend on the choice of the Riemannian metric. If $a \in M$ is a nondegenerate critical point of $\Phi: M \to \mathbb{R}$, then the local stability properties of the gradient flow $\dot{x} = \text{grad } \Phi(x)$ around a do not change with the Riemannian metric. However, if a is a degenerate critical point, then the qualitative picture of the local phase portrait of the gradient around a may well change with the Riemannian metric. See Shafer (1980) for results in this direction.

For an example of a gradient flow where the solutions converge to a set of equilibrium points rather than to single equilibria, see Palis and de Melo (1982). A recent book on differential geometric methods written from an applications oriented point of view is Doolin and Martin (1990), with background material on linear quadratic optimal control problems, the Riccati equation and flows on Grassmann manifolds. Background material on the matrix Riccati equation can be found in Reid (1972) and Anderson and Moore (1990). An analysis of the Riccati equation from a Lie theoretic perspective is given by Hermann (1979), Hermann and Martin (1982). A complete phase portrait analysis of the matrix Riccati equation has been obtained by Shayman (1986).

Structural stability properties of the matrix Riccati equation were studied by Schneider (1973), Bucy (1975). Finite escape time phenomena of the matrix Riccati differential equation are well studied in the literature, see Martin (1981) for a recent reference.

Numerical matrix eigenvalue methods are analyzed from a Lie group theoretic perspective by Ammar and Martin (1985). There also the connection of the matrix Riccati equation with numerical methods for finding invariant subspaces of a matrix by the power method is made. An early reference for this is Parlett and Poole (1973).

For a statement and proof of the Courant-Fischer minmax principle see Golub and Van Loan (1989). A generalization of Theorem 3.9 is Wielandt's minmax principle on partial sums of eigenvalues. For a proof as well as applications to eigenvalue inequalities for sums of Hermitian matrices see Bhatia (1987). The eigenvalue inequalities appearing in Corollary 3.11 are the simplest of a whole series of eigenvalue inequalities for sums of Hermitian matrices. They include those of Cauchy-Weyl, Lidskii, and Thompson and Freede. For this we refer to Bhatia (1987) and Marshall and Olkin (1979).

For an early investigation of isospectral matrix flows as a tool for solving matrix eigenvalue problems we refer to Rutishauser (1954). Apparently his pioneering work has long been overlooked until its recent rediscovery by Watkins and Elsner (1989). The first more recent papers on dynamical systems applications to numerical analysis are Kostant (1979), Symes (1980, 1982), Deift, Nanda and Tomei (1983), Chu (1984, 1986), Watkins (1984) and Nanda (1985).

References

Anderson, B.D.O. and J.B. Moore (1990) *Optimal Control: Linear Quadratic Methods*, Prentice Hall, New York

Ammar, G.S. and C.F. Martin (1986) The geometry of matrix eigenvalue methods, *Acta Applicandae Math.* 5, 239-278.

Andronov, A.A. and L. Pontryagin (1937) *Systèmes grossiers*, C.R. (Dokl.) Acad. URSS 14, 247-251

Andronov, A.A., Vitt, A.A. and S.E. Khaikin (1987) *Theory of Oscillators*, Dover Publ., Inc., New York.

Auchmuty, G. (1991) Globally and rapidly convergent algorithms for symmetric eigen problems, *SIAM Journal of Matrix Analysis and Applications*, 12, 690-706.

1.6. STANDARD LEAST SQUARES GRADIENT FLOWS

Beattie, C. and D.W. Fox (1989) Localization criteria and containment for Rayleigh quotient iteration, *SIAM Journal of Matrix Analysis and Applications*, 10, 80-93.

Batterson, S. and J. Smilie (1989) The dynamics of Rayleigh quotient iteration, *SIAM Journal of Numerical Analysis*, 26, 624-636.

Bhatia, R. (1987) *Perturbation bounds for matrix eigenvalues*, Pitman Research Notes in Mathematics Series 162, Longman Scientific & Technical, Harlow.

Bott, R. (1954) Nondegenerate critical manifolds, *Ann. of Mathematics*, 60, 248-261.

Bott, R. (1982) Lectures on Morse theory, old and new, *Bulletin of the American Mathematical Society*, 7, 331-358.

Bucy, R.S. (1975) Structural stability for the Riccati equation, *SIAM J. on Optimization and Control* 13, 749-753.

Chu, M.T. (1984) The generalized Toda flow, the QR algorithm and the center manifold theory, *SIAM J. Alg. Disc. Methods*, 5, 187-201.

Chu, M.T. (1986) A differential equation approach to the singular value decomposition of bidiagonal matrices, *Linear Algebra and its Applications*, 80, 71-80

Chu, M.T. (1988) On the continuous realization of iterative processes, *SIAM Review*, 30, 375-387.

Courant, R. (1922) Zur Theorie der kleinen Schwingungen, *Zts. f. angew. Math. und Mech.*, 2, 278-285.

Deift, D., T. Nanda, and C. Tomei (1983) Ordinary differential equations and the symmetric eigenvalue problems, *SIAM J. Numer. Anal*, 20, 1-22.

de la Rocque Palis, G. (1978) Linearly induced vector fields and $\mathbb{R}^2$ actions on spheres, *Journal of Differential Geometry*, 13, 163-191.

Doolin, B.F. and C.F. Martin (1990) *Introduction to Differential Geometry for Engineers*, Marcel Dekker, Inc. New York

Fan, K. (1949) On a theorem of Weyl concerning eigenvalues of linear transformations 1, *Proc. Nat. Acad. Sci.*, U.S.A.,35, 652-655.

Fischer, E. (1905) Über quadratische Formen mit reelen Koeffizienten, *Monatsh. f. Math. und Phys.*, 16, 234-249.

Francis, J.G.F. (1961) The QR transformation, a unitary analogue to the LR transformation, *Comput. J.*, 4, 265-280.

Frankel, T. (1965) Critical submanifolds of the classical groups and Stiefel manifolds, *Differential and Combinatorial Topology*, (S.S Cairns, Ed.,) Princeton University Press, Princeton, N.J.

Golub, G.H. and W. Kahan (1965) Calculating the singular values and pseudo-inverse of a matrix, *SIAM J. Numerical Anal.*, Series B2, 205-224.

Golub, G.H. and C.F. Van Loan (1989) *Matrix Computations*. Second Edition. The John Hopkins University Press, Baltimore.

Henniart, G. (1983) Les inegalities de Morse, *Seminaire Bourbaki*, 617, 1-17.

Hermann, R. (1979) *Cartanian geometry, nonlinear waves, and control theory, Part A*, Interdisciplinary Mathematics, Vol. 20, Math. Sci. Press, Brookline MA 02146.

Hermann, R. and C.F. Martin (1982) Lie and Morse theory of periodic orbits of vector fields and matrix Riccati equations I: General Lie-theoretic methods, *Math. Systems Theory* 15, 277-284.

Hirsch, M.W. (1976) *Differential topology*, Graduate Text in Math, 33, Springer Verlag, New York.

Kostant, B. (1979) The Solution to a generalized Toda lattice and representation theory, *Advances in Mathematics* 34, 195 - 338.

Martin, C.F. (1981) Finite escape time for Riccati differential equations, *Systems and Control Letters* 1, 127-131.

Marshall, A.W. and I. Olkin (1979) *Inequalities: Theory of Majorization and its Applications*, Academic Press, New York.

Milnor, J. (1963) *Morse Theory*, Princeton University Press.

Nanda, T. (1985) Differential equations and the QR algorithm, *SIAM J. Numer. Anal.* 22, 310-321.

Oja, E. (1982) A simplified neuron model as a principal component analyzer, *J. of Math. Biology.* 15, 267-273.

Oja, E. (1990) Neural networks, principal components and subspaces, *Int. Journal of Neural Systems*, Vol. 1, No. 1, 61-68.

Palais, R.S. (1962) Morse theory on Hilbert manifolds, *Topology*, 2, 299-340.

Palis, J. and W. de Melo (1982) *Geometric Theory of Dynamical Systems*, Springer-Verlag, New York.

Parlett, B.N. and W.G. Poole (1973) A geometric theory for the QR, LU and Power iterations, *SIAM J. Numer. Anal.* 10, 389-412.

Peixoto, M.M. (1962) Structural stability on two-dimensional manifolds, *Topology* 1, 101-120.

Reid, W.T. (1972) *Riccati Differential Equations*, Academic Press, New York.

Riddell, R.C. (1984) Minimax problems on Grassmann manifolds. Sums of eigenvalues, *Advances in Mathematics*, 54, 107-109.

Rutishauser, H. (1954) Ein Infinitesimales Analogon zum Quotienten-Differenzen-Algorithmus, *Arch. Math.*, (Basel), 5, 132-137.

Rutishauser, H. (1958) Solution of eigenvalue problems with the LR-transformation, *National Bureau of Standards Applied Mathematics Series*, 49, 47-81.

Schneider, C.R. (1973) Global aspects of the matrix Riccati equation, *Math. System Theory* 1, 281-286.

Shafer, D. (1980) Gradient vectorfields near degenerate singularities, *Global Theory of Dynamical Systems* (Z. Nitecki and C. Robinson, Eds.), Lecture Notes in Mathematics 819, Springer-Verlag, Berlin, 410-417.

Shayman, M.A. (1986) Phase portrait of the matrix Riccati equation, *SIAM J. Control and Optimization*, 24, 1-65.

Smale, S. (1960) Morse inequalities for a dynamical system, *Bull. Amer. Math. Soc.* 66, 43-49.

Smale, S. (1961) On gradient dynamical systems, *Annals of Mathematics* 74, 199-206.

Symes, W.W. (1980) Hamiltonian Group Actions and Integrable Systems, *Physica* 1D, 339-374.

Symes, W.W. (1982) The QR algorithm and scattering for the finite nonperiodic Toda lattice, *Physica.* 4D, 275-280.

Thom, R. (1949) Sur une partition en cellules associee anne fonction sur une varite, *C.R. Acad. Sci. Paris*, Series A-B, 228, 973-975.

von Neumann, J. (1937) Some matrix-inequalities and metrization of matricspaces, *Tomsk Univ. Rev.*, 1, 286-300. See also *John von Neumann Collected Works*, Vol. IV, (A.H. Taub Ed.) Pergamon Press, New York, 1962, 205-218.

Watkins, D.S. (1984) Isospectral Flows, *SIAM Rev.*, 26, 379-391.

Watkins, D. (1982) Understanding the QR algorithm, *SIAM Rev.* 24, 379-391.

Watkins, D. and L. Elsner (1989) Self-equivalent flows associated with the singular value decomposition, *SIAM J. Matrix Anal. Appl..* 10, 244-258.

Witten, E. (1982) Supersymmetry and Morse theory, *Journal of Diff. Geometry*, 17, 661-692.

Yan, W.Y, U. Helmke and J.B. Moore (1993) Global analysis of Oja's flow for neural networks, *IEEE Trans. on Neural Networks*, to appear.

Chapter 2

Double Bracket Isospectral Flows

2.1 Double Bracket Flows for Diagonalization

Brockett (1988) has introduced a new class of isospectral flows on the set of real symmetric matrices which have remarkable properties. He considers the ordinary differential equation

$$\dot{H}(t) = [H(t), [H(t), N]], \quad H(0) = H_0 \qquad (1.1)$$

where $[A, B] = AB - BA$ denotes the Lie bracket for square matrices and N is an arbitrary real symmetric matrix. We term this the *double bracket* equation. Brockett proves that (1.1) defines an isospectral flow which, under suitable assumptions on N, diagonalizes any symmetric matrix $H(t)$ for $t \to \infty$. Also, he shows that the flow (1.1) can be used to solve various combinatorial optimization tasks such as linear programming problems and the sorting of lists of real numbers. Further applications to the travelling salesman problem and to digital quantization of continuous-time signals have been described, see Brockett (1989), Brockett and Wong (1991). Note also the parallel efforts by Chu (1990), Driessel (1986), Chu and Driessel (1990) with applications to structured inverse eigenvalue problems and matrix least squares estimation.

The motivation for studying the double bracket equation (1.1) comes from the fact that it provides a solution to the following matrix least squares approximation problem.

Let $Q = \text{diag}(\lambda_1, \cdots, \lambda_n)$ with $\lambda_1 \geq \lambda_2 \geq \cdots \geq \lambda_n$ be a given real diagonal matrix and let

$$M(Q) = \{\Theta'Q\Theta \in \mathbb{R}^{n\times n} \mid \Theta\Theta' = I_n\} \qquad (1.2)$$

denote the set of all real symmetric matrices $H = \Theta'Q\Theta$ orthogonally equivalent to Q. Thus $M(Q)$ is the set of all symmetric matrices with eigenvalues $\lambda_1, \cdots, \lambda_n$. Given an arbitrary symmetric matrix $N \in \mathbb{R}^{n\times n}$, we consider the task of minimizing the matrix least squares index (distance)

$$\|N - H\|^2 = \|N\|^2 + \|H\|^2 - 2\operatorname{tr}(NH) \qquad (1.3)$$

of N to any $H \in M(Q)$ for the Frobenius norm $\|A\|^2 = \operatorname{tr}(AA')$. This is a least squares estimation problem with spectral constraints. Since $\|H\|^2 = \sum_{i=1}^n \lambda_i^2$ is constant for $H \in M(Q)$, the minimization of (1.3) is equivalent to the maximization of the trace functional $\phi(H) = \operatorname{tr}(NH)$ defined on $M(Q)$.

Heuristically, if N is chosen to be diagonal we would expect that the matrices $H_* \in M(Q)$ which minimize (1.3) are of the same form, i.e. $H_* = \pi Q \pi'$ for a suitable $n \times n$ permutation matrix. Since $M(Q)$ is a smooth manifold (Proposition 1.1 it seems natural to apply steepest descent techniques in order to determine the minima H_* of the distance function (1.3) on $M(Q)$.

This program will be carried out in this section. The intuitive meaning of equation (1.1) will be explained by showing that it is actually the gradient flow of the distance function (1.3) on $M(Q)$. Since $M(Q)$ is a homogeneous space for the Lie group $O(n)$ of orthogonal matrices, we first digress to the topic of Lie groups and homogeneous spaces before starting our analysis of the flow (1.1).

Digression: Lie Groups and Homogeneous Spaces A *Lie group G* is a group, which is also a smooth manifold, such that

$$G \times G \to G, \quad (x, y) \mapsto xy^{-1}$$

is a smooth map. Examples of Lie groups are

(i)   the general linear group $GL(n, \mathbb{R}) = \{T \in \mathbb{R}^{n\times n} | \det(T) \neq 0\}$

(ii) the special linear group $SL(n, \mathbb{R}) = \{T \in \mathbb{R}^{n\times n} | \det(T) = 1\}$

(iii) the orthogonal group $O(n) = \{T \in \mathbb{R}^{n\times n} | TT' = I_n\}$

(iv) the unitary group $U(n) = \{T \in \mathbb{C}^{n\times n} | TT^* = I_n\}$.

The groups $O(n)$ and $U(n)$ are compact Lie groups, i.e. Lie groups which are compact spaces. Also, $GL(n, \mathbb{R})$ and $O(n)$ have two connected components while $SL(n, \mathbb{R})$ and $U(n)$ are connected.

The tangent space $\mathcal{G} := T_e G$ of a Lie group G at the identity element $e \in G$ carries in a natural way the structure of a *Lie algebra*. The Lie algebras of the above examples of Lie groups are

2.1. DOUBLE BRACKET FLOWS FOR DIAGONALIZATION

(i) $\mathrm{gl}(n, \mathbb{R}) = \{X \in \mathbb{R}^{n \times n}\}$

(ii) $\mathrm{sl}(n, \mathbb{R}) = \{X \in \mathbb{R}^{n \times n} \mid \mathrm{tr}(X) = 0\}$

(iii) $\mathrm{skew}(n, \mathbb{R}) = \{X \in \mathbb{R}^{n \times n} \mid X' = -X\}$

(iv) $\mathrm{skew}(n, \mathbb{C}) = \{X \in \mathbb{C}^{n \times n} \mid X^* = -X\}$

In all of these cases the product structure on the Lie algebra is given by the *Lie bracket*

$$[X, Y] = XY - YX$$

of matrices X,Y.

A *Lie group action* of a Lie group G on a smooth manifold M is a smooth map

$$\sigma : G \times M \to M, \quad (g, x) \mapsto g \cdot x$$

satisfying for all $g, h \in G$, and $x \in M$

$$g \cdot (h \cdot x) = (gh) \cdot x, \quad e \cdot x = x.$$

The group action $\sigma : G \times M \to M$ is called *transitive* if there exists $x \in M$ such that every $y \in M$ satisfies $y = g \cdot x$ for some $g \in G$. A space M is called *homogeneous* if there exists a transitive G-action on M.

The *orbit* of $x \in M$ is defined by

$$\mathcal{O}(x) = \{g \cdot x \mid g \in G\}.$$

Thus the homogeneous spaces are the orbits of a group action. If G is a *compact* Lie group and $\sigma : G \times M \to M$ a Lie group action, then the orbits $\mathcal{O}(x)$ of σ are smooth, compact submanifolds of M.

Any Lie group action $\sigma : G \times M \to M$ induces an equivalence relation $\sim$ on M defined by

$$x \sim y \iff \text{There exists } g \in G \text{ with } y = g \cdot x$$

for $x, y \in M$. Thus the equivalence classes of $\sim$ are the orbits of $\sigma : G \times M \to M$. The *orbit space* of $\sigma : G \times M \to M$, denoted by $M/G = M/\sim$, is defined as the set of all equivalence classes of M, i.e.

$$M/G := \{\mathcal{O}(x) \mid x \in M\}.$$

Here M/G carries a natural topology, the quotient topology, which is defined as the finest topology on M/G such that the quotient map

$$\pi : M \to M/G, \quad \pi(x) := \mathcal{O}(x),$$

is continuous. Thus M/G is Hausdorff only if the orbits $\mathcal{O}(x), x \in M$, are closed subsets of M.

Given a Lie group action $\sigma : G \times M \to M$ and a point $X \in M$, the *stabilizer subgroup* of x is defined as

$$\text{Stab}(x) := \{g \in G \mid g \cdot x = x\}.$$

$\text{Stab}(x)$ is a closed subgroup of G and is also a Lie group.

For any subgroup $H \subset G$ the orbit space of the H-action $\alpha : H \times G \to G$, $(h, g) \mapsto gh^{-1}$, is the set of coset classes

$$G/H = \{g \cdot H \mid g \in G\}$$

which is a homogeneous space. If H is a closed subgroup of G, then G/H is a smooth manifold. In particular, $G/\text{Stab}(x), x \in M$, is a smooth manifold for any Lie group action $\sigma : G \times M \to M$.

Consider the smooth map

$$\sigma_x : G \to M, \quad g \mapsto g \cdot x,$$

for a given $x \in M$. Then the image $\sigma_x(G)$ of σ_x coincides with the G-orbit $\mathcal{O}(x)$ and σ_x induces a smooth bijection

$$\bar{\sigma}_x : G/\text{Stab}(x) \to \mathcal{O}(x).$$

This map is a diffeomorphism if G is a compact Lie group. For arbitrary non-compact Lie groups G, the map $\bar{\sigma}_x$ need not be a homeomorphism. Note that the topologies of $G/\text{Stab}(x)$ and $\mathcal{O}(x)$ are defined in a different way : $G/\text{Stab}(x)$ is endowed with the quotient topology while $\mathcal{O}(x)$ carries the relative subspace topology induced from M. If G is compact, then these topologies are homeomorphic via $\bar{\sigma}_x$. ∎

Returning to the study of the double bracket equation (1.1) as a gradient flow on the manifold $M(Q)$ of (1.2), let us first give a derivation of some elementary facts concerning the geometry of the set of symmetric matrices with fixed eigenvalues. Let

$$Q = \begin{bmatrix} \lambda_1 I_{n_1} & \cdots & 0 \\ \vdots & \ddots & \\ 0 & \cdots & \lambda_r I_{n_r} \end{bmatrix} \tag{1.4}$$

be a real diagonal $n \times n$ matrix with eigenvalues $\lambda_1 > \cdots > \lambda_r$ occurring with multiplicities $n_1, \cdots, n_r$, so that $n_1 + \cdots + n_r = n$.

Proposition 1.1 $M(Q)$ *is a smooth, connected, compact manifold of dimension*

$$\dim M(Q) = \frac{1}{2}\left(n^2 - \sum_{i=1}^{r} n_i^2\right).$$

2.1. DOUBLE BRACKET FLOWS FOR DIAGONALIZATION

Proof The proof uses some elementary facts and the terminology from differential geometry which are summarized in the above digression on Lie groups and homogeneous spaces. Let $O(n)$ denote the compact Lie group of all orthogonal matrices $\Theta \in \mathbb{R}^{n \times n}$. We consider the smooth Lie group action $\sigma : O(n) \times \mathbb{R}^{n \times n} \to \mathbb{R}^{n \times n}$ defined by $\sigma(\Theta, H) = \Theta H \Theta'$. Thus $M(Q)$ is an orbit of the group action σ and is therefore a compact manifold. The stabilizer subgroup $\text{Stab}(Q) \subset O(n)$ of $Q \in \mathbb{R}^{n \times n}$ is defined as $\text{Stab}(Q) = \{\Theta \in O(n) \mid \Theta Q \Theta' = Q\}$. Since $\Theta Q = Q\Theta$ if and only if $\Theta = \text{diag}(\Theta_1, \cdots, \Theta_r)$ with $\Theta_i \in O(n_i)$, we see that

$$\text{Stab}(Q) = O(n_1) \times \cdots \times O(n_r) \subset O(n) . \tag{1.5}$$

Therefore $M(Q) \cong O(n)/\text{Stab}(Q)$ is diffeomorphic to the homogeneous space

$$M(Q) \cong O(n)/O(n_1) \times \cdots \times O(n_r) , \tag{1.6}$$

and

$$\begin{aligned}
\dim M(Q) &= \dim O(n) - \sum_{i=1}^{r} \dim O(n_i) \\
&= \frac{n(n-1)}{2} - \frac{1}{2} \sum_{i=1}^{r} n_i(n_i - 1) \\
&= \frac{1}{2}\left(n^2 - \sum_{i=1}^{r} n_i^2\right) .
\end{aligned}$$

To see the connectivity of $M(Q)$ note that the subgroup $SO(n) = \{\Theta \in O(n) \mid \det\Theta = 1\}$ of $O(n)$ is connected. Now $M(Q)$ is the image of $SO(n)$ under the continuous map $f : SO(n) \to M(Q)$, $f(\Theta) = \Theta Q \Theta'$, and therefore also connected. This completes the proof. ∎

We need the following description of the tangent spaces of $M(Q)$.

Lemma 1.2 *The tangent space of $M(Q)$ at $H \in M(Q)$ is*

$$T_H M(Q) = \{[H, \Omega] \mid \Omega' = -\Omega \in \mathbb{R}^{n \times n}\} . \tag{1.7}$$

Proof Consider the smooth map $\sigma_H : O(n) \to M(Q)$ defined by $\sigma_H(\Theta) = \Theta H \Theta'$. Note that σ_H is a submersion and therefore it induces a surjective map on tangent spaces. The tangent space of $O(n)$ at the $n \times n$ identity matrix I is $T_I O(n) = \{\Omega \in \mathbb{R}^{n \times n} \mid \Omega' = -\Omega\}$ (Appendix C) and the derivative of σ_H at I is the surjective linear map

$$\begin{aligned}
D\sigma_H|_I : T_I O(n) &\to T_H M(Q) , \\
\Omega &\mapsto \Omega H - H\Omega .
\end{aligned} \tag{1.8}$$

This proves the result. ∎

We now state and prove our main result on the double bracket equation (1.1). The result in this form is due to Bloch, Brockett and Ratiu (1990).

Theorem 1.3 *Let $N \in \mathbb{R}^{n \times n}$ be symmetric, and Q satisfy (1.4).*
(a) The differential equation(1.1),

$$\dot{H} = [H, [H, N]], \quad H(0) = H'(0) = H_0,$$

defines an isospectral flow on the set of all symmetric matrices $H \in \mathbb{R}^{n \times n}$.

(b) There exists a Riemannian metric on $M(Q)$ such that (1.1) is the gradient flow $\dot{H} = \text{grad} f_N(H)$ of the function $f_N : M(Q) \to \mathbb{R}, f_N(H) = -\frac{1}{2}\|N - H\|^2$.

(c) The solutions of (1.1) exist for all $t \in \mathbb{R}$ and converge to a connected component of the set of equilibria points. The set of equilibria points H_∞ of (1.1) is characterized by $[H_\infty, N] = 0$, i.e. $NH_\infty = H_\infty N$.

(d) Let $N = \text{diag}(\mu_1, \cdots, \mu_n)$ with $\mu_1 > \cdots > \mu_n$. Then every solution $H(t)$ of (1.1) converges for $t \to \pm\infty$ to a diagonal matrix $\pi \text{diag}(\lambda_1, \cdots, \lambda_n)\pi' = \text{diag}(\lambda_{\pi(1)}, \cdots, \lambda_{\pi(n)})$, where $\lambda_1, \cdots, \lambda_n$ are the eigenvalues of H_0 and π is a permutation matrix.

(e) Let $N = \text{diag}(\mu_1, \cdots, \mu_n)$ with $\mu_1 > \cdots > \mu_n$ and define the function $f_N : M(Q) \to \mathbb{R}, f_N(H) = -\frac{1}{2}\|N - H\|^2$. The Hessian of $f_N : M(Q) \to \mathbb{R}$ at each critical point is nonsingular. For almost all initial conditions $H_0 \in M(Q)$ the solution $H(t)$ of (1.1) converges to Q as $t \to \infty$ with an exponential bound on the rate of convergence. For the case of distinct eigenvalues $\lambda_i \neq \lambda_j$, then the linearization at a critical point H_∞ is

$$\dot{\xi}_{ij} = -(\lambda_{\pi(i)} - \lambda_{\pi(j)})(\mu_i - \mu_j)\xi_{ij}, \quad i > j, \tag{1.9}$$

where $H_\infty = \text{diag}(\lambda_{\pi(1)}, \cdots, \lambda_{\pi(n)})$, for π a permutation matrix.

Proof To prove (a) note that $[H, N] = HN - NH$ is skew-symmetric if H, N are symmetric. Thus the result follows immediately from Lemma 1.4.1. It follows that every solution $H(t), t$ belonging to an interval in $\mathbb{R}$, of (1.1) satisfies $H(t) \in M(H_0)$. By Proposition 1.1, $M(H_0)$ is compact and therefore $H(t)$ exists for all $t \in \mathbb{R}$. Consider the time derivative of the t function

$$f_N(H(t)) = -\frac{1}{2}\|N - H(t)\|^2 = -\frac{1}{2}\|N\|^2 - \frac{1}{2}\|H_0\|^2 + \text{tr}(NH(t))$$

and note that

$$\frac{d}{dt}f_N(H(t)) = \text{tr}((N\dot{H}(t))) = \text{tr}(N[H(t), [H(t), N]]) .$$

2.1. DOUBLE BRACKET FLOWS FOR DIAGONALIZATION

Since $\text{tr}(A[B,C]) = \text{tr}([A,B]C)$, and $[A,B] = -[B,A]$, then with A, B symmetric matrices, we have $[A,B]' = [B,A]$ and thus

$$\text{tr}(N[H,[H,N]]) = \text{tr}([N,H][H,N]) = \text{tr}([H,N]'[H,N]).$$

Therefore

$$\frac{d}{dt} f_N(H(t)) = \|[H(t), N]\|^2 \ . \tag{1.10}$$

Thus $f_N(H(t))$ increases monotonically. Since $f_N : M(H_0) \to \mathbb{R}$ is a continuous function on the compact space $M(H_0)$, then f_N is bounded from above and below. It follows that $f_N(H(t))$ converges to a finite value and its time derivative must go to zero. Therefore every solution of (1.1) converges to a connected component of the set of equilibria points as $t \to \infty$. By (1.10), the set of equilibria of (1.1) is characterized by $[N, H_\infty] = 0$. This proves (c). Now suppose that $N = \text{diag}(\mu_1, \cdots, \mu_n)$ with $\mu_1 > \cdots > \mu_n$. By (c), the equilibria of (1.1) are the symmetric matrices $H_\infty = (h_{ij})$ which commute with N and thus

$$\mu_i h_{ij} = \mu_j h_{ij} \text{ for } i, j = 1, \cdots, n \ . \tag{1.11}$$

Now (1.11) is equivalent to $(\mu_i - \mu_j) h_{ij} = 0$ for $i, j = 1, \cdots, n$ and hence, by assumption on N, to $h_{ij} = 0$ for $i \neq j$. Thus H_∞ is a diagonal matrix with the same eigenvalues as H_0 and it follows that $H_\infty = \pi \text{diag}(\lambda_1, \cdots, \lambda_n) \pi'$. Therefore (1.1) has only a finite number of equilibrium points. We have already shown that every solution of (1.1) converges to a connected component of the set of equilibria and thus to a single equilibrium H_∞. This completes the proof of (d).

In order to prove (b) we need some preparation. For any $H \in M(Q)$ let $\sigma_H : O(n) \to M(Q)$ be the smooth map defined by $\sigma_H(\Theta) = \Theta' H \Theta$. The derivative at I_n is the surjective linear map

$$D\sigma_H|_I : T_I O(n) \to T_H M(Q) ,$$
$$\Omega \mapsto H\Omega - \Omega H \ .$$

Let $\text{skew}(n)$ denote the set of all real $n \times n$ skew-symmetric matrices. The kernel of $D\sigma_H|_I$ is then

$$K = \ker D\sigma_H|_I = \{\Omega \in \text{skew}(n) \mid H\Omega = \Omega H\} \ . \tag{1.12}$$

Let us endow the vector space $\text{skew}(n)$ with the standard inner product defined by $(\Omega_1, \Omega_2) \mapsto \text{tr}(\Omega_1' \Omega_2)$. The orthogonal complement of K in $\text{skew}(n)$ then is

$$K^\perp = \{Z \in \text{skew}(n) \mid \text{tr}(Z'\Omega) = 0 \ \forall \ \Omega \in K\} \ . \tag{1.13}$$

Since
$$\operatorname{tr}([N,H]'\Omega) = -\operatorname{tr}(N[H,\Omega]) = 0$$
for all $\Omega \in K$, then $[N,H] \in K^\perp$ for all symmetric matrices N.

For any $\Omega \in \operatorname{skew}(n)$ we have the unique decomposition
$$\Omega = \Omega_H + \Omega^H \tag{1.14}$$
with $\Omega_H \in K$ and $\Omega^H \in K^\perp$.

The gradient of a function on a smooth manifold M is only defined with respect to a fixed Riemannian metric on M (see Appendix C.9, C.10). To define the gradient of the distance function $f_N : M(Q) \to \mathbb{R}$ we therefore have to specify which Riemannian metric on $M(Q)$ we consider. Now to define a Riemannian metric on $M(Q)$ we have to define an inner product on each tangent space $T_H M(Q)$ of $M(Q)$. By Lemma 1.2, $D\sigma_H|_I : \operatorname{skew}(n) \to T_H M(Q)$ is surjective with kernel K and hence induces an isomorphism of $K^\perp \subset \operatorname{skew}(n)$ onto $T_H M(Q)$. Thus to define an inner product on $T_H M(Q)$ it is equivalent to define an inner product on $K^\perp$. We proceed as follows:

Define for tangent vectors $[H,\Omega_1], [H,\Omega_2] \in T_H M(Q)$
$$\langle\!\langle [H,\Omega_1], [H,\Omega_2] \rangle\!\rangle := \operatorname{tr}((\Omega_1^H)' \Omega_2^H), \tag{1.15}$$

where Ω_i^H, $i = 1,2$, are defined by (1.14). This defines an inner product on $T_H M(Q)$ and in fact a Riemannian metric on $M(Q)$. We refer to this as the normal Riemannian metric on $M(Q)$.

The gradient of $f_N : M(Q) \to \mathbb{R}$ with respect to this Riemannian metric is the unique vector field $\operatorname{grad} f_N$ on $M(Q)$ which satisfies the condition

$$\begin{aligned}(i) \quad & \operatorname{grad} f_N(H) \in T_H M(Q) \text{ for all } H \in M(Q) \\ (ii) \quad & Df_N|_H([H,\Omega]) = \langle\!\langle \operatorname{grad} f_N(H), [H,\Omega] \rangle\!\rangle, \end{aligned} \tag{1.16}$$

for all $[H,\Omega] \in T_H M(Q)$. By Lemma 1.2, condition (1.16) implies that for all $H \in M(Q)$
$$\operatorname{grad} f_N(H) = [H,X] \tag{1.17}$$
for some skew-symmetric matrix X (which possibly depends on H). By computing the derivative of f_N at H we find
$$Df_N|_H([H,\Omega]) = \operatorname{tr}(N[H,\Omega]) = \operatorname{tr}([H,N]'\Omega).$$
Thus (1.16) implies
$$\begin{aligned}\operatorname{tr}([H,N]'\Omega) &= \langle\!\langle \operatorname{grad} f_N(H), [H,\Omega] \rangle\!\rangle \\ &= \langle\!\langle [H,X], [H,\Omega] \rangle\!\rangle \\ &= \operatorname{tr}((X^H)'\Omega^H)\end{aligned} \tag{1.18}$$

2.1. DOUBLE BRACKET FLOWS FOR DIAGONALIZATION

for all $\Omega \in \text{skew}(n)$.

Since $[H, N] \in K^\perp$ we have $[H, N] = [H, N]^H$ and therefore $\text{tr}([H, N]'\Omega)$ $= \text{tr}([H, N]'\Omega^H)$. By (1.17) therefore

$$X^H = [H, N] . \qquad (1.19)$$

This shows

$$\begin{aligned} \text{grad} f_N(H) &= [H, X] = [H, X^H] \\ &= [H, [H, N]] . \end{aligned}$$

This completes the proof of (b).

For a proof that the Hessian of $f_N : M(Q) \to \mathbb{R}$ is nonsingular at each critical point, without assumption on the multiplicities of the eigenvalues of Q, we refer to Duistermaat, Kolk and Varadarajan (1983). Here we only treat the generic case where $n_1 = \cdots = n_n = 1$, i.e. $Q = \text{diag}(\lambda_1, \cdots, \lambda_n)$ with $\lambda_1, \cdots, \lambda_n$.

The linearization of the flow (1.1) on $M(Q)$ around any equilibrium point $H_\infty \in M(Q)$ where $[H_\infty, N] = 0$ is given by

$$\dot{\xi} = H_\infty(\xi N - N\xi) - (\xi N - N\xi)H_\infty , \qquad (1.20)$$

where $\xi \in T_{H_\infty} M(Q)$ is in the tangent space of $M(Q)$ at $H_\infty = \text{diag}(\lambda_{\pi(1)}, \cdots, \lambda_{\pi(n)})$, for π a permutation matrix. By Lemma 1.2, $\xi = [H_\infty, \Omega]$ holds for a skew-symmetric matrix Ω. Equivalently, in terms of the matrix entries, we have $\xi = (\xi_{ij})$, $\Omega = (\Omega_{ij})$ and $\xi_{ij} = (\lambda_{\pi(i)} - \lambda_{\pi(j)})\Omega_{ij}$. Consequently, (1.20) is equivalent to the decoupled set of differential equations.

$$\dot{\xi}_{ij} = -(\lambda_{\pi(i)} - \lambda_{\pi(j)})(\mu_i - \mu_j)\xi_{ij}$$

for $i, j = 1, \cdots, n$. Since $\xi_{ij} = \xi_{ji}$ the tangent space $T_{H_\infty} M(Q)$ is parametrized by the coordinates ξ_{ij} for $i > j$. Therefore the eigenvalues of the linearization (1.20) are $-(\lambda_{\pi(i)} - \lambda_{\pi(j)})(\mu_i - \mu_j)$ for $i > j$ which are all nonzero. Thus the Hessian is nonsingular. Similarly, $\pi = I_n$ is the unique permutation matrix for which $(\lambda_{\pi(i)} - \lambda_{\pi(j)})(\mu_i - \mu_j) > 0$ for all $i > j$. Thus $H_\infty = Q$ is the unique critical point of $f_N : M(Q) \to \mathbb{R}$ for which the Hessian is negative definite.

The union of the unstable manifolds (see Appendix B.11) of the equilibria points $H_\infty = \text{diag}(\lambda_{\pi(1)}, \cdots, \lambda_{\pi(n)})$ for $\pi \neq I_n$ forms a closed subset of $M(Q)$ of co-dimension at least one. (Actually, the co-dimension turns out to be exactly one). Thus the domain of attraction for the unique stable equilibrium $H_\infty = Q$ is an open and dense subset of $M(Q)$. This completes the proof of (e), and the theorem. ∎

Remarks. 1. In the generic case, where the ordered eigenvalues λ_i, μ_i of H_0, N are distinct, then the property $||N - H_0||^2 \geq ||N - H_\infty||^2$ leads to the *Wielandt-Hoffman inequality* $||N - H_0||^2 \geq \sum_{i=1}^n (\mu_i - \lambda_i)^2$. Since the eigenvalues of a matrix depend continuously on the coefficients, this last inequality holds in general, thus establishing the Wielandt-Hoffman inequality for all symmetric matrices N, H.

2. In part (c) of Theorem 1.3 it is stated that every solution of (1.1) converges to a connected component of the set of equilibria points. Actually Duistermaat, Kolk and Varadarajan (1983) have shown that $f_N : M(Q) \to \mathbb{R}$ is always a Morse-Bott function. Thus, using Proposition 1.3.3, it follows that any solution of (2.1.1) is converging to a single equilibrium point rather than to a set of equilibrium points.

3. If $Q = \begin{bmatrix} I_k & 0 \\ 0 & 0 \end{bmatrix}$ is the standard rank k projection operator, then the distance function $f_N: M(Q) \to \mathbb{R}$ is a Morse-Bott function on the Grassmann manifold $\text{Grass}_\mathbb{R}(k, n)$. This function then induces a Morse-Bott function on the Stiefel manifold $St(k, n)$, which coincides with the Rayleigh quotient $R_N: St(k, n) \to \mathbb{R}$. Therefore the Rayleigh quotient is seen as a Morse-Bott function on the Stiefel manifold.

4. The role of the matrix N for the double bracket equation is that of a parameter which guides the equation to a desirable final state. In a diagonalization exercise one would choose N to be a diagonal matrix while the matrix to be diagonalized would enter as the initial condition of the double bracket flow. Then in the generic situation, the ordering of the diagonal entries of N will force the eigenvalues of H_0 to appear in the same order.

5. It has been shown that the gradient flow of the least squares distance function $f_N: M(Q) \to \mathbb{R}$ with respect to the normal Riemannian metric is the double bracket flow. This is no longer true if other Riemannian metrices are considered. For example, the gradient of f_N with respect to the *induced* Riemannian metric is given by a very complicated formula in N and H which makes it hard to analyze. This is the main reason why the somewhat more complicated normal Riemannian metric is chosen. ∎

Flows on Orthogonal Matrices The double bracket flow (1.1) provides a method to compute the eigenvalues of a symmetric matrix H_0. What about the eigenvectors of H_0? Following Brockett, we show that suitable gradient flows evolving on the group of orthogonal matrices converge to the various eigenvector basis of H_0.

Let $O(n)$ denote the Lie group of $n \times n$ real orthogonal matrices and let N and H_0 be $n \times n$ real symmetric matrices. We consider the smooth

2.1. DOUBLE BRACKET FLOWS FOR DIAGONALIZATION

potential function

$$\phi : O(n) \to \mathbb{R}, \quad \phi(\Theta) = \operatorname{tr}(N\Theta' H_0 \Theta) \tag{1.21}$$

defined on $O(n)$. Note that

$$\| N - \Theta' H_0 \Theta \|^2 = \| N \|^2 - 2\phi(\Theta) + \| H_0 \|^2,$$

so that maximizing $\phi(\Theta)$ over $O(n)$ is equivalent to minimizing the least squares distances $\| N - \Theta' H_0 \Theta \|^2$ of N to $\Theta' H_0 \Theta \in M(H_0)$.

We need the following description of the tangent spaces of $O(n)$.

Lemma 1.4 *The tangent space of $O(n)$ at $\Theta \in O(n)$ is*

$$T_\Theta O(n) = \{\Theta \Omega \mid \Omega' = -\Omega \in \mathbb{R}^{n \times n}\}. \tag{1.22}$$

Proof Consider the smooth diffeomorphism $\lambda_\Theta : O(n) \to O(n)$ defined by left multiplication with Θ, i.e.. $\lambda_\Theta(\psi) = \Theta \psi$ for $\psi \in O(n)$. The tangent space of $O(n)$ at the $n \times n$ identity matrix I is $T_I O(n) = \{\Omega \in \mathbb{R}^{n \times n} | \Omega' = -\Omega\}$; i.e.. is equal to the Lie algebra of skew-symmetric matrices.

The derivative of λ_Θ at I is the linear isomorphism between tangent spaces

$$D(\lambda_\Theta)|_I : T_I O(n) \to T_\Theta O(n), \quad \Omega \mapsto \Theta \Omega. \tag{1.23}$$

The result follows. ∎

The standard Euclidean inner product $\langle A, B \rangle = \operatorname{tr}(A'B)$ on $\mathbb{R}^{n \times n}$ induces an inner product on each tangent space $T_\Theta O(n)$, defined by

$$\langle \Theta \Omega_1, \Theta \Omega_2 \rangle = \operatorname{tr}((\Theta \Omega_1)'(\Theta \Omega_2)) = \operatorname{tr}(\Omega_1' \Omega_2). \tag{1.24}$$

This defines a Riemannian matrix on $O(n)$ to which we refer to as the *induced Riemannian metric* on $O(n)$. As an aside for readers with a background in Lie group theory, this metric coincides with the Killing form on the Lie algebra of $O(n)$, up to a constant scaling factor.

Theorem 1.5 *Let $N, H_0 \in \mathbb{R}^{n \times n}$ be symmetric. Then:*

(a) The differential equation

$$\boxed{\dot{\Theta}(t) = H_0 \Theta(t) N - \Theta(t) N \Theta'(t) H_0 \Theta(t), \Theta(0) \in O(n)} \tag{1.25}$$

induces a flow on the Lie group $O(n)$ of orthogonal matrices. Also (1.25) is the gradient flow $\dot{\Theta} = \nabla \phi(\Theta)$ of the trace function (1.21) on $O(n)$ for the induced Riemannian metric on $O(n)$.

(b) The solutions of (1.25) exist for all $t \in \mathbb{R}$ and converge to a connected component of the set of equilibria points $\Theta_\infty \in O(n)$. The set of equilibria points Θ_∞ of (1.25) is characterized by $[N, \Theta'_\infty H_0 \Theta_\infty] = 0$.

(c) Let $N = \operatorname{diag}(\mu_1, \ldots, \mu_n)$ with $\mu_1 > \ldots > \mu_n$ and suppose H_0 has distinct eigenvalues $\lambda_1 > \ldots > \lambda_n$. Then every solution $\Theta(t)$ of (1.25) converges for $t \to \infty$ to an orthogonal matrix Θ_∞ satisfying $H_0 = \Theta_\infty \operatorname{diag}(\lambda_{\pi(1)}, \ldots, \lambda_{\pi(n)}) \Theta'_\infty$ for a permutation matrix π. The columns of Θ_∞ are eigenvectors of H_0.

(d) Let $N = \operatorname{diag}(\mu_1, \ldots, \mu_n)$ with $\mu_1 > \ldots > \mu_n$ and suppose H_0 has distinct eigenvalues $\lambda_1 > \ldots > \lambda_n$. The Hessian of $\phi : O(n) \to \mathbb{R}$ at each critical point is nonsingular. For almost all initial conditions $\Theta_0 \in O(n)$ the solution $\Theta(t)$ of (1.25) converges exponentially fast to an eigenbasis Θ_∞ of H_0, such that $H_0 = \Theta_\infty \operatorname{diag}(\lambda_1, \ldots, \lambda_n) \Theta'_\infty$ holds.

Proof The derivative of $\phi : O(n) \to \mathbb{R}, \phi(\Theta) = \operatorname{tr}(N\Theta' H_0 \Theta)$, at $\Theta \in O(n)$ is the linear map on the tangent space $T_\Theta O(n)$ defined by

$$\begin{aligned} D\phi|_\Theta(\Theta\Omega) &= \operatorname{tr}(N\Theta' H_0 \Theta\Omega - N\Omega\Theta' H_0 \Theta) \\ &= \operatorname{tr}([N, \Theta' H_0 \Theta]\Omega) \end{aligned} \quad (1.26)$$

for $\Omega' = -\Omega$. Let $\nabla\phi(\Theta)$ denote the gradient of ϕ at $\Theta \in O(n)$, defined with respect to the induced Riemannian metric on $O(n)$. Thus $\nabla\phi(\Theta)$ is characterized by

$$\begin{aligned} &(i) \quad \nabla\phi(\Theta) \in T_\Theta O(n) \\ &(ii) \quad D\phi|_\phi(\Theta\Omega) = <\nabla\phi(\Theta), \Theta\Omega> = \operatorname{tr}(\nabla\phi(\Theta)'\Theta\Omega) \end{aligned} \quad (1.27)$$

for all skew-symmetric matrices $\Omega \in \mathbb{R}^{n\times n}$. By (1.26) and (1.27) then

$$\operatorname{tr}((\Theta' \nabla \phi(\Theta))'\Omega) = \operatorname{tr}([N, \Theta' H_0 \Theta]\Omega)) \quad (1.28)$$

for all $\Omega' = -\Omega$ and thus, by skew symmetry of $\Theta' \nabla \phi(\Theta)$ and $[N, \Theta' H_0 \Theta]$, we have

$$\Theta' \nabla \phi(\Theta) = -[N, \Theta' H_0 \Theta] . \quad (1.29)$$

Therefore (1.25) is the gradient flow of $\phi : O(n) \to \mathbb{R}$, proving (a).

By compactness of $O(n)$ and by the convergence properties of gradient flows, the solutions of (1.25) exist for all $t \in \mathbb{R}$ and converge to a connected component of the set of equilibria points $\Theta_\infty \in O(n)$, corresponding to a fixed value of the trace function $\phi : O(n) \to \mathbb{R}$. Also, $\Theta_\infty \in O(n)$ is an equilibrium print of (1.25) if and only if $\Theta_\infty[N, \Theta'_\infty H_0 \Theta_\infty] = 0$, i.e.. as Θ_∞ is invertible, if and only if $[N, \Theta'_\infty H_0 \Theta_\infty] = 0$. This proves (b).

To prove (c) we need a lemma.

2.1. DOUBLE BRACKET FLOWS FOR DIAGONALIZATION 57

Lemma 1.6 Let $N = \text{diag}(\mu_1, \ldots, \mu_n)$ with $\mu_1 > \ldots > \mu_n$. Let $\psi \in O(n)$ with $H_0 = \psi \text{diag}(\lambda_1 I_{n_1}, \ldots, \lambda_r I_{n_r}) \psi'$ and $\sum_{i=1}^r n_i = n$. Then the set of equilibria points of (1.25) is characterized as

$$\Theta_\infty = \psi D \pi \qquad (1.30)$$

where $D = \text{diag}(D_1, \ldots, D_r) \in O(n_1) \times \ldots \times O(n_r), D_i \in O(n_i), i = 1, \ldots, r,$ and π is an $n \times n$ permutation matrix.

Proof By (b), Θ_∞ is an equilibrium point if and only if $[N, \Theta'_\infty H_0 \Theta_\infty] = 0$. As N has distinct diagonal entries this condition is equivalent to $\Theta'_\infty H_0 \Theta_\infty$ being a diagonal matrix Λ. This is clearly true for matrices of the form (1.30). Now the diagonal elements of Λ are the eigenvalues of H_0 and therefore $\Lambda = \pi' \text{diag}(\lambda_1 I_{n_1}, \ldots, \lambda_r I_{n_r}) \pi$ for an $n \times n$ permutation matrix π. Thus

$$\begin{aligned}\text{diag}(\lambda_1 I_{n_1}, \ldots, \lambda_r I_{n_r}) &= \pi \Theta'_\infty H_0 \Theta_\infty \pi' \\ &= \pi \Theta'_\infty \psi \text{diag}(\lambda_1 I_{n_1}, \ldots, \lambda_r I_{n_r}) \psi' \Theta_\infty \pi'.\end{aligned} \qquad (1.31)$$

Thus $\psi' \Theta_\infty \pi'$ commutes with $\text{diag}(\lambda_1 I_{n_1}, \ldots, \lambda_r I_{n_r})$. But any orthogonal matrix which commutes with $\text{diag}(\lambda_1 I_{n_1}, \ldots, \lambda_r I_{n_r})$, $\lambda_1 > \ldots > \lambda_r$, is of the form $D = \text{diag}(D_1, \ldots, D_r)$ with $D_i \in O(n_i), i = 1, \ldots, r$, orthogonal. The result follows. ∎

It follows from this lemma and the genericity conditions on N and H_0 in (c) that the equilibria of (1.25) are of the form $\Theta_\infty = \psi D \pi$, where $D = \text{diag}(\pm 1, \ldots, \pm 1)$ is an arbitrary sign matrix and π is a permutation matrix. In particular there are exactly $2^n n!$ equilibrium points of (1.25). As this number is finite, part (b) implies that every solution $\Theta(t)$ converges to an element $\Theta_\infty \in O(n)$ with $\Theta'_\infty H_0 \Theta_\infty = \pi' \text{diag}(\lambda_1, \ldots, \lambda_n) \pi = \text{diag}(\lambda_{\pi(1)}, \ldots, \lambda_{\pi(n)})$.

In particular, the column vectors of Θ_∞ are eigenvectors of H_0. This completes the proof of (c).

To prove (d) we linearize the flow (1.25) around each equilibrium point. The linearization of (1.25) at $\Theta_\infty = \psi D \pi$, where $D = \text{diag}(d_1, \ldots, d_n), d_i \in \{\pm 1\}$, is

$$\dot\xi = -\Theta_\infty [N, \xi' H_0 \Theta_\infty + \Theta'_\infty H_0 \xi] \qquad (1.32)$$

for $\xi = \Theta_\infty \Omega \in T_{\Theta_\infty} O(n)$. Thus (1.32) is equivalent to

$$\dot\Omega = -[N, [\Theta'_\infty H_0 \Theta_\infty, \Omega]] \qquad (1.33)$$

on the linear space $\text{skew}(n)$ of skew-symmetric $n \times n$ matrices Ω. As

$$\Theta'_\infty H_0 \Theta_\infty = \text{diag}(\lambda_{\pi(1)}, \ldots, \lambda_{\pi(n)}),$$

this is equivalent, in terms of the matrix entries of $\Omega = (\Omega_{ij})$, to the decoupled set of linear differential equations

$$\dot{\Omega}_{ij} = -(\lambda_{\pi(i)} - \lambda_{\pi(j)})(\mu_i - \mu_j)\Omega_{ij}, \quad i > j. \tag{1.34}$$

From this it follows immediately that the eigenvalues of the linearization at θ_∞ are nonzero. Furthermore, they are all negative if and only if $(\lambda_{\pi(1)}, \ldots, \lambda_{\pi(n)})$ and $(\mu_1, \ldots, \mu_n)$ are similarly ordered, that is if and only if $\pi = I_n$. Arguing as in the proof of Theorem 1.3, the union of the unstable manifolds of the critical points $\Theta_\infty = \psi D\pi$ with $\pi \neq I_n$ is a closed subset of $O(n)$ of co-dimension at least 1. Its complement thus forms an open and dense subset of $O(n)$. It is the union of the domains of attractions for the 2^n locally stable attractors $\Theta_\infty = \psi D$, with $D = \text{diag}(\pm 1, \ldots, \pm 1)$ arbitrary. This completes the proof of (d). ∎

Remark In part (b) of the above theorem it has been stated that every solution of (1.24) converges to a connected component of the set of equilibria points. Again, it can be shown that $\phi : O(n) \to \mathbb{R}, \phi(\Theta) = \text{tr}(N\Theta' H_0 \Theta)$, is a Morse-Bott function. Thus, using Proposition 1.3.6, any solution of (1.24) is converging to an equilibrium point rather than a set of equilibria. ∎

It is easily verified that $\Theta(t)$ given from (1.25) implies that

$$H(t) = \Theta'(t) H_0 \Theta(t) \tag{1.35}$$

is a solution of the double bracket equation (1.1). In this sense, the double bracket equation is seen as a projected gradient flow from $O(n)$. A final observation is that in maximizing $\phi(\Theta) = \text{tr}(N\Theta' H\Theta)$ over $O(n)$ in the case of generic matrices N, H_0 as in Theorem 1.5 part(d), then of the $2^n n!$ possible equilibria $\Theta_\infty = \psi D\pi$, the 2^n maxima of $\phi : O(n) \to \mathbb{R}$ with $\pi = I_n$ maximize the sum of products of eigenvalues $\sum_{i=1}^n \lambda_{\pi(i)} \mu_i$. This ties in with a classical result that to maximize $\sum_{i=1}^n \lambda_{\pi(i)} \mu_i$ there must be a "similar" ordering; see Hardy, Littlewood and Polya (1952).

Problems **1.** For a nonsingular positive definite symmetric matrix A, show that the solutions of the equation $\dot{H} = [H, A[H, N]A]$ converges to the set of H_∞ satisfying $[H_\infty, N] = 0$. Explore the convergence properties in the case where A is diagonal.

2. Let $A, B \in \mathbb{R}^{n \times n}$ be symmetric. For any integer $i \in \mathbb{N}$ define $\text{ad}_A^i(B)$ recursively by $\text{ad}_A(B) := AB - BA, \text{ad}_A^i(B) = \text{ad}_A(\text{ad}_A^{i-1}(B)), i \geq 2$. Prove that $\text{ad}_A^i(B) = 0$ for some $i \geq 1$ implies $\text{ad}_A(B) = 0$. Deduce that for $i \geq 1$ arbitrary

$$\dot{H} = [H, \text{ad}_H^i(N)]$$

has the same equilibria as (1.1).

3. Let $\text{diag}(H) = \text{diag}(h_{11}, \ldots, h_{nn})$ denote the diagonal matrix with diagonal entries identical to those of H. Consider the problem of minimizing the distance function $g \colon M(Q) \to \mathbb{R}$ defined by $g(H) = \|H - \text{diag}(H)\|^2$. Prove that the gradient flow of g with respect to the normal Riemannian metric is

$$\dot H = [H, [H, \text{diag}(H)]].$$

Show that the equilibrium points satisfy $[H_\infty, \text{diag}(H_\infty)] = 0$.

4. Let $N \in \mathbb{R}^{n \times n}$ be symmetric. Prove that the gradient flow of the function $H \mapsto \text{tr}(NH^2)$ on $M(Q)$ with respect to the normal Riemannian metric is

$$\dot H = [H, [H^2, N]]$$

Investigate the convergence properties of the flow. Generalize to $\text{tr}(NH^m)$ for $m \in \mathbb{N}$ arbitrary!

Main Points of Section An eigenvalue/eigenvector decomposition of a real symmetric matrix can be achieved by minimizing a matrix least squares distance function via a gradient flow on the Lie group of orthogonal matrices with an appropriate Riemannian metric. The distance function is a smooth Morse-Bott function.

The isospectral double bracket flow on homogeneous spaces of symmetric matrices converges to a diagonal matrix consisting of the eigenvalues of the initial condition. A specific choice of Riemannian metric allows a particularly simple form of the gradient. The convergence rate is exponential, with stability properties governed by a linearization of the equations around the critical points.

2.2 Toda Flows and the Riccati Equation

The Toda Flow An important issue in numerical analysis is to exploit special structures of matrices to develop faster and more reliable algorithms. Thus, for example, in eigenvalue computations an initial matrix is often first transformed into Hessenberg form and the subsequent operations are performed on the set of Hessenberg matrices. In this section we study this issue for the double bracket equation. For appropriate choices of the parameter matrix N, the double bracket flow (1.1) restricts to a flow on certain subclasses of symmetric matrices. We treat only the case of symmetric Hessenberg matrices. These are termed *Jacobi matrices* and are banded, being tri-diagonal, symmetric matrices $H = (h_{ij})$ with $h_{ij} = 0$ for $|i - j| \geq 2$.

Lemma 2.1 Let $N = \text{diag}(1, 2, \cdots, n)$ and let H be a Jacobi matrix. Then $[H, [H, N]]$ is a Jacobi matrix, and (1.1) restricts to a flow on the set of isospectral Jacobi matrices.

Proof The (i,j)-th entry of $B = [H, [H, N]]$ is

$$b_{ij} = \Sigma_{k=1}^{n}(2k - i - j)h_{ik}h_{kj} .$$

Hence for $|i - j| \geq 3$ and $k = 1, \cdots, n$, then $h_{ik} = 0$ or $h_{kj} = 0$ and therefore $b_{ij} = 0$. Suppose $j = i + 2$. Then for $k = i + 1$

$$b_{ij} = (2(i+1) - i - j)h_{i,i+1}h_{i+1,i+2} = 0 .$$

Similarly for $i = j + 2$. This completes the proof. ∎

Actually for $N = \text{diag}(1, 2, \cdots, n)$ and H a Jacobi matrix

$$HN - NH = H_u - H_\ell , \qquad (2.1)$$

where

$$H_u = \begin{bmatrix} h_{11} & h_{12} & & 0 \\ 0 & h_{22} & \ddots & \\ & \ddots & \ddots & h_{n-1,n} \\ 0 & & 0 & h_{nn}, \end{bmatrix}$$

and

$$H_\ell = \begin{bmatrix} h_{11} & 0 & \cdots & 0 \\ h_{21} & h_{22} & \ddots & \vdots \\ \vdots & \ddots & \ddots & 0 \\ 0 & \cdots & h_{n,n-1} & h_{nn} \end{bmatrix}$$

denote the upper triangular and lower triangular parts of H respectively. (This fails if H is not Jacobi; why?) Thus, for the above special choice of N, the double bracket flow induces the isospectral flow on Jacobi matrices

$$\boxed{\dot{H} = [H, H_u - H_\ell].} \qquad (2.2)$$

The differential equation (2.2) is called the *Toda flow*. This connection between the double bracket flow (1.1 and the Toda flow has been first observed by Bloch (1990). For a thorough study of the Toda lattice equation we refer to Kostant (1979). Flaschka (1974) and Moser (1975) have given the following Hamiltonian mechanics interpretation of (2.2).

Consider the system of n idealized mass points $x_1 < \cdots < x_n$ on the real axis.

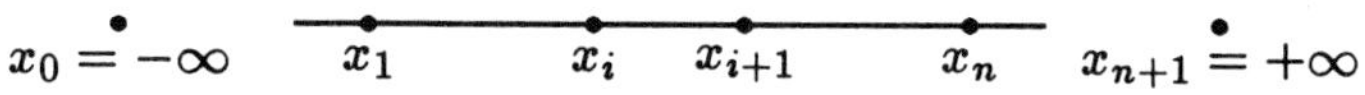

2.2. TODA FLOWS AND THE RICCATI EQUATION

$$x_0 = -\infty \quad x_1 \quad x_i \quad x_{i+1} \quad x_n \quad x_{n+1} = +\infty$$

Figure 2.2.1: Mass points on the real axis

Let us suppose that the potential energy of this system of points $x_1, \cdots, x_n$ is given by

$$V(x_1, \cdots, x_n) = \Sigma_{k=1}^{n} e^{x_k - x_{k+1}}$$

$x_{n+1} = +\infty$, while the kinetic energy is given as usual by $\frac{1}{2}\Sigma_{k=1}^{n}\dot{x}_k^2$. Thus the total energy of the system is given by the Hamiltonian

$$H(x,y) = \frac{1}{2}\Sigma_{k=1}^{n} y_k^2 + \Sigma_{k=1}^{n} e^{x_k - x_{k+1}}, \tag{2.3}$$

with $y_k = \dot{x}_k$ the momentum of the k-th mass point. Thus the associated Hamiltonian system is described by

$$\begin{aligned} \dot{x}_k &= \frac{\partial H}{\partial y_k} = y_k, \\ \dot{y}_k &= -\frac{\partial H}{\partial x_k} = e^{x_{k-1} - x_k} - e^{x_k - x_{k+1}}, \end{aligned} \tag{2.4}$$

for k=1,...,n. In order to see the connection between (2.2) and (2.4) we introduce the new set of coordinates (this trick is due to Flaschka).

$$a_k = \frac{1}{2}e^{(x_k - x_{k+1})/2}, \quad b_k = \frac{1}{2}y_k, \quad k = 1, \cdots, n. \tag{2.5}$$

Then with

$$H = \begin{bmatrix} b_1 & a_1 & & 0 \\ a_1 & b_2 & \ddots & \\ & \ddots & \ddots & a_{n-1} \\ 0 & & a_{n-1} & b_n \end{bmatrix},$$

the Jacobi matrix defined by a_k, b_k, it is easy to verify that (2.4) holds if and only if H satisfies the Toda lattice equation (2.2).

Let $\text{Jac}(\lambda_1, \cdots, \lambda_n)$ denote the set of $n \times n$ Jacobi matrices with eigenvalues $\lambda_1 \geq \cdots \geq \lambda_n$. The geometric structure of $\text{Jac}(\lambda_1, \cdots, \lambda_n)$ is rather complicated and not completely understood. However Tomei (1984) has shown that for distinct eigenvalues $\lambda_1 > \cdots > \lambda_n$ the set $\text{Jac}(\lambda_1, \cdots, \lambda_n)$ is a smooth, $(n-1)$-dimensional compact manifold. Moreover, Tomei (1984) has determined the Euler characteristic of $\text{Jac}(\lambda_1, \cdots, \lambda_n)$. He shows that for $n = 3$ the isospectral set $\text{Jac}(\lambda_1, \lambda_2, \lambda_3)$ for $\lambda_1 > \lambda_2 > \lambda_3$ is a compact Riemann surface of genus two. With these remarks in mind, the following result is an immediate consequence of Theorem 1.3.

Corollary 2.2 *(a) The Toda flow (2.2) is an isospectral flow on the set of real symmetric Jacobi matrices.*

(b) The solution $H(t)$ of (2.2) exists for all $t \in \mathbb{R}$ and converges to a diagonal matrix as $t \to \pm\infty$.

(c) Let $N = \text{diag}(1, 2, \cdots, n)$ and $\lambda_1 > \cdots > \lambda_n$. The Toda flow on the isospectral manifold $\text{Jac}(\lambda_1, \cdots, \lambda_n)$ is the gradient flow for the least squares distance function $f_N : \text{Jac}(\lambda_1, \cdots, \lambda_n) \to \mathbb{R}, f_N(H) = -\frac{1}{2}\|N - H\|^2$.

Remark In part (c) of the above Corollary, the underlying Riemannian metric on $\text{Jac}(\lambda_1, \cdots, \lambda_n)$ is just the restriction of the Riemannian metric $\langle\ ,\ \rangle$ appearing in Theorem 1.3 from $M(Q)$ to $\text{Jac}(\lambda_1, \cdots, \lambda_n)$. ∎

Connection with the Riccati equation There is also an interesting connection of the double bracket equation (1.1) with the Riccati equation. For any real symmetric $n \times n$ matrix

$$Q = \text{diag}(\lambda_1 I_{n_1}, \cdots, \lambda_r I_{n_r}) \qquad (2.6)$$

with distinct eigenvalues $\lambda_1 > \cdots > \lambda_r$ and multiplicities $n_1, \cdots, n_r$, with $\sum_{i=1}^r n_i = n$, let

$$M(Q) = \{\Theta' Q \Theta \mid \Theta' \Theta = I_n\} \qquad (2.7)$$

denote the isospectral manifold of all symmetric $n \times n$ matrices H which are orthogonally similar to Q. The geometry of $M(Q)$ is well understood, in fact, $M(Q)$ is known to be diffeomorphic to a flag manifold. By the isospectral property of the double bracket flow, it induces by restriction a flow on $M(Q)$ and therefore on a flag manifold. It seems difficult to analyze this induced flow in terms of the intrinsic geometry of the flag manifold; see however Duistermaat, Kolk and Varadarajan (1983). We will therefore restrict ourselves to the simplest nontrivial case $r = 2$. But first let us digress on the topic of flag manifolds.

Digression: Flag manifolds A *flag* in $\mathbb{R}^n$ is an increasing sequence of subspaces $V_1 \subset \cdots \subset V_r \subset \mathbb{R}^n$ of $\mathbb{R}^n, 1 \leq r \leq n$. Thus for $r = 1$ a flag is just a linear subspace of $\mathbb{R}^n$. Flags $V_1 \subset \cdots \subset V_n \subset \mathbb{R}^n$ with $\dim V_i = i, i = 1, \cdots, n$, are called *complete*, and flags with $r < n$ are called *partial*.

Given any sequence $(n_1, \cdots, n_r)$ of nonnegative integers with $n_1 + \cdots + n_r \leq n$, the *flag manifold* $\text{Flag}(n_1, \cdots, n_r)$ is defined as the set of all flags $(V_1, \cdots, V_r)$ of vector spaces with $V_1 \subset \cdots \subset V_r \subset \mathbb{R}^n$ and $\dim V_i = n_1 + \cdots + n_i, i = 1, \cdots, r$. $\text{Flag}(n_1, \cdots, n_r)$ is a smooth, connected, compact manifold. For $r = 1, \text{Flag}(n_1) =$

2.2. TODA FLOWS AND THE RICCATI EQUATION

Grass$_{\mathbb{R}}(n_1, n)$ is just the Grassmann manifold of n_1-dimensional linear subspaces of $\mathbb{R}^n$. In particular, for $n = 2$ and $n_1 = 1$, Flag(1) = $\mathbb{R}\mathbb{P}^1$ is the real projective line and is thus homeomorphic to the circle $S^1 = \{(x, y) \in \mathbb{R}^2 \mid x^2 + y^2 = 1\}$.

Let $Q = \text{diag}(\lambda_1 I_{n_1}, \cdots, \lambda_r I_{n_r})$ with $\lambda_1 > \cdots > \lambda_r$ and $n_1 + \cdots + n_r = n$ and let $M(Q)$ be defined by (1.2). Using (1.6) it can be shown that $M(Q)$ is diffeomorphic to the flag manifold Flag$(n_1, \cdots, n_r)$. This *isospectral picture* of the flag manifold will be particularly useful to us. In particular for $r = 2$ we have a diffeomorphism of $M(Q)$ with the real Grassmann manifold Grass$_{\mathbb{R}}(n_1, n)$ of n_1-dimensional linear subspaces of $\mathbb{R}^n$. This is in harmony with the fact, mentioned earlier in Chapter 1, that for $Q = \text{diag}(I_k, 0)$ the real Grassmann manifold Grass$_{\mathbb{R}}(k, n)$ is diffeomorphic to the set $M(Q)$ of rank k symmetric projection operators of $\mathbb{R}^n$.

Explicitly, to any orthogonal matrix $n \times n$ matrix

$$\Theta = \begin{bmatrix} \Theta_1 \\ \vdots \\ \Theta_r \end{bmatrix}$$

with $\Theta_i \in \mathbb{R}^{n_i \times n}, i = 1, \cdots, r$, and $n_1 + \cdots + n_r = n$, we associate the flag of vector spaces

$$V_\Theta := (V_1(\Theta), \cdots, V_r(\Theta)) \in \text{Flag}(n_1, \cdots, n_r).$$

Here $V_i(\Theta), i = 1, \cdots, r$, is defined as the $(n_1 + \cdots + n_i)$-dimensional vector space in $\mathbb{R}^n$ which is generated by the row vectors of the sub-matrix $\begin{bmatrix} \Theta_1 \\ \vdots \\ \Theta_i \end{bmatrix}$ This defines a map

$$\begin{aligned} f : M(Q) &\to \text{Flag}(n_1, \cdots, n_r) \\ \Theta' Q \Theta &\mapsto V_\Theta. \end{aligned}$$

Note that, if $\Theta' Q \Theta = \hat{\Theta}' Q \hat{\Theta}$ then $\hat{\Theta} = \psi \Theta$ for an orthogonal matrix ψ which satisfies $\psi' Q \psi = Q$. But this implies that $\psi = \text{diag}(\psi_1, \cdots, \psi_r)$ is block diagonal and therefore $V_{\hat{\Theta}} = V_\Theta$, so that f is well defined.

Conversely, $V_{\hat{\Theta}} = V_\Theta$ implies $\hat{\Theta} = \psi \Theta$ for an orthogonal matrix $\psi = \text{diag}(\psi_1, \cdots, \psi_r)$. Thus the map f is well-defined and injective. It is easy to see that f is in fact a smooth bijection. The inverse of f is

$$\begin{aligned} f^{-1} : \text{Flag}(n_1, \cdots, n_r) &\to M(Q) \\ V = (V_1, \cdots, V_r) &\mapsto \Theta'_V Q \Theta_V \end{aligned}$$

where

$$\Theta_V = \begin{bmatrix} \Theta_1 \\ \vdots \\ \Theta_r \end{bmatrix} \in O(n)$$

and Θ_i is any orthogonal basis of the orthogonal complement $V_{i-1}^\perp \cap V_i$ of V_{i-1} in V_i; $i = 1, \cdots, r$. It is then easy to check that f^{-1} is smooth and thus $f : M(Q) \to \text{Flag}(n_1, \cdots, n_r)$ is a diffeomorphism. ∎

With the above background material on flag manifolds, let us proceed with the connection of the double bracket equation to the Riccati equation. Let $\text{Grass}_\mathbb{R}(k, n)$ denote the *Grassmann manifold* of k-dimensional linear subspaces of $\mathbb{R}^n$. Thus $\text{Grass}_\mathbb{R}(k, n)$ is a compact manifold of dimension $k(n - k)$. For $k = 1$, $\text{Grass}_\mathbb{R}(1, n) = \mathbb{R}P^{n-1}$ is the $(n - 1)$-dimensional projective space of lines in $\mathbb{R}^n$ (see digression on projective spaces of Chapter 1.2).

For
$$Q = \begin{bmatrix} I_k & 0 \\ 0 & 0 \end{bmatrix},$$
$M(Q)$ coincides with the set of all rank k symmetric projection operators H of $\mathbb{R}^n$:
$$H' = H, \quad H^2 = H, \quad \text{rank } H = k, \tag{2.8}$$
and we have already shown (see digression) that $M(Q)$ is diffeomorphic to the Grassmann manifold $\text{Grass}_\mathbb{R}(k, n)$. The following result is a generalization of this observation.

Lemma 2.3 *For $Q = \text{diag}(\lambda_1 I_k, \lambda_2 I_{n-k})$, $\lambda_1 > \lambda_2$, the isospectral manifold $M(Q)$ is diffeomorphic to the Grassmann manifold $\text{Grass}_\mathbb{R}(k, n)$.*

Proof To any orthogonal $n \times n$ matrix
$$\Theta = \begin{bmatrix} \Theta_1 \\ \Theta_2 \end{bmatrix}, \quad \Theta_1 \in \mathbb{R}^{k \times n}, \quad \Theta_2 \in \mathbb{R}^{(n-k) \times n},$$
we associate the k dimensional vector-space $V_{\Theta_1} \subset \mathbb{R}^n$, which is generated by the k orthogonal row vectors of Θ_1. This defines a map
$$\begin{aligned} f : M(Q) &\to \text{Grass}_\mathbb{R}(k, n), \\ \Theta'Q\Theta &\mapsto V_{\Theta_1}. \end{aligned} \tag{2.9}$$
Note that, if $\Theta'Q\Theta = \hat{\Theta}'Q\hat{\Theta}$ then $\hat{\Theta} = \psi\Theta$ for an orthogonal matrix ψ which satisfies $\psi'Q\psi = Q$. But this implies that $\psi = \text{diag}(\psi_1, \psi_2)$ and therefore $V_{\hat{\Theta}_1} = V_{\Theta_1}$ and f is well defined. Conversely, $V_{\hat{\Theta}_1} = V_{\Theta_1}$ implies $\hat{\Theta} = \psi\Theta$ for an orthogonal matrix $\psi = \text{diag}(\psi_1, \psi_2)$. Thus (2.9) is injective. It is easy to see that f is a bijection and a diffeomorphism. In fact, as $M(Q)$ is the set of rank k symmetric projection operators, $f(H) \in \text{Grass}_\mathbb{R}(k, n)$ is the image of $H \in M(Q)$. Conversely let $X \in \mathbb{R}^{n \times k}$ be such that the columns of X generate a k-dimensional linear subspace $V \subset \mathbb{R}^n$. Then
$$P_X := X(X'X)^{-1}X'$$

2.2. TODA FLOWS AND THE RICCATI EQUATION

is the Hermitian projection operator onto V and

$$f^{-1}: \text{Grass}_{\mathbb{R}}(k,n) \to M(Q)$$

$$\text{column span}(X) \mapsto X(X'X)^{-1}X'$$

is the inverse of f. It is obviously smooth and a diffeomorphism.

∎

Every $n \times n$ matrix $N \in \mathbb{R}^{n \times n}$ induces a flow on the Grassmann manifold

$$\Phi_N : \mathbb{R} \times \text{Grass}_{\mathbb{R}}(k,n) \to \text{Grass}_{\mathbb{R}}(k,n)$$

defined by

$$\Phi_N(t,V) = e^{tN} \cdot V, \tag{2.10}$$

where $e^{tN} \cdot V$ denotes the image of the k-dimensional subspace $V \subset \mathbb{R}^n$ under the invertible linear transformation $e^{tN} : \mathbb{R}^n \to \mathbb{R}^n$. We refer to (2.10) as the *flow* on $\text{Grass}_{\mathbb{R}}(k,n)$ which is *linearly induced by N*.

We have already seen in Chapter 1, Section 2, that linearly induced flows on the Grassmann manifold $\text{Grass}_{\mathbb{R}}(k,n)$ correspond to the *matrix Riccati equation*

$$\dot{K} = A_{21} + A_{22}K - KA_{11} - KA_{12}K \tag{2.11}$$

on $\mathbb{R}^{(n-k) \times k}$, where A is partitioned as

$$A = \begin{bmatrix} A_{11} & A_{12} \\ A_{21} & A_{22} \end{bmatrix}.$$

Theorem 2.4 *Let $N \in \mathbb{R}^{n \times n}$ be symmetric and $Q = \text{diag}(\lambda_1 I_k, \lambda_2 I_{n-k})$, $\lambda_1 > \lambda_2$, for $1 \leq k \leq n-1$. The double bracket flow (1.1) is equivalent via the map f, defined by (2.9), to the flow on the Grassmann manifold $\text{Grass}_{\mathbb{R}}(k,n)$ linearly induced by $(\lambda_1 - \lambda_2)N$. It thus induces the Riccati equation (2.11) with generating matrix $A = (\lambda_1 - \lambda_2)N$.*

Proof Let $H_0 = \Theta_0' Q \Theta_0 \in M(Q)$ and let $H(t) \in M(Q)$ be a solution of (1.1) with $H(0) = H_0$. We have to show that for all $t \in \mathbb{R}$

$$f(H(t)) = e^{t(\lambda_1 - \lambda_2)N} V_0, \tag{2.12}$$

where f is defined by (2.9) and $V_0 = V_{\Theta_0}$. By Theorem 1.5, $H(t) = \Theta(t)' Q \Theta(t)$ where $\Theta(t)$ satisfies the gradient flow on $O(n)$

$$\dot{\Theta} = \Theta(\Theta' Q \Theta N - N \Theta' Q \Theta). \tag{2.13}$$

Hence for $X = \Theta'$

$$\dot{X} = NXQ - XQX'NX. \tag{2.14}$$

Let $X(t), t \in \mathbb{R}$, be any orthogonal matrix solution of (2.14). (Note that orthogonality of $X(t)$ holds automatically in case of orthogonal initial conditions.) Let $S(t) \in \mathbb{R}^{n \times n}$, $S(0) = I_n$, be the unique matrix solution of the linear time-varying system

$$\dot{S} = X(t)'NX(t)((\lambda_1 - \lambda_2)S - QS) + QX(t)'NX(t)S . \qquad (2.15)$$

Let $S_{ij}(t)$ denote the (i,j)-block entry of $S(t)$. Suppose $S_{21}(t_0) = 0$ for some $t_0 \in \mathbb{R}$. A straightforward computation using (2.15) shows that then also $\dot{S}_{21}(t_0) = 0$. Therefore (2.15) restricts to a time-varying flow on the subset of block upper triangular matrices. In particular, the solution $S(t)$ with $S(0) = I_n$ is block upper triangular for all $t \in \mathbb{R}$

$$S(t) = \begin{bmatrix} S_{11}(t) & S_{12}(t) \\ 0 & S_{22}(t) \end{bmatrix} .$$

Lemma 2.5 *For any solution $X(t)$ of (2.14) let*

$$S = \begin{bmatrix} S_{11} & S_{12} \\ 0 & S_{22} \end{bmatrix}$$

be the solution of (2.15) with $S(0) = I_n$. Then

$$X(t) \cdot S(t) = e^{t(\lambda_1 - \lambda_2)N} \cdot X(0) . \qquad (2.16)$$

Proof Let $Y(t) = X(t) \cdot S(t)$. Then

$$\begin{aligned} \dot{Y} &= \dot{X}S + X\dot{S} = NXQS - XQX'NXS \\ &\quad + NX((\lambda_1 - \lambda_2)S - QS) + XQX'NXS \\ &= (\lambda_1 - \lambda_2)NY . \end{aligned}$$

Let $Z(t) = e^{t(\lambda_1 - \lambda_2)N} \cdot X(0)$. Now Y and Z both satisfy the linear differential equation $\dot{\xi} = (\lambda_1 - \lambda_2)N\xi$ with identical initial condition $Y(0) = Z(0) = X(0)$. Thus $Y(t) = Z(t)$ for all $t \in \mathbb{R}$ and the lemma is proved. ∎

We now have the proof of Theorem 2.4 in our hands. In fact, let $V(t) = f(H(t)) \in \text{Grass}_{\mathbb{R}}(k, n)$ denote the vector space which is generated by the first k column vectors of $X(t)$. By the above lemma

$$V(t) = e^{t(\lambda_1 - \lambda_2)N} V(0)$$

which completes the proof. ∎

2.2. TODA FLOWS AND THE RICCATI EQUATION

One can use Theorem 2.4 to prove results on the dynamic Riccati equation arising in linear optimal control, see Anderson and Moore (1990). Let

$$N = \begin{bmatrix} A & -BB' \\ -C'C & -A' \end{bmatrix} \quad (2.17)$$

be the Hamiltonian matrix associated with a linear system

$$\dot{x} = Ax + Bu, \quad y = Cx \quad (2.18)$$

where $A \in \mathbb{R}^{n \times n}, B \in \mathbb{R}^{n \times m}$ and $C \in \mathbb{R}^{p \times n}$. If $m = p$ and $(A, B, C) = (A', C', B')$ then N is symmetric and we can apply Theorem 2.4.

Corollary 2.6 *Let $(A, B, C) = (A', C', B')$ be a symmetric controllable and observable realization. The Riccati equation*

$$\boxed{\dot{K} = -KA - A'K + KBB'K - C'C} \quad (2.19)$$

extends to a gradient flow on $\mathrm{Grass}_{\mathbb{R}}(n, 2n)$ given by the double bracket equation (1.1) under (2.19). Every solution in $\mathrm{Grass}_{\mathbb{R}}(n, 2n)$ converges to an equilibrium point. Suppose

$$N = \begin{bmatrix} A & -BB' \\ -C'C & -A' \end{bmatrix}$$

has distinct eigenvalues. Then (2.19) has $\binom{2n}{n}$ equilibrium points in the Grassmannian $\mathrm{Grass}_{\mathbb{R}}(n, 2n)$, exactly one of which is asymptotically stable.

Proof By Theorem 2.4 the double bracket flow on $M(Q)$ for $Q = \begin{bmatrix} I_k & 0 \\ 0 & 0 \end{bmatrix}$ is equivalent to the flow on the Grassmannian $\mathrm{Grass}_{\mathbb{R}}(n, 2n)$ which is linearly induced by N. If N has distinct eigenvalues, then the double bracket flow on $M(Q)$ has $\binom{2n}{n}$ equilibrium points with exactly one being asymptotically stable. Thus the result follows immediately from Theorem 1.3. ∎

Remarks 1. A transfer function $G(s) = C(sI - A)^{-1}B$ has a symmetric realization $(A, B, C) = (A', C', B')$ if and only if $G(s) = G(s)'$ and the Cauchy index of $G(s)$ is equal to the McMillan degree; Youla and Tissi (1966), Brockett and Skoog (1991). Such transfer functions arise frequently in circuit theory.

2. Of course the second part of the theorem is a well-known fact from linear optimal control theory. The proof here, based on the properties of the double

bracket flow, however, is new and offers a rather different approach to the stability properties of the Riccati equation.

3. A possible point of confusion arising here might be that in some cases the solutions of the Riccati equation (2.19) might not exist for all $t \in \mathbb{R}$ (finite escape time) while the solutions to the double bracket equation always exist for all $t \in \mathbb{R}$. One should keep in mind that Theorem 2.4 only says that the double bracket flow on $M(Q)$ is equivalent to a linearly induced flow on $\text{Grass}_{\mathbb{R}}(n, 2n)$. Now the Riccati equation (2.19) is the vector field corresponding to the linear induced flow on an open coordinate chart of $\text{Grass}_{\mathbb{R}}(n, 2n)$ and thus, up to the change of variables described by the diffeomorphism $f : M(Q) \to \text{Grass}_{\mathbb{R}}(n, 2n)$, coincides on that open coordinate chart with the double bracket equation. Thus the double bracket equation on $M(Q)$ might be seen as an extension or completion of the Riccati vector field (2.19).

4. For $Q = \begin{bmatrix} I_k & 0 \\ 0 & 0 \end{bmatrix}$ any $H \in M(Q)$ satisfies $H^2 = H$. Thus the double bracket equation on $M(Q)$ becomes the special Riccati equation on $\mathbb{R}^{n \times n}$

$$\boxed{\dot{H} = HN + NH - 2HNH}$$

which is shown in Theorem 2.4 to be equivalent to the *general* matrix Riccati equation (for N symmetric) on $\mathbb{R}^{(n-k) \times k}$. ∎

Finally we consider the case $k = 1$, i.e. the associated flow on the projective space $\mathbb{RP}^{n-1}$. Thus let $Q = \text{diag}(1, 0, \cdots, 0)$. Any $H \in M(Q)$ has a representation $H = x \cdot x'$ where $x' = (x_1, \cdots, x_n)$, $\sum_{j=1}^{n} x_j^2 = 1$, and x is uniquely determined up to multiplication by ± 1. The double bracket flow on $M(Q)$ is equivalent to the gradient flow of the standard Morse function

$$\Phi(x) = \frac{1}{2}\text{tr}(Nxx') = \frac{1}{2}\sum_{i,j=1}^{n} n_{ij}x_i x_j .$$

on $\mathbb{RP}^{n-1}$ see Milnor (1963). Moreover, the Lie bracket flow (1.25) on $O(n)$ induces, for Q as above, the Rayleigh quotient flow on the $(n-1)$ sphere S^{n-1} of Chapter 1, Section 3.

Problems 1. Prove that every solution of

$$\dot{H} = AH + HA - 2HAH, H(0) = H_0, \qquad (*)$$

has the form

$$H(t) = e^{tA}H_0(I_n - H_0 + e^{2tA}H_0)^{-1}e^{tA}, \ t \in \mathbb{R}$$

2.3. RECURSIVE LIE-BRACKET BASED DIAGONALIZATION

2. Show that the spectrum (i.e. the set of eigenvalues) $\sigma(H(t))$ of any solution is given by

$$\sigma(H(t)) = \sigma(H_0(e^{-2tA}(I_n - H_0) + H_0)).$$

3. Show that for any solution $H(t)$ of (*) also $G(t) = I_n - H(-t)$ solves (*).

4. Derive similar formulas for time-varying matrices $A(t)$.

Main Points of Section The double bracket equation $\dot{H} = [H,[H,N]]$ with $H(0)$ a Jacobi matrix and $N = \text{diag}(1,2,\cdots,n)$ preserves the Jacobi property in $H(t)$ for all $t \geq 0$. In this case the double bracket equation is the *Toda flow* $\dot{H} = [H, H_u - H_\ell]$ which has interpretations in Hamiltonian mechanics.

For the case of symmetric matrices N and $Q = \text{diag}(\lambda_1 I_k, \lambda_2 I_{n-k})$ with $\lambda_1 > \lambda_2$, the double bracket flow is equivalent to the flow on the Grassmann manifold $\text{Grass}_{\mathbb{R}}(k,n)$ linearly induced by $(\lambda_1 - \lambda_2)N$. This in turn is equivalent to a Riccati equation (2.11) with generating matrix $A = (\lambda_1 - \lambda_2)N$. This result gives an interpretation of certain Riccati equations of linear optimal control as gradient flows.

For the special case $Q = \text{diag}(I_k, 0, \cdots, 0)$, then $H^2 = H$ and the double bracket equation becomes the special Riccati equation $\dot{H} = HN + NH - 2HNH$. When $k = 1$, then $H = xx'$ with x a column vector and the double bracket equation is induced by $\dot{x}=(N - x'Nx I_n)x$, which is the Rayleigh quotient gradient flow of $\text{tr}(Nxx')$ on the sphere S^{n-1}.

2.3 Recursive Lie-Bracket Based Diagonalization

Now that the double bracket flow with its rich properties has been studied, it makes sense to ask whether or not there are corresponding recursive versions. Indeed there are. For some ideas and results in this direction we refer to Chu (1992), Brockett (1993), and Moore, Mahony, and Helmke (1993). In this section we study the following recursion, termed the Lie-bracket recursion,

$$\boxed{H_{k+1} = e^{-\alpha[H_k,N]} H_k e^{\alpha[H_k,N]}, \quad H_0 = H_0', \quad k \in \mathbb{N}} \qquad (3.1)$$

for arbitrary symmetric matrices $H_0 \in \mathbb{R}^{n \times n}$, and some suitably small scalar α, termed a step size scaling factor. A key property of the recursion (3.1) is that it is isospectral. This follows since $e^{\alpha[H_k,N]}$ is orthogonal, as indeed is any e^A where A is skew symmetric.

To motivate this recursion (3.1), observe that H_{k+1} is also the solution at time $t = \alpha$ to a linear matrix differential equation initialized by H_k as follows

$$\frac{d\bar{H}}{dt} = [\bar{H}, [H_k, N]], \quad \bar{H}(0) = H_k ,$$
$$H_{k+1} = \bar{H}(\alpha) .$$

For small t, and thus α small, $\bar{H}(t)$ appears to be a solution close to that of the corresponding double bracket flow $H(t)$ of $\dot{H} = [H, [H, N]], H(0) = H_k$. This suggests, that for step-size scaling factors not too large, or not decaying to zero too rapidly, that the recursion (3.1) should inherit the exponential convergence rate to desired equilibria of the continuous-time double bracket equation. Indeed, in some applications this piece-wise constant, linear, and isospectral differential equation may be more attractive to implement than a nonlinear matrix differential equation.

Our approach in this section is to optimize in some sense the α selections in (3.1) according to the potential of the continuous-time gradient flow. The subsequent bounding arguments are similar to those developed by Brockett (1993), see also Moore, Mahony and Helmke (1993). In particular, for the potential function

$$f_N(H_k) = -\frac{1}{2}\|N - H_k\|^2 = \operatorname{tr}(NH_k) - \frac{1}{2}\|N\| - \frac{1}{2}\|H_0\| \quad (3.2)$$

we seek to maximize at each iteration its increase

$$\triangle f_N(H_k, \alpha) := f_N(H_{k+1}) - f_N(H_k) = \operatorname{tr}(N(H_{k+1} - H_k)) \quad (3.3)$$

Lemma 3.1 *The constant step-size selection*

$$\alpha = \frac{1}{4\|H_0\| \cdot \|N\|} \quad (3.4)$$

satisfies $\triangle f_N(H_k, \alpha) > 0$ *if* $[H_k, N] \neq 0$.

Proof Let $H_{k+1}(\tau) = e^{-\tau[H_k,N]} H_k e^{\tau[H_k,N]}$ be the $k+1^{\text{th}}$ iteration of (3.1) for an arbitrary step-size scaling factor $\tau \in \mathbb{R}$. It is easy to verify that

$$\frac{d}{d\tau} H_{k+1}(\tau) = [H_{k+1}(\tau), [H_k, N]]$$
$$\frac{d^2}{d\tau^2} H_k + 1(\tau) = [[H_{k+1}(\tau), [H_k, N]], [H_k, N]].$$

Applying Taylor's theorem, then since $H_{k+1}(0) = H_k$

$$H_{k+1}(\tau) = H_k + \tau[H_k, [H_k, N]] + \tau^2 R_2(\tau),$$

2.3. RECURSIVE LIE-BRACKET BASED DIAGONALIZATION 71

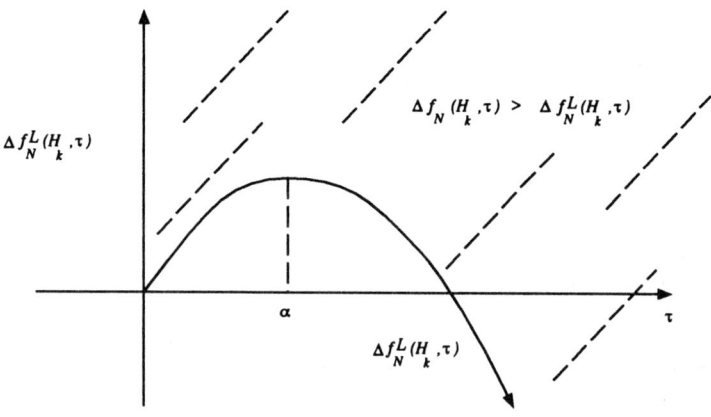

Figure 2.3.1: The lower bound on $\triangle f_N^L(H_k, \tau)$

where

$$\mathcal{R}_2(\tau) = \int_0^1 [[H_{k+1}(y\tau), [H_k, N]], [H_k, N]](1-y)dy. \quad (3.5)$$

Substituting into (3.3) gives, using matrix norm inequalities outlined in Appendix (A.6) and (A.9),

$$\begin{aligned}
\triangle f_N(H_k, \tau) &= \text{tr}(N(\tau[H_k, [H_k, N]] + \tau^2 \mathcal{R}_2(\tau))) \\
&= \tau\|[H_k, N]\|^2 + \tau^2 \text{tr}(N\mathcal{R}_2(\tau)) \\
&\geq \tau\|[H_k, N]\|^2 - \tau^2|\text{tr}(N\mathcal{R}_2(\tau))| \\
&\geq \tau\|[H_k, N]\|^2 - \tau^2\|N\| \cdot \|\mathcal{R}_2(\tau)\| \\
&\geq \tau\|[H_k, N]\|^2 - \tau^2\|N\| \cdot \\
&\quad \int_0^1 \|[[H_{k+1}(y\tau), [H_k, N]], [H_k, N]]\|(1-y)dy \\
&\geq \tau\|[H_k, N]\|^2 - 2\tau^2\|N\| \cdot \|H_0\| \cdot \|[H_k, N]\|^2 \\
&:= \triangle f_N^L(H_k, \tau). \quad (3.6)
\end{aligned}$$

Thus $\triangle f_N^L(H_k, \tau)$ is a lower bound for $\triangle f_N(H_k, \tau)$ and has the property that for sufficiently small $\tau > 0$, it is strictly positive, see Figure 2.3.1. Due to the explicit form of $\triangle f_N^L(H_k, \tau)$ in τ, it is immediately clear that if $[H_k, N] \neq 0$, then $\alpha = 1/(4\|H_0\|\|N\|)$ is the unique maximum of (3.6). Hence, $f_N(H_k, \alpha) \geq f_N^L(H_k, \alpha) > 0$ for $[H_k, N] \neq 0$. ∎

This lemma leads to the main result of the section.

Theorem 3.2 *Let $H_0 = H_0'$ be a real symmetric $n \times n$ matrix with eigenvalues $\lambda_1 \geq \ldots \geq \lambda_n$ and let $N = \text{diag}(\mu_1, \ldots, \mu_n)$, $\mu_1 > \ldots > \mu_n$. The*

Lie-bracket recursion (3.1), restated as

$$H_{k+1} = e^{-\alpha[H_k,N]} H_k e^{\alpha[H_k,N]}, \quad \alpha = 1/(4\|H_0\| \cdot \|N\|) \quad (3.7)$$

with initial condition H_0, has the following properties:

(a) The recursion is isospectral.

(b) If (H_k) is a solution of the Lie-bracket algorithm, then $f_N(H_k)$ of (3.2) is a strictly monotonically increasing function of $k \in \mathbb{N}$, as long as $[H_k, N] \neq 0$.

(c) Fixed points of the recursive equation are characterised by matrices $H_\infty \in M(H_0)$ such that $[H_\infty, N] = 0$, being exactly the equilibrium points of the double bracket equation (1.1).

(d) Let (H_k), $k = 1, 2, \ldots$, be a solution to the recursive Lie-bracket algorithm, then H_k converges to a matrix $H_\infty \in M(H_0)$, $[H_\infty, N] = 0$, an equilibrium point of the recursion.

(e) All equilibrium points of the recursive Lie-bracket algorithm are unstable, except for $Q = diag(\lambda_1, \ldots, \lambda_n)$ which is locally exponentially stable.

Proof To prove part (a), note that the Lie-bracket $[H, N]' = -[H, N]$ is skew-symmetric. As the exponential of a skew-symmetric matrix is orthogonal, (3.1) is just conjugation by an orthogonal matrix, and hence is an isospectral transformation. Part (b) is a direct consequence of Lemma 3.1.

For part (c) note that if $[H_k, N] = 0$, then by direct substitution into (3.1) we see $H_{k+l} = H_k$ for $l \geq 1$, and thus H_k is a fixed point. Conversely if $[H_k, N] \neq 0$, then from (3.6), $\Delta f_N(H_k, \alpha) \neq 0$, and thus $H_{k+1} \neq H_k$. In particular, fixed points of (3.1) are equilibrium points of (1.1), and furthermore, from Theorem 1.3, these are the only equilibrium points of (1.1).

To show part (d), consider the sequence H_k generated by the recursive Lie-bracket algorithm for a fixed initial condition H_0. Observe that part b) implies that $f_N(H_k)$ is strictly monotonically increasing for all k where $[H_k, N] \neq 0$. Also, since f_N is a continuous function on the compact set $M(H_0)$, then f_N is bounded from above, and $f_N(H_k)$ will converge to some $f_\infty \leq 0$ as $k \to \infty$. As $f_N(H_k) \to f_\infty$ then $\Delta f_N(H_k, \alpha^c) \to 0$.

For some small positive number ε, define an open set $D_\varepsilon \subset M(H_0)$ consisting of all points of $M(H_0)$, within an ϵ-neighborhood of some equilibrium point of (3.1). The set $M(H_0) - D_\varepsilon$ is a closed, compact subset of $M(H_0)$, on which the matrix function $H \mapsto \|[H, N]\|$ does not vanish. As a consequence, the difference function (3.3) is continuous and strictly positive on $M(H_0) - D_\varepsilon$, and thus, is under bounded by some strictly positive number $\delta_1 > 0$. Moreover, as $\Delta f_N(H_k, \alpha) \to 0$, there exists a $K = K(\delta_1)$ such that

2.3. RECURSIVE LIE-BRACKET BASED DIAGONALIZATION

for all $k > K$ then $0 \leq \Delta f_N(H_k, \alpha) < \delta_1$. This ensures that $H_k \in D_\varepsilon$ for all $k > K$. In other words, (H_k) is converging to some subset of possible equilibrium points.

From Theorem 1.3 and the imposed genericity assumption on N, it is known that the double bracket equation (1.1) has only a finite number of equilibrium points. Thus H_k converges to a finite subset in $M(H_0)$. Moreover, $[H_k, N] \to 0$ as $k \to \infty$. Therefore $\|H_{k+1} - H_k\| \to 0$ as $k \to \infty$. This shows that (H_k) must converge to an equilibrium point, thus completing the proof of part (d).

To establish exponential convergence, note that since α is constant, the map
$$H \mapsto e^{-\alpha[H,N]} H e^{\alpha[H,N]}$$
is a smooth recursion on all $M(H_0)$, and we may consider the linearization of this map at an equilibrium point $\pi' Q \pi$. The linearization of this recursion, expressed in terms of $\Xi_k \in T_{\pi'Q\pi}M(H_0)$, is

$$\Xi_{k+1} = \Xi_k - \alpha[(\Xi_k N - N\Xi_k)\pi'Q\pi - \pi'Q\pi(\Xi_k N - N\Xi_k)]. \tag{3.8}$$

Thus for the elements of $\Xi_k = (\xi_{ij})_k$ we have

$$(\xi_{ij})_{k+1} = [1 - \alpha(\lambda_{\pi(i)} - \lambda_{\pi(j)})(\mu_i - \mu_j)](\xi_{ij})_k, \text{ for } i, j = 1, \cdots, n. \tag{3.9}$$

The tangent space $T_{\pi'Q\pi}M(H_0)$ at $\pi'Q\pi$ consists of those matrices $\Xi = [\pi'Q\pi, \Omega]$ where $\Omega \in \text{skew}(n)$, the class of skew symmetric matrices. Thus, the matrices Ξ are linearly parametrized by their components ξ_{ij}, where $i < j$, and $\lambda_{\pi(i)} \neq \lambda_{\pi(j)}$. As this is a linearly independent parametrisation, the eigenvalues of the linearization (3.8) can be read directly from the linearization (3.9), and are $1 - \alpha(\lambda_{\pi(i)} - \lambda_{\pi(j)})(\mu_i - \mu_j)$, for $i < j$ and $\lambda_{\pi(i)} \neq \lambda_{\pi(j)}$. From classical stability theory for discrete-time linear dynamical systems, (3.8) is asymptotically stable if and only if all eigenvalues have absolute value less than 1. Equivalently, (3.8) is asymptotically stable if and only if
$$0 < \alpha(\lambda_{\pi(i)} - \lambda_{\pi(j)})(\mu_i - \mu_j) < 2$$
for all $i < j$ with $\lambda_{\pi(i)} \neq \lambda_{\pi(j)}$. This condition is only satisfied when $\pi = I$ and consequently $\pi'Q\pi = Q$. Thus the only possible stable equilibrium point for the recursion is $H_\infty = Q$. Certainly $(\lambda_i - \lambda_j)(\mu_i - \mu_j) < 4\|N\|_2 \|H_0\|_2$. Also since $\|N\|_2 \|H_0\|_2 < 2\|N\| \, \|H_0\|$ we have $\alpha < \frac{1}{2\|N\|_2 \|H_0\|_2}$. Therefore $\alpha(\lambda_i - \lambda_j)(\mu_i - \mu_j) < 2$ for all $i < j$ which establishes exponential stability of (3.8). This completes the proof. ∎

Remarks 1. In the nongeneric case where N has multiple eigenvalues, the proof techniques for parts (d) and (e) do not apply. All the results except convergence to a *single* equilibrium point remain in force.

2. It is difficult to characterise the set of exceptional initial conditions, for which the algorithm converges to some unstable equilibrium point $H_\infty \neq Q$. However, in the continuous-time case it is known that the unstable basins of attraction of such points are of zero measure in $M(H_0)$, see Section 2.1.

3. By using a more sophisticated bounding argument, a variable step size selection can be determined as

$$\alpha_k = \frac{1}{2\|[H_k, N]\|} \log\left(\frac{\|[H_k, N]\|^2}{\|H_0\| \cdot \|[N, [H_k, N]]\|} + 1\right) \quad (3.10)$$

Rigorous convergence results are given for this selection in Moore, Mahony, and Helmke (1993). The convergence rate is faster with this selection. ∎

Recursive Flows on Orthogonal Matrices The associated recursions on the orthogonal matrices corresponding to the gradient flows (1.25) are

$$\boxed{\Theta_{k+1} = \Theta_k e^{\alpha_k [\Theta_k' H_0 \Theta_k, N]}, \quad \alpha = 1/(4\|H_0\| \cdot \|N\|)} \quad (3.11)$$

where Θ_k is defined on $O(n)$ and α is a general step-size scaling factor. Thus $H_k = \Theta_k' H_0 \Theta_k$ is the solution of the Lie-bracket recursion (3.1). Precise results on (3.11) are now stated and proved for generic H_0 and constant step-size selection, although corresponding results are established in Moore, Mahony and Helmke (1993) for the variable step-size scaling factor (3.10).

Theorem 3.3 *Let $H_0 = H_0'$ be a real symmetric $n \times n$ matrix with distinct eigenvalues $\lambda_1 > \ldots > \lambda_n$. Let $N \in \mathbb{R}^{n \times n}$ be $\mathrm{diag}(\mu_1, \ldots, \mu_n)$ with $\mu_1 > \ldots > \mu_n$. Then the recursion (3.11) referred to as the associated orthogonal Lie-bracket algorithm, has the following properties:*

(a) A solution Θ_k, $k = 1, 2, \ldots$, to the associated orthogonal Lie-bracket algorithm remains orthogonal.

(b) Let $f_{N,H_0} : O(n) \to \mathbb{R}$, $f_{N,H_0}(\Theta) = -\frac{1}{2}\|\Theta' H_0 \Theta - N\|^2$ be a function on $O(n)$. Let Θ_k, $k = 1, 2, \ldots$, be a solution to the associated orthogonal Lie-bracket algorithm. Then $f_{N,H_0}(\Theta_k)$ is a strictly monotonically increasing function of $k \in \mathbb{N}$, as long as $[\Theta_k' H_0 \Theta_k, N] \neq 0$.

(c) Fixed points of the recursive equation are characterised by matrices $\Theta \in O(n)$ such that

$$[\Theta' H_0 \Theta, N] = 0.$$

There are exactly $2^n n!$ such fixed points.

(d) Let Θ_k, $k = 1, 2, \ldots$, be a solution to the associated orthogonal Lie-bracket algorithm, then Θ_k converges to an orthogonal matrix Θ_∞, satisfying $[\Theta_\infty' H_0 \Theta_\infty, N] = 0$.

2.3. RECURSIVE LIE-BRACKET BASED DIAGONALIZATION 75

(e) *All fixed points of the associated orthogonal Lie-bracket algorithm are strictly unstable, except those 2^n points $\Theta_* \in O(n)$ such that*

$$\Theta'_* H_0 \Theta_* = Q,$$

where $Q = \text{diag}(\lambda_1, \ldots, \lambda_n)$. Such points Θ_* are locally exponentially asymptotically stable and $H_0 = \Theta_* Q \Theta'_*$ is the eigenspace decomposition of H_0.

Proof Part (a) follows directly from the orthogonal nature of $e^{\alpha[\Theta'_k H_0 \Theta_k, N]}$. Let $g : O(n) \to M(H_0)$ be the matrix valued function $g(\Theta) = \Theta' H_0 \Theta$. Observe that g maps solutions $(\Theta_k \mid k \in \mathbb{N})$ of (3.11) to solutions $(H_k \mid k \in \mathbb{N})$ of (3.1).

Consider the potential $f_{N,H}(\Theta_k) = -\frac{1}{2}\|\Theta'_k H_0 \Theta_k - N\|^2$, and the potential $f_N = -\frac{1}{2}\|H_k - N\|^2$. Since $g(\Theta_k) = H_k$ for all $k = 1, 2, \ldots$, then $f_{N,H_0}(\Theta_k) = f_N(g(\Theta_k))$ for $k = 1, 2 \ldots$. Thus $f_N(H_k) = f_N(g(\Theta_k)) = f_{N,H_0}(\Theta_k)$ is strictly monotonically increasing for $[H_k, N] = [g(\Theta_k), N] \neq 0$, and part (b) follows.

If Θ_k is a fixed point of the associated orthogonal Lie-bracket algorithm with initial condition Θ_0, then $g(\Theta_k)$ is a fixed point of the Lie-bracket algorithm. Thus, from Theorem 3.2, $[g(\Theta_k), N] = [\Theta'_k H_0 \Theta_k, N] = 0$. Moreover, if $[\Theta'_k H_0 \Theta_k, N] = 0$ for some given $k \in \mathbb{N}$, then by inspection $\Theta_{k+l} = \Theta_k$ for $l = 1, 2, \ldots$, and Θ_k is a fixed point of the associated orthogonal Lie-bracket algorithm. A simple counting argument shows that there are precisely $2^n n!$ such points and part (c) is established.

To prove (d), note that since $g(\Theta_k)$ is a solution to the Lie-bracket algorithm, it converges to a limit point $H_\infty \in M(H_0)$, $[H_\infty, N] = 0$, by Theorem 3.2. Thus Θ_k must converge to the pre-image set of H_∞ via the map g. The genericity condition on H_0 ensures that the set generated by the pre-image of H_∞ is a finite disjoint set. Since $\|[g(\Theta_k), N]\| \to 0$ as $k \to \infty$, then $\|\Theta_{k+1} - \Theta_k\| \to 0$ as $k \to \infty$. From this convergence of Θ_k follows.

To prove part (e), observe that, due to the genericity condition on H_0, the dimension of $O(n)$ is the same as the dimension of $M(H_0)$. Thus g is locally a diffeomorphism on $O(n)$, and taking a restriction of g to such a region, the local stability structure of the equilibria are preserved under the map g^{-1}. Thus, all fixed points of the associated orthogonal Lie-bracket algorithm are locally unstable except those that map via g to the unique locally asymptotically stable equilibrium of the Lie-bracket recursion. ∎

Simulations A simulation has been included to demonstrate the recursive schemes developed. The simulation deals with a real symmetric 7×7 initial condition, H_0, generated by an arbitrary orthogonal similarity transformation of matrix $Q = \text{diag}(1, 2, 3, 4, 5, 6, 7)$. The matrix N was chosen to be

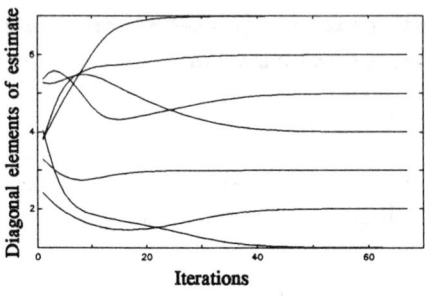 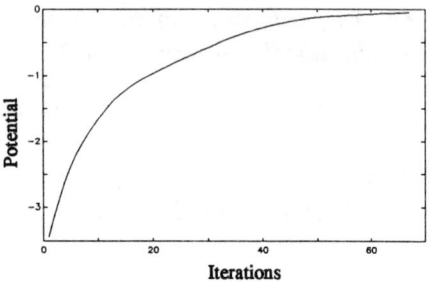

Figure 2.3.2: The recursive Lie-bracket scheme

Figure 2.3.3: The potential function $f_N(H_k)$

diag$(1,2,3,4,5,6,7)$ so that the minimum value of f_N occurs at Q such that $f_N(Q) = 0$. Figure 2.3.2 plots the diagonal entries of H_k at each iteration and demonstrates the asymptotic convergence of the algorithm. The exponential behaviour of the curves appears at around iteration 30, suggesting that this is when the solution H_k enters the locally exponentially attractive domain of the equilibrium point Q. Figure 2.3.3 shows the evolution of the potential $f_N(H_k) = -\frac{1}{2}\|H_k - N\|^2$, demonstrating its monotonic increasing properties and also displaying exponential convergence after iteration 30.

Computational Considerations It is worth noting that an advantage of the recursive Lie-bracket scheme over more traditional linear algebraic schemes for the same tasks, is the presence of a step-size α and the arbitrary target matrix N. The focus in this section is only on the step size selection. The challenge remains to devise more optimal (possibly time-varying) target matrix selection schemes for improved performance. This suggests an application of optimal control theory.

It is also possible to consider alternatives of the recursive Lie-bracket scheme which have improved computational properties. For example, consider a (1,1) Padé approximation to the matrix exponential

$$e^{\alpha[H_k,N]} \approx \frac{2I - \alpha[H_k,N]}{2I + \alpha[H_k,N]}.$$

Such an approach has the advantage that, as $[H_k, N]$ is skew symmetric, then the Padé approximation will be orthogonal, and will preserve the isospectral nature of the Lie-bracket algorithm. Similarly, an (n,n) Padé approximation of the exponential for any n will also be orthogonal.

Actually Newton methods involving second order derivatives can be devised to give local quadratic convergence. These can be switched in to the Lie-bracket recursions here as appropriate, making sure that at each iteration the potential function increases. The resulting schemes are then much more

2.3. RECURSIVE LIE-BRACKET BASED DIAGONALIZATION

competitive with commercial diagonalization packages than the purely linear methods of this section. Of course there is the possibility that quadratic convergence can also be achieved using shift techniques, but we do not explore this fascinating territory here.

Another approach is to take just an Euler iteration,

$$H_{k+1} = H_k + \alpha[H_k, [H_k, N]],$$

as a recursive algorithm on $\mathbb{R}^{n \times n}$. A scheme such as this is similar in form to approximating the curves generated by the recursive Lie-bracket scheme by straight lines. The approximation will not retain the isospectral nature of the Lie-bracket recursion, but this fact may be overlooked in some applications, because it is computationally inexpensive. We can not recommend this scheme, or higher order versions, except in the neighborhood of an equilibrium.

Main Points of Section   In this section we have proposed a numerical scheme for the calculation of double bracket gradient flows on manifolds of similar matrices. Step-size selections for such schemes has been discussed and results have been obtained on the nature of equilibrium points and on their stability properties. As a consequence, the schemes proposed in this section could be used as a computational tool with known bounds on the total time required to make a calculation. Due to the computational requirements of calculating matrix exponentials these schemes may not be useful as a direct numerical tool in traditional computational environments, however, they provide insight into discretising matrix flows such as is generated by the double bracket equation.

Notes for Chapter 2

As we have seen in this chapter, the isospectral set $M(Q)$ of symmetric matrices with fixed eigenvalues is a homogeneous space and the least squares distance function $f_N: M(Q) \to \mathbb{R}$ is a smooth Morse-Bott function. Quite a lot is known about the critical points of such functions and there is a rich mathematical literature on Morse theory developed for functions defined on homogeneous spaces.

If Q is a rank k projection operator, then $M(Q)$ is a Grassmannian. In this case the trace function $H \mapsto \text{tr}(NH)$ is the classical example of a Morse function on Grassmann manifolds. See Milnor (1963) for the case where $k = 1$ and Wu (1965), Hangan (1968) for a slightly different construction of a Morse function for arbitrary k. For a complete characterization of the critical points of the trace functional on classical groups, the Stiefel manifold and Grassmann manifolds, with applications to the topology of these spaces, we refer to Frankel (1965) and Shayman (1982). For a complete analysis of the critical points and their Morse indices of the trace function on more general classes of homogeneous spaces we refer to Hermann (1962, 1963, 1964), Takeuchi (1965) and Duistermaat, Kolk and Varadarajan (1983). For results on gradient flows of certain least squares functions defined on infinite dimensional homogeneous spaces arising in physics we refer to the important work of Atiyah and Bott (1982), Pressley (1982) and Pressley and Segal (1986). The work of Byrnes and Willems (1986) contains an interesting application of moment map techniques from symplectic geometry to total least squares estimation.

The geometry of varieties of isospectral Jacobi matrices has been studied by Tomei (1984) and Davis (1987). The set $\text{Jac}(\lambda_1, \ldots, \lambda_n)$ of Jacobi matrices with distinct eigenvalues $\lambda_1 > \ldots > \lambda_n$ is shown to be a smooth compact manifold. Furthermore, an explicit construction of the universal covering space is given. The report of Driessel (1987 a) contains another elementary proof of the smoothness of $\text{Jac}(\lambda_1, \ldots, \lambda_n)$. Also the tangent space of $\text{Jac}(\lambda_1, \ldots, \lambda_n)$ at a Jacobi matrix L is shown to be the vector space of all Jacobi matrices ζ of the form $\zeta = L\Omega - \Omega L$, where Ω is skew symmetric. Closely related to the above is the work by de Mari and Shayman (1988) and de Mari, Procesi and Shayman (1992) on Hessenberg varieties; i.e. varieties of invariant flags of a given Hessenberg matrix. Furthermore there are interesting connections with torus varieties; for this we refer to Gelfand and Serganova (1987).

The double bracket equation (1.1) and its properties were first studied by Brockett (1988); see also Chu and Driessel (1990). The simple proof of the Wielandt-Hoffman inequality via the double bracket equation is due to Chu and Driessel (1990). A systematic analysis of the double bracket flow (1.2) on adjoint orbits of compact Lie groups appears in Bloch, Brockett and Ratiu (1990, 1992). For an application to subspace learning see Brockett (1991a). In Brockett (1989b) it is shown that the double bracket equation can simulate any finite automaton. Least squares matching problems arising in computer vision and pattern analysis are tackled via double bracket-like equations in Brockett (1989a). An interesting connection exists between the double bracket flow (1.1) and a fundamental equation arising in micromagnetics. The Landau-Lifshitz equation on the two-sphere is a nonlinear diffusion equation which, in the absence of diffusion terms, becomes equivalent to the double bracket equation. Stochastic versions of the double bracket flow are studied by Colonius and Kliemann (1990).

A thorough study of the Toda lattice equation with interesting links to representation theory has been made by Kostant (1979). For the Hamiltonian mechanics interpretation of the QR-algorithm and the Toda flow see Flaschka (1974, 1975), Moser (1975), and Bloch (1990). For connections of the Toda flow with scaling actions on spaces of rational functions

2.3. RECURSIVE LIE-BRACKET BASED DIAGONALIZATION

in system theory see Byrnes (1978), Brockett and Krishnaprasad (1980) and Krishnaprasad (1979). An interesting interpretation of the Toda flow from a system theoretic viewpoint is given in Brockett and Faybusovich (1991); see also Faybusovich (1989). Numerical analysis aspects of the Toda flow have been treated by Symes (1980, 1982), Chu (1984 a, 1984 b), Deift, Nanda and Tomei (1983), and Shub and Vasquez (1987). Expository papers are Watkins (1984), Chu (1984 b). The continuous double bracket flow $\dot{H} = [H, [H, \operatorname{diag} H]]$ is related to the discrete Jacobi method for diagonalization. For a phase portrait analysis of this flow see Driessel (1987 b). See also Wilf (1981) and Golub and Van Loan (1983), Section 8.4 "Jacobi methods", for a discussion on the Jacobi method.

For the connection of the double bracket flow to Toda flows and flows on Grassmannians much of the initial work was done by Bloch (1990, 1991) and then by Bloch, Brockett and Ratiu (1990) and Bloch, Flaschka and Ratiu (1990). The connection to the Riccati flow was made explicit in Helmke (1991) and independently observed by Faybusovich. The paper of Faybusovich (1992) contains a complete description of the phase portrait of the Toda flow and the corresponding QR algorithm, including a discussion of structural stability properties. In Faybusovich (1989) the relationship between QR-like flows and Toda-like flows is described. Monotonicity properties of the Toda flow are discussed in Lagarias (1991). A VLSI type implementation of the Toda flow by a nonlinear lossless electrical network is given by Paul, Hueper and Nossek (1992).

Infinite dimensional versions of the Toda flow with applications to sorting of function values are in Brockett and Bloch (1989), Deift, Li and Tomei (1985), Bloch, Brockett, Kodama and Ratiu (1989). See also the closely related work by Bloch (1985a, 1987) and Bloch and Byrnes (1986).

Numerical integration schemes of ordinary differential equations on manifolds are presented by Crouch and Grossmann (1991), Crouch, Grossmann and Yan (1992a, 1992b) and Ge-Zhong and Marsden (1988). Discrete-time versions of some classical integrable systems are analyzed by Moser and Veselov (1991). For complexity properties of discrete integrable systems see Arnold (1990) and Veselov (1992). The recursive Lie-bracket diagonalization algorithm (3.1) is analyzed in detail in Moore, Mahony and Helmke (1993). Related results appear in Brockett (1993) and Chu (1992). Step-size selections for discretizing the double bracket flow also appear in the recent PhD thesis of Smith (1993).

References

Anderson, B.D.O. and J.B. Moore (1990) *Optimal control: linear quadratic methods*, Prentice Hall, New York.

Arnold, V.I. (1990) Dynamics of intersections, *Analysis et cetera* (P. Rabinowich and E. Zehnder, Eds.) Academic Press, New York.

Atiyah, M. and R. Bott (1982) On the Yang-Mills equations over Riemann surfaces, *Phil. Trans. Roy. Soc. London A*, 308, 523 - 615.

Bloch, A.M. (1985a) A completely integrable Hamiltonian system associated with line fitting in complex vector spaces, *Bull. Amer. Math. Soc.*, 12, 250 - 254.

Bloch, A.M. (1985b) Estimation, principal components and Hamiltonian systems, *Systems and Control Letters*, 6, 1-15.

Bloch, A.M. and C.I. Byrnes (1986) An infinite-dimensional variational problem arising in estimation theory, *Algebraic and Geometric Methods in Nonlinear Control Theory*, (M. Fliess and H. Hazewinkel, Eds.), Reidel Publ.

Bloch, A.M. (1987) An infinite-dimensional Hamiltonian system on projective Hilbert space, *Trans. Amer. Math. Soc.*, 302, 787-796.

Bloch, A.M., R.W. Brockett, Y. Kodama and R. Ratiu (1989) Spectral equations for the long wave limit of the Toda lattice equations, *Hamiltonian systems, Transformation groups and Spectral Transform Methods* (J. Harnad and J.E. Marsden, Eds.)

Bloch, A.M., (1990) Steepest descent, linear programming and Hamiltonian flows, *Contemporary Math.* 114, 77-88.

Bloch, A.M., H. Flaschka and T. Ratiu (1990) A convexity theorem for isospectral sets of Jacobi matrices in a compact Lie algebra, *Duke Math J.* 61, 41-65.

Bloch, A.M. (1991) The Kähler structure of the total least squares problem, Brockett's steepest descent equations and constrained flows, *Realization and Modelling in System Theory* (M. Kaashoek, J.W. van Schuppen and A.C.M. Ran, Eds.), Birkhäuser, Boston.

Bloch, A.M., R.W. Brockett and T.Ratiu (1990) A new formulation of the generalized Toda lattice equations and their fixed point analysis via the moment map, *Bull. Amer. Math. Soc.*,23, 447-456..

Bloch, A.M., R.W. Brockett and T.Ratiu (1992) Completely integrable gradient flows, *Communications in Mathematical Physics*, 23, 47-456.

Bott, R. and H. Samelson (1958) Application of the theory of Morse to symmetric spaces, *Amer. J. Math.* 80, 964 - 1029

Brockett, R.W. (1989a) Least squares matching problems, *Linear Algebra Appl.*, 122-124, 761-777.

Brockett, R.W. (1989b) Smooth dynamical systems which realize arithmetical and logical operations, *Three Decades of Mathematical System Theory*, 135, Lecture Notes in Control and Information Sciences, Springer-Verlag, 19-30.

Brockett, R.W. (1988) Dynamical systems that sort lists and solve linear programming problems, *Proc. 27th IEEE Conference on Decision and Control, Austin, TX*, 779-803. See also *Linear Algebra Appl.* 146 (1991)., 79-91.

Brockett, R.W. (1991) Dynamical systems that learn subspaces, *Mathematical System Theory - The Influence of Kalman*, (A.C. Antoulas, Ed.), 579-592.

Brockett, R.W. (1993) Differential geometry and the design of gradient algorithms, *Proceedings of Symposia in Pure Mathematics*, 54, 69-91.

Brockett, R.W. and A.M. Bloch (1989) Sorting with the dispersionless limit of the Toda lattice, *Hamiltonian Systems, Transformation Groups and Spectral Transform Methods* (J. Harnad and J.E. Marsden, Eds.), 103-112.

Brockett, R.W. and L.E. Faybusovich (1991) Toda flows, inverse spectral transform and realization theory, *Systems and Control Letters*, 16, 779-803.

Brockett, R.W. and P.S. Krishnaprasad (1980) A scaling theory for linear systems, *IEEE Transactions on Automatic Control* AC-25, 197-207.

Brockett, R.W. and R.A. Skoog (1971) A new perturbation theory for the synthesis of nonlinear networks, SIAM-AMS Proc. Vol. III, *Mathematical Aspects of Electrical Network Analysis* Providence.R.I. 17-33.

Byrnes, C.I. (1978) On certain families of rational functions arising in dynamics, *Proc. IEEE*, 1002-1006.

Byrnes, C.I. and J.C. Willems (1986) Least-squares estimation, linear programming and momentum: A geometric parametrization of local minima, *IMA J. of Math. Control and Inf.*, 3, 103-118.

Chu, M.T. (1984a) On the global convergence of the Toda Lattice for real normal matrices and its application to the eigenvalue problems, *SIAM J. Math. Anal*, 15, 98-104

Chu, M.T. (1984b) The generalized Toda flow, the QR algorithm and the center manifold theory, *SIAM J. Disc. Meth.*, 5, 187-201.

Chu, M.T. and K.R.Driessel (1990) The projected gradient method for least squares matrix approximations with spectral constraints, *SIAM J. Numer. Anal.* 27, 1050-1060.

Chu, M.T. (1992a) Matrix differential equations: A continuous realization process for linear algebra problems, *Nonlinear Analysis, TMA* 18, 1125-1146.

Chu, M.T. (1992b) Numerical methods for inverse singular value problems, *SIAM J. Numer. Anal.* 29, 885-903.

Colonius, F. and W. Kliemann (1990) Linear control semigroup acting on projective space, Report No. 224, Universität Augsburg.

Crouch, P.E. and R. Grossman (1991) Numerical integration of ordinary differential equations on manifolds, *J. of Nonlinear Science*, to appear.

Crouch, P.E., R. Grossmann and Y. Yan (1992a) On the numerical integration of the dynamic attitude equations, *Proc. of Conference on Decision and Control*, Tuscon, Arizona, 1497-1501.

Crouch, P.E., R. Grossman and Y. Yan (1992b) A third order Runge-Kutta algorithm on a manifold, preprint

Davis, M.W. (1987) Some aspherical manifolds, *Duke Math. Journal*, 5, 105-139.

Deift, P., L.C. Li and C. Tomei (1985) Toda flows with infinitely many variables, *J. of Functional Analysis* 64, 358-402.

Deift, P., T. Nanda and C. Tomei (1983) Ordinary differential equations for the symmetric eigenvalue problem, *SIAM J. Numer. Anal.*, 20, 1-22.

de Mari, F. and M.A. Shayman (1988) Generalized Eulerian numbers and the topology of the Hessenberg variety of a matrix, *Acta Applicandae Math.* 12, 213-235.

de Mari, F., Procesi, C. and M.A. Shayman (1992) Hessenberg varieties, *Transactions of the American Mathematical Society* 332, 529-534.

Driessel, K.R. (1986) On isospectral gradient flow-solving matrix eigen- problems using differential equations, *Inverse Problems* (J.R. Cannon and U. Hornung, Eds.) ISNM 77, Birkhäuser Publ. 69 - 91.

Driessel, K.R. (1987 a) On isospectral surfaces in the space of symmetric tridiagonal matrices, *Technical Report #* 544, Department of Mathematical Sciences, Clemson University.

Driessel, K.R. (1987 b) On finding the eigenvalues and eigenvectors of a matrix by means of an isospectral gradient flow, *Technical Report #* 541, Department of Mathematical Sciences, Clemson University.

Duistermaat, J.J., J.A.C. Kolk and V.S. Varadarajan (1983) Functions, flow and oscillatory integrals on flag manifolds and conjugacy classes in real semisimple Lie groups, *Compositio Math.* 49, 309-398.

Faybusovich, L.E. (1989) QR-type factorizations, the Yang-Baxter equations, and an eigenvalue problem of control theory, *Linear Algebra and its Appl.* 122/123/124, 943 - 971

Faybusovich, L. (1992) Toda flows and isospectral manifolds, *Proc. American Mathematical Society* 115, 837-847.

Flaschka, H. (1974) The Toda lattice, I, *Phys. Rev. B* 9, 1924-1925.

Flaschka, H. (1975) Discrete and periodic illustrations of some aspects of the inverse methods, in dynamical system, theory and applications, J. Moser, ed., *Lecture Notes in Physics*, 38 Springer-Verlag, Berlin.

Frankel, T. (1963) Critical submanifolds of the classical groups and Steifel manifolds, *Differential and Combinatorial Topology*, (S.S. Cairns, Ed.,) Princeton University Press, Princeton, N.J.

Fried, D. (1986) The cohomology of an isospectral flow, *Proc. Amer. Math. Soc.*, 98, 363-368.

Gelfand, I.M. and V.V. Serganova (1987) Combinatorial geometries and torus strata on homogeneous compact manifolds, *Russian Math. Surveys* 42, 133-168.

Golub, G.H. and C.F. Van Loan (1989) *Matrix Computations*, Second Edition, The John Hopkins University Press, Baltimore.

Ge-Zhong and J.E. Marsden (1988) Lie-Poisson Hamilton-Jacobi theory and Lie-Poisson integration, *Phys. Lett.*, A-133, 134 - 139.

Hangan, T. (1968) A Morse function on Grassmann manifolds, *Journal Diff. Geometry* 2, 363 - 367.

Hardy, G.H., Littlewood, J.E. and G. Polya (1952) *Inequalities*, 2nd edition. Cambridge University Press, Cambridge, England.

Helmke, U. (1991) Isospectral flows on symmetric matrices and the Riccati equation, *Systems and Control Letters*, 16, 159-166.

Hermann, R. (1962) Geometric aspects of potential theory in the bounded symmetric domains, *Math. Annalen*, 148, 349 - 366

Hermann, R. (1963) Geometric aspects of potential theory in the symmetric bounded domains II, *Math. Ann.*, 151, 143 - 149

Hermann, R. (1964) Geometric aspects of potential theory in symmetric spaces III, *Math. Annalen*, 153, 384 - 394.

Kostant, B. (1979) The Solution to a generalized Toda lattice and representation theory, *Advances in Mathematics* 34, 195 - 338.

Krishnaprasad, P.S. (1979) Symplectic mechanics and rational functions, *Richerche Automat.* 10, 107-135.

Lagarias, J.U. (1991) Monotonicity properties of the Toda flow, the QR flow and subspace iteration, *SIAM Journal of Matrix Analysis and Applications*, 12, 449-462.

Milnor, J. (1963) *Morse Theory*, Ann. of Math. Studies, 51, Princeton University Press, Princeton, N.J.

Moore, J.B., R.E. Mahony, and U. Helmke (1993) Numerical gradient algorithms for eigenvalues and singular value decomposition, *SIAM Journal of Matrix Analysis*, to appear.

Moser, J. (1975) Finitely many mass points on the line under the influence of an exponential potential - An integrable system, in J. Moser, Ed., *Dynamic Systems Theory and Applications* Springer-Verlag, Berlin - New York. 467-497.

Moser, J. and A.P. Veselov (1991) Discrete versions of some classical integrable systems and factorization of matrix polynomials, *Commun. Math. Phys.* 139, 217-243.

Nakamura, Y. (1988) Fractional transformation group induced by QR factorization and linear prediction problems, *Systems and Control Letters*, 10, 181 - 184

Palis, J. and W. de Melo (1982) *Geometric Theory of Dynamical Systems*, Springer-Verlag, New York.

Paul, St., K. Hueper and J.A. Nossek (1992) A class of nonlinear lossless dynamical systems, *Archiv für Elektronik und Übertragungstechnik* 46, 219-227.

Pressley, A.N. (1982) The energy flow on the loop space of a compact Lie group, *J. London Math. Soc.*, 26, 557-566.

Pressley, A.N. and G. Segal (1986) *Loop Groups*, Oxford Mathematical Monographs, Oxford.

Schneider, C.R. (1973) Global aspects of the matrix Riccati equation, *Math. Systems Theory* 7, 281-286.

Shayman, M.A. (1982) Morse functions for the classical groups and Grassmann manifolds, unpublished manuscript.

Shaymann, M.A. (1986) Phase portrait of the matrix Riccati equation, *SIAM J. Control Optim.* 24, 1-65.

Shub, M. and A.T. Vasquez (1987) Some linearly induced Morse-Smale systems, the QR algorithm and the Toda lattice, in: *The Legacy of Sonya Kovaleskaya* (L. Keen, ed.) Contemporary Mathematics 64, A.M.S., Providence, 181-194.

Smith, S.T. (1993) *Geometric optimization methods for adaptive filtering*, PhD Thesis, Harvard University.

Symes, W.W. (1980) Systems of the Toda type, inverse spectral problems, and representation theory, *Inventiones Mathematicae* 59, 13-51.

Symes, W.W. (1982) The QR algorithm and scattering for the finite non periodic Toda lattice, *Physica* 4D, 275-280.

Takeuchi, M. (1965) Cell decompositions and Morse equalities on certain symmetric spaces, *J. Fac. of Sci. Univ. of Tokyo*, 12, 81 - 192.

Tomei, C. (1984) The topology of isospectral manifolds of tri-diagonal matrices, *Duke Math. Journal* 51, 981-996.

Veselov, A.P. (1992) Growth and integrability in the dynamics of mappings, *Commun. Math. Phys.* 145, 181-193.

Watkins, D.S. (1984) Isospectral flows, *SIAM Review* 26, 379-391.

Wilf, H.S. (1981) An algorithm-inspired proof of the spectral theorem in E^n, *Amer. Math. Monthly* 88, 49-50.

Wu, W.T. (1965) On critical sections of convex bodies, *Sci. Sinica* 14, 1721 - 1728

Youla, D.C. and P. Tissi (1966) N-port synthesis via reactance extraction - Part I, *IEEE Intern. Convention Record*, 183-205.

Chapter 3

Singular Value Decomposition

3.1 SVD via Double Bracket Flows

Many numerical methods used in application areas such as signal processing, estimation, and control are based on the singular value decomposition (SVD) of matrices. The SVD is widely used in least squares estimation, systems approximations, and numerical linear algebra.

In this chapter, the double bracket flow of the previous chapter is applied to the singular value decomposition (SVD) task. In the following section, a first principles derivation of these self-equivalent flows is given to clarify the geometry of the flows. This work is also included as a means of re-enforcing for the reader the technical approach developed in the previous chapter. The work is based on Helmke and Moore (1992). Note also the parallel efforts of Chu and Driessel (1990) and Smith (1991).

Singular Values via Double Bracket Flows Recall, that the singular values of a matrix $H \in \mathbb{R}^{m \times n}$ with $m \geq n$ are defined as the nonnegative square roots σ_i of the eigenvalues of $H'H$, i.e. $\sigma_i(H) = \lambda_i^{\frac{1}{2}}(H'H)$ for $i = 1, \ldots, n$. Thus, for calculating the singular values σ_i of $H \in \mathbb{R}^{m \times n}$, $m \geq n$, let us first consider, perhaps naively, the diagonalization of $H'H$ or HH' by the double bracket isospectral flows evolving on the vector space of real symmetric $n \times n$ and $m \times m$ matrices.

$$\frac{d}{dt}(H'H) = [H'H, [H'H, N_1]],$$
$$\frac{d}{dt}(HH') = [HH', [HH', N_2]]. \tag{1.1}$$

The equilibrium points of these equations are the solutions of

$$[(H'H)_\infty, N_1] = 0, \quad [(HH')_\infty, N_2] = 0. \tag{1.2}$$

Thus if

$$N_1 = \mathrm{diag}(\mu_1,\ldots,\mu_n) \in \mathbb{R}^{n\times n}$$
$$N_2 = \mathrm{diag}(\mu_1,\ldots,\mu_n,0,\ldots,0) \in \mathbb{R}^{m\times m} \tag{1.3}$$

for $\mu_1 > \cdots \mu_n > 0$, then from Theorem 2.1.3 these equilibrium points are characterised by

$$\begin{aligned}(H'H)_\infty &= \pi_1 \,\mathrm{diag}\,(\sigma_1^2,\ldots,\sigma_n^2)\pi_1', \\ (HH')_\infty &= \pi_2 \,\mathrm{diag}\,(\sigma_1^2,\ldots,\sigma_n^2,0,\ldots,0)\pi_2'.\end{aligned} \tag{1.4}$$

Here σ_i^2 are the squares of the singular values of $H(0)$ and π_1, π_2 are permutation matrices. Thus we can use equations (1.1) as a means to compute, asymptotically, the singular values of $H_0 = H(0)$.

A disadvantage of using the flow (1.1) for the singular value decomposition is the need to "square" a matrix, which as noted in Chapter 1, Section 5, may lead to numerical problems. This suggests that we be more careful in the application of the double bracket equations.

We now present another direct approach to the SVD task based on the double-bracket isospectral flow of the previous chapter. This approach avoids the undesirable squarings of H appearing in (1.1). First recall, from Chapter 1, Section 5, that for any matrix $H \in \mathbb{R}^{m\times n}$ with $m \geq n$, there is an associated symmetric matrix $\widehat{H} \in \mathbb{R}^{(m+n)\times(m+n)}$ as

$$\widehat{H} := \begin{bmatrix} 0 & H \\ H' & 0 \end{bmatrix}. \tag{1.5}$$

The crucial fact is that the eigenvalues of $\widehat{H}$ are given by $\pm\sigma_i, i = 1,\cdots,n$, and possibly zero, where σ_i are the singular values of H.

Theorem 1.1 *With the definitions*

$$\widehat{H}(t) = \begin{bmatrix} 0 & H(t) \\ H'(t) & 0 \end{bmatrix}, \quad \widehat{N} = \begin{bmatrix} 0 & N \\ N' & 0 \end{bmatrix},$$

$$\widehat{H}_0 = \begin{bmatrix} 0 & H_0 \\ H_0' & 0 \end{bmatrix}, \tag{1.6}$$

where $N \in \mathbb{R}^{m\times n}$, *then* $\widehat{H}(t)$ *satisfies the double bracket isospectral flow on* $\mathbb{R}^{(m+n)\times(m+n)}$

$$\frac{d\widehat{H}(t)}{dt} = [\widehat{H}(t), [\widehat{H}(t), \widehat{N}]], \quad \widehat{H}(0) = \widehat{H}_0, \tag{1.7}$$

3.1. SVD VIA DOUBLE BRACKET FLOWS

if and only if $H(t)$ satisfies the self-equivalent flow on $\mathbb{R}^{m \times n}$

$$\boxed{\begin{aligned} \dot{H} &= H(H'N - N'H) - (HN' - NH')H, \quad H(0) = H_0 \\ &= H\{H, N\} - \{H', N'\}H. \end{aligned}} \quad (1.8)$$

using a mild generalization of the Lie-bracket notation

$$\{X, Y\} = X'Y - Y'X = \{Y, X\}' = -\{Y, X\}. \quad (1.9)$$

The solutions $H(t)$ of (1.8) exist for all $t \in \mathbb{R}$. Moreover, the solutions of (1.8) converge, in either time direction, to the limiting solutions H_∞ satisfying

$$H'_\infty N - N'H_\infty = 0, \quad H_\infty N' - NH_\infty = 0. \quad (1.10)$$

Proof Follows by substitution of (1.6) into (1.7), and by applying Theorem 2.1.3 for double bracket flows. ∎

Remarks 1. The matrix $\widehat{N}$ is of course not diagonal and, more important, 0 is a multiple eigenvalue of $\widehat{N}$ for $m \neq n$. Thus certain of the stability results of Theorem 2.1.3 do not apply immediately to the flow (1.8). However, similar proof techniques as used in deriving the stability results in Theorem 2.1.3 do apply. This is explained in the next section.

2. The gradient flow associated with the maximization of $\text{tr}(\widehat{N}\widehat{\Theta}'\widehat{H}_0\widehat{\Theta})$ on the manifold of orthogonal $(m+n) \times (m+n)$ matrices $\widehat{\Theta}$, is given by Theorem 2.1.5 as

$$\frac{d}{dt}\widehat{\Theta}(t) = \widehat{\Theta}(t)[\widehat{\Theta}'(t)\widehat{H}_0\widehat{\Theta}(t), \widehat{N}] \quad (1.11)$$

If $\widehat{\Theta} = \begin{bmatrix} V & 0 \\ 0 & U \end{bmatrix}$, (1.11) is easily seen to be equivalent to

$$\begin{aligned} \dot{U} &= U\{V'H_0U, N\}, \\ \dot{V} &= V\{(V'H_0U)', N'\}. \end{aligned} \quad (1.12)$$

This gradient flow on $O(m) \times O(n)$ allows the determination of the left and right eigenvectors of H_0. Its convergence properties are studied in the next section. ∎

Problems 1. Verify that the flow $\dot{H} = H[H'H, N_1] - [HH', N_2]H$ for $H(t)$ implies (1.1) for $H(t)H'(t)$ and $H'(t)H(t)$.

2. Flesh out the details of the proof of Theorem 2.3, and verify the claims in Remark 2 following this proof.

Main Points of Section The singular value decomposition of a matrix $H_0 \in \mathbb{R}^{m \times n}$ can be obtained by applying the double bracket flow to the symmetric $(m+n) \times (m+n)$ matrix (1.6). The resulting flow can be simplified as a self equivalent flow (1.8) for SVD.

3.2 A Gradient Flow Approach to SVD

In this section, we parallel the analysis of the double bracket isospectral flow on symmetric matrices with a corresponding theory for self-equivalent flows on rectangular matrices $H \in \mathbb{R}^{m \times n}$, for $m \geq n$. A matrix $H_0 \in \mathbb{R}^{m \times n}$ has a *singular value decomposition* (SVD)

$$H_0 = V\Sigma U' \tag{2.1}$$

where U, V are real $n \times n$ and $m \times m$ orthogonal matrices satisfying $VV' = I_m$, $UU' = I_n$ and

$$\Sigma = \begin{bmatrix} \operatorname{diag}(\sigma_1 I_{n_1}, \ldots, \sigma_r I_{n_r}) \\ 0_{(m-n) \times n} \end{bmatrix} = \begin{bmatrix} \Sigma_1 \\ 0 \end{bmatrix}, \quad \sum_{i=1}^{r} n_i = n, \tag{2.2}$$

with $\sigma_1 > \ldots > \sigma_r \geq 0$ being the singular values of H_0. We will approach the SVD task by showing that it is equivalent to a certain norm minimization problem.

Consider the class $S(\Sigma)$ of all real $m \times n$ matrices H having singular values $(\sigma_1, \ldots \sigma_r)$ occurring with multiplicities $(n_1, \ldots, n_r)$ as

$$S(\Sigma) = \{V'\Sigma U \in \mathbb{R}^{m \times n} | \ U \in O(n), V \in O(m)\}, \tag{2.3}$$

with $O(n)$ the group of all real orthogonal $n \times n$ matrices. Thus $S(\Sigma)$ is the compact set of all $m \times n$ matrices H which are orthogonally equivalent to Σ via (2.1). Given an arbitrary matrix $N \in \mathbb{R}^{m \times n}$, we consider the task of minimizing the least squares distance

$$\|H - N\|^2 = \|H\|^2 + \|N\|^2 - 2\operatorname{tr}(N'H) \tag{2.4}$$

of N to any $H \in S(\Sigma)$. Since the Frobenius norm $\|H\|^2 = \sum_{i=1}^{r} n_i \sigma_i^2$ is constant for $H \in S(\Sigma)$, the minimization of (2.4) is equivalent to the maximization of the inner product function $\phi(H) = 2\operatorname{tr}(N'H)$, defined on $S(\Sigma)$. Heuristically, if N is chosen to be of the form

$$N = \begin{bmatrix} N_1 \\ 0_{(m-n) \times n} \end{bmatrix}, \quad N_1 = \operatorname{diag}(\mu_1, \ldots, \mu_n) \in \mathbb{R}^{n \times n}, \tag{2.5}$$

we would expect the minimizing matrices $H_* \in S(\Sigma)$ of (2.4) to be of the same form. i.e. $H_* = \begin{bmatrix} \pi & 0 \\ 0 & I \end{bmatrix} \Sigma \pi' S$ for a suitable $n \times n$ permutation matrix π and a sign matrix $S = \operatorname{diag}(\pm 1, \ldots, \pm 1)$. In fact, if $N_1 = \operatorname{diag}(\mu_1, \ldots, \mu_n)$ with $\mu_1 > \ldots > \mu_n > 0$ then we would expect the minimizing H_* to be equal to Σ. Since $S(\Sigma)$ turns out to be a smooth manifold (Proposition 2.1) it seems natural to apply steepest descent techniques in order to determine the minima H_* of the distance function (2.4) on $S(\Sigma)$.

3.2. A GRADIENT FLOW APPROACH TO SVD

To achieve the SVD we consider the induced function on the product space $O(n) \times O(m)$ of orthogonal matrices

$$\phi : O(n) \times O(m) \longrightarrow \mathbb{R},$$

$$\phi(U, V) = 2\mathrm{tr}(NV'H_0 U), \qquad (2.6)$$

defined for fixed arbitrary real matrices $N \in \mathbb{R}^{n \times m}, H_0 \in S(\Sigma)$. This leads to a coupled gradient flow

$$\begin{aligned}\dot{U}(t) &= \nabla_U \phi(U(t), V(t)), & U(0) &= U_0 \in O(n), \\ \dot{V}(t) &= \nabla_V \phi(U(t), V(t)), & V(0) &= V_0 \in O(m),\end{aligned} \qquad (2.7)$$

on $O(n) \times O(m)$. These turn out to be the flows (1.12) as shown subsequently.

Associated with the gradient flow (2.7) is a flow on the set of real $m \times n$ matrices H, derived from $H(t) = V'(t) H_0 U(t)$. This turns out to be a self-equivalent flow (i.e. a singular value preserving flow) (2.4) on $S(\Sigma)$ which, under a suitable genericity assumption on N, converges exponentially fast, for almost every initial condition $H(0) = H_0 \in S(\Sigma)$, to the global minimum H_* of (2.4).

We now proceed with a formalization of the above norm minimization approach to SVD. First, we present a phase portrait analysis of the self-equivalent flow (1.8) on $S(\Sigma)$. The equilibrium points of (1.8) are determined and, under suitable genericity assumptions on N and Σ, their stability properties are determined. A similar analysis for the gradient flow (2.7) is given.

Self-equivalent Flows on Real Matrices We start with a derivation of some elementary facts concerning varieties of real matrices with prescribed singular values.

Proposition 2.1 $S(\Sigma)$ *is a smooth, compact, manifold of dimension*

$$\dim S(\Sigma) = \begin{cases} n(m-1) - \sum_{i=1}^{r} \frac{n_i(n_i-1)}{2} & \text{if } \sigma_r > 0, \\ n(m-1) - \sum_{i=1}^{r} \frac{n_i(n_i-1)}{2} - n_r(\frac{n_r-1}{2} + (m-n)) & \text{if } \sigma_r = 0. \end{cases}$$

Proof $S(\Sigma)$ is an orbit of the compact Lie group $O(m) \times O(n)$, acting on $\mathbb{R}^{m \times n}$ via the equivalence action:

$$\eta : (O(m) \times O(n)) \times \mathbb{R}^{m \times n} \longrightarrow \mathbb{R}^{m \times n}, \quad \eta((V, U), H) = V'HU.$$

See digression on Lie groups and homogeneous spaces. Thus $S(\Sigma)$ is a homogeneous space of $O(m) \times O(n)$ and therefore a compact manifold. The stabilizer group $\mathrm{Stab}(\Sigma) = \{(V,U) \in O(m) \times O(n)|\ V'\Sigma U = \Sigma\}$ of Σ is the set of all pairs of block-diagonal orthogonal matrices (V,U)

$$U = \mathrm{diag}(U_{11},\ldots,U_{rr})\,,\quad V = \mathrm{diag}(V_{11},\ldots,V_{r+1,r+1})\,,$$
$$U_{ii} = V_{ii} \in O(n_i)\,,\quad i=1,\ldots,r\,,\quad V_{r+1,r+1} \in O(m-n) \qquad (2.8)$$

if $\sigma_r > 0$, and if $\sigma_r = 0$ then

$$U = \mathrm{diag}(U_{11},\ldots,U_{rr})\,,\quad V = \mathrm{diag}(V_{11},\ldots,V_{rr})\,,$$
$$U_{ii} = V_{ii} \in O(n_i)\,,\quad i=1,\ldots,r-1\,,$$
$$U_{rr} \in O(n_r)\,,\quad V_{rr} \in O(m-n+n_r)\,.$$

Hence there is an isomorphism of homogeneous spaces $S(\Sigma) \cong (O(m) \times O(n))/\mathrm{Stab}(\Sigma)$ and

$$\begin{aligned}\dim S(\Sigma) &= \dim(O(m) \times O(n)) - \dim \mathrm{Stab}(\Sigma)\,, \\ &= \frac{m(m-1)}{2} + \frac{n(n-1)}{2} - \dim \mathrm{Stab}(\Sigma)\,.\end{aligned}$$

Moreover,

$$\dim \mathrm{Stab}(\Sigma) = \begin{cases} \sum_{i=1}^{r} \frac{n_i(n_i-1)}{2} + \frac{(m-n)(m-n-1)}{2} & \text{if } \sigma_r > 0\,, \\ \sum_{i=1}^{r} \frac{n_i(n_i-1)}{2} + \frac{(m-n+n_r)(m-n+n_r-1)}{2} & \text{if } \sigma_r = 0\,. \end{cases}$$

The result follows. ∎

Remark Let $Q = \begin{bmatrix} I_n \\ 0_{(m-n)\times n} \end{bmatrix}$. Then $S(Q)$ is equal to the (compact) Stiefel manifold $St(n,m)$ consisting of all $X \in \mathbb{R}^{m \times n}$ with $X'X = I_n$. In particular, for $m=n$, $S(Q)$ is equal to the orthogonal group $O(n)$. Thus the manifolds $S(\Sigma)$ are a generalization of the compact Stiefel manifolds $St(n,m)$, appearing in Chapter 1.3. ∎

Let

$$\Sigma_0 = \left[\begin{array}{ccc} \mu_1 I_{n_1} & \cdots & 0 \\ \vdots & \ddots & \vdots \\ 0 & \cdots & \mu_r I_{n_r} \\ \hline & 0_{(m-n)\times n} & \end{array}\right]$$

be given with $\mu_1, \ldots, \mu_r$ arbitrary real numbers. Thus $\Sigma_0 \in S(\Sigma)$ if and only if $\{|\mu_1|,\ldots,|\mu_r|\} = \{\sigma_1,\ldots,\sigma_r\}$. We need the following description of the tangent space of $S(\Sigma)$ at Σ_0.

Lemma 2.2 *Let Σ be defined by (2.2) and Σ_0 as above, with $\sigma_r > 0$. Then the tangent space $T_{\Sigma_0} S(\Sigma)$ of $S(\Sigma)$ at Σ_0 consists of all block partitioned real*

3.2. A GRADIENT FLOW APPROACH TO SVD

$m \times n$ matrices

$$\xi = \begin{bmatrix} \xi_{11} & \cdots & \xi_{1r} \\ \vdots & & \\ \xi_{r+1,1} & \cdots & \xi_{r+1,r} \end{bmatrix}$$

$\xi_{ij} \in \mathbb{R}^{n_i \times n_j}$, $i,j = 1, \ldots, r$; $\xi_{r+1,j} \in \mathbb{R}^{(m-n) \times n_j}$, $j = 1, \ldots, r$

with $\xi'_{ii} = -\xi_{ii}$, $i = 1, \ldots, r$.

Proof Let skew(n) denote the Lie algebra of $O(n)$, i.e. skew(n) consists of all skew-symmetric $n \times n$ matrices and is identical with tangent space of $O(n)$ at the identity matrix I_n, see Appendix C. The tangent space $T_{\Sigma_0} S(\Sigma)$ is the image of the $\mathbb{R}$-linear map

$$\mathcal{L} : \text{skew}(m) \times \text{skew}(n) \to \mathbb{R}^{m \times n} ,$$

$$(X, Y) \mapsto -X\Sigma_0 + \Sigma_0 Y , \qquad (2.9)$$

i.e. of the Fréchet derivative of $\eta : O(m) \times O(n) \to \mathbb{R}$, $\eta(U, V) = V'\Sigma_0 U$, at (I_m, I_n).

It follows that the image of $\mathcal{L}$ is contained in the $\mathbb{R}$-vector space Ξ consisting of all block partitioned matrices ξ as in the lemma. Thus it suffices to show that both spaces have the same dimension. By Prop. 3.3.1

$$\begin{aligned} \dim \text{ image } (\mathcal{L}) &= \dim S(\Sigma) \\ &= nm - \sum_{i=1}^{r} \frac{n_i(n_i + 1)}{2} . \end{aligned}$$

From (2.9) also dim $\Xi = nm - \sum_{i=1}^{r} \frac{n_i(n_i+1)}{2}$ and the result follows. ∎

Recall from Chapter 1, Section 5, that a differential equation (or a flow)

$$\dot{H}(t) = f(t, H(t)) \qquad (2.10)$$

defined on the vector space of real $m \times n$ matrices $H \in \mathbb{R}^{m \times n}$ is called *self-equivalent* if every solution $H(t)$ of (2.10) is of the form

$$H(t) = V'(t) H(0) U(t) \qquad (2.11)$$

with orthogonal matrices $U(t) \in O(n), V(t) \in O(m), U(0) = I_n, V(0) = I_m$. Recall that self-equivalent flows have a simple characterization given by Lemma 1.5.1.

The following theorem gives explicit examples of self-equivalent flows on $\mathbb{R}^{m \times n}$.

Theorem 2.3 *Let $N \in \mathbb{R}^{m \times n}$ be arbitrary and $m \geq n$*

(a) The differential equation (1.8) repeated as

$$\boxed{\dot{H} = H(H'N - N'H) - (HN' - NH')H\ ,\ H(0) = H_0}\qquad(2.12)$$

defines a self-equivalent flow on $\mathbb{R}^{m \times n}$.

(b) The solutions $H(t)$ of (2.12) exist for all $t \in \mathbb{R}$ and $H(t)$ converges to an equilibrium point H_∞ of (2.12) as $t \to +\infty$. The set of equilibria points H_∞ of (2.12) is characterized by

$$H'_\infty N = N' H_\infty\ ,\ N H'_\infty = H_\infty N'. \qquad(2.13)$$

(c) Let N be defined as in (2.5), with $\mu_1, \ldots, \mu_n$ real and $\mu_i + \mu_j \neq 0$ of all $i, j = 1, \ldots, n$. Then every solution $H(t)$ of (2.12) converges for $t \to \pm\infty$ to an extended diagonal matrix H_∞ of the form

$$H_\infty = \left[\begin{array}{ccc} \lambda_1 & & 0 \\ & \ddots & \\ 0 & & \lambda_n \\ \hline & 0_{(m-n)\times n} & \end{array}\right] \qquad(2.14)$$

with $\lambda_1, \ldots, \lambda_n$ real numbers.

(d) If $m = n$ and $N = N'$ is symmetric (or $N' = -N$ is skew-symmetric), (2.12) restricts to the isospectral flow

$$\boxed{\dot{H} = [H, [H, N]]} \qquad(2.15)$$

evolving on the subset of all symmetric (respectively skew-symmetric) $n \times n$ matrices $H \in \mathbb{R}^{n \times n}$

Proof For every (locally defined) solution $H(t)$ of (2.12), the matrices $C(t) = H'(t)N(t) - N'(t)H(t)$ and $D(t) = N(t)H'(t) - H(t)N'(t)$ are skew-symmetric. Thus (a) follows immediately from Lemma 1.5.1. For any $H(0) = H_0 \in \mathbb{R}^{m \times n}$ there exist orthogonal matrices $U \in O(n), V \in O(m)$ with $V'H_0U = \Sigma$ as in (2.1) where $\Sigma \in \mathbb{R}^{m \times n}$ is the extended diagonal matrix given by (2.2) and $\sigma_1 > \ldots > \sigma_r \geq 0$ are the singular values of H_0 with corresponding multiplicities $n_1, \ldots, n_r$. By (a), the solution $H(t)$ of (2.12) evolves in the compact set $S(\Sigma)$ and therefore exists for all $t \in \mathbb{R}$.

Recall that $\{H, N\}' = \{N, H\} = -\{H, N\}$, then the time derivative of the Lyapunov-type function $\operatorname{tr}(N'H(t))$ is

$$2\frac{d}{dt}\operatorname{tr}(N'H(t)) = \operatorname{tr}(N'\dot{H}(t) + \dot{H}(t)'N)$$

3.2. A GRADIENT FLOW APPROACH TO SVD

$$\begin{aligned}
&= \operatorname{tr}(N'H\{H,N\} - N'\{H',N'\}H - \{H,N\}H'N + H'\{H',N'\}N) \\
&= \operatorname{tr}(\{H,N\}\{H,N\}' + \{H',N'\}\{H',N'\}') \\
&= (\|\{H,N\}\|^2 + \|\{H',N'\}\|^2) \,. \quad (2.16)
\end{aligned}$$

Thus $\operatorname{tr}(N'H(t))$ increases monotonically. Since $H \mapsto \operatorname{tr}(N'H)$ is a continuous function defined on the compact space $S(\Sigma)$, then $\operatorname{tr}(N'H)$ is bounded from below and above. Therefore (2.16) must go to zero as $t \to +\infty$ (and indeed for $t \to -\infty$). It follows that every solution $H(t)$ of (2.12) converges to an equilibrium point and the set of equilibria of (2.12) is characterized by $\{H_\infty, N\} = \{H'_\infty, N'\} = 0$, or equivalently (2.13). This proves (b). To prove (c) recall that (2.5) partitions N as $\begin{bmatrix} N_1 \\ 0 \end{bmatrix}$ with $N_1 = N'_1$. Now partition H_∞ as $H_\infty = \begin{bmatrix} H_1 \\ H_2 \end{bmatrix}$. This is an equilibrium point of (2.12) if and only if

$$H'_1 N_1 = N_1 H_1, \; N_1 H'_1 = H_1 N_1, \; N_1 H'_2 = 0 \,. \quad (2.17)$$

Since N_1 is nonsingular, (2.17) implies $H_2 = 0$. Also since $N_1 = N'_1$, then $N_1(H_1 - H'_1) = (H'_1 - H_1)N_1$ and therefore with $H_1 = (h_{ij}|i,j = 1,\ldots,n)$

$$(\mu_i + \mu_j)(h_{ij} - h_{ji}) = 0$$

for all $i,j = 1,\ldots,n$. By assumption $\mu_i + \mu_j \neq 0$ and therefore $H_1 = H'_1$ is symmetric. Thus under (2.17) $N_1 H_1 = H_1 N_1$, which implies that H_1 is real diagonal.

To prove (d) we note that for symmetric (or skew-symmetric) matrices N, H, equation (2.12) is equivalent to the double bracket equation (2.15). The isospectral property of (2.15) now follows readily from Lemma 1.5.1, applied for $D = -C$. This completes the proof of Theorem 2.3. ∎

Remark Let $N' = U'_0(N_1, 0_{n \times (m-n)})V_0$ be the singular value decomposition of N' with real diagonal N_1. Using the change of variables $H \mapsto V'_0 H U_0$ we can always assume without loss of generality that N' is equal to $(N_1, 0_{n \times (m-n)})$ with N_1 real diagonal. ∎

From now on we will always assume that Σ and N are given from (2.2) and (2.5).

An important property of the self-equivalent flow (2.12) is that it is a gradient flow. To see this, let

$$\widehat{S}(\Sigma) := \{\widehat{H} \in M(\widehat{\Sigma}) \mid H \in S(\Sigma)\}$$

denote the subset of $(m+n) \times (m+n)$ symmetric matrices $\widehat{H} \in M(\widehat{\Sigma})$ which are of the form (1.6). Here $M(\widehat{\Sigma})$ is the isospectral manifold defined by $\{\Theta\hat{\Sigma}\Theta' \in \mathbb{R}^{(m+n)\times(m+n)} | \Theta \in O(m+n)\}$. The map

$$i : S(\Sigma) \to M(\hat{\Sigma}), \ i(H) = \hat{H}$$

defines a diffeomorphism of $S(\Sigma)$ onto its image $\hat{S}(\Sigma) = i(S(\Sigma))$. By Lemma 2.2 the double bracket flow (1.7) has $\hat{S}(\Sigma)$ as an invariant submanifold.

Endow $M(\hat{\Sigma})$ with the normal Riemannian metric defined in Chapter 2. Thus (1.7) is the gradient flow of the least squares distance function $f_N : M(\hat{\Sigma}) \to \mathbb{R}$, $f_N(H) = -\frac{1}{2}\|N - H\|^2$, with respect to the normal Riemannian metric. By restriction, the normal Riemannian metric on $M(\hat{\Sigma})$ induces a Riemannian metric on the submanifold $\hat{S}(\Sigma)$; see Appendix C. Therefore, using the diffeomorphism $i : S(\Sigma) \to \hat{S}(\Sigma)$, a Riemannian metric on $M(\hat{\Sigma})$ induces a Riemannian on $S(\Sigma)$. We refer to this as the *normal Riemannian metric on $S(\Sigma)$*. Since $\|\hat{N} - \hat{H}\|^2 = 2\|N - H\|^2$, Lemma 2.2 implies that the gradient of $F_N : S(\Sigma) \to \mathbb{R}$, $F_N(H) = -\|N - H\|^2$, with respect to this normal Riemannian metric on $S(\Sigma)$ is given by (1.8). We have thus proved

Corollary 2.4 *The differential equation (2.12) is the gradient flow of the distance function $F_N : S(\Sigma) \to \mathbb{R}$, $F_N(H) = -\|N - H\|^2$, with respect to the normal Riemannian metric on $S(\Sigma)$.*

Since self-equivalent flows do not change the singular values or their multiplicities, (2.12) restricts to a flow on the compact manifold $S(\Sigma)$. The following result is an immediate consequence of Theorem 2.3 (a) - (c).

Corollary 2.5 *The differential equation (2.12) defines a flow on $S(\Sigma)$. Every equilibrium point H_∞ on $S(\Sigma)$ is of the form*

$$H_\infty = \begin{bmatrix} \pi\Sigma_1\pi'S \\ 0 \end{bmatrix} \quad (2.18)$$

where π is an $n \times n$ permutation matrix and $S = \text{diag}(s_1, \ldots, s_n)$, $s_i \in \{-1, 1\}$, $i = 1, \ldots, n$, is a sign matrix.

We now analyze the local stability properties of the flow (2.12) on $S(\Sigma)$ around each equilibrium point. The linearization of the flow (2.12) on $S(\Sigma)$ around any equilibrium point $H_\infty \in S(\Sigma)$ is given by

$$\dot{\xi} = H_\infty(\xi'N - N'\xi) - (\xi N' - N\xi')H_\infty \quad (2.19)$$

where $\xi \in T_{H_\infty}S(\Sigma)$ is the tangent space of $S(\Sigma)$. By Lemma 2.2

$$\xi = \begin{bmatrix} \xi_{11} & \cdots & \xi_{1r} \\ \vdots & & \vdots \\ \xi_{r+1,1} & \cdots & \xi_{r+1,r} \end{bmatrix} = \begin{bmatrix} \xi_1 \\ \xi_2 \end{bmatrix}$$

3.2. A GRADIENT FLOW APPROACH TO SVD

with $\xi_1 = (\xi_{ij}) \in \mathbb{R}^{n \times n}, \xi'_{ii} = -\xi_{ii}$, and $\xi_2 = [\xi_{r+1,1}, \ldots, \xi_{r+1,r}] \in \mathbb{R}^{(m-n) \times n}$.

Using (2.18), (2.19) is equivalent to the decoupled system of equations

$$\begin{aligned} \dot{\xi}_1 &= \pi \Sigma_1 \pi' S(\xi'_1 N_1 - N'_1 \xi_1) - (\xi_1 N_1 - N_1 \xi'_1) \pi \Sigma_1 \pi' S \\ \dot{\xi}_2 &= -\xi_2 N_1 \pi \Sigma_1 \pi' S \end{aligned} \qquad (2.20)$$

In order to simplify the subsequent analysis we now assume that the singular values of $\Sigma = \begin{bmatrix} \Sigma_1 \\ 0 \end{bmatrix}$ are distinct, i.e

$$\Sigma_1 = \text{diag}(\sigma_1, \ldots, \sigma_n), \quad \sigma_1 > \ldots > \sigma_n > 0. \qquad (2.21)$$

Furthermore, we assume that $N = \begin{bmatrix} N_1 \\ 0_{(m-n) \times n} \end{bmatrix}$ with

$$N_1 = \text{diag}(\mu_1, \ldots, \mu_n), \quad \mu_1 > \ldots > \mu_n > 0.$$

Then $\pi \Sigma_1 \pi' S = \text{diag}(\lambda_1, \ldots, \lambda_n)$ with

$$\lambda_i = s_i \sigma_{\pi(i)}, \quad i = 1, \ldots, n, \qquad (2.22)$$

and $\xi_{ii} = -\xi_{ii}$ are zero for $i = 1, \ldots, n$. Then (2.20) is equivalent to the system for $i, j = 1, \ldots, n$,

$$\begin{aligned} \dot{\xi}_{ij} &= -(\lambda_i \mu_i + \lambda_j \mu_j)\xi_{ij} + (\lambda_i \mu_j + \lambda_j \mu_i)\xi_{ji}, \\ \dot{\xi}_{n+1,j} &= -\mu_j \lambda_j \xi_{n+1,j}. \end{aligned} \qquad (2.23)$$

This system of equations is equivalent to the system of equations

$$\begin{aligned} \frac{d}{dt} \begin{bmatrix} \xi_{ij} \\ \xi_{ji} \end{bmatrix} &= \begin{bmatrix} -(\lambda_i \mu_i + \lambda_j \mu_j) & (\lambda_i \mu_j + \lambda_j \mu_i) \\ (\lambda_i \mu_j + \lambda_j \mu_i) & -(\lambda_i \mu_i + \lambda_j \mu_j) \end{bmatrix} \begin{bmatrix} \xi_{ij} \\ \xi_{ji} \end{bmatrix}, i < j \\ \dot{\xi}_{n+1,j} &= -\mu_j \lambda_j \xi_{n+1,j}. \end{aligned} \qquad (2.24)$$

The eigenvalues of

$$\begin{bmatrix} -(\lambda_i \mu_i + \lambda_j \mu_j) & \lambda_i \mu_j + \lambda_j \mu_i \\ \lambda_i \mu_j + \lambda_j \mu_i & -(\lambda_i \mu_i + \lambda_j \mu_j) \end{bmatrix}$$

are easily seen to be $\{-(\lambda_i + \lambda_j)(\mu_i + \mu_j), -(\lambda_i - \lambda_j)(\mu_i - \mu_j)\}$ for $i < j$.

Let H_∞ be an equilibrium point of the flow (2.12) on the manifold $S(\Sigma)$. Let n_+ and n_- denote the number of positive real eigenvalues and negative real eigenvalues respectively of the linearization (2.19). The *Morse index* $\text{ind} H_\infty$ of (2.20) at H_∞ is defined by

$$\text{ind}(H_\infty) = n_- - n_+. \qquad (2.25)$$

By (2.21), (2.22) the eigenvalues of the linearization (2.23) are all nonzero real numbers and therefore

$$n_+ + n_- = \dim S(\Sigma) = n(m-1)$$
$$n_- = \frac{1}{2}(n(m-1) + \text{ind }(H_\infty)) \qquad (2.26)$$

This sets the stage for the key stability results for the self-equivalent flows (2.12).

Corollary 2.6 *Let $\sigma_1 > \ldots > \sigma_n > 0$ and let $(\lambda_1,\ldots,\lambda_n)$ be defined by (2.22). Then (2.12) has exactly $2^n.n!$ equilibrium points on $S(\Sigma)$. The linearization of the flow (2.12) on $S(\Sigma)$ at $H_\infty = \begin{bmatrix} \pi\Sigma_1\pi'S \\ 0 \end{bmatrix}$ is given by (2.23) and has only nonzero real eigenvalues. The Morse index at H_∞ is*

$$\text{ind }(H_\infty) = (m-n)\sum_{i=1}^n s_i + \sum_{i<j} \text{sign }(s_i\sigma_{\pi(i)} - s_j\sigma_{\pi(j)})$$
$$+ \sum_{i<j} \text{sign }(s_i\sigma_{\pi(i)} + s_j\sigma_{\pi(j)}). \qquad (2.27)$$

In particular, $\text{ind}(H_\infty) = \pm \dim S(\Sigma)$ if and only if $s_1 = \ldots = s_n = \pm 1$ and $\pi = I_n$. Also, $H_\infty = \Sigma$ is the uniquely determined asymptotically stable equilibrium point on $S(\Sigma)$.

Proof It remains to prove (2.27). By (2.23) and as $\mu_i > 0$

$$\text{ind }(H_\infty) = (m-n)\sum_{i=1}^n \text{sign }\mu_i\lambda_i$$
$$+ \sum_{i<j}(\text{sign }(\mu_i - \mu_j)(\lambda_i - \lambda_j) + \text{sign }(\mu_i + \mu_j)(\lambda_i + \lambda_j))$$
$$= (m-n)\sum_{i=1}^n \text{sign }\lambda_i + \sum_{i<j} \text{sign }(\lambda_i - \lambda_j)$$
$$+ \sum_{i<j} \text{sign }(\lambda_i + \lambda_j)$$

with $\lambda_i = s_i\sigma_{\pi(i)}$. The result follows. ∎

Remark From (2.23) and Corollary 2.6 it follows that, for almost all initial conditions on $S(\Sigma)$, the solutions of (2.12) on $S(\Sigma)$ approach the

3.2. A GRADIENT FLOW APPROACH TO SVD

attractor Σ exponentially fast. The rate of exponential convergence to Σ is equal to
$$\rho = \min\{\mu_n \sigma_n, \min_{i<j}(\mu_i - \mu_j)(\sigma_i - \sigma_j)\}. \tag{2.28}$$

∎

A Gradient Flow on Orthogonal Matrices Let $N' = (N_1, 0_{n \times (m-n)})$ with $N_1 = \text{diag}(\mu_1, \ldots, \mu_n)$ real diagonal and $\mu_1 > \ldots > \mu_n > 0$. Let H_0 be a given real $m \times n$ matrix with singular value decomposition
$$H_0 = V_0' \Sigma U_0 \tag{2.29}$$
where $U_0 \in O(n), V_0 \in O(m)$ and Σ is given by (2.2). For simplicity we always assume that $\sigma_r > 0$. With these choices of N, H_0 we consider the smooth function on the product space $O(n) \times O(m)$ of orthogonal groups
$$\phi : O(n) \times O(m) \longrightarrow \mathbb{R}, \quad \phi(U, V) = 2\text{tr}(N'V'H_0U). \tag{2.30}$$

We always endow $O(n) \times O(m)$ with its standard Riemannian metric $\langle\,,\,\rangle$ defined by $\langle(\Omega_1, \Omega_2), (\widehat{\Omega}_1, \widehat{\Omega}_2)\rangle = \text{tr}(\Omega_1 \widehat{\Omega}_1') + \text{tr}(\Omega_2 \widehat{\Omega}_2')$ for $(\Omega_1, \Omega_2), (\widehat{\Omega}_1, \widehat{\Omega}_2) \in T_{(U,V)}(O(n) \times O(m))$. Thus $\langle\,,\,\rangle$ is the Riemannian metric induced by the imbedding $O(n) \times O(m) \subset \mathbb{R}^{n \times n} \times \mathbb{R}^{m \times m}$, where $\mathbb{R}^{n \times n} \times \mathbb{R}^{m \times m}$, is equipped with its standard symmetric inner product. The Riemannian metric $\langle\,,\,\rangle$ on $O(n) \times O(m)$ in the case where $m = n$ coincides with the Killing form, up to a constant scaling factor.

Lemma 2.7 *The gradient flow of $\phi : O(n) \times O(m) \longrightarrow \mathbb{R}$ (with respect to the standard Riemannian metric) is*

$$\boxed{\begin{aligned} \dot{U} &= U\{V'H_0U, N\}, & U(0) &= U_0 \in O(n), \\ \dot{V} &= V\{U'H_0'V, N'\}, & V(0) &= V_0 \in O(m). \end{aligned}} \tag{2.31}$$

The solutions $(U(t), V(t))$ of (2.31) exist for all $t \in \mathbb{R}$ and converge to an equilibrium point $(U_\infty, V_\infty) \in O(n) \times O(m)$ of (2.31) for $t \longrightarrow \pm\infty$.

Proof We proceed as in Chapter 2, Section 1. The tangent space of $O(n) \times O(m)$ at (U, V) is given by
$$T_{(U,V)}(O(n) \times O(m))$$
$$= \{(U\Omega_1, V\Omega_2) \in \mathbb{R}^{n \times n} \times \mathbb{R}^{m \times m} \mid \Omega_1 \in \text{skew}(n), \Omega_2 \in \text{skew}(m)\}.$$

The Riemannian matrix on $O(n) \times O(m)$ is given by the inter product $\langle\,,\,\rangle$ on the tangent space $T_{(U,V)}(O(n) \times O(m))$ defined for $(A_1, A_2), (B_1, B_2) \in T_{(U,V)}(O(n) \times O(m))$ by
$$\langle(A_1, A_2), (B_1, B_2)\rangle = \text{tr}(A_1 B_1') + \text{tr}(A_2 B_2')$$

for $(U, V) \in O(n) \times O(m)$.

The derivative of $\phi : O(n) \times O(m) \to \mathbb{R}$, $\phi(U, V) = 2\mathrm{tr}(N'V'H_0 U)$, at an element (U, V) is the linear map on the tangent space $T_{(U,V)}(O(n) \times O(m))$ defined by

$$\begin{aligned} D\phi|_{(U,V)}(U\Omega_1, V\Omega_2) &= 2\mathrm{tr}\left(N'V'H_0 U\Omega_1 - N'\Omega_2 V'H_0 U\right) \\ &= \mathrm{tr}\left[(N'V'H_0 U - U'H_0'VN)\Omega_1 + (NU'H_0'V - V'H_0 U N')\Omega_2\right] \\ &= \mathrm{tr}\left(\{V'H_0 Y, N\}'\Omega_1 + \{U'H_0'V, N'\}'\Omega_2\right) \end{aligned} \quad (2.32)$$

for skew-symmetric matrices $\Omega_1 \in \mathrm{skew}(n), \Omega_2 \in \mathrm{skew}(m)$. Let

$$\nabla \phi = \begin{pmatrix} \nabla_U \phi \\ \nabla_V \phi \end{pmatrix}$$

denote the gradient vector field of ϕ, defined with respect to the above Riemannian metric on $O(n) \times O(m)$. Thus $\nabla \phi(U, V)$ is characterised by

(i) $\nabla \phi(U, V) \in T_{(U,V)}(O(n) \times O(m))$

(ii) $D\phi|_{(U,V)}(U\Omega_1, V\Omega_2) = \langle \nabla \phi(U, V), (U\Omega_1, V\Omega_2) \rangle$
$= \mathrm{tr}\left(\nabla_U \phi(U, V)' U\Omega_1\right) + \mathrm{tr}\left(\nabla_V \phi(U, V)' V\Omega_2\right)$

for all $(\Omega_1, \Omega_2) \in \mathrm{skew}(n) \times \mathrm{skew}(m)$. Combining (2.32) and (i), (ii) we obtain

$$\begin{aligned} \nabla_U \phi &= U\{V'H_0 U, N\} \\ \nabla_V \phi &= V\{U'H_0'V, N'\}. \end{aligned}$$

Thus (2.32) is the gradient flow $\dot{U} = \nabla_U \phi(U, V)$, $\dot{V} = \nabla_V \phi(U, V)$ of ϕ.

By compactness of $O(n) \times O(m)$ the solutions of (2.31) exist for all $t \in \mathbb{R}$. Moreover, by the properties of gradient flows as summarized in the digression on convergence of gradient flows, Chapter 1.3, the ω-limit set of any solution $(U(t), V(t))$ of (2.31) is contained in a connected component of the intersection of the set of equilibria (U_∞, V_∞) of (2.31) with some level set of $\phi : O(n) \times O(m) \to \mathbb{R}$.

It can be shown that $\phi : O(n) \times O(m) \to \mathbb{R}$ is a Morse-Bott function. Thus, by Proposition 1.3.6, any solution $(U(t), V(t))$ of (2.32) converges to an equilibrium point. ∎

The following result describes the equilibrium points of (2.31).

3.2. A GRADIENT FLOW APPROACH TO SVD

Lemma 2.8 Let $\sigma_r > 0$ and $(U_0, V_0) \in O(n) \times O(m)$ as in (2.29). A pair $(U, V) \in O(n) \times O(m)$ is an equilibrium point of (2.31) if and only if

$$U = U_0' \operatorname{diag}(U_1, \ldots, U_r)\pi' S,$$
$$V = V_0' \operatorname{diag}(U_1, \ldots, U_r, U_{r+1}) \begin{bmatrix} \pi' & 0 \\ 0 & I_{m-n} \end{bmatrix}, \quad (2.33)$$

where $\pi \in \mathbb{R}^{n \times n}$ is a permutation matrix, $S = \operatorname{diag}(s_1, \ldots, s_n)$, $s_i \in \{-1, 1\}$, is a sign-matrix and $(U_1, \ldots, U_r, U_{r+1}) \in O(n_1) \times \ldots \times O(n_r) \times O(n-m)$ are orthogonal matrices.

Proof From (2.31) it follows that the equilibria (U, V) are characterised by $\{V'H_0U, N'\} = 0$, $\{U'H_0'V, N'\} = 0$. Thus, by Corollary 2.5, (U, V) is an equilibrium point of (2.31) if and only if

$$V'H_0U = \begin{bmatrix} \pi\Sigma_1\pi'S \\ 0 \end{bmatrix}. \quad (2.34)$$

Thus the result follows from the form of the stabilizer group given in (2.8). ∎

Remark In fact the gradient flow (1.12) is related to the self-equivalent flow (1.8). Let $(U(t), V(t))$ be a solution of (1.12). Then $H(t) = V(t)'H_0U(t) \in S(\Sigma)$ is a solution of the self-equivalent flow (1.8) since

$$\begin{aligned}
\dot{H} &= V'H_0\dot{U} + \dot{V}'H_0U \\
&= V'H_0U\{V'H_0U, N\} - \{(V'H_0U)', N'\}V'H_0U \\
&= H\{H, N\} - \{H', N'\}H.
\end{aligned}$$

Note that by (2.33) the equilibria (U_∞, V_∞) of the gradient flow satisfy

$$\begin{aligned}
V_\infty' H_0 U_\infty &= \begin{bmatrix} \pi & 0 \\ 0 & I_{m-n} \end{bmatrix} \Sigma \pi' S \\
&= \begin{bmatrix} \pi\Sigma_1\pi'S \\ 0 \end{bmatrix}.
\end{aligned}$$

Thus, up to permutations and possible sign factors, the equilibria of (1.12) just yield the singular value decomposition of H_0. ∎

In order to relate the local stability properties of the self-equivalent flow (1.8) on $S(\Sigma)$ to those of the gradient flow (2.31) we consider the smooth function

$$f : O(n) \times O(m) \longrightarrow S(\Sigma)$$

defined by

$$f(U, V) = V'H_0U. \quad (2.35)$$

By (2.8), $f(U,V) = f(\widehat{U}, \widehat{V})$ if and only if there exists orthogonal matrices $(U_1, \ldots, U_r, U_{r+1}) \in O(n_1) \times \ldots \times O(n_r) \times O(m-n)$ with

$$\begin{aligned} \widehat{U} &= U_0' \text{ diag } (U_1, \ldots, U_r) U_0 \cdot U , \\ \widehat{V} &= V_0' \text{ diag } (U_1, \ldots, U_r, U_{r+1}) V_0 \cdot V . \end{aligned} \quad (2.36)$$

and $(U_0, V_0) \in O(n) \times O(m)$ as in (2.29). Therefore the fibres of (2.35) are all diffeomorphic to $O(n_1) \times \ldots \times O(n_r) \times O(m-n)$. For π an $n \times n$ permutation matrix and $S = \text{diag}(\pm 1, \ldots, \pm 1)$ an arbitrary sign-matrix let $C(\pi, S)$ denote the submanifold of $O(n) \times O(m)$ which consists of all pairs (U, V)

$$\begin{aligned} U &= U_0' \text{ diag } (U_1, \ldots, U_r) \pi' S , \\ V &= V_0' \text{ diag } (U_1, \ldots, U_{r+1}) \begin{bmatrix} \pi & 0 \\ 0 & I_{m-n} \end{bmatrix} , \end{aligned}$$

with $(U_1, \ldots, U_r, U_{r+1}) \in O(n_1) \times \ldots \times O(n_r) \times O(m-n)$ arbitrary. Thus, by Lemma 2.8, the union

$$C = \bigcup_{(\pi, S)} C(\pi, S) \quad (2.37)$$

of all $2^n \cdot n!$ sets $C(\pi, S)$ is equal to the set of equilibria of (2.31). That is, the behaviour of the self-equivalent flow (1.8) around an equilibrium point $\begin{bmatrix} \pi \Sigma_1 \pi' S \\ 0 \end{bmatrix}$ now is equivalent to the behaviour of the solutions of the gradient flow (2.31) as they approach the invariant submanifold $C(\pi, S)$. This is made more precise in the following theorem. For any equilibrium point $(U_\infty, V_\infty) \in C$ let $W^s(U_\infty, V_\infty) \subset O(n) \times O(m)$ denote the stable manifold. See Appendix C.11. Thus $W^s(U_\infty, V_\infty)$ is the set of all initial conditions (U_0, V_0) such that the corresponding solution $(U(t), V(t))$ of (1.12) converges to (U_∞, V_∞) as $t \longrightarrow +\infty$.

Theorem 2.9 *Suppose the singular values $\sigma_1, \ldots, \sigma_n$ of H_0 are pairwise distinct with $\sigma_1 > \ldots > \sigma_n > 0$. Then $W^s(U_\infty, V_\infty)$ is an immersed submanifold in $O(n) \times O(m)$. Convergence in $W^s(U_\infty, V_\infty)$ to (U_∞, V_∞) is exponentially fast. Outside of a union of invariant submanifolds of codimension ≥ 1, every solution of the gradient flow (1.12) converges to the submanifold $C(I_n, I_n)$.*

Proof By the stable manifold theorem, Appendix C.11, $W^s(U_\infty, V_\infty)$ is an immersed submanifold. The stable manifold $W^s(U_\infty, V_\infty)$ of (1.12) at (U_∞, V_∞) is mapped diffeomorphically by $f : O(n) \times O(m) \longrightarrow S(\Sigma)$ onto the stable manifold of (1.8) at $H_\infty = V_\infty' H_0 U_\infty$. The second claim

3.2. A GRADIENT FLOW APPROACH TO SVD

follows since convergence on stable manifolds is always exponential. For any equilibrium point $(U_\infty, V_\infty) \in C(I_n, I_n)$

$$H_\infty = V'_\infty H_0 U_\infty = \Sigma$$

is the uniquely determined exponentially stable equilibrium point of (1.8) and its stable manifold $W^s(H_\infty)$ is the complement of a union Γ of submanifolds of $S(\Sigma)$ of codimension ≥ 1. Thus

$$W^s(U_\infty, V_\infty) = O(n) \times O(m) - f(\Gamma)$$

is dense in $O(n) \times O(m)$ with codim $f^{-1}(\Gamma) \geq 1$. The result follows. ∎

Problems 1. Verify that $\phi : O(n) \times O(m) \to \mathbb{R}$, $\phi(U,V) = \text{tr}(N'V'H_0 U)$, is a Morse-Bott function. Here

$$N = \begin{bmatrix} N_1 \\ O_{(m-n) \times n} \end{bmatrix}, \quad N_1 = \text{diag}(\mu_1, \ldots, \mu_n), \quad \mu_1 > \ldots > \mu_n > 0.$$

2. Let $A, B \in \mathbb{R}^{n \times n}$ with singular values $\sigma_1 \geq \ldots \geq \sigma_n$ and $\tau_1 \geq \ldots \geq \tau_n$ respectively. Show that the maximum value of the trace function $\text{tr}(AUBV)$ on $O(n) \times O(n)$ is

$$\max_{U,V \in O(n)} \text{tr}(AUBV) = \sum_{i=1}^{n} \sigma_i \tau_i.$$

3. Same notation as above. Show that

$$\max_{U,V \in SO(n)} \text{tr}(AUBV) = \sum_{i=1}^{n-1} \sigma_i \tau_i + (\text{sign}(\det(AB))) \sigma_n \tau_n$$

Main Points of Section The results on finding singular value decompositions via gradient flows, depending on the viewpoint, are seen to be both a generalization and specialization of the results of Brockett on the diagonalization of real symmetric matrices. The work ties in nicely with continuous time interpolations of the classical discrete-time QR algorithm by means of self-equivalent flows.

Notes for Chapter 3

There is an enormous literature on the singular value decomposition (SVD) and its applications within numerical linear algebra, statistics, signal processing and control theory. Standard applications of the SVD include those for solving linear equations and least squares problems; see Lawson and Hansen (1974). A strength of the singular value decomposition lies in the fact that it provides a powerful tool for solving ill-conditioned matrix problems; see Hansen (1990) and the references therein. A lot of the current research on the SVD in numerical analysis has been pioneered and is based on the work of Golub and his school. Discussions on the SVD can be found in almost any modern textbook on numerical analysis. The book of Golub and Van Loan (1989) is an excellent source of information on the SVD. A thorough discussion on the SVD and its history can also be found in Klema and Laub (1980).

A basic numerical method for computing the SVD of a matrix is the algorithm by Golub and Reinsch (1970). For an SVD updating method applicable to time-varying matrices we refer to Moonen, van Dooren and Vandewalle (1990). A neural network algorithm for computing the SVD has been proposed by Sirat (1991).

In the statistical literature, neural network theory and signal processing, the subject of matrix diagonalization and SVD is often referred to as principal component analysis or the Karhunen-Loeve expansion/transform method; see Hotelling (1933, 1935), Dempster (1969), Oja (1990) and Fernando and Nicholson (1980). Applications of the singular value decomposition within control theory and digital signal processing have been pioneered by Mullis and Roberts (1976) and B.C. Moore (1981). Since then the SVD has become an invaluable tool for system approximation and model reduction theory; see e.g. Glover (1984). Applications of the SVD for rational matrix valued functions to feedback control can be found in Hung and MacFarlane (1982).

Early results on the singular value decomposition, including a characterization of the critical points of the trace function (3.2.10) as well as a computation of the associated Hessian, are due to von Neumann (1937). The idea of relating the singular values of a matrix A to the eigenvalues of the symmetric matrix $\begin{bmatrix} 0 & A' \\ A & 0 \end{bmatrix}$ is a standard trick in linear algebra which is probably due to Wielandt. Inequalities for singular values and eigenvalues of a possibly nonsymmetric, square matrix include those of Weyl, Polya, Horn and Fan and can be found in Gohberg and Krein (1969) and Bhatia (1987).

Generalizations of the Frobenius norm are unitarily invariant matrix norms. These satisfy $\|UAV\| = \|A\|$ for all orthogonal (or unitary) matrices U, V and arbitrary matrices A. A characterization of unitarily invariant norms has been obtained by von Neumann (1937). The appropriate generalization of the SVD for square invertible matrices to arbitrary Lie groups is the Cartan decomposition. Basic textbooks on Lie groups and Lie algebras are Bröcker and tom Dieck (1985) and Humphreys (1972). The Killing form is a symmetric bilinear form on a Lie algebra. It is nondegenerate if the Lie algebra is semisimple.

An interesting application of Theorem 2.3 is to the Wielandt-Hoffman inequality for singular values of matrices. For this we refer to Chu and Driessel (1990); see also Bhatia (1987). For an extension of the results of Chapter 3 to complex matrices see Helmke and Moore (1992). Parallel work is that of Chu and Driessel (1990) and Smith (1991). The analysis in Chu and Driessel (1990) is not entirely complete and they treat only the generic case of simple, distinct singular values. The crucial observation that (2.12) is the gradient flow for the least squares cost function (2.4) on $S(\Sigma)$ is due to Smith (1991). His proof, however, is slightly different from the one presented here.

For an analysis of a recursive version of the self-equivalent flow (2.12) we refer to Moore, Mahony and Helmke (1993). It is shown that the iterative scheme

$$H_{k+1} = e^{-\alpha_k \{H'_k, N'\}} H_k e^{\alpha_k \{H_k, N\}} \quad , k \in \mathbb{N}_0,$$

with initial condition H_0 and step-size selection $\alpha_k := 1/(8\|H_0\| \cdot \|N\|)$ defines a self-equivalent recursion on $S(H_0)$ such that every solution $(H_k | k \in \mathbb{N}_0)$ converges to the same set of equilibria points as for the continuous flow (2.12). Moreover, under suitable genericity assumption, the convergence is exponentially fast, see also Brockett (1993).

For extensions of the singular value decomposition to several matrices and restricted versions we refer to Zha (1989 a, 1989 b) and de Moor and Golub (1989).

References

Bhatia, R. (1987) *Perturbation bounds for matrix eigenvalues*, Pitman Research Notes in Mathematics Series 162, Longman Scientific & Technical, New York.

Bröcker, Th. and T. tom Dieck (1985) *Representations of Compact Lie Groups*, Springer-Verlag, New York.

Brockett, R.W. (1993) Differential geometry and the design of gradient algorithms, *Proceedings of Symposia in Pure Mathematics*, 54, 69-91.

Chu, M.T. (1986) A differential equation approach to the singular value decomposition of bidiagonal matrices, *Linear Algebra and its Appl.*, 80, 71-80.

Chu, M.T. and Driessel, K.R., (1990) The projected gradient method for least squares approximations with spectral constraints, *SIAM J. Numer. Anal.*, 27 1050 - 1060.

de Moor, B.L.R., and G.H. Golub (1989) The restricted singular value decomposition: Properties and applications, *Technical Report* NA-89-03, Department of Computer Science, Stanford University, Stanford.

Dempster, A.P. (1969) *Elements of Continuous Multivariable Analysis*, Reading, MA: Addison-Wesley.

Fernando, K.V.M. and H. Nicholson (1980) Karhunen-Loeve expansion with reference to singular value decomposition and separation of variables, *Proc. IEE*, part D, vol. 127, 204-206.

Glover, K. (1984) All optimal Hankel-norm approximations of linear multivariable systems and their L^∞-error bounds, *International Journal of Control* 39, 1115-1193.

Gohberg, I.C. and M.G. Krein (1969) *Introduction to the Theory of Linear Nonselfadjoint operators*, Transl. Math. Monographs, Vol. 18, Amer. Math. Soc., Providence, R.I.

Golub, G.H. and C. Reinsch (1970) Singular value decomposition and least squares solutions, *Numer. Math.* 14, 403-420.

Golub, G.H. and C.F. Van Loan (1989) *Matrix Computations*, Second Edition, The John Hopkins University Press, Baltimore.

Hansen, P.C. (1990) Truncated singular value decomposition solutions to discrete ill-posed problems with ill-determined numerical rank, *SIAM J. Sci. Stat. Comp.* 11, 503-518.

Helmke, U. and J.B. Moore (1992) Singular value decomposition via gradient and self equivalent flows, *Linear Algebra and its Appl.* 69, 223-248.

Hotelling, H. (1933) Analysis of a complex of statistical variables into principal components, *J. Educ. Psych.* 24, 417-441, 498-520.

Hotelling, H. (1935) Simplified calculation of principal components, *Psychometrika* 1, 27-35.

Humphreys, J.E. (1972) *Introduction to Lie Algebras and Representation Theory*, Springer-Verlag, New York.

Hung, Y.S. and A.G.J. MacFarlane (1982) *Multivariable Feedback: A Quasi-Classical Approach*, Lecture Notes in Control and Information Sciences, Vol. 40, Springer, Berlin.

Klema, V.C. and A.J. Laub (1980) The singular value decomposition: Its computation and some applications, *IEEE Transactions on Automatic Control* AC-25, 164-176.

Lawson, C.L. and R.J. Hanson (1974) *Solving Least Squares Problems*, Prentice-Hall, Englewood Cliffs, N.J.

Moonen, M., van Dooren, P. and J. Vandewalle (1990) SVD updating for tracking slowly time-varying systems. A parallel implementation, *Signal Processing, Scattering and Operator Theory, and Numerical Methods* (M.A. Kaashoek, J.H. van Schuppen and A.C.M. Ran, Eds.), Birkhäuser, Boston, 487-494.

Moore, B.C. (1981) Principal component analysis in linear systems: controllability, observability, and model reduction, *IEEE Transactions on Automatic Control* AC-26, 17-32.

Moore, J.B., R.E. Mahony, and U. Helmke (1993) Numerical gradient algorithms for eigenvalues and singular value decompositions, *SIAM Journal of Matrix Analysis*, to appear.

Mullis, C.T. and R.A. Roberts (1976) Synthesis of minimum roundoff noise fixed point digital filters, *IEEE Transactions on Circuits and Systems*, CAS-23, 551-562.

Sirat, J.A. (1991) A fast neural algorithm for principal component analysis and singular value decompositions, *International Journal of Neural Systems* 2, 147-155.

Smith, S.T., (1991) Dynamical systems that perform the singular value decomposition, *Systems and Control Letters* 16, 319-328.

von Neumann, J. (1937) Some matrix inequalities and metrization of matrix space, *Tomsk Univ. Rev.* 1, 286-300.

Watkins, D.S. and Elsner, L., (1988) Self-similar flows, *Linear Algebra and its Appl.* 110, 213-242.

Watkins, D.S. and Elsner, L., (1989) Self-equivalent flows associated with the singular value decomposition, *SIAM J. Matrix Anal. Appl.*, 10, 244-258.

Zha, H. (1989a) A numerical algorithm for computing the restricted singular value decomposition of matrix triplets, *Scientific Report* 89-1, Konrad-Zuse Zentrum für Informationstechnik, Berlin.

Zha, H. (1989b) Restricted singular value decomposition of matrix triplets, *Scientific Report* 89-2, Konrad-Zuse Zentrum für Informationstechnik, Berlin.

Chapter 4

Linear Programming

4.1 The Role of Double Bracket Flows

In Chapter 2, the double bracket equation

$$\dot{H}(t) = [H(t), [H(t), N]], \quad H(0) = H_0 \quad (1.1)$$

for a real diagonal matrix N, and its recursive version, is presented as a scheme for diagonalizing a real symmetric matrix. Thus with a generic initial condition $H(0) = H_0$ where H_0 is real symmetric, $H(t)$ converges to a diagonal matrix H_∞, with its diagonal elements ordered according to the ordering in the prespecified diagonal matrix N.

If a generic non-diagonal initial matrix $H_0 = \Theta' \Sigma_0 \Theta$ is chosen, where Σ_0 is diagonal with a different ordering property than N, then the flow thus allows a re-ordering, or *sorting*. Now in a linear programming exercise, one can think of the vertices of the associated compact convex constraint set as needing an ordering to find the one vertex which has the greatest cost. The cost of each vertex can be entered in a diagonal matrix N. Then the double bracket equation, or its recursive form, can be implemented to achieve, from an initial $H_0 = H_0'$ with one nonzero eigenvalue, a diagonal H_∞ with one nonzero diagonal element. This nonzero element then "points" to the corresponding diagonal element of N, having the maximum value, and so the maximum cost vertex is identified.

The optimal solution in linear programming via a double bracket flow is achieved from the interior of the constraint set, and is therefore similar in spirit to the celebrated *Khachian's and Karmarkar's algorithms*. Actually, there is an essential difference between Karmarkar's and Brockett's algorithms due to the fact that the optimizing trajectory in Brockett's approach is constructed using a gradient flow evolving on a higher dimensional smooth manifold. An important property of this gradient flow and its recursive version is that it converges exponentially fast to the optimal solution. However, we caution that the computational effort in evaluating the vertex costs entered into N, is itself a computational cost of order m, where m is the number of vertices.

Starting from the seminal work of Khachian and Karmarkar, the development of interior point algorithms for linear programming is currently making fast progress. The purpose of this chapter is not to cover the main recent achievements in this field, but rather to show the connection between interior points flows for linear programming and matrix least squares estimation. Let us mention that there are also exciting connections with *artificial neural network theory*, as can be seen from the work of, e.g., Pyne (1956), Chua and Lin (1984), Tank and Hopfield (1985). A connection of linear programming with completely integrable Hamiltonian systems is made in the important work of Bayer and Lagarias (1989), Bloch (1990) and Faybusovich (1991).

In this chapter we first formulate the double bracket algorithm for linear programming and then show that it converges exponentially fast to the optimal solution. Explicit bounds for the rate of convergence are given as well as for the time needed for the trajectory produced to enter an ε-neighborhood of the optimal solution. In the special case where the convex constraint set is the standard simplex, Brockett's equation, or rather its simplification studied here, is shown to induce an interior point algorithm for sorting. The algorithm in this case is formally very similar to Karmarkar's interior point flow and this observation suggests the possibility of common generalizations of these algorithms. Very recently, Faybusovich (1991) have proposed a new class of interior point flows and for linear programming which, in the case of a standard simplex, coincide with the flow studied here. These interior point flows are studied in Section 3.

The Linear Programming Problem Let $C(v_1, \cdots, v_m) \subset \mathbb{R}^n$ denote the convex hull of m vectors $v_1, \cdots, v_m \in \mathbb{R}^n$. Given the compact convex set $C(v_1, \cdots, v_m) \subset \mathbb{R}^n$ and a row vector $c' = (c_1, \cdots, c_n) \in \mathbb{R}^n$ the linear programming problem then asks to find a vector $x \in C(v_1, \cdots, v_m)$ which maximizes $c' \cdot x$. Of course, the optimal solution is a vertex point v_{i_*} of the constraint set $C(v_1, \cdots, v_m)$.

Since $C(v_1, \cdots, v_m)$ is not a smooth manifold in any reasonable sense it is not possible to apply in the usual way steepest descent gradient methods in order to find the optimum. One possible way to circumvent such technical difficulties would be to replace the constraint set $C(v_1, \cdots, v_m)$ by some suitable compact manifold M so that the optimization takes place on M instead of $C(v_1, \cdots, v_m)$. A mathematically convenient way here would be

4.1. THE ROLE OF DOUBLE BRACKET FLOWS

to construct a suitable resolution space for the singularities of $C(v_1, \cdots, v_m)$. This is the approach taken by Brockett.

Let $\Delta_{m-1} = \{(\eta_1, \cdots, \eta_m) \in \mathbb{R}^m \mid \eta_i \geq 0, \sum_{i=1}^{m} \eta_i = 1\}$ denote the standard $(m-1)$-dimensional simplex in $\mathbb{R}^m$. Let

$$T = (v_1, \cdots, v_m) \tag{1.2}$$

be the real $n \times m$ matrix whose column vectors are the vertices $v_1, \cdots, v_m$. Thus $T : \mathbb{R}^m \to \mathbb{R}^n$ maps the simplex $\Delta_{m-1} \subset \mathbb{R}^m$ linearly onto the set $C(v_1, \cdots, v_m)$. We can thus use T in order to replace the constraint set $C(v_1, \cdots, v_m)$ by the standard simplex Δ_{m-1}.

Suppose that we are to solve the linear programming problem consisting of maximizing $c'x$ over the compact convex set of all $x \in C(v_1, \cdots, v_m)$. Brockett's recipe to solve the problem is this (cf. Brockett (1988), Theorem 6).

Theorem 1.1 *Let N be the real $m \times m$ matrix defined by*

$$N = \mathrm{diag}(c'v_1, \cdots, c'v_m), \tag{1.3}$$

and assume that the genericity condition $c'v_i \neq c'v_j$ for $i \neq j$ holds. Let $Q = \mathrm{diag}(1, 0, \cdots, 0) \in \mathbb{R}^{m \times m}$. Then for almost all orthogonal matrices $\Theta \in O(m)$ the solution $H(t)$ of the differential equation

$$\boxed{\begin{aligned}\dot{H} &= [H, [H, N]] \\ &= H^2 N + N H^2 - 2HNH, \quad H(0) = \Theta' Q \Theta\end{aligned}} \tag{1.4}$$

converges as $t \to \infty$ to a diagonal matrix $H_\infty = \mathrm{diag}(0, \cdots, 0, 1, 0, \cdots, 0)$, with the entry 1 being at position i_* so that $x = v_{i_*}$ is the optimal vertex of the linear programming problem.

In fact, this is an immediate consequence of the remark following Theorem 2.1.3. Thus the optimal solution of a linear programming problem can be obtained by applying the linear transformation T to a vector obtained from the diagonal entries of the stable limiting solution of (1.4). Brockett's method, while theoretically appealing, has however a number of shortcomings.

First, it works with a huge overparametrization of the problem. The differential equation (1.4) evolves on the $\frac{1}{2}m(m+1)$-dimensional vector space of real symmetric $m \times m$ matrices H, while the linear programming problem is set up in the n-dimensional space $C(v_1, \cdots, v_m)$. Of course, usually n will be much smaller than $\frac{1}{2}m(m+1)$.

Second, convergence to H_∞ is guaranteed only for a generic choice of orthogonal matrices. No explicit description of this generic set of initial data is given.

Finally the method requires the knowledge of the values of the cost functional at all vertex points in order to define the matrix N.

Clearly the last point is the most critical one and therefore Theorem 1.1 should be regarded only as a theoretical approach to the linear programming problem. To overcome the first two difficulties we proceed as follows.

Let $S^{m-1} = \{(\xi_1, \cdots, \xi_m) \in \mathbb{R}^m \mid \sum_{i=1}^m \xi_i^2 = 1\}$ denote the set of unit vectors of $\mathbb{R}^m$. We consider the polynomial map

$$f : S^{m-1} \to \triangle_{m-1}$$

defined by

$$f(\xi_1, \cdots, \xi_m) = (\xi_1^2, \cdots, \xi_m^2). \tag{1.5}$$

By composing f with the map $T : \triangle_{m-1} \to C(v_1, \cdots, v_m)$ we obtain a real algebraic map

$$\pi_T = T \circ f : S^{m-1} \to C(v_1, \cdots, v_m),$$

$$(\xi_1, \cdots, \xi_m) \mapsto T \begin{pmatrix} \xi_1^2 \\ \vdots \\ \xi_m^2 \end{pmatrix}. \tag{1.6}$$

The linear programming task is to maximize the restriction of the linear functional

$$\lambda : C(v_1, \cdots, v_m) \to \mathbb{R}$$

$$x \mapsto c'x \tag{1.7}$$

over $C(v_1, \cdots, v_m)$. The idea now is to consider instead of the maximization of (1.7) the maximization of the induced smooth function $\lambda \circ \pi_T : S^{m-1} \to \mathbb{R}$ on the sphere S^{m-1}. Of course, the function $\lambda \circ \pi_T : S^{m-1} \to \mathbb{R}$ has the form as that of a Rayleigh quotient, so that we can apply our previous theory developed in Chapter 1.4. Let S^{m-1} be endowed with the Riemannian metric defined by

$$\langle \xi, \eta \rangle = 2\xi'\eta, \quad \xi, \eta \in T_x S^{m-1}. \tag{1.8}$$

Theorem 1.2 *Let N be defined by (1.3) and assume the genericity condition*

$$c'(v_i - v_j) \neq 0 \text{ for all } i \neq j. \tag{1.9}$$

a) *The gradient vector-field of $\lambda \circ \pi_T$ on S^{m-1} is*

$$\boxed{\dot{\xi} = (N - \xi'N\xi I)\xi,} \tag{1.10}$$

4.1. THE ROLE OF DOUBLE BRACKET FLOWS

$|\xi|^2 = 1$. Also, (1.10) has exactly 2^m equilibrium points given by the standard basis vectors $\pm e_1, \cdots, \pm e_m$ of $\mathbb{R}^m$.

b) The eigenvalues of the linearization of (1.10) at $\pm e_i$ are

$$c'(v_1 - v_i), \cdots, c'(v_{i-1} - v_i), \; c'(v_{i+1} - v_i), \; \cdots, \; c'(v_m - v_i), \quad (1.11)$$

and there is a unique index $1 \leq i_* \leq m$ such that $\pm e_{i_*}$ is asymptotically stable.

c) Let $X \cong S^{m-2}$ be the smooth codimension-one submanifold of S^{m-1} defined by

$$X = \{(\xi_1, \cdots, \xi_m) \mid \xi_{i_*} = 0\}. \quad (1.12)$$

With the exception of initial points contained in X, every solution $\xi(t)$ of (1.10) converges exponentially fast to the stable attractor $[\pm e_{i_*}]$. Moreover, $\pi_T(\xi(t))$ converges exponentially fast to the optimal solution $\pi_T(\pm e_{i_*}) = v_{i_*}$ of the linear programming problem, with a bound on the rate of convergence

$$|\pi_T(\xi(t)) - v_{i_*}| \leq \text{const } e^{-\mu(t-t_0)},$$

where

$$\mu = \min_{j \neq i_*} |c'(v_j - v_{i_*})|. \quad (1.13)$$

Proof   For convenience of the reader we repeat the argument for the proof, taken from Chapter 1 Section 4. For any diagonal $m \times m$ matrix $N = \text{diag}(n_1, \cdots, n_m)$ consider the Rayleigh quotient $\varphi : \mathbb{R}^m - \{0\} \to \mathbb{R}$ defined by

$$\varphi(x_1, \cdots, x_m) = \frac{\sum_{i=1}^m n_i x_i^2}{\sum_{i=1}^m x_i^2}. \quad (1.14)$$

A straightforward computation shows that the gradient of φ at a unit vector $x \in S^{m-1}$ is

$$\nabla \varphi(x) = (N - x'Nx)x.$$

Furthermore, if $n_i \neq n_j$ for $i \neq j$, then the critical points of the induced map $\varphi : S^{m-1} \to \mathbb{R}$ are, up to a sign $\pm$, the standard basis vectors, i.e. $\pm e_1, \cdots, \pm e_m$. This proves (a). The Hessian of $\varphi : S^{m-1} \to \mathbb{R}$ at $\pm e_i$ is readily computed as

$$H_\varphi(\pm e_i) = \text{diag}(n_1 - n_i, \cdots, n_{i-1} - n_i, n_{i+1} - n_i, \cdots, m_m - n_i).$$

Let i_* be the unique index such that $n_{i_*} = \max_{1 \leq j \leq m} n_j$. Thus $H_\varphi(\pm e_i) < 0$ if and only if $i = i_*$ which proves (b). Let $X \cong S^{m-2}$ be the closed submanifold of S^{m-1} defined by (1.12). Then $S^{m-1} - X$ is equal to the stable manifold of $\pm e_{i_*}$ and X is equal to the union of the stable manifolds of the other equilibrium points $\pm e_i, i \neq i_*$. The result follows. ∎

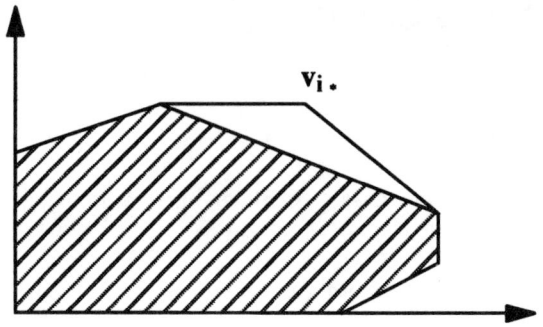

Figure 4.1.1: The convex set $C(v_1, \cdots, v_m)$

Remarks 1. From (1.12)

$$\pi_T(X) = C(v_1, \cdots, v_{i_*-1}, v_{i_*+1}, \cdots, v_m) \ . \tag{1.15}$$

Thus if $n = 2$ and $C(v_1, \cdots, v_m) \subset \mathbb{R}^2$ is the convex set, illustrated by the following Figure 4.1.1 with optimal vertex point v_{i_*}.

then the shaded region describes the image $\pi_T(X)$ of the set of exceptional initial conditions. The gradient flow (1.10) on the $(m-1)$-space S^{m-1} is identical with the Rayleigh quotient gradient flow; see Chapter 1 Section 2.

2. Moreover, the flow (1.4) with $Q = \text{diag}(1, 0, \cdots, 0)$ is equivalent to the double bracket gradient flow (1.10) on the isospectral set $M(Q) = \mathbb{R}\mathbb{P}^{m-1}$. In fact, with $H = \xi\xi'$ and $\xi(t)$ a solution of (1.10) we have $H^2 = H$ and

$$\begin{aligned}\dot{H} &= \dot{\xi}\xi' + \xi\dot{\xi}' = (N - \xi'N\xi I)\xi\xi' + \xi\xi'(N - \xi'N\xi I) \\ &= NH + HN - 2HNH = [H, [H, N]].\end{aligned} \tag{1.16}$$

Also, (1.10) has an interesting interpretation in neural network theory; see Oja (1982).

3. For any (skew-) symmetric matrix $\Omega \in \mathbb{R}^{m \times m}$

$$\boxed{\dot{\xi} = (N + \Omega - \xi'(N + \Omega)\xi I_m)\xi} \tag{1.17}$$

defines a flow on S^{m-1}. If $\Omega = -\Omega'$ is skew-symmetric, the functional $\varphi(\xi) = \xi'(N + \Omega)\xi = \xi'N\xi$ is not changed by Ω and therefore has the same critical points. Thus, while (1.17) is not the gradient flow of φ (if $\Omega = -\Omega'$), it can still be of interest for the linear programming problem. If $\Omega = \Omega'$ is symmetric, (1.17) is the gradient flow of $\psi(\xi) = \xi'(N + \Omega)\xi$ on S^{m-1}. ∎

4.1. THE ROLE OF DOUBLE BRACKET FLOWS

Interior Point Flows on the Simplex Brockett's equation (1.4) and its simplified version (1.10) both evolve on a high-dimensional manifold M so that the projection of the trajectories into the polytope leads to a curve which approaches the optimum from the interior. In the special case where the polytope is the $(m-1)$-simplex $\triangle_{m-1} \subset \mathbb{R}^m$, (1.10) actually leads to a flow evolving in the simplex such that the optimum is approached from all trajectories starting in the interior of the constraint set. Such algorithms are called *interior point algorithms*, an example being the celebrated *Karmarkar algorithm* (1984), or the algorithm proposed by Khachian (1979). Here we like to take such issues a bit further.

In the special case where the constraint set is the standard simplex $\triangle_{m-1} \subset \mathbb{R}^m$, then equation (1.10) on S^{m-1} becomes

$$\dot{\xi}_i = (c_i - \sum_{j=1}^m c_j \xi_j^2)\xi_i, \quad i=1,\cdots,m, \qquad (1.18)$$

$\sum_{i=1}^m \xi_i^2 = 1$. Thus with the substitution $x_i = \xi_i^2, i=1,\cdots,m$, we obtain

$$\dot{x}_i = 2(c_i - \sum_{j=1}^m c_j x_j)x_i, \quad i=1,\cdots,m, \qquad (1.19)$$

$x_i \geq 0, \sum_{i=1}^m x_i = 1$. Since $\sum_{i=1}^m \dot{x}_i = 0$, (1.19) is a flow on the simplex $\triangle_{m-1}$. The set $X \subset S^{m-1}$ of exceptional initial conditions is mapped by the quadratic substitution $x_i = \xi_i^2, i=1,\cdots,m$, onto the boundary $\partial \triangle_{m-1}$ of the simplex. Thus Theorem 1.2 implies

Corollary 1.3 *Equation (1.19) defines a flow on $\triangle_{m-1}$. Every solution $x(t)$ with initial condition $x(0)$ in the interior of $\triangle_{m-1}$ converges to the optimal solution e_{i_*} of the linear programming problem: Maximize $c'x$ over $x \in \triangle_{m-1}$. The exponential rate of convergence is given by*

$$|x(t) - e_{i_*}| \leq \text{constant } e^{-2\mu t}, \quad \mu = \min_{j \neq i_*}(c_{i_*} - c_j).$$

Remarks 1. Equation (1.19) is a *Volterra-Lotka* type of equation and thus belongs to a well studied class of equations in population dynamics; cf. Schuster, Sigmund and Wolff (1978), Zeeman (1980).

2. If the interior $\overset{\circ}{\triangle}_{m-1}$ of the simplex is endowed with the Riemannian metric defined by

$$\langle\langle \xi, \eta \rangle\rangle = \sum_{i=1}^m \frac{\xi_i \eta_i}{x_i}, \quad x = (x_1, \cdots, x_m) \in \overset{\circ}{\triangle}_{m-1},$$

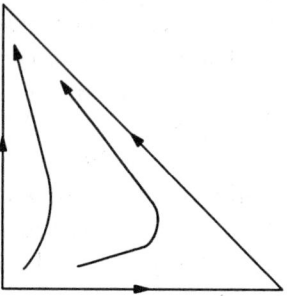

Figure 4.1.2: Phase portrait of (1.19)

then (1.19) is actually (up to an irrelevant factor by 2) the gradient flow of the linear functional $x \mapsto c'x$ on $\overset{\circ}{\triangle}_{m-1}$, see Figure 4.2.1.

3. Karmarkar (1990) has analyzed a class of interior point flows which are the continuous-time versions of the discrete-time algorithm described in Karmarkar (1984). In the case of the standard simplex, Karmarkar's equation turns out to be

$$\dot{x}_i = (c_i x_i - \sum_{j=1}^{m} c_j x_j^2) x_i \, , \quad i = 1, \cdots, m, \tag{1.20}$$

$x_i \geq 0, \sum_{i=1}^{m} x_i = 1$. This flow is actually the gradient flow for the quadratic cost function $\sum_{j=1}^{m} c_j x_j^2$ rather than for the linear cost function $\sum_{j=1}^{m} c_j x_j$. Thus Karmakar's equation solves a quadratic optimization problem on the simplex. A more general class of equations would be

$$\boxed{\dot{x}_i = [c_i f_i(x_i) - \sum_{j=1}^{m} c_j f_j(x_j) x_j] x_i \, , \quad i = 1, \cdots, m,}$$
(1.21)

with $f_j : [0, 1] \to \mathbb{R}$ monotonically increasing C^1 functions, see Section 2. Incidentally, the Karmarkar flow (1.20) is just a special case of the equations studied by Zeeman (1980). ∎

The following result estimates the time a trajectory of (1.19) needs in order to reach an ε-neighborhood of the optimal vertex.

Proposition 1.4 *Let $0 < \varepsilon < 1$ and $\mu = \min_{j \neq i_*}(c_{i_*} - c_j)$. Then for any initial condition $x(0)$ in the interior of $\triangle_{m-1}$ the solution $x(t)$ of (1.19) is contained in an ε-neighborhood of the optimum vertex e_{i_*} if $t \geq t_\varepsilon$ where*

$$t_\varepsilon = \frac{|\log(\min_{1 \leq i \leq m} x_i(0)) \varepsilon^2 / 2|}{2\mu} . \tag{1.22}$$

4.1. THE ROLE OF DOUBLE BRACKET FLOWS

Proof We first introduce a lemma.

Lemma 1.5 *Every solution $x(t)$ of (1.19) is of the form*

$$x(t) = \frac{e^{2tN}x(0)}{\langle e^{2tN}x(0)\rangle}, \qquad (1.23)$$

where $N = \mathrm{diag}(c_1, \cdots, c_m)$ and $\langle e^{2tN}x(0)\rangle = \sum_{j=1}^{m} e^{2tc_j}x_j(0)$.

Proof In fact, by differentiating the right hand side of (1.23) one sees that both sides satisfy the same conditions (1.19) with identical initial conditions. Thus (1.23) holds. ∎

Using (1.23) one has

$$\begin{aligned}\|x(t) - e_{i_*}\|^2 &= \|x(t)\|^2 - 2\frac{\langle e^{2tN}x(0), e_{i_*}\rangle}{\langle e^{2tN}x(0)\rangle} + 1 \\ &\leq 2 - 2\frac{e^{2tc_{i_*}}x_{i_*}(0)}{\langle e^{2tN}x(0)\rangle}.\end{aligned} \qquad (1.24)$$

Now

$$\sum_{j=1}^{m} e^{2t(c_j - c_{i_*})}x_j(0) \leq e^{-2\mu t} + x_{i_*}(0), \qquad (1.25)$$

and thus

$$\|x(t) - e_{i_*}\|^2 \leq 2 - 2(1 + e^{-2\mu t}x_{i_*}(0)^{-1})^{-1}.$$

Therefore $\|x(t) - e_{i_*}\| \leq \varepsilon$ if

$$(1 + e^{-2\mu t}x_{i_*}(0)^{-1})^{-1} > 1 - \frac{\varepsilon^2}{2},$$

i.e., if

$$t \geq \left|\frac{\log(\frac{\varepsilon^2 x_{i_*}(0)}{2-\varepsilon^2})}{2\mu}\right|.$$

From this the result easily follows. ∎

Note that for the initial condition $x(0) = \frac{1}{m}(1, \cdots, 1)$ the estimate (1.22) becomes

$$t_\varepsilon \geq \frac{|\log \frac{\varepsilon^2}{2m}|}{2\mu}, \qquad (1.26)$$

and (1.26) gives an effective lower bound for (1.22) valid for all $x(0) \in \overset{\circ}{\triangle}_{m-1}$.

We can use Proposition 1.4 to obtain an explicit estimate for the time needed in either Brockett's flow (1.4) or for (1.10) that the projected interior point trajectory $\pi_T(x(t))$ enters an ε-neighborhood of the optimal solution.

Thus let N, T be defined by (1.2), (1.3) with (1.9) understood and let v_{i_*} denote the optimal solution for the linear programming problem of maximizing $c'x$ over the convex set $C(v_1, \cdots, v_m)$.

Theorem 1.6 *Let $0 < \varepsilon < 1$, $\mu = \min_{j \neq i_*}(c'v_{i_*} - c'v_j)$, and let X be defined by (1.12). Then for all initial conditions $\xi(0) \in S^{m-1} - X$ the projected trajectory $\pi_T(\xi(t)) \in C(v_1, \cdots, v_m)$ of the solution $\xi(t)$ of (1.10) is in an ε-neighborhood of the optimal vertex v_{i_*} for all $t \geq t_\varepsilon$ with*

$$t_\varepsilon = \frac{|\log \frac{(\varepsilon/\|T\|)^2}{2m}|}{2\mu}. \tag{1.27}$$

Proof By (1.6), $\pi_T(\xi(t)) = Tx(t)$ where $x(t) = (\xi_1(t)^2, \cdots, \xi_m(t)^2)$ satisfies

$$\dot{x}_i = 2(c'v_i - \sum_{j=1}^{m} c'v_j x_j)x_i, \quad i = 1, \cdots, m.$$

Proposition 4.1.1 implies $\|x(t) - e_{i_*}\| < \frac{\varepsilon}{\|T\|}$ for $t \geq \frac{|\log \frac{(\varepsilon/\|T\|)^2}{2m}|}{2\mu}$ and hence

$$\|\pi_T(\xi(t)) - v_{i_*}\| \leq \|T\| \cdot \|x(t) - e_{i_*}\| < \varepsilon.$$

∎

Main Points of the Section The "gradient method" for linear programming consists in the following program:

(i) To find a smooth compact manifold M and a smooth map $\pi : M \to \mathbb{R}^n$, which maps M onto the convex constraint set C.

(ii) To solve the gradient flow of the smooth function $\lambda \circ \pi : M \to \mathbb{R}$ and determine its stable equilibria points.

In the cases discussed here we had $M = S^{m-1}$ and $\pi : S^{m-1} \to \mathbb{R}^n$ was the Rayleigh quotient formed as the composition of the linear map $T : \Delta_{m-1} \to C(v_1, \cdots, v_m)$ with the smooth map from $S^{m-1} \to \Delta_{m-1}$ defined by (1.5).

Likewise, interior point flows for linear programming evolve in the interior of the constraint set.

4.2 Interior Point Flows on a Polytope

In the previous sections interior point flows for optimizing a linear functional on a simplex were considered. Here, following the pioneering work

4.2. INTERIOR POINT FLOWS ON A POLYTOPE

of Faybusovich (1991a,1992), we extend our previous results by considering interior point gradient flows for a cost function defined on the interior of an arbitrary polytope. Polytopes, or compact convex subsets of $\mathbb{R}^n$, can be parametrized in various ways. A standard way of describing such polytopes $\mathcal{C}$ is as follows.

Mixed Equality-Inequality Constraints For any $x \in \mathbb{R}^n$ we write $x \geq 0$ if all coordinates of x are non-negative. Given $A \in \mathbb{R}^{m \times n}$ with $\text{rk} A = m < n$ and $b \in \mathbb{R}^m$ let

$$\mathcal{C} = \{x \in \mathbb{R}^n \mid x \geq 0, \, Ax = b\}$$

be the convex constraint set of our optimization problem. The *interior* of $\mathcal{C}$ is the smooth manifold defined by

$$\overset{\circ}{\mathcal{C}} = \{x \in \mathbb{R}^n \mid x > 0, \, Ax = b\}$$

In the sequel we assume that $\overset{\circ}{\mathcal{C}}$ is a nonempty subset of $\mathbb{R}^n$ and $\mathcal{C}$ is compact. Thus $\mathcal{C}$ is the closure of $\overset{\circ}{\mathcal{C}}$. The optimization task is then to optimize (i.e. minimize or maximize) the cost function $\phi : \mathcal{C} \to \mathbb{R}$ over $\mathcal{C}$. Here we assume that $\phi : \mathcal{C} \to \mathbb{R}$ is the restriction of a smooth function $\phi : \mathbb{R}^n \to \mathbb{R}$. Let

$$\nabla \phi(x) = (\frac{\partial \phi}{\partial x_1}(x), \cdots, \frac{\partial \phi}{\partial x_n}(x))' \qquad (2.1)$$

denote the usual gradient vector of ϕ in $\mathbb{R}^n$.

For $x \in \mathbb{R}^n$ let

$$D(x) = \text{diag}(x_1, \cdots, x_n) \in \mathbb{R}^{n \times n}. \qquad (2.2)$$

For any $x \in \overset{\circ}{\mathcal{C}}$ let $T_x(\overset{\circ}{\mathcal{C}})$ denote the tangent space of $\overset{\circ}{\mathcal{C}}$ at x. Thus $T_x(\overset{\circ}{\mathcal{C}})$ coincides with the tangent space of the affine subspace of $\{x \in \mathbb{R}^n \mid Ax = b\}$ at x:

$$T_x(\overset{\circ}{\mathcal{C}}) = \{\xi \in \mathbb{R}^n \mid A\xi = 0\}, \qquad (2.3)$$

that is, with the kernel of A; cf. Chapter 1.6. For any $x \in \overset{\circ}{\mathcal{C}}$, the diagonal matrix $D(x)$ is positive definite and thus

$$\langle\langle \xi, \eta \rangle\rangle := \xi' D(x)^{-1} \eta, \quad \xi, \eta \in T_x(\overset{\circ}{\mathcal{C}}), \qquad (2.4)$$

defines a positive definite inner product on $T_x(\overset{\circ}{\mathcal{C}})$ and, in fact, a Riemannian metric on the interior $\overset{\circ}{\mathcal{C}}$ of the constraint set.

The gradient $\text{grad}\phi$ of $\phi : \overset{\circ}{\mathcal{C}} \to \mathbb{R}$ with respect to the Riemannian metric $\langle\langle \, , \, \rangle\rangle$, at $x \in \overset{\circ}{\mathcal{C}}$ note that $\text{grad}\phi(x)$, $x \in \overset{\circ}{\mathcal{C}}$ is characterized by the properties

(i) $\mathrm{grad}\phi(x) \in T_x(\overset{\circ}{\mathcal{C}})$

(ii) $\langle\langle\mathrm{grad}\phi(x),\xi\rangle\rangle = \nabla\phi(x)'\xi$

for all $\xi \in T_x(\overset{\circ}{\mathcal{C}})$. Here (i) is equivalent to $A \cdot \mathrm{grad}\phi(x) = 0$ while (ii) is equivalent to

$$(D(x)^{-1}\mathrm{grad}\phi(x) - \nabla\phi(x))'\xi = 0 \quad \forall\, \xi \in \ker A$$
$$\iff D(x)^{-1}\mathrm{grad}\phi(x) - \nabla\phi(x) \in (\ker A)^{\perp} = \mathrm{Im}(A')$$
$$\iff D(x)^{-1}\mathrm{grad}\phi(x) - \nabla\phi(x) = A'\lambda \tag{2.5}$$

for a uniquely determined $\lambda \in \mathbb{R}^m$.

Thus

$$\mathrm{grad}\phi(x) = D(x)\nabla\phi(x) + D(x)A'\lambda. \tag{2.6}$$

Multiplying both sides of this equation by A and noting that $A \cdot \mathrm{grad}\phi(x) = 0$ and $AD(x)A' > 0$ for $x \in \overset{\circ}{\mathcal{C}}$ we obtain

$$\lambda = -(AD(x)A')^{-1}AD(x)\nabla\phi(x).$$

Thus

$$\mathrm{grad}\phi(x) = [I_n - D(x)A'(AD(x)A')^{-1}A]D(x)\nabla\phi(x) \tag{2.7}$$

is the gradient of ϕ at an interior point $x \in \overset{\circ}{\mathcal{C}}$. We have thus proved

Theorem 2.1 *The gradient flow $\dot{x} = \mathrm{grad}\phi(x)$ of $\phi: \overset{\circ}{\mathcal{C}} \to \mathbb{R}$ with respect to the Riemannian metric (2.4) on the interior of the constraint set is*

$$\boxed{\dot{x} = (I_n - D(x)A'(AD(x)A')^{-1}A)D(x)\nabla\phi(x).} \tag{2.8}$$

We refer to (2.8) as the *Faybusovich flow* on $\overset{\circ}{\mathcal{C}}$. Let us consider a few special cases of this result. If $A = (1,\cdots,1) \in \mathbb{R}^{1\times m}$ and $b = 1$, the constraint set $\mathcal{C}$ is just the standard $(m-1)$-dimensional simplex Δ_{m-1}. The gradient flow (2.8) then simplifies to

$$\dot{x} = (D(\nabla\phi(x)) - x'\nabla\phi(x)I_n)x,$$

that is, to

$$\boxed{\dot{x}_i = \left(\frac{\partial\phi}{\partial x_i} - \sum_{j=1}^{n}x_j\frac{\partial\phi}{\partial x_j}\right)x_i\,, \quad i=1,\cdots,n.} \tag{2.9}$$

If $\phi(x) = c'x$ then (2.9) is equivalent to the interior point flow (1.19) on Δ_{m-1} (up to the constant factor 2). If

$$\phi(x) = \sum_{j=1}^{n} f_j(x_j)$$

4.2. INTERIOR POINT FLOWS ON A POLYTOPE

then (2.8) is equivalent to

$$\dot{x}_i = (f'_i(x_i) - \sum_{j=1}^n x_j f'_j(x_j))x_i, \quad i = 1, \cdots, n,$$

i.e. to (1.21). In particular, Karmarkar's flow (1.20) on the simplex is thus seen as the gradient flow of the quadratic cost function

$$\phi(x) = \frac{1}{2} \sum_{j=1}^n c_j x_j^2$$

on Δ_{m-1}.

As another example, consider the least squares cost function on the simplex $\phi : \Delta_{m-1} \to \mathbb{R}$ defined by

$$\phi(x) = \frac{1}{2}\|Fx - g\|^2$$

Then $\nabla\phi(x) = F'(Fx - g)$ and the gradient flow (2.8) is equivalent to

$$\dot{x}_i = ((F'Fx)_i - (F'g)_i - x'F'Fx + x'F'g)x_i \quad i = 1, \cdots, n \qquad (2.10)$$

In particular, for $F = I_n$, (2.9) becomes

$$\dot{x}_i = (x_i - g_i - \|x\|^2 + x'g)x_i, \quad i = 1, \cdots, n \qquad (2.11)$$

If $g \notin \Delta_{m-1}$, then the gradient flow (2.11) in $\overset{\circ}{\Delta}_{m-1}$ converges to the best approximation of $g \in \mathbb{R}^n$ by a vector on the boundary $\partial\Delta_{m-1}$ of Δ_{m-1}; see Figure 4.2.1.

Finally, for C arbitrary and $\phi(x) = c'x$ the gradient flow (2.8) on $\overset{\circ}{C}$ becomes

$$\dot{x} = [D(c) - D(x)A'(AD(x)A')^{-1}AD(c)]x \qquad (2.12)$$

which is identical with the gradient flow for linear programming proposed and studied extensively by Faybusovich (1991). Independently, this flow was also studied by Herzel, Recchioni and Zirilli (1991). Faybusovich (1991a, 1992) has also presented a complete phase portrait analysis of (2.12). He shows that the solutions $x(t) \in \overset{\circ}{C}$ converge exponentially fast to the optimal vertex of the linear programming problem.

Inequality Constraints   A different description of the constraint set is as

$$C = \{x \in \mathbb{R}^n \mid Ax \geq b\} \qquad (2.13)$$

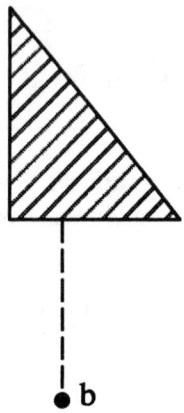

Figure 4.2.1: Best approximation of b on a simplex

where $A \in \mathbb{R}^{m \times n}, b \in \mathbb{R}^m$. Here we assume that $x \mapsto Ax$ is injective, that is $\mathrm{rk} A = n \leq m$. Consider the injective map

$$F : \mathbb{R}^n \to \mathbb{R}^m \ , \ F(x) = Ax - b$$

Then

$$C = \{x \in \mathbb{R}^n \mid F(x) \geq 0\}$$

and

$$\overset{\circ}{C} = \{x \in \mathbb{R}^n \mid F(x) > 0\}$$

In this case the tangent space $T_x(\overset{\circ}{C})$ of $\overset{\circ}{C}$ at x is unconstrained and thus consists of all vectors $\xi \in \mathbb{R}^n$. A Riemannian metric on $\overset{\circ}{C}$ is defined by

$$\begin{aligned}\langle\langle \xi, \eta \rangle\rangle &:= DF|_x(\xi)' D(F(x))^{-1} DF|_x(\eta) \\ &= \xi' A' \mathrm{diag}(Ax - b)^{-1} A \eta\end{aligned} \qquad (2.14)$$

for $x \in \overset{\circ}{C}$ and $\xi, \eta \in T_x(\overset{\circ}{C}) = \mathbb{R}^n$. Here $DF|_x : T_x(\overset{\circ}{C}) \to T_{F(x)} \mathbb{R}^m$ is the derivative or tangent map of F at x and $D(F(x))$ is defined by (2.2). The reader may verify that (2.14) does indeed define a positive definite inner product on $\mathbb{R}^n$, using the injectivity of A. It is now rather straightforward to compute the gradient flow $\dot{x} = \mathrm{grad}\phi(x)$ on $\overset{\circ}{C}$ with respect to this Riemannian metric on $\overset{\circ}{C}$. Here $\phi : \mathbb{R}^n \to \mathbb{R}$ is a given smooth function.

Theorem 2.2 *The gradient flow $\dot{x} = \mathrm{grad}\phi(x)$ of a smooth function $\phi : \overset{\circ}{C} \to \mathbb{R}$ with respect to the Riemannian metric (2.14) on $\overset{\circ}{C}$ is*

$$\boxed{\dot{x} = (A' \mathrm{diag}(Ax - b)^{-1} A)^{-1} \nabla \phi(x)} \qquad (2.15)$$

4.2. INTERIOR POINT FLOWS ON A POLYTOPE

where $\nabla \phi(x) = (\frac{\partial \phi}{\partial x_1}, \cdots, \frac{\partial \phi}{\partial x_n})'$

Proof The gradient is defined by the characterization

(i) $\mathrm{grad}\phi(x) \in T_x(\overset{\circ}{\mathcal{C}})$

(ii) $\langle\langle \mathrm{grad}\phi(x), \xi \rangle\rangle = \nabla \phi(x)'\xi$ for all $\xi \in T_x(\overset{\circ}{\mathcal{C}})$.

Since $T_x(\overset{\circ}{\mathcal{C}}) = \mathbb{R}^n$ we only have to consider property (ii) which is equivalent to

$$\mathrm{grad}\phi(x)' A' \mathrm{diag}(Ax - b)^{-1} A\xi = \nabla \phi(x)'\xi$$

for all $\xi \in \mathbb{R}^n$. Thus

$$\mathrm{grad}\phi(x) = (A' \mathrm{diag}(Ax - b)^{-1} A)^{-1} \nabla \phi(x).$$

∎

Let us consider some special cases. If $\phi(x) = c'x$, as in the linear programming case, then (2.15) is equivalent to

$$\dot{x} = (A' \mathrm{diag}(Ax - b)^{-1} A)^{-1} c. \tag{2.16}$$

If we assume further that A is invertible, i.e. that we have exactly n inequality constraints, then (2.16) becomes the *linear* differential equation

$$\dot{x} = A^{-1} \mathrm{diag}(Ax - b)(A')^{-1} c. \tag{2.17}$$

Let us consider next the case of a n-dimensional simplex domain $\mathcal{C} = \{x \in \mathbb{R}^n \mid x \geq 0, \sum_{i=1}^n x_i \leq 1\}$, now regarded as a subset of $\mathbb{R}^n$. Here $m = n + 1$ and

$$A = \begin{bmatrix} I_n \\ -e' \end{bmatrix}, \quad b = \begin{bmatrix} 0 \\ \vdots \\ 0 \\ -1 \end{bmatrix}$$

with $e' = (1, \cdots, 1) \in \mathbb{R}^n$. Thus

$$\mathrm{diag}(Ax - b)^{-1} = \mathrm{diag}(x_1^{-1}, \cdots, x_n^{-1}, (1 - \sum_{i=1}^n x_i)^{-1}).$$

and

$$A' \mathrm{diag}(Ax - b)^{-1} A = \mathrm{diag}(x)^{-1} + (1 - \sum_{1}^n x_i)^{-1} e e'.$$

Using the matrix inversion lemma, see Appendix A, gives

$$\begin{aligned}(A'\operatorname{diag}(Ax-b)^{-1}A)^{-1} &= (D(x)^{-1} + (1-\Sigma x_i)^{-1}ee')^{-1} \\ &= D(x) - D(x)e(e'D(x)e + (1-\Sigma x_i))^{-1}e'D(x) \\ &= (I - D(x)ee')D(x).\end{aligned}$$

Thus the gradient flow (2.17) in this case is

$$\begin{aligned}\dot{x} &= (I - D(x)ee')D(x)c \\ &= (D(c) - xc')x \\ &= (D(c) - c'xI_n)x.\end{aligned}$$

in harmony with Corollary 1.3.

Problems 1. Let $A \in \mathbb{R}^{m \times n}$, rk $A = m < n$, $B \in \mathbb{R}^{m \times n}$ and consider the convex subset of positive semidefinite matrices

$$\mathcal{C} = \{P \in \mathbb{R}^{n \times n} \mid P = P' \geq 0,\ AP = B\}$$

with nonempty interior

$$\overset{\circ}{\mathcal{C}} = \{P \in \mathbb{R}^{n \times n} \mid P = P' > 0,\ AP = B\}.$$

Show that

$$\langle\langle \xi, \eta \rangle\rangle := \operatorname{tr}(P^{-1}\xi P^{-1}\eta),\ \xi, \eta \in T_P(\overset{\circ}{\mathcal{C}}),$$

defines a Riemannian metric on $\overset{\circ}{\mathcal{C}}$.

2. Let $W_1, W_2 \in \mathbb{R}^{n \times n}$ be symmetric matrices. Prove that the gradient flow of the cost function $\phi : \overset{\circ}{\mathcal{C}} \to \mathbb{R}$, $\phi(P) = \operatorname{tr}(W_1 P + W_2 P^{-1})$ with respect to the above Riemannian metric on $\overset{\circ}{\mathcal{C}}$ is

$$\dot{P} = (I_n - PA'(APA')^{-1}A)(PW_1P - W_2).$$

4.3 Recursive Linear Programming/Sorting

Our aim in this section is to obtain a recursive version of Brockett's linear programming/sorting scheme (1.4). In particular, we base our algorithms on the recursive Lie-bracket algorithm of Chapter 2 Section 3, as originally presented in Moore, Mahony and Helmke (1993).

Theorem 1.2 gives an analysis of the Rayleigh quotient gradient flow on the sphere S^{m-1},

$$\dot{\xi} = (N - \xi'N\xi I)\xi,\ \xi(0) = \xi_0 \tag{3.1}$$

4.3. RECURSIVE LINEAR PROGRAMMING/SORTING

and shows how this leads to a solution to the linear programming/sorting problem. As pointed out in the remarks following Theorem 1.2, any solution $\xi(t)$ of (3.1) induces a solution to the isospectral double bracket flow

$$\dot{H} = [H,[H,N]], \quad H(0) = H_0 = \xi_0\xi_0'$$

via the quadratic substitution $H = \xi\xi'$. Similarly, if (ξ_k) is a solution to the recursion

$$\xi_{k+1} = e^{-\alpha[\xi_k\xi_k',N]}\xi_k, \quad \alpha = \frac{1}{4\|N\|}, \quad \xi_0 \in S^{m-1}, \tag{3.2}$$

evolving on S^{m-1}, then $(H_k) = (\xi_k\xi_k')$ is a solution to the recursive Lie-bracket scheme

$$H_{k+1} = e^{-\alpha[H_k,N]}H_k e^{\alpha[H_k,N]}, \quad \alpha = \frac{1}{4\|N\|}, \quad \xi_0 \in S^{m-1}. \tag{3.3}$$

We now study (3.2) as a recursive solution to the linear programming/sorting problem.

Theorem 3.1 Linear Programming/Sorting Algorithm. *Consider the maximization of $c'x$ over the polytope $C(v_1,\cdots,v_m)$, where $x, c, v_i \in \mathbb{R}^n$. Assume the genericity condition $c'v_i \neq c'v_j$ holds for all $i \neq j$. Let $N = $ diag $(c'v_1,\cdots,c'v_m)$. Then:*

(a) *The solutions (ξ_k) of the recursion (3.2) satisfy*

1) $\xi_k \in S^{m-1}$ *for all $k \in \mathbb{N}$.*

2) *There are exactly 2^m fixed points corresponding to $\pm e_1,\ldots,\pm e_m$, the standard basis vectors of $\mathbb{R}^m$.*

3) *All fixed points are unstable, except for $\pm e_{i_*}$ with $c'v_{i_*} = \max_{i=1,\ldots,m}(c'v_i)$, which is exponentially stable.*

4) *The recursive algorithm (3.2) acts to monotonically increase the cost*

$$r_N(\xi_k) = \xi_k' N \xi_k.$$

(b) *Let $p_0 = \sum_{i=1}^m \eta_i v_i \in C$, $\eta_i \geq 0$, $\eta_1 + \ldots + \eta_m = 1$, be arbitrary and define $\zeta_0 = (\sqrt{\eta_1},\ldots,\sqrt{\eta_m})$. Let $\xi_k = (\xi_{k,1},\ldots,\xi_{k,m})$ be given by (3.2). Then the sequence of points*

$$p_k = \sum_{i=1}^m \xi_{k,i} v_i \tag{3.4}$$

converges exponentially fast to the unique optimal vertex of the polytope C for generic initial conditions ξ_0.

Proof The proof of part (a) is an immediate consequence of Theorem 2.3.3, once it is observed that ξ_k is the first column vector of an orthogonal matrix solution $\Theta_k \in O(n)$ to the double bracket recursion

$$\Theta_{k+1} = e^{-\alpha[\Theta_k Q \Theta'_k, N]} \Theta_k.$$

Here $Q = \text{diag}(1, 0, \ldots, 0)$ and $\xi_0 = \Theta_0 e_1$.

Part (b) follows immediately from part (a). ∎

Remarks 1. By choosing $p_0 = \frac{1}{m}\sum_{i=1}^{m} v_i \in C$ as the central point of the polytope and setting $\xi_0 = (\frac{1}{\sqrt{m}}, \ldots, \frac{1}{\sqrt{m}})$, it is guaranteed that the sequence of interior points p_k defined by (3.4) converges to the optimal vertex.

2. Of course, it is not our advise to use the above algorithm as a practical method for solving linear programming problems. In fact, the same comments as for the continuous time double bracket flow (1.4) apply here.

3. It should be noted that a computationally simple form of (3.2) exists which does not require the calculation of the matrix exponential.

$$\begin{aligned} \xi_{k+1} &= \left(\cos(\alpha y_k) - \xi'_k N \xi_k \frac{\sin(\alpha y_k)}{y_k}\right)\xi_k + \frac{\sin(\alpha y_k)}{y_k} N \xi_k \\ y_k &= (x'_k N^2 x_k - (x'_k N x_k)^2)^{\frac{1}{2}} \end{aligned} \quad (3.5)$$

4. Another alternative to the exponential term $e^{-\alpha_k(\xi_k \xi'_k N - N \xi_k \xi'_k)} = e^A$ is a (1,1) Padé approximation $\frac{2I+A}{2I-A}$ which preserves the orthogonal nature of the recursion ξ_k and reduces the computational cost of each step of the algorithm.

5. Although the linear programming/sorting algorithm evolves in the interior of the polytope to search for the optimal vertex, it is not strictly a dynamical system defined on the polytope C, but rather a dynamical system with output variables on C. It is a projection of the recursive Rayleigh quotient algorithm from the sphere S^{m-1} to the polytope C.

It remains as a challenge to find isospectral-like recursive interior point algorithms similar to Karmarkar's algorithm evolving on C. Actually, one interesting case where the desired algorithm is available as a recursive interior point algorithm, is the special case where the polytope is the standard simplex Δ_{m-1}, see Problem 2 at the end of the section.

6. An attraction of this algorithm, is that it can deal with time-varying or noisy data. Consider a process in which the vertices of a linear polytope are given by a sequence of vectors $\{v_i(k)\}$ where $k = 1, 2, \ldots$, and similarly the

4.3. RECURSIVE LINEAR PROGRAMMING/SORTING

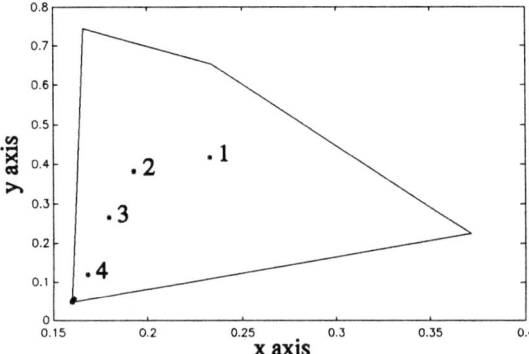

Figure 4.3.1: The recursion occurring inside the polytope $\mathcal{C}$

cost vector $c = c(k)$ is also a function of k. Then the only difference is the change in target matrix at each step as

$$N(k) = \text{diag}(c'(k)v_1(k), \ldots, c'(k)v_m(k)).$$

Now since such a scheme is based on a gradient flow algorithm for which the convergence rate is exponential, it can be expected that for slowly time-varying data, the tracking response of the algorithm should be reasonable, and there should be robustness to noise. ∎

Simulations Two simulations were run to give an indication of the nature of the recursive interior-point linear programming algorithm. To illustrate the phase diagram of the recursion in the polytope $\mathcal{C}$, a linear programming problem simulation with 4 vertices embedded in $\mathbb{R}^2$ is shown in Figure 4.3.1. This figure underlines the interior point nature of the algorithm.

The second simulation in Figure 4.3.2 is for 7 vectors embedded in $\mathbb{R}^7$. This indicates how the algorithm behaves for time-varying data. Here the cost vector $c \in \mathbb{R}^n$ has been chosen as a time varying sequence $c = c(k)$ where $\frac{c(k+1)-c(k)}{c(k)} \sim 0.1$. For each step the optimum direction e_{i_*} is calculated using a standard sorting algorithm and then the norm of the difference between the estimate $||\xi(k) - e_{i_*}||_2$ is plotted. Note that since the cost vector $c(k)$ is changing the optimal vertex may change abruptly while the algorithm is running. In this simulation such a jump occurs at iteration 10. It is interesting to note that for the iterations near to the jump the change in $\xi(k)$ is small indicating that the algorithm slows down when the optimal solution is in doubt.

Problems 1. Verify (3.5).

2. Verify that the Rayleigh quotient algorithm (3.5) on the sphere S^{m-1}

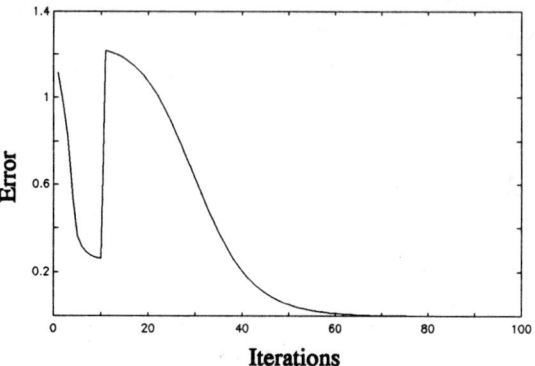

Figure 4.3.2: Response to a time-varying cost function

induces the following interior point algorithm on the simplex Δ_{m-1}. Let $c = (c_1, \ldots, c_m)'$. For any solution $\xi_k = (\xi_{k,1}, \ldots, \xi_{k,m}) \in S^{m-1}$ of (3.5) set $x_k = (x_{k,1}, \ldots, x_{k,m}) := (\xi_{k,1}^2, \ldots, \xi_{k,m}^2) \in \Delta_{m-1}$. Then (x_k) satisfies the recursion

$$x_{k+1} = \left((\cos(\alpha y_k) - c'x_k \frac{\sin(\alpha y_k)}{y_k}) I + \frac{\sin(\alpha y_k)}{y_k} N \right)^2 x_k$$

with $\alpha = \frac{1}{4\|N\|}$ and $y_k = \sqrt{\sum_{i=1}^{m} c_i^2 x_{k,i} - (\sum_{i=1}^{m} c_i x_{k,i})^2}$.

3. Show that every solution (x_k) which starts in the interior of the simplex converges exponentially to the optimal vertex e_{i_*} with $c_{i_*} = \max_{i=1,\ldots,m} c_i$.

Main Points of Section The linear programming/sorting algorithm presented in this section is based on the recursive Lie-bracket scheme in the case where H_0 is a rank one projection operator. There are computationally simpler forms of the algorithm which do not require the calculation of the matrix exponential.

An attraction of the exponentially convergent algorithms, such as is described here, is that they can be modified to be tracking algorithms which can the track time-varying data or filter noisy data. In such a situation the algorithms presented here provide simple computational schemes which will track the optimal solution, and yet be robust.

Notes for Chapter 4

As textbooks on optimization theory and nonlinear programming we mention Luenberger (1969, 1973). Basic monographies on linear programming, the simplex method and combinatorial optimization are Dantzig (1963), Grötschel, Lovász and Schrijver (1988). The book Werner (1992) also contains a discussion of Karmarkar's method.

Sorting algorithms are discussed in Knuth (1973). For a collection of articles on interior point methods we refer to Lagarias and Todd (1990) as well as to the Special Issue of Linear Algebra and its Applications (1991), Volume 151. Convergence properties of interior point flows are investigated by Meggido and Shub (1989). Classical papers on interior point methods are those of Khachian (1979) and Karmarkar (1984). The paper of Karmarkar (1990) contains an interesting interpretation of the (discrete-time) Karmarkar algorithm as the discretization of a continuous time interior point flow. Variable step-size selections are made with respect to the curvature of the interior point flow trajectories. For related work we refer to Sonnevend and Stoer (1990), Sonnevend, Stoer and Zhao (1990, 1991).

The idea of using isospectral matrix flows to solve combinatorial optimization tasks such as sorting and linear programming is due to Brockett (1991); see also Bloch (1990) and Helmke (1993) for additional information. For further applications to combinatorial assignment problems and least squares matching problems see Brockett and Wong (1991), Brockett (1989). Starting from the work of Brockett (1991), a systematic approach to interior point flows for linear programming has been developed in the important pioneering work of Faybusovich.

Connections of linear programming with completely integrable Hamiltonian systems are made in Bayer and Lagarias (1989), Bloch (1990) and Faybusovich (1991 a, b, 1992). For a complete phase portrait analysis of the interior point gradient flow (2.12) we refer to Faybusovich (1991 a, 1992). Quadratic convergence to the optimum via a discretization of (2.12) is established in Herzel et al. (1991).

An important result from linear algebra which is behind Brockett's approach to linear programming is the Schur-Horn theorem. The Schur-Horn theorem states that the set of vectors formed by the diagonal entries of Hermitian matrices with eigenvalues $\lambda_1, \ldots, \lambda_n$ coincides with the convex polytope with vertices $(\lambda_{\pi(1)}, \ldots, \lambda_{\pi(n)})$, where π varies over all $n!$ permutations of $1, \ldots, n$; Schur (1923), Horn (1953). A Lie group theoretic generalization of this result is due to Kostant (1973). For deep connections of such convexity results with symplectic geometry we refer to Atiyah (1982, 1983); see also Byrnes and Willems (1986), Bloch and Byrnes (1986).

There is an interesting connection of interior point flows with population dynamics. In population dynamics interior point flows on the standard simplex naturally arise. The book of Akin (1979) contains a systematic analysis of interior point flows on the standard simplex of interest in population dynamics. Akin refers to the Riemannian metric (2.4) on the interior of the standard simplex as the Shahshahani metric. See also the closely related work of Smale (1976), Schuster, Sigmund and Wolff (1978) and Zeeman (1980) on Volterra-Lotka equations for predator-prey models in population dynamics. For background material on the Volterra-Lotka equation see Hirsch and Smale (1974).

Highly interconnected networks of coupled nonlinear artificial neurons can have remarkable computational abilities. For exciting connections of optimization with artificial neural networks we refer to Pyne (1956), Chua and Lin (1984), Hopfield and Tank (1985), Tank and Hopfield (1985). Hopfield (1982, 1984) has shown that certain well behaved differential equations on a multidimensional cube may serve as a model for associative memories. In the pioneering work Hopfield and Tank (1985) show that such dynamical systems are capable of solving complex optimization problems such as the Travelling Salesman Problem. See also Peterson and Soederberg (1989) for related work. For an analysis of the Hopfield model we refer to Aiyer et al. (1990). For further work we refer to Yuille (1990) and the references therein. An open problem for future research is to find possible connections between the interior point flows, as described in this chapter, and Hopfield type neural networks for linear programming.

References

Aiyer, S.V.B., Niranjan, M. and F. Fallside (1990) A theoretical investigation into the performance of the Hopfield model, *IEEE Transactions on Neural Networks* 1, 204-215.

Akin, E. (1979) *The geometry of population genetics*, Lecture Notes in Biomathematics **31**, Springer Verlag, Berlin.

Atiyah, M.F. (1982) Convexity and commuting Hamiltonians, *Bull. London Math Soc.*, 16, 1-15.

Atiyah, M.F. (1983) Angular momentum, convex polyhedra and algebraic geometry, *Proceedings of the Edinburgh Math. Soc.*, 26, 121-138.

Bayer, D.A. and J.C.Lagarias (1989) The nonlinear geometry of linear programming I, II, *Trans. Amer. Math. Soc.*, 314, 499-580.

Bloch, A.M. (1986) An infinite-dimensional variational problem arising in estimation theory, *Algebraic and Geometric Methods in Nonlinear Control Theory* (M. Fliess and M. Hazewinkel, Eds.), Reidel Publ.

Bloch, A.M. (1990) Steepest descent, linear programming and Hamiltonian flows, *Contemporary Math.* 114, 77-88.

Brockett, R.W. (1989) Least squares matching problems, *Linear Algebra Appl.*, 122-124, 761-777.

Brockett, R.W. (1991) Dynamical systems that sort lists, diagonalize matrices and solve linear programming problems, *Linear Algebra Appl.*, 146, 79-91.

Brockett, R.W. and W.S.Wong (1991) A gradient flow for the assignment problem, *New Trends in System Theory*, Progress in System and Control Theory, (G. Conte, A.M. Perdon and B. Wyman, Eds.), Birkhäuser, 170-177.

Brockett, R.W. and A.M.Bloch (1989) Sorting with the dispersionless limit of the Toda Lattice, in *Hamiltonian Systems, Transformation Groups and Spectral Transform Methods*, CRM, (J. Harnad and J.E. Marsden, Eds.) Université de Montréal, Montréal, Canada, 103-112.

Byrnes, C.I. and J.C.Willems (1986) Least-squares estimation, linear programming and momentum: A geometric parameterization of local minima, *IMA J. Math. Control and Info*, 3, 103-118.

Chua, L.O. and G.N.Lin (1984) Nonlinear programming without computation, *IEEE Trans. Circuits and Systems*, 31, 182-188.

Dantzig, G.B. (1963) *Linear Programming and Extensions*, Princeton University Press, Princeton, N.J.

Deift, P., T.Nanda and C.Tomei (1983) Ordinary differential equations for the symmetric eigenvalue problem, *SIAM J. Numer. Anal.*, 20, 1-22.

Dikin, I.I. (1967) Iterative solutions of problems of linear and quadratic programming, *Soviet Math. Dokl.*, 8, 674-675.

Faybusovich, L.E. (1991a) Dynamical systems which solve optimization problems with linear constraints, *IMA J. Math. Control and Info.*, 8, 135-149.

Faybusovich, L.E. (1991b) Hamiltonian structure of dynamical systems which solve linear programming problems, *Physica*, D 53, 217-232.

Faybusovich, L.E. (1991c) Interior print methods and entropy, *Proc of IEEE Conf. on Decision and Control*, Brighton, UK, 2094-2095.

Faybusovich, L.E. (1992) Dynamical systems that solve linear programming problems, *Proc. of IEEE Conference on Decision and Control*, Tuscon, Arizona, 1626-1631.

Grötschel, M., L. Lovász and A. Schrijver (1988) *Geometric algorithms and combinatorial optimization*, Springer-Verlag, Berlin.

Helmke, U. (1993) Isospectral flows and linear programming, *Journal of Australian Mathematical Soc.*, Series B, 34, 495-510.

Helmke, U. (1991) Isospectral flows on symmetric matrices and the Riccati equation, *Systems and Control Letters*, 16, 159-166.

Herzel, S., M.C.Recchioni and F.Zirilli (1991) A quadratically convergent method for linear programming, *Linear Algebra Appl.*, 151, 255-290.

Hirsch, M.W. and S. Smale (1974) *Differential Equations, Dynamical Systems, and Linear Algebra*, Academic Press, New York

Hopfield, J.J. (1982) Neural networks and physical systems with emergent collective computational abilities, *Proc. Nat. Acad. Sci. USA* 79, 2554.

Hopfield, J.J. (1984) Neurons with graded response have collective computational properties like those of two-state neurons, *Proc. National Academy of Sciences USA* 81, 3088-3092.

Hopfield, J.J. and D.W. Tank (1985) Neural computation of decisions in optimization problems, *Biolog. Cybern.* 52, 1-25.

Horn, R. (1953) Doubly stochastic matrices and the diagonal of a rotation matrix, *American Journal of Mathematics*, 76, 620-630.

Karmarkar, N. (1984) A new polynomial time algorithm for linear programming, *Combinatorica*, 4, 373-395.

Karmarkar, N. (1990) Riemannian geometry underlying interior point methods for linear programming, *Contemporary Mathematics* 114, AMS, 51-75.

Khachian, L.G. (1979) A polynomial algorithm in linear programming, *Doklady Akademit Nauk SSSR* 244S, 1093-1096, translated in *Soviet Mathematics Doklady* (1979) 201, 191-194.

Knuth, D.E. (1973) *The art of computer programming, Vol. 3: Sorting and Searching*, Addison-Wesley, Reading.

Kostant, (1973) On convexity, the Weyl Group and the Iwasawa decomposition, *Ann. Sci. Ecole Norm. Sup.*, 6, 413-455.

Lagarias, J. and M.J. Todd (Eds.) (1990) *Mathematical Development Arising from Linear Programming*, Providence, R.I., Contemporary Math. 114, AMS.

Luenberger, D.G. (1969) *Optimization by Vector Space Methods*, John Wiley, New York.

Luenberger, D.G. (1973) *Introduction to linear and nonlinear programming*, Addison-Wesley Publ. Co., Reading.

Mahony, R.E. and J.B.Moore (1992) Recursive interior-point linear programming based on Lie-Brockett flows, *Proc. of Int. Conf. on Optimization Techniques and Applications*, Singapore.

Meggido, N. and M. Shub (1989) Boundary behaviour of interior-point algorithms in linear programming, *Math. Oper. Research* 14, 97-146.

Moore, J.B., R.E. Mahony and U. Helmke (1993) Numerical gradient algorithms for eigenvalue and singular value decomposition, *SIAM J. of Matrix Analysis*, to appear.

Oja, E. (1982) A simplified neuron model as a principal component analyzer, *J. of Math. Biology*, 15, 267-273.

Peterson C. and B. Soederberg (1989) A new method for mapping optimization problems onto neural networks, *International Journal of Neural Systems* 1, 3-22.

Pyne, I.B. (1956) Linear programming on an analogue computer, *Trans. AIEE*, Part I, 75, 139-143.

Schur, I (1923) Über eine Klasse von Mittelbildungen mit Anwendungen auf die Determinanten Theorie, *Sitzungsber. der Berliner Math. Gesellschaft.* 22, 9-20.

Schuster, P., K.Sigmund and R. Wolff (1978) Dynamical systems under constant organisation I: Topological analysis of a family of nonlinear differential equations, *Bull. Math. Biophys.*, 40, 743-769.

Smale, S. (1976) On the differential equations of species in competition, *Journal of Mathematical Biology* 3, 5-7.

Sonnevend, G. and J. Stoer (1990) Global ellipsoidal approximations and homotopy methods for solving convex programs, *Applied Mathematics and Optimization* 21, 139-165.

Sonnevend, G., Stoer, J. and G. Zhao (1990) On the complexity of following the central path of linear programs by linear extrapolation, *Methods of Operations Research* 62, 19-31.

Sonnevend, G., Stoer, J. and G. Zhao (1991) On the complexity of following the central path of linear programs by linear extrapolation II, *Mathematical Programming* (Series B).

Stone, R.E. and C.A.Tovey (1991) The simplex and projective scaling algorithms as iteratively reweighted least squares methods, *SIAM Review* 33, 220-237.

Tank, D.W. and J.J.Hopfield (1985) Simple 'neural' optimization networks: an A/D converter, signal decision circuit and a linear programming circuit, *IEEE Transactions, Circuits and Systems*, CAS-33, 533-541, preprint.

Werner, J. (1992) *Numerische Mathematik 2*, Vieweg Verlag, Braunschweig.

Yuille, A.L. (1990) Generalized deformable models, statistical physics and matching problems, *Neural Computation* 2, 1-24.

Zeeman, E.C. (1980) Population dynamics from game theory, in *Global Theory of Dynamical Systems*, (Z. Nitecki and C. Robinson, Eds.), Lecture Notes in Mathematics 819, Springer Verlag Berlin, 471-497.

Chapter 5

Approximation and Control

5.1 Approximations by Lower Rank Matrices

In this chapter, we analyse three further matrix least squares estimation problems. The first is concerned with the task of approximating matrices by lower rank ones. This has immediate relevance to total least squares estimation and representations by linear neural networks. The section also serves as a prototype for linear system approximation. In the second, short section we introduce dynamical systems achieving the polar decomposition of a matrix. This has some contact with Brockett's (1989) investigation on matching problems of interest in computer vision and image detection. Finally the last section deals with inverse eigenvalue problems arising in control theory. This section also demonstrates the potential of the dynamical systems approach to optimize feedback controllers.

In this section, we study the problem of approximating finite-dimensional linear operators by lower rank linear operators. A classical result from matrix analysis, the Eckart-Young-Mirsky Theorem, see Corollary 1.10, states that the best approximation of a given $M \times N$ matrix A by matrices of smaller rank is given by a truncated singular value decomposition of A. Since the set $M(n, M \times N)$ of real $M \times N$ matrices X of fixed rank n is a smooth manifold, see Proposition 1.8, we thus have an optimization problem for the smooth Frobenius norm distance function

$$f_A : M(n, M \times N) \to \mathbb{R}, \quad f_A(X) = ||A - X||^2.$$

This problem is equivalent to the total linear least squares problem, see Golub and van Loan (1980). The fact that we have here assumed that the rank of X is precisely n instead of being less than or equal to n is no restriction in generality, as we will later show that any best approximant $\widehat{X}$ of A of rank $\leq n$ will automatically satisfy rank $\widehat{X} = n$.

The critical points and, in particular, the global minima of the distance function $f_A(X) = ||A - X||^2$ on manifolds of fixed rank symmetric and rectangular matrices are investigated. Also, gradient flows related to the minimization of the constrained distance function $f_A(X)$ are studied. Similar results are presented for symmetric matrices. For technical reasons it is convenient to study the symmetric matrix case first and then to deduce the corresponding results for rectangular matrices. This work is based on Helmke and Shayman (1993) and Helmke, Prechtel and Shayman (1993).

Approximations by Symmetric Matrices. Let $S(N)$ denote the set of all $N \times N$ real symmetric matrices. For integers $1 \leq n \leq N$ let

$$S(n, N) = \{X \in \mathbb{R}^{N \times N} | \ X' = X, \text{rank} X = n\} \tag{1.1}$$

denote the set of real symmetric $N \times N$ matrices of rank n. Given a fixed real symmetric $N \times N$ matrix A we consider the distance function

$$f_A : S(n, N) \to \mathbb{R} \ , \ X \mapsto ||A - X||^2 \tag{1.2}$$

where $||X||^2 = \text{tr}(XX')$ is the Frobenius norm. We are interested in finding the critical points and local and global minima of f_A, i.e. the best rank n symmetric approximants of A. The following result summarizes some basic geometric properties of the set $S(n, N)$.

Proposition 1.1 *(a) $S(n, N)$ is a smooth manifold of dimension $\frac{1}{2}n(2N - n + 1)$ and has $n + 1$ connected components*

$$S(p, q; N) = \{X \in S(n, N) | \ \text{sig} \ X = p - q\} \tag{1.3}$$

where $p, q \geq 0, p + q = n$ and sig denotes signature. The tangent space of $S(n, N)$ at an element X is

$$T_X S(n, N) = \{\Delta X + X \Delta' | \ \Delta \in \mathbb{R}^{N \times N}\} \tag{1.4}$$

(b) The topological closure of $S(p, q; N)$ in $\mathbb{R}^{N \times N}$ is

$$\overline{S(p, q; N)} = \bigcup_{p' \leq p, q' \leq q} S(p', q'; N) \tag{1.5}$$

and the topological closure of $S(n, N)$ in $\mathbb{R}^{N \times N}$ is

$$\overline{S(n, N)} = \bigcup_{i=0}^{n} S(i, N) \tag{1.6}$$

Proof The function which associates to every $X \in S(n, N)$ its signature is locally constant. Thus $S(p, q; N)$ is open in $S(n, N)$ with $S(n, N) =$

5.1. APPROXIMATIONS BY LOWER RANK MATRICES

$\bigcup_{p+q=n,\, p,q\geq 0} S(p,q;N)$. It follows that $S(n,N)$ has at least $n+1$ connected components.

Any element $X \in S(p,q;N)$ is of the form $X = \gamma Q \gamma'$ where $\gamma \in GL(N, \mathbb{R})$ is invertible and $Q = \text{diag}(I_p, -I_q, 0) \in S(p,q;N)$. Thus $S(p,q;N)$ is an orbit of the congruence group action $(\gamma, X) \mapsto \gamma X \gamma'$ of $GL(N, \mathbb{R})$ on $S(n,N)$. It follows that $S(p,q;N)$ with $p,q \geq 0$, $p+q = n$, is a smooth manifold and therefore also $S(n,N)$ is. Let $GL^+(N, \mathbb{R})$ denote the set of invertible $N \times N$ matrices γ with $\det \gamma > 0$. Then $GL^+(N, \mathbb{R})$ is a connected subset of $GL(N, \mathbb{R})$ and $S(p,q;N) = \{\gamma Q \gamma' \mid \gamma \in GL^+(N, \mathbb{R})\}$. Consider the smooth surjective map

$$\eta : GL(N, \mathbb{R}) \to S(p,q;N), \quad \eta(\gamma) = \gamma Q \gamma'. \tag{1.7}$$

Then $S(p,q;N)$ is the image of the connected set $GL^+(N, \mathbb{R})$ under the continuous map γ and therefore $S(p,q;N)$ is connected. This completes the proof that the sets $S(p,q;N)$ are precisely the $n+1$ connected components of $S(n,N)$. Furthermore, the derivative of η at $\gamma \in GL(N, \mathbb{R})$ is the linear map $D\eta(\gamma) : T_\gamma GL(N, \mathbb{R}) \to T_X S(p,q;N), X = \gamma Q \gamma'$, defined by $D\eta(\gamma)(\triangle) = \triangle X + X \triangle'$ and maps $T_\gamma GL(N, \mathbb{R}) \cong \mathbb{R}^{N \times N}$ surjectively onto $T_X S(p,q;N)$. Since X and $p,q \geq 0$, $p+q = n$, are arbitrary this proves (1.4). Let

$$\triangle = \begin{bmatrix} \triangle_{11} & \triangle_{12} & \triangle_{13} \\ \triangle_{21} & \triangle_{22} & \triangle_{23} \\ \triangle_{31} & \triangle_{32} & \triangle_{33} \end{bmatrix}$$

be partitioned according to the partition $(p,q,N-n)$ of N. Then $\triangle \in \ker D\eta(I)$ if and only if $\triangle_{13} = 0, \triangle_{23} = 0$ and the $n \times n$ submatrix

$$\begin{bmatrix} \triangle_{11} & \triangle_{12} \\ \triangle_{21} & \triangle_{22} \end{bmatrix}$$

is skew-symmetric. A simple dimension count thus yields $\dim \ker D\eta(I) = \frac{1}{2}n(n-1) + N(N-n)$ and therefore

$$\dim S(p,q;N) = N^2 - \dim \ker D\eta(I) = \frac{1}{2}n(2N - n + 1).$$

This completes the proof of (a). The proof of (b) is left as an exercise to the reader. ∎

Theorem 1.2 (a) *Let $A \in \mathbb{R}^{n \times n}$ be symmetric and let*

$$N_+ = \dim \, Eig^+(A) \,, \quad N_- = \dim \, Eig^-(A) \tag{1.8}$$

be the numbers of positive and negative eigenvalues of A, respectively. The critical points X of the distance function $f_A : S(n,N) \to \mathbb{R}$ are characterized by $AX = XA = X^2$.

(b) *If A has N distinct eigenvalues $\lambda_1 > \cdots > \lambda_N$, and $A = \Theta\mathrm{diag}(\lambda_1, \cdots, \lambda_N)\Theta'$ for $\Theta \in O(N)$, then the restriction of the distance function $f_A : S(p,q;N) \to \mathbb{R}$ has exactly*

$$\binom{N_+}{p}\cdot\binom{N_-}{q} = \frac{N_+!N_-!}{p!(N_+-p)!q!(N_--q)!}$$

critical points. In particular, f_A has critical points in $S(p,q;N)$ if and only if $p \leq N_+$ and $q \leq N_-$. The critical points $X \in S(p,q;N)$ of f_A with $p \leq N_+, q \leq N_-$, are characterized by

$$X = \Theta\mathrm{diag}(x_1,\cdots,x_N)\Theta' \tag{1.9}$$

with

$$x_i = 0 \text{ or } x_i = \lambda_i \ , \ i = 1,\cdots,N \tag{1.10}$$

and exactly p of the x_i are positive and q are negative.

Proof Without loss of generality, we may assume that $A = \mathrm{diag}(\lambda_1, \cdots, \lambda_N)$ with $\lambda_1 \geq \cdots \geq \lambda_N$. A straightforward computation shows that the derivative of $f_A : S(n,N) \to \mathbb{R}$ at X is the linear map $Df_A(X) : T_X S(n,N) \to \mathbb{R}$ defined by

$$\begin{aligned}Df_A(X)(\Delta X + X\Delta') &= -2\mathrm{tr}((A-X)(\Delta X + X\Delta')) \\ &= -4\mathrm{tr}(X(A-X)\Delta) \end{aligned} \tag{1.11}$$

for all $\Delta \in \mathbb{R}^{N \times N}$. Therefore $X \in S(n,N)$ is a critical point of f_A if and only if $X^2 = XA$. By symmetry of X then $XA = AX$. This proves (a). Now assume that $\lambda_1 > \cdots > \lambda_N$. From $X^2 = AX = XA$, and since A has distinct eigenvalues, X must be diagonal. Thus the critical points of f_A are the diagonal matrices $X = \mathrm{diag}(x_1,\cdots,x_N)$ with $(A-X)X = 0$ and therefore $x_i = 0$ or $x_i = \lambda_i$ for $i = 1,\cdots,N$. Also, $X \in S(p,q;N)$ if and only if exactly p of the x_i are positive and q are negative. Consequently, using the standard symbol for the binomial coefficient, there are

$$\binom{N_+}{p}\cdot\binom{N_-}{q}$$

critical points of f_A in $S(p,q;N)$, characterized by (1.9). This completes the proof. ∎

Example Let $N = 2$, $n = 1$ and $A = \begin{bmatrix} a & b \\ b & c \end{bmatrix}$. The variety of rank ≤ 1 real symmetric 2×2 matrices is a cone $\{X = \begin{bmatrix} x & y \\ y & z \end{bmatrix} \mid xz - y^2\}$ depicted in Figure 5.1.1.

5.1. APPROXIMATIONS BY LOWER RANK MATRICES

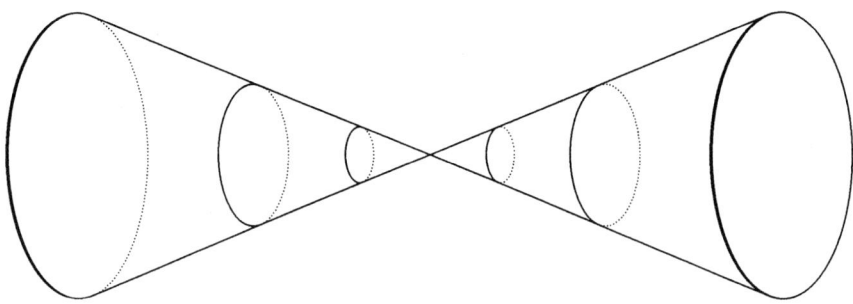

Figure 5.1.1: Real symmetric matrices of rank ≤ 1

The function $f_A : S(1,2) \to \mathbb{R}$ has generically 2 local minimum, if $\text{sig}(A) = 0$, and 1 local = global minimum, if $A > 0$ or $A < 0$. No other critical points exist. ∎

Theorem 1.2 has the following immediate consequence.

Corollary 1.3 *Let* $A = \Theta \text{diag}(\lambda_1, \cdots, \lambda_N) \Theta'$ *with* $\Theta \in O(N)$ *a real orthogonal* $N \times N$ *matrix and* $\lambda_1 \geq \cdots \geq \lambda_N$. *A minimum* $\hat{X} \in S(p,q;N)$ *for* $f_A : S(p,q;N) \to \mathbb{R}$ *exists if and only if* $p \leq N_+$ *and* $q \leq N_-$. *One such minimizing* $\hat{X} \in S(p,q;N)$ *is given by*

$$\hat{X} = \Theta \text{diag}(\lambda_1, \cdots, \lambda_p, 0, \cdots, 0, \lambda_{N-q+1}, \cdots, \lambda_N)\Theta' \quad (1.12)$$

and the minimum value of $f_A : S(p,q;N) \to \mathbb{R}$ *is* $\sum_{i=p+1}^{N-q} \lambda_i^2$. $\hat{X} \in S(p,q;N)$ *given by (1.12) is the unique minimum of* $f_A : S(p,q;N) \to \mathbb{R}$ *if* $\lambda_p > \lambda_{p+1}$ *and* $\lambda_{N-q} > \lambda_{N-q+1}$.

The next result shows that a best symmetric approximant of A of rank $\leq n$ with $n < \text{rank} A$ necessarily has rank n. Thus, for $n < \text{rank} A$, the global minimum of the function $f_A : \overline{S(n,N)} \to \mathbb{R}$ is always an element of $S(n,N)$. Recall that the singular values of $A \in \mathbb{R}^{N \times N}$ are the nonnegative square roots of the eigenvalues of AA'.

Proposition 1.4 *Let* $A \in \mathbb{R}^{N \times N}$ *be symmetric with singular values* $\sigma_1 \geq \cdots \geq \sigma_N$ *and let* $n \in \mathbb{N}$ *be any integer with* $n < \text{rank } A$. *There exists* $\hat{X} \in S(n,N)$ *which minimizes* $f_A : S(n,N) \to \mathbb{R}$. *Moreover, if* $\hat{X} \in \mathbb{R}^{N \times N}$ *is any symmetric matrix which satisfies*

$$\begin{aligned} \text{rank } \hat{X} &\leq n \\ \|A - \hat{X}\| &= \inf\{\|A - X\| \mid X \in S(i,N), 0 \leq i \leq n\} \end{aligned} \quad (1.13)$$

then necessarily rank $\hat{X} = n$. In particular, the minimum value

$$\mu_n(A) = \min\{f_A(X) \mid X \in S(n, N)\} \tag{1.14}$$

coincides with the minimum value $\min\{f_A(X) \mid X \in S(i, N), 0 \leq i \leq n\}$. One has

$$\|A\|^2 = \mu_0(A) > \mu_1(A) > \cdots > \mu_r(A) = 0, \quad r = \mathrm{rank} A, \tag{1.15}$$

and for $0 \leq n \leq r$

$$\mu_n(A) = \sum_{i=n+1}^{N} \sigma_i^2. \tag{1.16}$$

Proof By Proposition 1.1

$$\overline{S(n, N)} = \bigcup_{i=0}^{n} S(i, N)$$

is a closed subset of $\mathbb{R}^{N \times N}$ and therefore $f_A : \overline{S(n, N)} \to \mathbb{R}$ is proper. Since any proper continuous function $f : X \to Y$ has a closed image $f(X)$ it follows that a minimizing $\hat{X} \in \overline{S(n, N)}$ for $f_A : \overline{S(n, N)} \to \mathbb{R}$ exists:

$$\|A - \hat{X}\|^2 = \min\{f_A(X) | X \in \overline{S(n, N)}\}$$

Suppose $\mathrm{rk}(\hat{X}) < n$. Then for $\varepsilon \in \mathbb{R}$ and $b \in \mathbb{R}^N, \|b\| = 1$, arbitrary we have $\mathrm{rk}(\hat{X} + \varepsilon bb') \leq n$ and thus

$$\begin{aligned} \|A - \hat{X} - \varepsilon bb'\|^2 &= \|A - \hat{X}\|^2 - 2\varepsilon \mathrm{tr}((A - \hat{X})bb') + \varepsilon^2 \\ &\geq \|A - \hat{X}\|^2. \end{aligned}$$

Thus for all $\varepsilon \in \mathbb{R}$ and $b \in \mathbb{R}^N, \|b\| = 1$,

$$\varepsilon^2 \geq 2\varepsilon \mathrm{tr}((A - \hat{X})bb').$$

This shows $\mathrm{tr}((A - \hat{X})bb') = 0$ for all $b \in \mathbb{R}^N$. Hence $A = \hat{X}$, contrary to assumption. Therefore $\hat{X} \in S(n, N)$ and $\mu_0(A) > \cdots > \mu_r(A) = 0$ for $r = \mathrm{rk}(A)$. Thus (1.16) follows immediately from Corollary 1.3. ∎

We now show that the best symmetric approximant of a symmetric matrix $A \in \mathbb{R}^{N \times N}$ in the Frobenius norm is in general uniquely determined.

Theorem 1.5 Let $A = \Theta \mathrm{diag}(\lambda_1, \cdots, \lambda_N)\Theta'$ with $\Theta \in O(N)$ real orthogonal $N \times N$ matrix and $\lambda_1^2 \geq \cdots \geq \lambda_n^2 > \lambda_{n+1}^2 \geq \cdots \geq \lambda_N^2$. Then

$$\hat{X} = \Theta \mathrm{diag}(\lambda_1, \cdots, \lambda_n, 0, \cdots, 0)\Theta' \in S(n, N) \tag{1.17}$$

is the uniquely determined best symmetric approximant of A of rank $\leq n$.

5.1. APPROXIMATIONS BY LOWER RANK MATRICES

Proof We have
$$||A - \hat{X}||^2 = \sum_{i=n+1}^{N} \lambda_i^2 = \mu_n(A)$$

by (1.16), and thus $\hat{X} \in S(n, N)$ is a global minimum of $f_A : S(n, N) \to \mathbb{R}$. By Proposition 1.4 every symmetric best approximant of A of rank $\leq n$ has rank n. Let $X_0 \in S(n, N)$ denote any minimum of $f_A : S(n, N) \to \mathbb{R}$. Then X_0 is a critical point of f_A and therefore, by Theorem 1.2, is of the form

$$X_0 = \Theta \text{diag}(x_1, \cdots, x_n)\Theta'$$

with $x_i = \lambda_i$ for indices i satisfying $1 \leq i_1 < \cdots < i_n \leq N$ and $x_i = 0$ otherwise. For $I = \{i_1, \cdots, i_n\}$ the minimal distance $||A - X_0||^2 = \sum_{i \notin I} \lambda_i^2$ coincides with $\mu_n(A) = \sum_{i=n+1}^{N} \lambda_i^2$ if and only if $I = \{1, \cdots, n\}$, i.e. if and only if $X_0 = \Theta \text{diag}(\lambda_1, \cdots, \lambda_n, 0, \cdots, 0)\Theta'$. ∎

An important problem in linear algebra is that of finding the best positive semidefinite symmetric approximant of a given symmetric $N \times N$ matrix A. By Corollary 1.3 we have

Corollary 1.6 *Let $A = \Theta \text{diag}(\lambda_1, \cdots, \lambda_N)\Theta'$ with $\Theta \in O(N)$ real orthogonal and $\lambda_1 \geq \cdots \geq \lambda_n > \lambda_{n+1} \geq \cdots \geq \lambda_N, \lambda_n > 0$. Then*

$$\hat{X} = \Theta \text{diag}(\lambda_1, \cdots, \lambda_n, 0, \cdots, 0)\Theta' \in S(n, N) \quad (1.18)$$

is the unique positive semidefinite symmetric best approximant of A of rank $\leq n$.

In particular
$$\hat{X} = \Theta \text{diag}(\lambda_1, \cdots, \lambda_{N_+}, 0, \cdots, 0)\Theta' \quad (1.19)$$

is the uniquely determined best approximant of A in the class of positive semidefinite symmetric matrices. This implies the following result due to Higham (1988).

Corollary 1.7 *Let $A \in \mathbb{R}^{N \times N}$ and let $B = \frac{1}{2}(A + A')$ be the symmetric part of A. Let $B = UH$ be the polar decomposition ($UU' = I_N$, $H = H' \geq 0$). Then $\hat{X} = \frac{1}{2}(B + H)$ is the unique positive semidefinite approximant of A in the Frobenius norm.*

Proof For any symmetric matrix X we have $||A - X||^2 = ||B - X||^2 + \frac{1}{2}(||A||^2 - \text{tr}(A^2))$. Let $B = \Theta \text{diag}(\lambda_1, \cdots, \lambda_N)\Theta'$ and let $B = UH$ be the polar decomposition of B. Then

$$\text{diag}(\lambda_1, \cdots, \lambda_N) = (\Theta'U\Theta) \cdot (\Theta'H\Theta)$$

is a polar decomposition of diag $(\lambda_1, \cdots, \lambda_N)$ and thus

$$H = \Theta \text{diag}(|\lambda_1|, \cdots, |\lambda_N|)\Theta'$$

and

$$U = \Theta \text{diag}(I_{N_+}, -I_{N-N_+})\Theta',$$

where N_+ is the number of positive eigenvalues of $\frac{1}{2}(A + A')$. Note that $H = (B^2)^{\frac{1}{2}}$ is uniquely determined. Therefore, by Corollary 1.6

$$\frac{1}{2}(B + H) = \frac{1}{2}(U + I)H = \Theta \text{diag}(I_{N_+}, 0)\Theta' H \quad (1.20)$$
$$= \Theta \text{diag}(\lambda_1, \cdots, \lambda_{N_+}, 0, \cdots, 0)\Theta' \quad (1.21)$$

is the uniquely determined best approximant of B in the class of positive semidefinite symmetric matrices. The result follows. ∎

Approximations by Rectangular Matrices Here we address the related classical issue of approximating a real rectangular matrix by matrices of lower rank.

For integers $1 \leq n \leq \min(M, N)$ let

$$M(n, M \times N) = \{X \in \mathbb{R}^{M \times N} | \text{rank} X = n\} \quad (1.22)$$

denote the set of real $M \times N$ matrices of rank n. Given $A \in \mathbb{R}^{M \times N}$ the approximation task is to find the minimum and, more generally, the critical points of the distance function

$$F_A : M(n, M \times N) \to \mathbb{R}, \quad F_A(X) = ||A - X||^2 \quad (1.23)$$

for the Frobenius norm $||X||^2 = \text{tr}(XX')$ on $\mathbb{R}^{M \times N}$.

Proposition 1.8 $M(n, M \times N)$ *is a smooth and connected manifold of dimension* $n(M + N - n)$, *if* $\max(M, N) > 1$. *The tangent space of* $M(n, M \times N)$ *at an element X is*

$$T_X M(n, M \times N) = \{\Delta_1 X + X \Delta_2 \mid \Delta_1 \in \mathbb{R}^{M \times M}, \Delta_2 \in \mathbb{R}^{N \times N}\}. \quad (1.24)$$

Proof Let $Q \in M(n, M \times N)$ be defined by $Q = \begin{bmatrix} I_n & 0 \\ 0 & 0 \end{bmatrix}$ Since every $X \in M(n, M \times N)$ is congruent to Q by the congruence action $((\gamma_1, \gamma_2), X) \mapsto \gamma_1 X \gamma_2^{-1}, \gamma_1 \in GL(M, \mathbb{R}), \gamma_2 \in GL(N, \mathbb{R})$, the set $M(n, M \times N)$ is an orbit of this smooth real algebraic Lie group action of $GL(M) \times GL(N)$ on $\mathbb{R}^{M \times N}$ and therefore a smooth manifold; see Appendix C. Here γ_1, γ_2 may be chosen to have positive determinant. Thus $M(n, M \times N)$ is the image of the connected subset $GL^+(M) \times GL^+(N)$ of the continuous (and in fact smooth) map $\pi : GL(M) \times GL(N) \to \mathbb{R}^{M \times N}$, $\pi(\gamma_1, \gamma_2) = \gamma_1 Q \gamma_2^{-1}$, and hence is also connected. The derivative of π at (γ_1, γ_2) is the linear map on the tangent space $T_{(\gamma_1, \gamma_2)}(GL(M) \times GL(N)) \cong \mathbb{R}^{M \times M} \times \mathbb{R}^{N \times N}$ defined by

$$D\pi(\gamma_1, \gamma_2)((\Delta_1, \Delta_2)) = \Delta_1(\gamma_1 Q \gamma_2^{-1}) - (\gamma_1 Q \gamma_2^{-1})\Delta_2$$

and $T_X M(n, M \times N)$ is the image of this map. Finally, the dimension result follows from a simple parameter counting. In fact, from the Schur complement formula, see Horn and Johnson (1990), any $M \times N$ matrix of rank n

$$X = \begin{bmatrix} X_{11} & X_{12} \\ X_{21} & X_{22} \end{bmatrix} = \begin{bmatrix} X_{11} & 0 \\ X_{21} & I \end{bmatrix} \begin{bmatrix} I & X_{11}^{-1} X_{12} \\ 0 & X_{22} - X_{21} X_{11}^{-1} X_{12} \end{bmatrix}$$

with $X_{11} \in \mathbb{R}^{n \times n}$ invertible, $X_{21} \in \mathbb{R}^{(M-n) \times n}$, $X_{12} \in \mathbb{R}^{n \times (N-n)}$ satisfies $X_{22} = X_{21} X_{11}^{-1} X_{12}$ and thus depends on $n^2 + n(M + N - 2n) = n(M + N - n)$ independent parameters. This completes the proof. ∎

The general approximation problem for rectangular matrices can be reduced to the approximation problem for symmetric matrices, using the same symmetrisation trick as in Chapter 3. To this end, define for $A, X \in \mathbb{R}^{M \times N}$

$$\hat{A} = \begin{bmatrix} 0 & A \\ A' & 0 \end{bmatrix}, \quad \hat{X} = \begin{bmatrix} 0 & X \\ X' & 0 \end{bmatrix}. \tag{1.25}$$

Thus $\hat{A}$ and $\hat{X}$ are $(M+N) \times (M+N)$ symmetric matrices. If $A \in \mathbb{R}^{M \times N}$ has singular values $\sigma_1, \cdots, \sigma_k$, $k = \min(M, N)$, then the eigenvalues of $\hat{A}$ are $\pm \sigma_1, \cdots, \pm \sigma_k$, and possibly 0. By (1.25) we have a smooth injective imbedding

$$M(n, M \times N) \to S(2n, M+N), \quad X \mapsto \hat{X} \tag{1.26}$$

with $2\|A - X\|^2 = \|\hat{A} - \hat{X}\|^2$. It is easy to check that, for $X \in M(n, M \times N)$, $\hat{X}$ is a critical point (or a minimum) for $f_{\hat{A}} : S(2n, M+N) \to \mathbb{R}$ if and only if X is a critical point (or a minimum) for $F_A : M(n, M \times N) \to \mathbb{R}$. Thus the results for the symmetric case all carry over to results on the function $F_A : M(n, M \times N) \to \mathbb{R}$, and the next result follows. Recall, see Chapter 3, that an $M \times N$ matrix A has a singular value decomposition

$$A = \Theta_1 \Sigma \Theta_2, \quad \text{and} \quad \Sigma = \begin{bmatrix} \sigma_1 & & & & 0 \\ & \ddots & & & \vdots \\ & & \sigma_k & 0 \\ 0 & \cdots & & 0 & 0 \end{bmatrix} \in \mathbb{R}^{M \times N}$$

where Θ_1, Θ_2 are orthogonal matrices.

Let $\Sigma_n, n \leq k$, be obtained from Σ by setting $\sigma_{n+1}, \cdots, \sigma_k$ equal to zero.

Theorem 1.9 Let $A = \Theta_1 \Sigma \Theta_2$ be the singular value decomposition of $A \in \mathbb{R}^{M \times N}$ with singular values $\sigma_1 \geq \cdots \geq \sigma_k > 0$, $1 \leq k \leq \min(M, N)$.

(a) The critical points of $F_A : M(n, M \times N) \to \mathbb{R}$, $F_A(X) = \|A - X\|^2$, are characterized by $(A - X)X' = 0, X'(A - X) = 0$.

(b) $F_A : M(n, M \times N) \to \mathbb{R}$ has a finite number of critical points if and only if $M = N$ and A has M distinct singular values.

(c) If $\sigma_n > \sigma_{n+1}, n \leq k$, there exists a unique global minimum $X_{\min}$ of $F_A : M(n, M \times N) \to \mathbb{R}$ which is given by

$$X_{\min} = \Theta_1 \Sigma_n \Theta_2. \tag{1.27}$$

As an immediate Corollary we obtain the following classical theorem of Eckart and Young (1936).

Corollary 1.10 Let $A = \Theta_1 \Sigma \Theta_2$ be the singular value decomposition of $A \in \mathbb{R}^{M \times N}$, $\Theta_1 \Theta_1' = I_M$, $\Theta_2 \Theta_2' = I_N$, $\Sigma = \begin{bmatrix} \operatorname{diag}(\sigma_1, \cdots, \sigma_k) & 0 \\ 0 & 0 \end{bmatrix} \in \mathbb{R}^{M \times N}$ with $\sigma_1 \geq \cdots \geq \sigma_k > 0$. Let $n \leq k$ and $\sigma_n > \sigma_{n+1}$. Then with

$$\Sigma_n = \begin{bmatrix} \operatorname{diag}(\sigma_1, \cdots, \sigma_n, 0, \cdots, 0) & 0 \\ 0 & 0 \end{bmatrix} \in \mathbb{R}^{M \times N}$$

$X_{\min} = \Theta_1 \Sigma_n \Theta_2$ is the unique $M \times N$ matrix of rank n which minimizes $\|A - X\|^2$ over the set of matrices of rank less than or equal to n.

Gradient Flows In this subsection we develop a gradient flow approach to find the critical points of the distance functions $f_A : S(n, N) \to \mathbb{R}$ and $F_A : M(n, M \times N) \to \mathbb{R}$. We first consider the symmetric matrix case.

By Proposition 1.1 the tangent space of $S(n, N)$ at an element X is the vector space

$$T_X S(n, N) = \{\Delta X + X \Delta' \mid \Delta \in \mathbb{R}^{N \times N}\}.$$

For $A, B \in \mathbb{R}^{N \times N}$ we define

$$\{A, B\} = AB + B'A' \tag{1.28}$$

which, by the way, coincides with the product defined on $\mathbb{R}^{N \times N}$, when $\mathbb{R}^{N \times N}$ is considered as a Jordan algebra. Thus the tangent space $T_X S(n, N)$ is the image of the linear map

$$\pi_X : \mathbb{R}^{N \times N} \to \mathbb{R}^{N \times N}, \quad \Delta \mapsto \{\Delta, X\} \tag{1.29}$$

5.1. APPROXIMATIONS BY LOWER RANK MATRICES

while the kernel of π_X is

$$\ker \pi_X = \{\Delta \in \mathbb{R}^{N \times N} \mid \Delta X + X \Delta' = 0\} \quad (1.30)$$

Taking the orthogonal complement $(\ker \pi_X)^\perp = \{Z \in \mathbb{R}^{N \times N} \mid \text{tr}(Z'\Delta) = 0 \ \forall \Delta \in \ker \pi_X\}$, with respect to the standard inner product on $\mathbb{R}^{N \times N}$

$$\langle A, B \rangle = \text{tr}(A'B), \quad (1.31)$$

yields the isomorphism of vector spaces

$$(\ker \pi_X)^\perp \cong \mathbb{R}^{N \times N} / \ker \pi_X \cong T_X S(n, N). \quad (1.32)$$

We have the orthogonal decomposition of $\mathbb{R}^{N \times N}$

$$\mathbb{R}^{N \times N} = \ker \pi_X \oplus (\ker \pi_X)^\perp$$

and hence every element $\Delta \in \mathbb{R}^{N \times N}$ has a unique decomposition

$$\Delta = \Delta_X + \Delta^X \quad (1.33)$$

where $\Delta_X \in \ker \pi_X$ and $\Delta^X \in (\ker \pi_X)^\perp$.

Given any pair of tangent vectors $\{\Delta_1, X\}, \{\Delta_2, X\}$ of $T_X S(n, M)$ we define

$$\langle\langle \{\Delta_1, X\}, \{\Delta_2, X\} \rangle\rangle := 4\text{tr}((\Delta_1^X)' \Delta_2^X). \quad (1.34)$$

It is easy to show that $\langle\langle \, , \, \rangle\rangle$ defines a nondegenerate symmetric bilinear form on $T_X S(n, N)$ for each $X \in S(n, N)$. In fact, $\langle\langle \, , \, \rangle\rangle$ defines a Riemannian metric of $S(n, N)$. We refer to $\langle\langle \, , \, \rangle\rangle$ as the *normal Riemannian metric* on $S(n, N)$.

Theorem 1.11 *Let $A \in \mathbb{R}^{N \times N}$ be symmetric.*

(a) The gradient flow of $f_A : S(n, N) \to \mathbb{R}$ with respect to the normal Riemannian metric $\langle\langle \, , \, \rangle\rangle$ is

$$\boxed{\dot{X} = -\text{grad} f_A(X) = \{(A - X)X, X\} = (A - X)X^2 + X^2(A - X)} \quad (1.35)$$

(b) For any $X(0) \in S(n, N)$ the solution $X(t) \in S(n, N)$ of (1.35) exists for all $t \geq 0$.

(c) Every solution $X(t) \in S(n, N)$ of (1.35) converges to an equilibrium point X_∞ characterized by $X(X - A) = 0$. Also, X_∞ has rank less than or equal to n.

Proof The gradient of f_A with respect to the normal metric is the uniquely determined vector field on $S(n,N)$ characterized by

$$Df_A(X)(\{\Delta, X\}) = \langle\langle \mathrm{grad} f_A(X), \{\Delta, X\}\rangle\rangle$$
$$\mathrm{grad} f_A(X) = \{\Omega, X\} \in T_X S(n, N) \tag{1.36}$$

for all $\Delta \in \mathbb{R}^{N \times N}$ and some (unique) $\Omega \in (\ker \pi_X)^\perp$. A straightforward computation shows that the derivative of $f_A : S(n,N) \to \mathbb{R}$ at X is the linear map defined on $T_X S(n,N)$ by

$$\begin{aligned} Df_A(X)(\{\Delta, X\}) &= 2\mathrm{tr}(X\{\Delta, X\} - A\{\Delta, X\}) \\ &= 4\mathrm{tr}(((X-A)X)'\Delta). \end{aligned} \tag{1.37}$$

Thus (1.36) is equivalent to

$$\begin{aligned} 4\mathrm{tr}(((X-A)X)'\Delta) &= \langle\langle \mathrm{grad} f_A(X), \{\Delta, X\}\rangle\rangle \\ &= \langle\langle \{\Omega, X\}, \{\Delta, X\}\rangle\rangle \\ &= 4\mathrm{tr}((\Omega^X)'\Delta^X) \\ &= 4\mathrm{tr}(\Omega' \Delta^X) \end{aligned} \tag{1.38}$$

since $\Omega \in (\ker \pi_X)^\perp$ implies $\Omega = \Omega^X$. For all $\Delta \in \ker \pi_X$

$$\mathrm{tr}(X(X-A)\Delta) = \frac{1}{2}\mathrm{tr}((X-A)(\Delta X + X\Delta')) = 0$$

and therefore $(X-A)X \in (\ker \pi_X)^\perp$. Thus

$$\begin{aligned} 4\mathrm{tr}(((X-A)X)'\Delta) &= 4\mathrm{tr}(((X-A)X)'(\Delta_X + \Delta^X)) \\ &= 4\mathrm{tr}(((X-A)X)'\Delta^X)) \end{aligned}$$

and (1.38) is equivalent to

$$\Omega = (X-A)X$$

Thus

$$\mathrm{grad} f_A(X) = \{(X-A)X, X\} \tag{1.39}$$

which proves (a).

For (b) note that for any $X \in S(n,N)$ we have $\{X, X(X-A)\} \in T_X S(n,N)$ and thus (1.35) is a vector field on $S(n,N)$. Thus for any initial condition $X(0) \in S(n,N)$ the solution $X(t)$ of (1.35) satisfies $X(t) \in S(n,N)$ for all t for which $X(t)$ is defined. It suffices therefore to show the existence of solutions of (1.35) for all $t \geq 0$. To this end consider any solution $X(t)$ of (1.35). By

$$\begin{aligned} \frac{d}{dt} f_A(X(t)) &= 2\mathrm{tr}((X-A)\dot{X}) \\ &= 2\mathrm{tr}((X-A)\{(A-X)X, X\}) \\ &= -4\mathrm{tr}((A-X)^2 X^2) \\ &= -4\|(A-X)X\|^2 \end{aligned} \tag{1.40}$$

5.1. APPROXIMATIONS BY LOWER RANK MATRICES

(since $\operatorname{tr}(A\{B,C\}) = 2\operatorname{tr}(BCA)$ for $A = A'$). Thus $f_A(X(t))$ decreases monotonically and the equilibria points of (1.35) are characterized by $(X - A)X = 0$. Also,

$$\|A - X(t)\| \leq \|A - X(0)\|$$

and $X(t)$ stays in the compact set $\{X \in \overline{S(n,N)} \mid \|A-X\| \leq \|A-X(0)\|\}$. By the closed orbit lemma, see Appendix C, the closure $\overline{S(n,N)}$ of each orbit $S(n,N)$ is a union of orbits $S(i,N)$ for $0 \leq i \leq n$. Since the vector field is tangent to all $S(i,N)$, $0 \leq i \leq n$, the boundary of $S(n,N)$ is invariant under the flow. Thus $X(t)$ exists for all $t \geq 0$ and the result follows. ∎

Remarks **1.** An important consequence of Theorem 1.11 is that the differential equation on the vector space of symmetric $N \times N$ matrices

$$\boxed{\dot{X} = X^2(A - X) + (A - X)X^2} \tag{1.41}$$

is rank preserving, i.e. rank $X(t) =$ rank $X(0)$ for all $t \geq 0$, and therefore also signature preserving. Also, $X(t)$ always converges in the spaces of symmetric matrices to some symmetric matrix $X(\infty)$ as $t \to \infty$ and hence $\operatorname{rk}X(\infty) \leq \operatorname{rk}X(0)$. Here $X(\infty)$ is a critical point of $f_A : S(n,N) \to \mathbb{R}, n \leq \operatorname{rk}X(0)$.

2. The gradient flow (1.35) of $f_A : S(n,N) \to \mathbb{R}$ can also be obtained as follows.

Consider the task of finding the gradient for the function $g_A : GL(N) \to \mathbb{R}$ defined by

$$g_A(\gamma) = \|A - \gamma'Q\gamma\|^2 \tag{1.42}$$

where $Q \in S(n,N)$. Let $\langle\langle \xi, \eta \rangle\rangle = 4\operatorname{tr}((\gamma^{-1}\xi)'(\gamma^{-1}\eta))$ for $\xi, \eta \in T_\gamma GL(N)$. It is now easy to show that

$$\dot{\gamma} = \gamma\gamma'Q\gamma(\gamma'Q\gamma - A) \tag{1.43}$$

is the gradient flow of $g_A : GL(N) \to \mathbb{R}$ with respect to the Riemannian metric $\langle\langle \, , \, \rangle\rangle$ on $GL(N)$. In fact, the gradient $\operatorname{grad}(g_A)$ with respect to $\langle\langle \, , \, \rangle\rangle$ is characterized by

$$\begin{aligned}\operatorname{grad}g_A(\gamma) &= \gamma \cdot \Omega \\ Dg_A(\gamma)(\xi) &= \langle\langle \operatorname{grad}g_A(\gamma), \xi \rangle\rangle \\ &= 4\operatorname{tr}(\Omega'\gamma^{-1}\xi)\end{aligned} \tag{1.44}$$

for all $\xi \in T_\gamma GL(N)$. The derivative of g_A on $T_\gamma GL(N)$ is $(X = \gamma'Q\gamma)$

$$Dg_A(\gamma)(\xi) = -4\operatorname{tr}((A - X)\gamma'Q\xi)$$

and hence

$$\Omega'\gamma^{-1} = (X - A)\gamma'Q.$$

This proves (1.43). It is easily verified that $X(t) = \gamma(t)'Q\gamma(t)$ is a solution of (1.35), for any solution $\gamma(t)$ of (1.43). ■

We now turn to the task of determining the gradient flow of $F_A: M(n, M \times N) \to \mathbb{R}$. A differential equation $\dot{X} = F(X)$ evolving on the matrix space $\mathbb{R}^{M \times N}$ is said to be *rank preserving* if the rank $\mathrm{rk} X(t)$ of every solution $X(t)$ is constant as a function of t. The following characterization is similar to that of isospectral flows.

Lemma 1.12 *Let $I \subset \mathbb{R}$ be an interval and let $A(t) \in \mathbb{R}^{M \times M}, B(t) \in \mathbb{R}^{N \times N}, t \in I$, be a continuous time-varying family of matrices. Then*

$$\dot{X}(t) = A(t)X(t) + X(t)B(t), \quad X(0) \in \mathbb{R}^{M \times N} \qquad (1.45)$$

is rank preserving. Conversely, every rank preserving differential equation on $\mathbb{R}^{M \times N}$ is of the form (1.45) for matrices $A(t), B(t)$.

Proof For any fixed $X \in \mathbb{R}^{M \times N}$ with rank $X = n$, and $n \leq \min(M, N)$ arbitrary, $A(t)X + XB(t) \in T_X M(n, M \times N)$. Thus (1.45) defines a time varying vector field on each subset $M(n, M \times N) \subset \mathbb{R}^{M \times N}$. Thus for any initial condition $X_0 \in M(n, M \times N)$ the solution $X(t)$ of (1.45) satisfies $X(t) \in M(n, M \times N)$ for all $t \in I$. Therefore (1.45) is rank preserving. Conversely, suppose $\dot{X} = F(X)$ is rank preserving. Then it defines a vector field on $M(n, M \times N)$ for any $1 \leq n \leq \min(M, N)$. By Proposition 1.8 therefore $F(X) = \Delta_1(X) \cdot X + X \cdot \Delta_2(X)$, $X \in M(n, M \times N)$, for $M \times M$ and $N \times N$ matrices Δ_1 and Δ_2. Setting $A(t) = \Delta_1(X(t))$, $B(t) = \Delta_2(X(t))$ completes the proof. ■

To obtain the gradient flow of the distance function $F_A : M(n, M \times N) \to \mathbb{R}$ in the general approximation problem we proceed as above. Let

$$i : M(n, M \times N) \to S(2n, M + N)$$

denote the imbedding defined by

$$i(X) = \hat{X} = \begin{bmatrix} 0 & X \\ X' & 0 \end{bmatrix}.$$

The gradient flow of $f_{\hat{A}} : S(2n, M + N) \to \mathbb{R}$ is

$$\dot{Z} = -(Z^2(Z - \hat{A}) + (Z - \hat{A})Z^2), \quad Z \in S(2n, M + N)$$

For $Z = \hat{X} = \begin{bmatrix} 0 & X \\ X' & 0 \end{bmatrix}$ the right hand side is simplified as

$$-\begin{bmatrix} 0 & XX'(X - A) + (X - A)X'X \\ (XX'(X - A) + (X - A)X'X)' & 0 \end{bmatrix}$$

5.1. APPROXIMATIONS BY LOWER RANK MATRICES

Thus the gradient flow (1.35) on $S(2n, M+N)$ of $f_{\tilde{A}}$ leaves the submanifold $i(M(n, M \times N)) \subset S(2n, M+N)$ invariant. The normal Riemannian metric of $S(2n, M+N)$ induces by restriction a Riemannian metric on $i(M(n, M \times N))$ and hence on $M(n, M \times N)$. We refer to this as the *normal Riemannian metric* of $M(n, M \times N)$. The above computation together with Theorem 1.11 then shows the following theorem.

Theorem 1.13 Let $A \in \mathbb{R}^{M \times N}$.

(a) The gradient flow of $F_A : M(n, M \times N) \to \mathbb{R}, F_A(X) = ||A - X||^2$, with respect to the normal Riemannian metric on $M(n, M \times N)$ is

$$\boxed{\dot{X} = -\mathrm{grad} F_A(X) = XX'(A-X) + (A-X)X'X} \quad (1.46)$$

(b) For any $X(0) \in M(n, M \times N)$ the solution $X(t)$ of (1.46) exists for all $t \geq 0$ and rank $X(t) = n$ for all $t \geq 0$.

(c) Every solution $X(t)$ of (1.46) converges to an equilibrium point satisfying $X_\infty(X'_\infty - A') = 0, X'_\infty(X_\infty - A) = 0$ and X_∞ has rank $\leq n$. ∎

A Riccati Flow The Riccati differential equation

$$\dot{X} = (A-X)X + X(A-X)$$

appears to be the simplest possible candidate for a rank preserving flow on $S(N)$ which has the same set of equilibria as the gradient flow (1.35). Moreover, the restriction of the right hand side of (1.35) on the subclass of projection operators, characterized by $X^2 = X$, coincides with the above Riccati equation. This motivates us to consider the above Riccati equation in more detail. As we will see, the situation is particularly transparent for positive definite matrices X.

Theorem 1.14 Let $A \in \mathbb{R}^{N \times N}$ be symmetric.

(a) The Riccati equation

$$\boxed{\dot{X} = (A-X)X + X(A-X)}, \quad X(0) \in S(n, N), \quad (1.47)$$

defines a rank preserving flow on $S(n, N)$.

(b) Assume A is invertible. Then the solutions $X(t)$ of (1.47) are given by

$$X(t) = e^{tA} X_0 [I_N + A^{-1}(e^{2At} - I_N)X_0]^{-1} e^{tA} \quad (1.48)$$

(c) For any positive semidefinite initial condition $X(0) = X(0) \geq 0$, the solution $X(t)$ of (1.47) exists for all $t \geq 0$ and is positive semidefinite.

(d) Every positive semidefinite solution $X(t) \in \overline{S^+(n,N)} = S(n,0;N)$ of (1.47) converges to a connected component of the set of equilibrium points, characterized by $(A - X_\infty)X_\infty = 0$. Also X_∞ is positive semidefinite and has rank $\leq n$. If A has distinct eigenvalues then every positive semidefinite solution $X(t)$ converges to an equilibrium point.

Proof The proof of (a) runs similarly to that of Lemma 1.12; see Problem. To prove (b) it suffices to show that $X(t)$ defined by (1.48) satisfies the Riccati equation. By differentiation of (1.48) we obtain

$$\dot{X}(t) = AX(t) + X(t)A - 2X(t)^2,$$

which shows the claim.

For (c) note that (a) implies that $X(t) \in S^+(n,N)$ for all $t \in [0, t_{max}[$. Thus it suffices to show that $X(t)$ exists for all $t \geq 0$; i.e. $t_{max} = \infty$. This follows from a simple Lyapunov argument. First, we note that the set $\overline{S^+(n,N)}$ of positive semidefinite matrices X of rank $\leq n$ is a closed subset of $S(N)$. Consider the distance function $f_A : S(N) \to \mathbb{R}_+$ defined by $f_A(X) = ||A - X||^2$. Thus f_A is a proper function of $S(N)$ and hence also on $\overline{S^+(n,N)}$. For every positive semidefinite solution $X(t)$, $t \in [0, t_{max}[$, let $X(t)^{1/2}$ denote the unique positive semidefinite symmetric square root. A simple computation shows

$$\begin{aligned}\frac{d}{dt}f_A(X(t)) &= -4\mathrm{tr}[(A - X(t))\dot{X}(t)] \\ &= -4||(A - X(t))X(t)^{\frac{1}{2}}||^2 \leq 0.\end{aligned}$$

Thus f_A is a Lyapunov function for (1.47), restricted to the class of positive semidefinite matrices, and equilibrium points $X_\infty \in \overline{S^+(n,N)}$ are characterized by $(A - X_\infty)X_\infty = 0$. In particular, $f_A(X(t))$ is a monotonically decreasing function of t and the solution $X(t)$ stays in the compact subset

$$\{X \in \overline{S^+(n,N)} \mid f_A(X) \leq f_A(X(0))\}.$$

Thus $X(t)$ is defined for all $t \geq 0$. By the closed orbit lemma, Appendix C, the boundary of $\overline{S^+(n,N)}$ is a union of orbits of the congruence action on $GL(N,\mathbb{R})$. Since the Riccati vectorfield is tangent to these orbits, the boundary is invariant under the flow. Thus $X(t) \in S^+(n,N)$ for all $t \geq 0$. By La Salle's principle of invariance, the ω-limit set of $X(t)$ is a connected component of the set of positive semidefinite equilibrium points. If A has distinct eigenvalues, then the set of positive semidefinite equilibrium points is finite. Thus the result follows. ∎

Remark The above proof shows that the least squares distance function $f_A(X) = ||A - X||^2$ is a Lyapunov function for the Riccati equation, evolving

5.1. APPROXIMATIONS BY LOWER RANK MATRICES

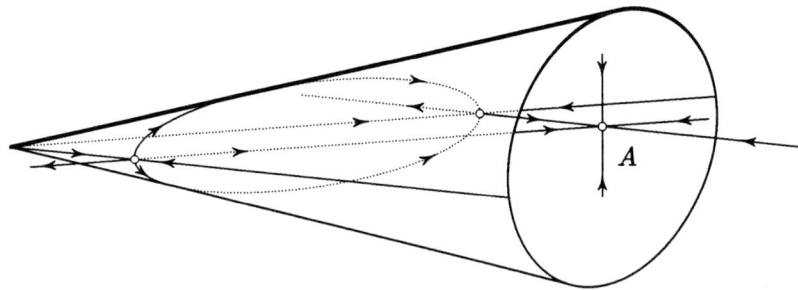

Figure 5.1.2: Riccati flow on positive semidefinite matrices

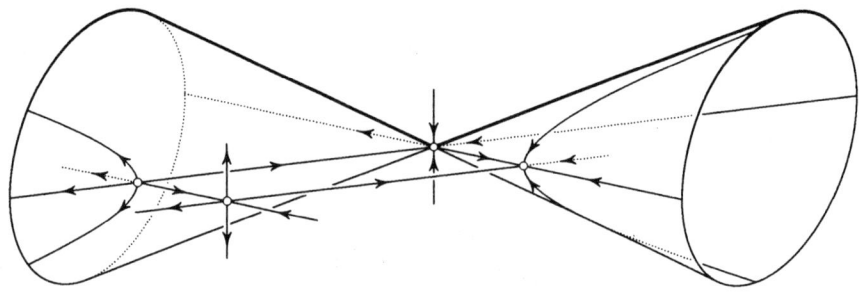

Figure 5.1.3: Riccati flow on $\overline{S(1,2)}$.

on the subset of positive semidefinite matrices X. In particular, the Riccati equation exhibits gradient-like behaviour if restricted to positive definite initial conditions. If X_0 is an indefinite matrix then also $X(t)$ is indefinite, and $f_A(X)$ is no longer a Lyapunov function for the Riccati equation. ∎

The following Figure 5.1.2 illustrates the phase portrait of the Riccati flow on $S(2)$ for $A = A'$ positive semidefinite. Only a part of the complete phase portrait is shown here, concentrating on the cone of positive semidefinite matrices in $S(2)$. There are two equilibrium points on $S(1,2)$. For the flow on $S(2)$, both equilibria are saddle points, one having 1 positive and 2 negative eigenvalues while the other one has 2 positive and 1 negative eigenvalues. The induced flow on $S(1,2)$ has one equilibrium as a local attractor while the other one is a saddle point. Figure 5.1.3 illustrates the phase portrait in the case where $A = A'$ is indefinite.

Problems 1. Let $I \subset \mathbb{R}$ be an interval and let $A(t) \in \mathbb{R}^{N \times N}$, $t \in I$, be a continuous family of matrices. Show that

$$\dot{X}(t) = A(t)X(t) + X(t)A(t)', X(0) \in S(n)$$

is a rank (and hence signature) preserving flow on $S(N)$. Prove that, conversely, any rank preserving vector field on $S(N)$ is of this form.

2. Let $A, B \in \mathbb{R}^{N \times N}$. What is the gradient flow of the total least squares function $F_{A,B}: M(n, N \times N) \to \mathbb{R}$, $F_{A,B}(X) = ||A - BX||^2$, with respect to the normal Riemannian metric on $M(n, N \times N)$? Characterize the equilibria!

Main Points of Section The approximation problem of a matrix by a lower rank one in the Frobenius norm is a further instance of a matrix least squares estimation problem. The general matrix case can be reduced to the approximation problem for symmetric matrices. Explicit formulas are given for the critical points and local minima in terms of the eigenspace decomposition. A certain Riemannian metric leads to a particularly simple expression of the gradient vector field. A remarkable property of the gradient vector field is that the solutions are rank preserving.

5.2 The Polar Decomposition

Every real $n \times n$ matrix A admits a decomposition

$$A = \Theta \cdot P \tag{2.1}$$

where P is positive semidefinite symmetric and Θ is an orthogonal matrix satisfying $\Theta\Theta' = \Theta'\Theta = I_n$. While $P = (A'A)^{\frac{1}{2}}$ is uniquely determined, Θ is uniquely determined only if A is invertible. The decomposition (2.1) is called the *polar decomposition* of A. While effective algebraic algorithms achieving the polar decomposition are well known, we are interested in finding dynamical systems achieving the same purpose.

Let $O(n)$ and $\mathcal{P}(n)$ denote the set of real orthogonal and positive definite symmetric $n \times n$ matrices respectively. In the sequel we assume for simplicity that A is invertible. Given $A \in \mathbb{R}^{n \times n}$, $\det(A) \neq 0$, we consider the smooth function

$$F_A : O(n) \times \mathcal{P}(n) \to \mathbb{R} \ , F_A(\Theta, P) = ||A - \Theta P||^2 \tag{2.2}$$

where $||A||^2 = \text{tr}(AA')$ is the Frobenius norm. Since $F(\Theta_0, P_0) = 0$ if and only if $A = \Theta_0 P_0$ we see that the global minimum of F corresponds to the polar decomposition. This motivates the use of a gradient flow for $F_A : O(n) \times \mathcal{P}(n) \to \mathbb{R}$ to achieve the polar decomposition. Of course, other approximation problems suggest themselves as well. Thus, if Θ is restricted to be the identity matrix, we have the problem studied in the previous section of finding the best positive definite approximant of a given matrix A. Similarly if P is restricted to be the identity matrix, then the question amounts to finding the best orthogonal matrix approximant of a given invertible matrix A. In this case $||A - \Theta||^2 = ||A||^2 - 2\text{tr}(A'\Theta) + n$ and

5.2. THE POLAR DECOMPOSITION

we have a least square matching problem, studied by Shayman (1982) and Brockett (1989).

Theorem 2.1 Let $A \in \mathbb{R}^{n \times n}, det(A) \neq 0$, and let $O(n)$ and $\mathcal{P}(n)$ be endowed with the constant Riemannian metric arising from the Euclidean inner product of $\mathbb{R}^{n \times n}$. The (minus) gradient flow of $F_A : O(n) \times \mathcal{P}(n) \to \mathbb{R}$ is

$$\begin{aligned} \dot{\Theta} &= \Theta P A' \Theta - AP \\ \dot{P} &= -2P + A'\Theta + \Theta'A . \end{aligned} \quad (2.3)$$

For every initial condition $(\Theta(0), P(0)) \in O(n) \times \mathcal{P}(n)$ the solution $(\Theta(t), P(t))$ of (2.3) exists for all $t \geq 0$ with $\Theta(t) \in O(n)$ and $P(t)$ symmetric (but not necessarily positive semidefinite). Every solution of (2.3) converges to an equilibrium point of (2.3) as $t \to +\infty$. The equilibrium points of (2.3) are $(\Theta_\infty, P_\infty)$ with $\Theta_\infty = \Theta_0 \Psi$, $\Psi \in O(n)$,

$$P_\infty = \frac{1}{2}(P_0 \Psi + \Psi' P_0)$$
$$\Psi(P_\infty P_0)\Psi = P_0 P_\infty \quad (2.4)$$

and $A = \Theta_0 P_0$ is the polar decomposition. For almost every initial condition $\Theta(0) \in O(n), P(0) \in \mathcal{P}(n)$, then $(\Theta(t), P(t))$ converges to the polar decomposition (Θ_0, P_0) of A.

Proof The derivative of $F_A : O(n) \times \mathcal{P}(n) \to \mathbb{R}$ at (Θ, P) is the linear map on the tangent space

$$\begin{aligned} DF(\Theta, P)(\Theta\Omega, S) &= -2tr(A'\Theta\Omega P) - 2tr(A'\Theta S) + 2tr(PS) \\ &= -tr((PA'\Theta - \Theta'AP)\Omega) - tr((A'\Theta - P - P + \Theta'A)S) \end{aligned}$$

for arbitrary matrices $\Omega' = -\Omega, S' = S$. Thus the gradient is $\nabla F_A = (\nabla_\Theta F_A, \nabla_P F_A)$ with

$$\begin{aligned} \nabla_\Theta F_A &= \Theta(\Theta'AP - PA'\Theta) = AP - \Theta PA'\Theta \\ \nabla_P F_A &= 2P - A'\Theta - \Theta'A. \end{aligned}$$

This is also the gradient flow of the extended function $\hat{F}_A : O(n) \times S(n) \to \mathbb{R}$, $\hat{F}_A(\Theta, S) = ||A - \Theta S||^2$, where $S(n)$ is the set of all real symmetric matrices. Since $\hat{F}_A : O(n) \times S(n) \to \mathbb{R}$ is proper, the existence of solutions (2.3) for $t \geq 0$ follows. Moreover, every solution converges to a critical point of $\hat{F}_A$ characterized by (1.31). ∎

A difficulty with the above ODE approach to the polar decomposition is that equation (2.3) for the polar part $P(t)$ does not in general evolve in the space of positive definite matrices. In fact, positive definiteness may be lost during the evolution of (2.3). For example, let $n = 1, a = 1$, and $\Theta_0 = -1$. Then $(\Theta_\infty, P_\infty) = (-1, -1)$. If, however, $A'\Theta_\infty + \Theta'_\infty A \geq 0$ then $P(t) > 0$ for all $t \geq 0$.

Research Problems More work is to be done in order to achieve a reasonable ODE method for polar decomposition. If $\mathcal{P}(n)$ is endowed with the normal Riemannian metric instead of the constant Euclidean one used in the above theorem, then the gradient flow on $O(n) \times \mathcal{P}(n)$ becomes

$$\begin{aligned} \dot{\Theta} &= \Theta(PA')\Theta - AP \\ \dot{P} &= 4[(\tfrac{1}{2}(A'\Theta + \Theta'A) - P)P^2 + P^2(\tfrac{1}{2}(A'\Theta + \Theta'A) - P))]. \end{aligned}$$

Show that this flow evolves on $O(n) \times \mathcal{P}(n)$, i.e. $\Theta(t) \in O(n)$, $P(t) \in \mathcal{P}(n)$ exists for all $t \geq 0$. Furthermore, for all initial conditions $\Theta(0) \in O(n)$, $P(0) \in \mathcal{P}(n)$, $(\Theta(t), P(t))$ converges (exponentially?) to the unique polar decomposition $\Theta_0 P_0 = A$ of A (for $t \to \infty$). Thus this seems to be the right gradient flow.

A different, somewhat simpler, *gradient-like* flow on $O(n) \times \mathcal{P}(n)$ which also achieves the polar decomposition is

$$\begin{aligned} \dot{\Theta} &= \Theta(PA')\Theta - AP \\ \dot{P} &= (\tfrac{1}{2}(A'\Theta + \Theta'A) - P)P + P(\tfrac{1}{2}(A'\Theta + \Theta'A) - P). \end{aligned}$$

Note that the equation for P now is a Riccati equation. Analyse these flows.

Main Points of Section The polar decomposition of a matrix is investigated from a matrix least squares point of view. Gradient flows converging to the orthogonal and positive definite factors in a polar decomposition are introduced. These flows are coupled Riccati equations on the orthogonal matrices and positive definite matrices, respectively.

5.3 Output Feedback Control

Feedback is a central notion in modern control engineering and systems theory. It describes the process of "feeding back" the output or state variables in a dynamical systems configuration through the input channels. Here we concentrate on output feedback control of linear dynamical systems.

Consider finite-dimensional linear dynamical systems of the form

$$\begin{aligned} \dot{x}(t) &= Ax(t) + Bu(t) \\ y(t) &= Cx(t). \end{aligned} \tag{3.1}$$

5.3. OUTPUT FEEDBACK CONTROL

Here $u(t) \in \mathbb{R}^m$ and $y(t) \in \mathbb{R}^p$ are the input and output respectively of the system while $x(t)$ is the state vector and $A \in \mathbb{R}^{n \times n}$, $B \in \mathbb{R}^{n \times m}$ and $C \in \mathbb{R}^{p \times n}$ are real matrices. We will use the matrix triple (A, B, C) to denote the dynamical system (3.1). Using output feedback the input $u(t)$ of the system is replaced by a new input

$$u(t) = Ky(t) + v(t) \qquad (3.2)$$

defined by a feedback gain matrix $K \in \mathbb{R}^{m \times p}$. Combining equations (3.1) and (3.2) yields the closed loop system

$$\begin{aligned} \dot{x}(t) &= (A + BKC)x(t) + Bv(t) \\ y(t) &= Cx(t) \end{aligned} \qquad (3.3)$$

An important task in linear systems theory is that of pole placement or *eigenvalue assignment* via output feedback. Thus for a given system (A, B, C) and a self-conjugate set $\{s_1, \ldots, s_n\} \subset \mathbb{C}$ one is asked to find a feedback matrix $K \in \mathbb{R}^{m \times p}$ such that $A + BKC$ has eigenvalues $s_1, \ldots s_n$. A simple dimension count shows that the condition $mp \geq n$ is necessary for the solvability of the problem; see also the notes for Chapter 5. In general, however, for $mp < n$, there is no choice of a feedback gain matrix K such that $A + BKC$ has prescribed eigenvalues. It is desirable in this case to find a feedback matrix that generates the best closed loop approximation to a specified eigenstructure. Thus, for a fixed matrix $F \in \mathbb{R}^{n \times n}$ with eigenvalues $s_1, \ldots, s_n$ the task is to find matrices $T \in GL(n, \mathbb{R})$ and $K \in \mathbb{R}^{m \times p}$ that minimize the distance $||F - T(A + BKC)T^{-1}||^2$. One would hope to find an explicit formulae for the optimal feedback gain that achieves the best approximation, however, the question appears to be too difficult to tackle directly. Thus, algorithmic solutions become important.

A natural generalization of eigenvalue or eigenstructure assignment of linear systems is that of *optimal system assignment*. Here the distance of a target system (F, G, H) to an output feedback orbit (see below) is minimized. In this section a gradient flow approach is developed to solve such output feedback optimization problems. More complicated problems, such as simultaneous eigenvalue assignment of several systems or eigenvalue assignment problems for systems with symmetries, can be treated in a similar manner. We regard this as an important aspect of the approach. Although no new output feedback pole placement theorems are proved here we believe that the new methodology introduced is capable of offering new insights into these difficult questions.

Gradient Flows on Output Feedback Orbits We begin with a brief description of the geometry of output feedback orbits. Two linear systems (A_1, B_1, C_1) and (A_2, B_2, C_2) are called *output feedback equivalent* if

$$(A_2, B_2, C_2) = (T(A_1 + B_1KC_1)T^{-1}, TB_1, C_1T^{-1}) \qquad (3.4)$$

holds for $T \in GL(n, \mathbb{R})$ and $K \in \mathbb{R}^{m \times p}$. Thus the system (A_2, B_2, C_2) is obtained from (A_1, B_1, C_1) using a linear change of basis $T \in GL(n, \mathbb{R})$ in the state space $\mathbb{R}^n$ and a feedback transformation $K \in \mathbb{R}^{m \times p}$. Observe that the set $GL(n, \mathbb{R}) \times \mathbb{R}^{m \times p}$ of feedback transformation is a Lie group under the operation $(T_1, K_1) \circ (T_2, K_2) = (T_1 T_2, K_1 + K_2)$. This group is called the *output feedback group*.

The Lie group $GL(n, \mathbb{R}) \times \mathbb{R}^{m \times p}$ acts on the vector space of all triples

$$L(n, m, p) = \{(A, B, C) \mid (A, B, C) \in \mathbb{R}^{n \times n} \times \mathbb{R}^{n \times m} \times \mathbb{R}^{p \times n}\} \quad (3.5)$$

via the Lie group action

$$\alpha : (GL(n, \mathbb{R}) \times \mathbb{R}^{m \times p}) \times L(n, m, p) \to L(n, m, p)$$
$$((T, K), (A, B, C)) \mapsto (T(A + BKC)T^{-1}, TB, CT^{-1}) \quad (3.6)$$

Thus, the orbits

$$\mathcal{F}(A, B, C) = \{(T(A + BKC)T^{-1}, TB, CT^{-1}) \mid (T, K) \in GL(n, \mathbb{R}) \times \mathbb{R}^{m \times p}\} \quad (3.7)$$

are the set of systems which are output feedback equivalent to (A, B, C). It is readily verified that the transfer function $G(s) = C(sI - A)^{-1}B$ associated with (A, B, C), Appendix B.1, changes under output feedback via the linear fractional transformation

$$G(s) \mapsto G_K(s) = (I_p - G(s)K)^{-1}G(s).$$

Let $G(s)$ be an arbitrary strictly proper $p \times m$ rational transfer function of McMillan degree n, see Appendix B for results on linear system theory. It is a consequence of Kalman's realization theorem that

$$\begin{aligned} \mathcal{F}_G &= \{(A, B, C) \in L(n, m, p) \mid C(sI_n - A)^{-1}B = (I_p - G(s)K)^{-1}G(s) \\ &\qquad \text{for some } K \in \mathbb{R}^{m \times p}\} \\ &= \mathcal{F}(A, B, C) \end{aligned}$$

coincides with the output feedback orbit $\mathcal{F}(A, B, C)$, for (A, B, C) controllable and observable.

Lemma 3.1 *Let $(A, B, C) \in L(n, m, p)$. Then*

(a) The output feedback orbit $\mathcal{F}(A, B, C)$ is a smooth submanifold of all triples $L(n, m, p)$.

(b) The tangent space of $\mathcal{F}(A, B, C)$ at (A, B, C) is

$$T_{(A,B,C)}\mathcal{F} = \{([X, A] + BLC, XB, -CX) \mid X \in \mathbb{R}^{n \times n}, L \in \mathbb{R}^{m \times p}\} \quad (3.8)$$

5.3. OUTPUT FEEDBACK CONTROL

Proof $\mathcal{F}(A, B, C)$ is an orbit of the real algebraic Lie group action (3.6) and thus, by Appendix C, is a smooth submanifold of $L(n, m, p)$. To prove (b) we consider the smooth map

$$\Gamma : GL(n, \mathbb{R}) \times \mathbb{R}^{m \times p} \to L(n, m, p)$$
$$(T, K) \mapsto (T(A + BKC)T^{-1}, TB, CT^{-1}).$$

The derivative of Γ at the identity element $(I_n, 0)$ of $GL(n, \mathbb{R}) \times \mathbb{R}^{m \times p}$ is the surjective linear map

$$D\Gamma|_{(I_n,0)} : T_{(I_n,0)}(GL(n, \mathbb{R}) \times \mathbb{R}^{m \times p}) \to T_{(A,B,C)}\mathcal{F}(A, B, C)$$
$$(X, L) \mapsto ([X, A] + BLC, XB, -CX)$$

for $X \in T_{I_n} GL(n, \mathbb{R})$, $L \in T_0 \mathbb{R}^{m \times p}$. The result follows. ∎

Let $(A, B, C), (F, G, H) \in L(n, m, p)$ and consider the potential

$$\Phi : \mathcal{F}(A, B, C) \to \mathbb{R}$$
$$\Phi(T(A + BKC)T^{-1}, TB, CT^{-1}) :=$$
$$\|T(A + BKC)T^{-1} - F\|^2 + \|TB - G\|^2 + \|CT^{-1} - H\|^2 \quad (3.9)$$

In order to determine the gradient flow of this distance function we must specify a Riemannian metric on $\mathcal{F}(A, B, C)$. The construction of the Riemannian metric that we consider parallels a similar development in Section 1 of this chapter. By Lemma (3.1), the tangent space $T_{(A,B,C)}\mathcal{F}(A, B, C)$ at an element (A, B, C) is the image of the linear map

$$\pi : \mathbb{R}^{n \times n} \times \mathbb{R}^{m \times p} \to L(n, m, p)$$

defined by

$$\pi(X, L) = ([X, A] + BLC, XB, -CX),$$

which has kernel

$$\ker \pi = \{(X, L) \in \mathbb{R}^{n \times n} \times \mathbb{R}^{m \times p} \mid ([X, A] + BLC, XB, -CX) = (0, 0, 0)\}.$$

Taking the orthogonal complement $(\ker \pi)^\perp$ with respect to the standard Euclidean inner product on $\mathbb{R}^{n \times n} \times \mathbb{R}^{m \times p}$

$$< (Z_1, M_1), (Z_2, M_2) > := \operatorname{tr}(Z_1 Z_2') + \operatorname{tr}(M_1 M_2')$$

yields the isomorphism of vector spaces

$$(\ker \pi)^\perp \approx T_{(A,B,C)}\mathcal{F}(A, B, C).$$

We have the orthogonal decomposition of $\mathbb{R}^{n \times n} \times \mathbb{R}^{m \times p}$

$$\mathbb{R}^{n \times n} \times \mathbb{R}^{m \times p} = \ker \pi \oplus (\ker \pi)^\perp$$

and hence every element has a unique decomposition

$$(X, L) = (X_\perp, L_\perp) + (X^\perp, L^\perp)$$

with $(X_\perp, L_\perp) \in \ker \pi$, $(X^\perp, L^\perp) \in (\ker \pi)^\perp$. Given any pair of tangent vectors $([X_i, A] + BL_iC, X_iB, -CX_i)$, $i = 1, 2$, of $T_{(A,B,C)}\mathcal{F}(A, B, C)$ we define

$$\ll ([X_1, A] + BL_1C, X_1B, -CX_1), ([X_2, A] + BL_2C, X_2B, -CX_2) \gg$$
$$:= 2\mathrm{tr}(X_1^\perp (X_2^\perp)') + 2\mathrm{tr}(L_1^\perp (L_2^\perp)')$$
(3.10)

It is easily shown the $\ll \cdot, \cdot \gg$ defines a nondegenerate symmetric bilinear form on $T_{(A,B,C)}\mathcal{F}(A, B, C)$. In fact, $\ll \cdot, \cdot \gg$ defines a Riemannian metric on $\mathcal{F}(A, B, C)$ which is termed the *normal Riemannian metric*.

Theorem 3.2 (System Assignment)   *Suppose $(A, B, C), (F, G, H) \in L(n, m, p)$.*

(a) The gradient flow $(\dot{A}, \dot{B}, \dot{C}) = -\mathrm{grad}\Phi(A, B, C)$ of $\Phi : \mathcal{F}(A, B, C) \to \mathbb{R}$ given by (3.9), with respect to the normal Riemannian metric is

$$\boxed{\begin{aligned}\dot{A} &= [A, [A - F, A'] + (B - G)B' - C'(C - H)] - BB'(A - F)C'C \\ \dot{B} &= -([A - F, A'] + (B - G)B' - C'(C - H))B \\ \dot{C} &= C([A - F, A'] + (B - G)B' - C'(C - H))\end{aligned}}$$
(3.11)

(b) Equilibrium points $(A_\infty, B_\infty, C_\infty)$ of (3.11) are characterised by

$$[A_\infty - F, A'_\infty] + (B_\infty - G)B'_\infty - C'_\infty(C_\infty - H) = 0 \quad (3.12)$$
$$B'_\infty(A_\infty - F)C'_\infty = 0 \quad (3.13)$$

(c) For every initial condition $(A(0), B(0), C(0)) \in \mathcal{F}(A, B, C)$ the solution $(A(t), B(t), C(t))$ of (3.11) exists for all $t \geq 0$ and remains in $\mathcal{F}(A, B, C)$.

(d) Every solution $(A(t), B(t), C(t))$ of (3.11) converges to a connected component of the set of equilibrium points $(A_\infty, B_\infty, C_\infty) \in \overline{\mathcal{F}(A, B, C)}$.

Proof   The derivative of $\Phi : \mathcal{F}(A, B, C) \to \mathbb{R}$ at (A, B, C) is the linear map on the tangent space $T_{(A,B,C)}\mathcal{F}(A, B, C)$ defined by

$$D\Phi|_{(A,B,C)}([X, A] + BLC, XB, -CX)$$
$$= 2\mathrm{tr}(([X, A] + BLC)(A' - F') + XB(B' - G') - (C' - H')CX)$$
$$= 2\mathrm{tr}(X([A, A' - F'] + B(B' - G') - (C' - H')C)$$
$$+ \mathrm{tr}(LC(A' - F')B).$$

5.3. OUTPUT FEEDBACK CONTROL

The gradient of Φ with respect to the normal metric is the uniquely determined vector field on $\mathcal{F}(A, B, C)$ characterised by $\text{grad}\Phi(A, B, C) = ([Z, A] + BMC, ZB, -CZ)$,

$$\begin{aligned} D\Phi|_{(A,B,C)}([X, A] + BLC, XB, -CX) \\ = \ll \text{grad}\Phi(A, B, C), ([X, A] + BLC, XB, -CX) \gg \\ = 2\text{tr}(Z^\perp (X^\perp)') + 2\text{tr}(M^\perp (L^\perp)') \end{aligned}$$

for all $(X, L) \in \mathbb{R}^{n \times n} \times \mathbb{R}^{m \times p}$ and some $(Z, M) \in \mathbb{R}^{n \times n} \times \mathbb{R}^{m \times p}$. Virtually the same argument as for the proof of Theorem 1.11 then shows that

$$\begin{aligned} Z^\perp &= ([A, A' - F'] + B(B' - G') - (C' - H')C)' & (3.14) \\ M^\perp &= (C(A' - F')B)'. & (3.15) \end{aligned}$$

Therefore (3.11) gives the (negative of the) gradient flow of Φ, which proves (a). Part (c) is an immediate consequence of (3.14) and (3.15). As $\Phi : L(n, m, p) \to \mathbb{R}$ is proper it restricts to a proper function $\Phi : \overline{\mathcal{F}(A, B, C)} \to \mathbb{R}$ on the topological closure of the output feedback orbit. Thus every solution $(A(t), B(t), C(t))$ of (3.11) stays in the compact set

$$\{(A_1, B_1, C_1) \in \overline{\mathcal{F}(A(0), B(0), C(0))} \mid \Phi(A_1, B_1, C_1) \leq \Phi(A(0), B(0), C(0))\}$$

and thus exists for all $t \geq 0$. By the orbit closure lemma, see Appendix C.8, the boundary of $\mathcal{F}(A, B, C)$ is a union of output feedback orbits. Since the gradient flow (3.11) is tangent to the feedback orbits $\mathcal{F}(A, B, C)$, the solutions are contained in $\mathcal{F}(A(0), B(0), C(0))$ for all $t \geq 0$. This completes the proof of (b). Finally, (d) follows from general convergence results for gradient flows, see Appendix C.12. ∎

Remarks 1. Although the solutions of the gradient flow (3.11) all converge to some equilibrium point $(A_\infty, B_\infty, C_\infty)$ as $t \to +\infty$, it may well be that such an equilibrium point is contained the boundary of the output feedback orbit. This seems reminiscent to the occurrence of *high gain output feedback* in pole placement problems; see however, Problem 6.

2. Certain symmetries of the realization (F, G, H) result in associated invariance properties of the gradient flow (3.11). For example, if $(F, G, H) = (F', H', G')$ is a symmetric realization, then by inspection the flow (3.11) on $L(n, m, p)$ induces a flow on the invariant submanifold of all symmetric realizations $(A, B, C) = (A', C', B')$. In this case the induced flow on $\{(A, B) \in \mathbb{R}^{n \times n} \times \mathbb{R}^{n \times m} \mid A = A'\}$ is

$$\boxed{\begin{aligned} \dot{A} &= [A, [A, F] + BG' - GB'] - BB'(A - F)BB' \\ \dot{B} &= -([A, F] + BG' - GB')B. \end{aligned}}$$

In particular, for $G = 0$, we obtain the extension of the double bracket flow

$$\boxed{\begin{aligned} \dot{A} &= [A,[A,F]] - BB'(A-F)BB' \\ \dot{B} &= -([A,F])B. \end{aligned}}$$

Observe that the presence of the feedback term $BB'(A-F)BB'$ in the equation for $A(t)$ here destroys the isospectral nature of the double bracket equation. ∎

Flows on the Output Feedback Group Of course there are also associated gradient flows achieving the optimal feedback gain K_∞ and state space coordinate transformation T_∞. Let $(A, B, C) \in L(n, m, p)$ be a given realization and let $(F, G, H) \in L(n, m, p)$ be a "target system". To find the optimal output feedback transformation of (A, B, C) which results in a best approximation of (F, G, H), we consider the smooth function

$$\phi : GL(n, \mathbb{R}) \times \mathbb{R}^{m \times p} \to \mathbb{R}$$
$$\phi(T, K) = \|T(A + BKC)T^{-1} - F\|^2 + \|TB - G\|^2 + \|CT^{-1} - H\|^2$$
(3.16)

on the feedback group $GL(n, \mathbb{R}) \times \mathbb{R}^{m \times p}$. Any tangent vector of $GL(n, \mathbb{R}) \times \mathbb{R}^{m \times p}$ at an element (T, K) is of the form (XT, L) for $X \in \mathbb{R}^{n \times n}$ and $L \in \mathbb{R}^{m \times p}$. In the sequel we endow $GL(n, \mathbb{R}) \times \mathbb{R}^{m \times p}$ with the *normal Riemannian metric* defined by

$$\ll (X_1 T, L_1), (X_2 T, L_2) \gg := 2\operatorname{tr}(X_1' X_2) + 2\operatorname{tr}(L_1' L_2) \quad (3.17)$$

for any pair of tangent vectors $(X_i T, L_i) \in T_{(T,K)}(GL(n, \mathbb{R}) \times \mathbb{R}^{m \times p})$, $i = 1, 2$.

Theorem 3.3 *Let $(A, B, C) \in L(n, m, p)$ and consider the smooth function $\phi : GL(n, \mathbb{R}) \times \mathbb{R}^{m \times p} \to \mathbb{R}$ defined by (3.16) for a target system $(F, G, H) \in L(n, m, p)$.*

(a) The gradient flow $(\dot{T}, \dot{K}) = -\operatorname{grad}\phi(T, K)$ of ϕ with respect to the normal Riemannian metric (3.17) on $GL(n, \mathbb{R}) \times \mathbb{R}^{m \times p}$ is

$$\boxed{\begin{aligned} \dot{T} &= -[T(A+BKC)T^{-1} - F, (T(A+BKC)T^{-1})'] \\ & \quad - (TB-G)B'T'T + (T')^{-1}C'(CT^{-1} - H)T \\ \dot{K} &= -B'T'(T(A+BKC)T^{-1} - F)(T')^{-1}C' \end{aligned}}$$
(3.18)

(b) The equilibrium points $(T_\infty, K_\infty) \in \mathbb{R}^{n \times n} \times \mathbb{R}^{n \times m}$ are characterised by

$$[T_\infty(A+BKC)T_\infty^{-1} - F, (T_\infty(A+BK_\infty C)T_\infty^{-1})'] =$$
$$(T_\infty')^{-1}C'(CT_\infty^{-1} - H) - (T_\infty B - G)B'T_\infty',$$
$$CT_\infty^{-1}(T_\infty^{-1}(T_\infty(A+BK_\infty C)T_\infty^{-1} - F)'T_\infty B = 0$$

5.3. OUTPUT FEEDBACK CONTROL

(c) Let $(T(t), K(t))$ be a solution of (3.18). Then $(A(t), B(t), C(t)) := (T(t)(A + BK(t)C)T(t)^{-1}, T(t)B, CT(t)^{-1})$ is a solution (3.11).

Proof The Fréchet-derivative of $\phi : GL(n, \mathbb{R}) \times \mathbb{R}^{m \times p} \to \mathbb{R}$ is the linear map on the tangent space defined by

$$\begin{aligned}
D\Phi|_{(T,K)}(XT, L) &= 2\text{tr}((T(A+BKC)T^{-1} - F)'([X, T(A+BKC)T^{-1}] \\
&\quad + TBLCT^{-1})) + 2\text{tr}((TB-G)'XTB \\
&\quad - (CT^{-1} - H)CT^{-1}X) \\
&= 2\text{tr}(X([T(A+BKC)T^{-1}, (T(A+BKC)T^{-1} - F)'] \\
&\quad + TB(TB-G)' - (CT^{-1} - H)'CT^{-1})) \\
&\quad + 2\text{tr}(L(CT^{-1}(T(A+BKC)T^{-1} - F)'TB))
\end{aligned}$$

Therefore the gradient vector $\text{grad}\phi(T, K)$ is

$$\text{grad}\phi(T, K) = \begin{pmatrix} [T(A+BKC)T^{-1} - F, (T(A+BKC)T^{-1})'] \\ + (TB-G)B'T'T - (T')^{-1}C'(CT^{-1} - H)T \\ B'T'(T(A+BKC)T^{-1} - F)(T')^{-1}C' \end{pmatrix}$$

This proves (a). Part (b) follows immediately from (3.18). For (c) note that (3.18) is equivalent to

$$\begin{aligned}
\dot{T} &= -[\mathcal{A}(t) - F, \mathcal{A}(t)']T - (\mathcal{B}(t) - G)\mathcal{B}(t)'T + \mathcal{C}(t)'(\mathcal{C} - H)T \\
\dot{K} &= \mathcal{B}(t)'(\mathcal{A}(t) - F)\mathcal{C}(t)',
\end{aligned}$$

where $\mathcal{A} = T(A + BKC)T^{-1}$, $\mathcal{B} = TB$ and $\mathcal{C} = CT^{-1}$. Thus,

$$\begin{aligned}
\dot{\mathcal{A}} &= [\dot{T}T^{-1}, \mathcal{A}] + \mathcal{B}\dot{K}\mathcal{C} \\
&= -[[\mathcal{A} - F, \mathcal{A}'] + (\mathcal{B} - G)\mathcal{B}' - \mathcal{C}'(\mathcal{C} - H), \mathcal{A}] - \mathcal{B}\mathcal{B}'(\mathcal{A} - F)\mathcal{C}'\mathcal{C} \\
\dot{\mathcal{B}} &= \dot{T}T^{-1}\mathcal{B} = -([\mathcal{A} - F, \mathcal{A}'] + (\mathcal{B} - G)\mathcal{B}' - \mathcal{C}'(\mathcal{C} - H))\mathcal{B} \\
\dot{\mathcal{C}} &= \mathcal{C}\dot{T}T^{-1} = \mathcal{C}([\mathcal{A} - F, \mathcal{A}'] + (\mathcal{B} - G)\mathcal{B}' - \mathcal{C}'(\mathcal{C} - H)),
\end{aligned}$$

which completes the proof of (c). ∎

Remarks 1. Note that the function $\phi : GL(n, \mathbb{R}) \times \mathbb{R}^{m \times p} \to \mathbb{R}$ is not necessarily proper. Therefore the existence of the solutions $(T(t), K(t))$ of (3.18) for all $t \geq 0$ is not guaranteed a-priori. In particular, finite escape time behaviour is not precluded.

2. A complete phase portrait analysis of (3.18) would be desirable but is not available. Such an understanding requires deeper knowledge of the geometry of the output feedback problem. Important question here are:

- Under which conditions are there finitely many equilibrium points of (3.18)?

- Existence of global minima for $\phi : GL(n, \mathbb{R}) \times \mathbb{R}^{m \times p} \to \mathbb{R}$?
- Are there any local minima of the cost function $\phi : GL(n, \mathbb{R}) \times \mathbb{R}^{m \times p} \to \mathbb{R}$ besides global minima? ∎

Optimal Eigenvalue Assignment For optimal eigenvalue assignment problems one is interested in minimizing the cost function

$$\varphi : GL(n, \mathbb{R}) \times \mathbb{R}^{m \times p} \to \mathbb{R}$$
$$\varphi(T, K) = \|T(A + BKC)T^{-1} - F\|^2 \qquad (3.19)$$

rather than (3.16). In this case the gradient flow $(\dot{T}, \dot{K}) = -\text{grad}\varphi(T, K)$ on $GL(n, \mathbb{R}) \times \mathbb{R}^{m \times p}$ is readily computed to be

$$\boxed{\begin{aligned} \dot{T} &= [(T(A + BKC)T^{-1})', T(A + BKC)T^{-1} - F]T \\ \dot{K} &= B'T'(F - T(A + BKC)T^{-1})(T')^{-1}C'. \end{aligned}} \qquad (3.20)$$

The following result is an immediate consequence of the proof of Theorem 3.2.

Corollary 3.4 (Eigenvalue Assignment) *Let $(A, B, C) \in L(n, m, p)$ and $F \in \mathbb{R}^{n \times n}$.*

(a) The gradient flow $(\dot{A}, \dot{B}, \dot{C}) = -\text{grad}\Psi(A, B, C)$ of $\Psi : \mathcal{F}(A, B, C) \to \mathbb{R}$, $\Psi(A, B, C) = \|A - F\|^2$, with respect to the normal Riemannian metric is

$$\boxed{\begin{aligned} \dot{A} &= [A, [A - F, A']] - BB'(A - F)C'C \\ \dot{B} &= -[A - F, A']B \\ \dot{C} &= C[A - F, A'] \end{aligned}} \qquad (3.21)$$

(b) Equilibrium points $(A_\infty, B_\infty, C_\infty)$ of (3.21) are characterized by

$$[A_\infty - F, A'_\infty] = 0 \qquad (3.22)$$
$$B'_\infty(A_\infty - F)C_\infty = 0 \qquad (3.23)$$

(c) For every initial condition $(A(0), B(0), C(0)) \in \mathcal{F}(A, B, C)$ the solution $(A(t), B(t), C(t))$ of (3.21) exists for all $t \geq 0$ and remains in $\mathcal{F}(A, B, C)$.

(d) Every solution $(A(t), B(t), C(t))$ of (3.21) converges to a connected component of the set of equilibrium points $(A_\infty, B_\infty, C_\infty) \in \overline{\mathcal{F}(A, B, C)}$.

In the special case where $(F, G, H) = (F', H', G')$ and also $(A, B, C) = (A', C', B')$ are symmetric realizations the equation (3.20) restricts to a gradient flow on $O(n) \times S(m)$, where $S(m)$ is the set of $m \times m$ symmetric

5.3. OUTPUT FEEDBACK CONTROL

matrices. In this case the flow (3.20) simplifies to

$$\dot{\Theta} = [F, \Theta(A + BKB')\Theta']\Theta, \quad \Theta_0 \in O(n)$$
$$\dot{K} = B'(\Theta'F\Theta - (A + BKB')B, \quad K_0 \in S(m)$$

The convergence properties of this flow is more easily analysed due to the compact nature of $O(n)$, see Mahony and Helmke (1993).

Problems 1. A flow on $L(n, m, p)$

$$\dot{A} = f(A, B, C)$$
$$\dot{B} = g(A, B, C)$$
$$\dot{C} = h(A, B, C)$$

is called *feedback preserving* if for all t the solutions $(A(t), B(t), C(t))$ are feedback equivalent to $(A(0), B(0), C(0))$, i.e.

$$\mathcal{F}(A(t), B(t), C(t)) = \mathcal{F}(A(0), B(0), C(0))$$

for all t and all $(A(0), B(0), C(0)) \in L(n, m, p)$. Show that every feedback preserving flow on $L(n, m, p)$ has the form

$$\dot{A}(t) = [A(t), \mathcal{L}(t)] + BM(t)C$$
$$\dot{B}(t) = \mathcal{L}(t)B(t)$$
$$\dot{C}(t) = -C\mathcal{L}(t)$$

for suitable matrix functions $t \mapsto \mathcal{L}(t)$, $t \mapsto M(t)$. Conversely, show that every such flow on $L(n, m, p)$ is feedback preserving.

2. Let $N(s)D(s)^{-1} = C(sI - A)^{-1}B$ be a coprime factorization of the transfer function $C(sI - A)^{-1}B$, where $N(s) \in \mathbb{R}[s]^{p \times m}$, $D(s) \in \mathbb{R}[s]^{m \times m}$. Show that the coefficients of the polynomial entries appearing in $N(s)$ are invariants for the flow (3.11). Use this to find n independent algebraic invariants for (3.11) when $m = p = 1$!

3. For any integer $N \in \mathbb{N}$ let $(A_1, B_1, C_1), \ldots, (A_N, B_N, C_N) \in L(n, m, p)$ be given. Prove that

$$\mathcal{F}((A_1, B_1, C_1), \ldots, (A_N, B_N, C_N)) :=$$
$$\{(T(A_1 + B_1KC_1)T^{-1}, TB_1, C_1T^{-1}),$$
$$\ldots, (T(A_N + B_NKC_N)T^{-1}, TB_N, C_NT^{-1}) \mid T \in GL(n, \mathbb{R}), K \in \mathbb{R}^{m \times p}\}$$

is a smooth submanifold of the N-fold product space $L(n, m, p) \times \ldots \times L(n, m, p)$.

4. Prove that the tangent space of $\mathcal{F}((A_1, B_1, C_1), \ldots, (A_N, B_N, C_N))$ at $(A_1, B_1, C_1), \ldots, (A_N, B_N, C_N)$ is equal to

$$\{([X, A_1] + B_1 L C_1, X B_1, -C X_1),$$
$$\ldots, ([X, A_N] + B_N L C_N, X B_N, -C X_N) \mid X \in \mathbb{R}^{n \times n}, L \in \mathbb{R}^{m \times p}\}.$$

5. Show that the gradient vector field of

$$\Psi((A_1, B_1, C_1), \ldots, (A_N, B_N, C_N)) := \sum_{i=1}^{N} \|A_i - F_i\|^2$$

with respect to the normal Riemannian metric on the orbit $\mathcal{F}((A_1, B_1, C_1), \ldots, (A_N, B_N, C_N))$ is

$$\dot{A}_i = [A_i, \sum_{j=1}^{N} [A_j - F_j, A_j']] - \sum_{j=1}^{N} B_j B_j' (A_j - F_j) C_j' C_j,$$

$$\dot{B}_i = -\sum_{j=1}^{N} [A_j - F_j, A_j'] B_i,$$

$$\dot{C}_i = C_i \sum_{j=1}^{N} [A_j - F_j, A_j'],$$

for $i = 1, \ldots N$. Characterize the equilibrium points.

6. Show that every solution $(T(t), K(t))$ of (3.18) satisfies the feedback gain bound

$$\|K(t) - K(0)\|^2 \leq \frac{1}{2} \phi(T(0), K(0)).$$

Main Points of Section The optimal eigenstructure assignment problem for linear systems (A, B, C) is that of finding a feedback equivalent system such that the resulting closed loop state matrix $T(A + BKC)T^{-1}$ is as close as possible to a prescribed target matrix F. A related task in control theory is that of eigenvalue assignment or pole placement. A generalisation of the eigenstructure assignment problem is that of optimal system assignment. Here the distance of a target system to an ouput feedback orbit is minimized. Gradient flows achieving the solution to these problems are proposed. For symmetric systems realizations these flows generalise the double bracket flows. The approach easily extends to more complicated situations, i.e. for systems with symmetries and simultaneous eigenvalue assignment tasks.

Notes for Chapter 5

For a thorough investigation of the total least squares problem we refer to Golub and Van Loan (1980). In de Moor and David (1992) it is argued that behind every least squares estimation problem there is an algebraic Riccati equation. This has been confirmed by de Moor and David (1992) for the total least squares problem. A characterization of the optimal solution for the total least squares problem in terms of a solution of an associated algebraic Riccati equation is obtained. Their work is closely related to that of Helmke (1991) and a deeper understanding of the connection between these papers would be desirable.

The total least squares problem amounts to fitting a k-dimensional vector space to a data set in n real or complex dimensions. Thus the problem can be reformulated as that of minimizing the total least squares distance function, which is a trace function $\text{tr}(AH)$ defined on a Grassmann manifold of Hermitian projection operators H. See Pearson (1901) for early results. Byrnes and Willems (1986) and Bloch (1985) have used the symplectic structure on a complex Grassmann manifold to characterize the critical points of the total least squares distance function. Bloch (1985, 1987) has analyzed the Hamiltonian flow associated with the function and described its statistical significance. In Bloch (1990) the Kähler structure of complex Grassmannians is used to derive explicit forms of the associated Hamiltonian and gradient flows. He shows that the gradient flow of the total least squares distance function coincides with Brockett's double bracket equation.

In Helmke, Prechtel and Shayman (1993) a phase portrait analysis of the gradient flow (1.46) and the Riccati flow on spaces of symmetric matrices is given. Local stability properties of the linearizations at the equilibria points are determined. A recursive numerical algorithm based on a variable step-size discretization is proposed. In Helmke and Shayman (1993) the critical point structure of matrix least squares distance functions defined on manifolds of fixed rank symmetric, skew-symmetric and rectangular matrices is obtained.

Eckart and Young (1936) have solved the problem of approximating a matrix by one of lower rank, if the distance is measured by the Frobenius norm. They show that the best approximant is given by a truncated singular value decomposition. Mirsky (1960) has extended their result to arbitrary unitarily invariant norms, i.e. matrix norms which satisfy $\|UXV\| = \|X\|$ for all unitary matrices U and V. A generalization of the Eckart-Young-Mirsky theorem has been obtained by Golub, Hoffman and Stewart (1987). In that paper also the connection between total least squares estimation and the minimization task for the least squares distance function $F_A \colon M(n, M \times N) \to \mathbb{R}, F_A(X) = \|A - X\|^2$, is explained. Matrix approximation problems for structured matrices such as Hankel or Toeplitz matrices are of interest in control theory and signal processing. Important papers are those of Adamyan, Arov and Krein (1971), who derive the analogous result to the Eckart-Young-Mirsky theorem for Hankel operators. The theory of Adamyan, Arov and Krein had a great impact on model reduction control theory, see Kung and Lin (1981).

Halmos (1972) has considered the task of finding a best symmetric positive semidefinite approximant P of a linear operator A for the 2-norm. He derives a formula for the distance $\|A - P\|_2$. A numerical bisection method for finding a nearest positive semidefinite matrix approximant is studied by Higham (1988). Higham (1988) also gives a simple formula for the best positive semidefinite approximant to a matrix A in terms of the polar decomposition of the symmetrization $(A + A')/2$ of A. Problems with matrix definiteness constraints are described in Fletcher (1985), Parlett (1978). For further results on matrix nearness problems we refer to Demmel (1987), Higham (1986) and Van Loan (1985).

Baldi and Hornik (1989) have provided a neural network interpretation of the total least squares problem. They prove that the least squares distance function (1.23) $F_A: M(n, N \times N) \to \mathbb{R}, F_A(X) = \|A - X\|^2$, has generically a unique local and global minimum. Moreover, a characterization of the critical points is obtained. Thus, equivalent results to those appearing in the first section of the chapter are obtained.

The polar decomposition can be defined for an arbitrary Lie group. It is then called the Iwasawa decomposition. For algorithms computing the polar decomposition we refer to Higham (1986).

Textbooks on feedback control and linear systems theory are those of Kailath (1980), Sontag (1990) and Wonham (1985). A celebrated result from linear systems theory is the pole-shifting theorem of Wonham (1967). It asserts that a state space system $(A, B) \in \mathbb{R}^{n \times n} \times \mathbb{R}^{n \times m}$ is controllable if and only if for every monic real polynomial $p(s)$ of degree n there exists a state feedback gain $K \in \mathbb{R}^{m \times n}$ such that the characteristic polynomial of $A + BK$ is $p(s)$. Thus controllability is a necessary and sufficient condition for eigenvalue assignment by state feedback.

The corresponding, more general, problem of eigenvalue assignment by constant gain output feedback is considerably harder and a complete solution is not known to this date. The question has been considered by many authors with important contributions by Brockett and Byrnes (1981), Byrnes (1989), Ghosh (1988), Hermann and Martin (1977), Kimura (1975), Rosenthal (1992), Wang (1991) and Willems and Hesselink (1978). Important tools for the solution of the problem are those from algebraic geometry and topology. The paper of Byrnes (1989) is an excellent survey of the recent

5.3. OUTPUT FEEDBACK CONTROL

developments on the output feedback problem.

Hermann and Martin (1977) have shown that the condition $mp \geq n$ is necessary for generic eigenvalue assignment by constant gain output feedback. A counter-example ($m = p = 2, n = 4$) of Willems and Hesselink (1978) shows that the condition $mp \geq n$ is in general not sufficient for generic eigenvalue assignment. Brockett and Bynres (1981) have shown that generic eigenvalue assignment is possible if $mp = n$ and an additional hypothesis on the degree of the pole-placement map is satisfied. Wang (1991) has shown that $mp > n$ is a sufficient condition for generic eigenvalue assignment via output feedback.

For results on simultaneous eigenvalue assignment of a finite number of linear systems by output feedback we refer to Ghosh (1988).

The development of efficient numerical methods for eigenvalue assignment by output feedback is a challenge. Methods from matrix calculus have been applied by Godbout and Jordan (1980) to determine gradient matrices for output feedback. In Mahony and Helmke (1993) gradient flows for optimal eigenvalue assignment for symmetric state space systems, of interest in circuit theory, are studied.

References

Adamyan, Arov and Krein (1971) Analytic properties of Schmidt pairs for a Hankel operator and the generalized Schur-Takagi problem, *Math. USSR Sbornik* 15, 31-73.

Baldi P. and K. Hornik (1989) Neural networks and principal component analysis: learning from examples without local minima, *Neural Networks* 2, 53-58.

Bloch, A.M. (1985) Estimation, principal components and Hamiltonian systems, *Systems and Control Letters* 3, 103-108.

Bloch, A.M. (1987) An infinite-dimensional Hamiltonian system on projective Hilbert space, *Transactions of the American Mathematical Society* 302, 787-796.

Bloch, A.M. (1990) The Kähler structure of the total least squares problem, Brockett's steepest descent equations, and constrained flows, *Realization and Modelling in Systems Theory* (M. Kaashoek, J.H. van Schuppen and A.C. Ran, Eds.), Birkhäuser Publ., Boston.

Brockett, R.W. (1989) Least squares matching problems, *Linear Algebra Appl.* 122-127, 701-777.

Brockett, R.W. and C.I. Byrnes (1981) Multivariable Nyquist criteria, root loci and pole placement: A geometric viewpoint, *IEEE Transactions on Automatic Control* AC-26, 271-284.

Byrnes, C.I. (1989) Pole assignment by output feedback, *Lecture Notes in Control and Information Sciences* 135, Springer-Verlag, Berlin, 31-78.

Byrnes, C.I. and J.C. Willems (1986) Least-squares estimation, linear programming and momentum: A geometric parametrization of local minima, *IMA J. of Math. Control and Inf.* 3, 103-118.

Demmel, J.W. (1987) On condition numbers and the distance to the nearest ill-posed problem, *Numer. Math.* 51, 251-289.

De Moor, B. and J. David (1992) Total linear least squares and the algebraic Riccati equation, *Systems and Control Letters* 5, 329-337.

Eckart, G. and Young, G. (1936) The approximation of one matrix by another of lower rank, *Psychometrika* 1. 221-218.

Fletcher, R. (1985) Semi-definite matrix constraints in optimization, *SIAM J. Control Optim.* 23, 493-513.

Ghosh, B.K. (1988) An approach to simultaneous system design. Part II: Nonswitching gain and dynamic feedback compensation by algebraic geometric methods, *SIAM J. Control and Optimization* 26, 919-963.

Godbout, L.F. and D. Jordan (1980) Gradient matrices for output feedback systems, *International Journal of Control* 32, 411-433.

Golub, G.H. and C. Van Loan (1980) An analysis of the total least squares problem, *SIAM J. Numer. Anal.* 17, 883-843.

Golub, G.H., A. Hoffman and G.W. Stewart (1987) A generalization of the Eckart-Young-Mirsky matrix approximation theorem, *Linear Algebra and its Appl.* 88/89, 317-327.

Halmos, P. (1972) Positive approximants of operators, *Indiana Univ. Math. U.* 21, 951-960.

Hayden, T.L. and Well, J. (1988) Approximation by matrices positive semidefinite on a subspace, *Linear Algebra and its Appl.* 109, 115-130.

Helmke, U. (1991) Isospectral flows on symmetric matrices and the Riccati equation, *Systems and Control Letters* 16, 159-166.

Helmke, U. and M.A. Shayman (1993) Critical points of matrix least squares distance functions, *Linear Algebra and its Applications*, to appear.

Helmke, U., Prechtel, M. and M.A. Shayman (1993) Riccati-like flows and matrix approximations, submitted to *Kybernetika*.

Hermann, R. and C.F. Martin (1977) Applications of algebraic geometry to system theory, Part I, *IEEE Transactions on Automatic Control* AC-22, 19-25.

Higham, N.J. (1986) Computing the polar decomposition - with applications, *SIAM J. Sci. Stast. Comput.* 7, 1160-1174.

Higham, N.J. (1988) Computing a nearest symmetric positive semidefinite matrix, *Linear Algebra and its Appl.* 103, 103-118.

Kailath, T. (1980) *Linear Systems*, Prentice Hall, Englewood Cliffs.

Kimura, H. (1975) Pole assignment by gain output feedback, *IEEE Transactions on Automatic Control* AC-20, 509-516.

Kung, S.Y. and D.W. Lin (1981) Optimal Hankel-norm model reductions: multivariable systems, *Transactions on Automatic Control* AC-26, 832-852.

Mahony, R.E. and U. Helmke (1993) System assignment and pole placement for symmetric realizations, preprint.

Mirsky, L. (1960) Symmetric gauge functions and unitarily invariant norms, *Quart. J. Math. Oxford* 11, 50-59.

Parlett, B.N. (1978) Progress in numerical analysis, *SIAM Rev.* 20, 443-456.

Pearson, K. (1901) On lines and planes of closest fit to points in space, *Phil. Mag.* 559-572.

Rosenthal, J. (1992) New results in pole assignment by real output feedback, *SIAM J. Control and Optimization* 30, 203-211.

Shayman, M.A. (1982) Morse functions for the classical groups and Grassmann manifolds, unpublished manuscript.

Sontag, E.D. (1990) *Mathematical Control Theory*, Springer-Verlag, New York.

van Loan, C.F. (1985) How near is a stable matrix to an unstable matrix? *Linear Algebra and its Role in Systems Theory* (B.N. Datta, Ed.), Contemporary Math. 47, Amer. Math. Soc., 465-478.

Wang, X. (1991) On output feedback via Grassmannians, *SIAM J. Control and Optimization* 29, 926-935.

Willems, J.C. and W.H. Hesselink (1978) Generic properties of the pole placement problem, *Proc. IFAC World Congress*, Helsinki.

Wonham, W.M. (1967) On pole assignment in multi-input controllable linear systems, *IEEE Transactions on Automatic Control* AC-12, 660-665.

Wonham, W.M. (1985) *Linear Multivariable Control*, Third Edition, Springer-Verlag, New York

Chapter 6

Balanced Matrix Factorizations

6.1 Introduction

The singular value decomposition of a finite-dimensional linear operator is a special case of the following more general matrix factorization problem:

Given a matrix $H \in \mathbb{R}^{k \times \ell}$ find matrices $X \in \mathbb{R}^{k \times n}$ and $Y \in \mathbb{R}^{n \times \ell}$ such that
$$H = XY. \tag{1.1}$$
A factorization (1.1) is called *balanced* if
$$X'X = YY' \tag{1.2}$$
holds and *diagonal balanced* if
$$X'X = YY' = D \tag{1.3}$$
holds for a diagonal matrix D. If $H = XY$ then $H = XT^{-1}TY$ for all invertible $n \times n$ matrices T, so that factorizations (1.1) are never unique. A coordinate basis transformation $T \in GL(n)$ is called *balancing*, and *diagonal balancing*, if (XT^{-1}, TY) is a balanced, and diagonal balanced factorization, respectively.

Diagonal balanced factorizations (1.3) with $n = \text{rank}(H)$ are equivalent to the singular value decomposition of H. In fact if (1.3) holds for $H = XY$ then with
$$U = D^{-\frac{1}{2}}Y, \quad V = XD^{-\frac{1}{2}}, \tag{1.4}$$
we obtain the singular value decomposition $H = VDU, UU' = I_n, V'V = I_n$.

A parallel situation arises in system theory, where factorizations $H = O \cdot R$ of a Hankel matrix H by the observability and controllability matrices O and R are crucial for realization theory. Realizations (A, B, C) are easily constructed from O and R. Balanced and diagonal balanced realizations then correspond to balanced and diagonal balanced factorizations of the Hankel matrix. While diagonal balanced realizations are an important tool for system approximation and model reduction theory, they are not always best for certain minimum sensitivity applications arising, for example in digital filtering. There is therefore the need to study both types of factorizations as developed in this chapter. System theory issues are studied in Chapters 7-9, based on the analysis of balanced matrix factorizations as developed in this chapter.

In order to come to grips with the factorization tasks (1.1) - (1.3) we have to study the underlying geometry of the situation.

Let

$$\mathcal{F}(H) = \{(X, Y) \in \mathbb{R}^{k \times n} \times \mathbb{R}^{n \times \ell} \mid H = XY\}, \quad n = \mathrm{rk}(H) \qquad (1.5)$$

denote the set of all minimal, that is full rank, factorizations (X, Y) of H. Applying a result from linear algebra, Appendix A.8, we have

$$\mathcal{F}(H) = \{(XT^{-1}, TY) \in \mathbb{R}^{k \times n} \times \mathbb{R}^{n \times \ell} \mid T \in GL(n)\} \qquad (1.6)$$

for any initial factorization $H = XY$ with $\mathrm{rk}(X) = \mathrm{rk}(Y) = n$. We are thus led to study the sets

$$\mathcal{O}(X, Y) := \{(XT^{-1}, TY) \in \mathbb{R}^{k \times n} \times \mathbb{R}^{n \times \ell} \mid T \in GL(n)\} \qquad (1.7)$$

for arbitrary n, k, l. The geometry of $\mathcal{O}(X, Y)$ and their topological closures is of interest for invariant theory and the necessary tools are developed in Sections 2 and 3. It is shown there that $\mathcal{O}(X, Y)$ and hence $\mathcal{F}(H)$ are smooth manifolds. Balanced and diagonal balanced factorizations are characterized as the critical points of the cost functions

$$\Phi_N : \mathcal{O}(X, Y) \to \mathbb{R}$$
$$\Phi_N(XT^{-1}, TY) = \mathrm{tr}(N(T')^{-1}X'XT^{-1}) + \mathrm{tr}(NTYY'T') \qquad (1.8)$$

in terms of a symmetric target matrix N, which is the identity matrix or a diagonal matrix, respectively. If N is the identity matrix, the function Φ_I, denoted Φ, has the appealing form

$$\Phi(XT^{-1}, TY) = \|XT^{-1}\|^2 + \|TY\|^2, \qquad (1.9)$$

for the Frobenius norm.

We are now able to apply gradient flow techniques in order to compute balanced and diagonal balanced factorizations. This is done in Sections 3-5.

6.2 Kempf-Ness Theorem

The purpose of this section is to recall an important recent result as well as the relevant terminology from invariant theory. The result is due to Kempf and Ness (1979) and plays a central role in our approach to the balanced factorization problem. For general references on invariant theory we refer to Kraft (1984), Dieudonné and Carrell (1971).

Digression on group actions Let $GL(n)$ denote the group of invertible real $n \times n$ matrices and let $O(n) \subset GL(n)$ denote the subgroup consisting of all real orthogonal $n \times n$ matrices Θ characterized by $\Theta\Theta' = \Theta'\Theta = I_n$.

A *group action* of $GL(n)$ on a finite dimensional real vector space V is a map

$$\alpha : GL(n) \times V \to V, \quad (g, x) \mapsto g \cdot x \qquad (2.1)$$

satisfying for all $g, h \in GL(n), x \in V$

$$g \cdot (h \cdot x) = (gh) \cdot x, \quad e \cdot x = x$$

where $e = I_n$ denotes the $n \times n$ identity matrix.

If the coordinates of $g \cdot x$ in (2.1) are polynomials in the coefficients of g, x and $\det(g)^{-1}$ then, (2.1) is called an *algebraic group action*, and $\alpha : GL(n) \times V \to V$ is called a *linear* algebraic group action if in addition the maps

$$\alpha_g : V \to V, \quad \alpha_g(x) = g \cdot x \qquad (2.2)$$

are linear, for all $g \in GL(n)$.

A typical example of a linear algebraic group action is the similarity action

$$\alpha : GL(n) \times \mathbb{R}^{n \times n} \to \mathbb{R}^{n \times n},$$

$$(S, A) \mapsto SAS^{-1} \qquad (2.3)$$

where the (i,j)-entry of SAS^{-1} is a polynomial in the coefficients of S, A and $(\det S)^{-1}$.

Given a group action (2.1) and an element $x \in V$, the *stabilizer* of x is the subgroup of $GL(n)$ defined by

$$\text{Stab}(x) = \{g \in GL(n) \mid g \cdot x = x\} \qquad (2.4)$$

The *orbit* of x is defined as the set $\mathcal{O}(x)$ of all $y \in V$ which are $GL(n)$-equivalent to x, that is

$$\mathcal{O}(x) = \{g \cdot x \mid g \in GL(n)\} \qquad (2.5)$$

We say that the orbit $\mathcal{O}(x)$ is *closed* if $\mathcal{O}(x)$ is a closed subset of V.

In the above example, the orbit $\mathcal{O}(A)$ of an $n \times n$ matrix A is just the set of all matrices B which are similar to A.

If $GL(n)/\text{Stab}(x)$ denotes the quotient space of $GL(n)$ by the stabilizer group $\text{Stab}(x)$ then $\mathcal{O}(x)$ is homeomorphic to $GL(n)/\text{Stab}(x)$. In fact, for any algebraic group action of $GL(n)$, both $\mathcal{O}(x)$ and $GL(n)/\text{Stab}(x)$ are smooth manifolds which are diffeomorphic to each other; see Appendix C. ∎

A positive definite inner product $\langle\ ,\ \rangle$ on the vector space V is called *orthogonally invariant* with respect to the group action (2.1) if

$$\langle \Theta \cdot x, \Theta \cdot y \rangle = \langle x, y \rangle \qquad (2.6)$$

holds for all $x, y \in V$ and $\Theta \in O(n)$. This induces an $O(n)$-invariant Hermitian norm on V defined by $||x||^2 = \langle x, x \rangle$.

Choose any such $O(n)$-invariant norm on V. For any given $x \in V$ we consider the distance or norm functions

$$\Phi : \mathcal{O}(x) \to \mathbb{R}, \quad \Phi(y) = ||y||^2 \qquad (2.7)$$

and

$$\phi_x : GL(n) \to \mathbb{R}, \quad \phi_x(g) = ||g \cdot x||^2 \qquad (2.8)$$

Note that $\phi_x(g)$ is just the square of the distance of the transformed vector $g \cdot x$ to the origin $0 \in V$.

We can now state the Kempf-Ness result, where the real version presented here is actually due to Slodowy (1989). We will not prove the theorem here, because we will prove in Chapter 6 a generalization of the Kempf-Ness theorem which is due to Azad and Loeb (1990).

Theorem 2.1 *Let $\alpha : GL(n) \times V \to V$ be a linear algebraic action of $GL(n)$ of a finite-dimensional real vector space V and let $||\cdot||$ be derived from an orthogonally invariant inner product on V. Then*

(a) The orbit $\mathcal{O}(x)$ is closed if and only if the norm function $\Phi : \mathcal{O}(x) \to \mathbb{R}$ (respectively $\phi_x : G \to \mathbb{R}$) has a global minimum (i.e. $||g_0 \cdot x|| = \inf_{g \in G} ||g \cdot x||$ for some $g_0 \in G$).

(b) Let $\mathcal{O}(x)$ be closed. Every critical point of $\Phi : \mathcal{O}(x) \to \mathbb{R}$ is a global minimum and the set of global minima of Φ is a single $O(n)$-orbit.

(c) Let $\mathcal{O}(x)$ be closed and let $C(\Phi) \subset \mathcal{O}(x)$ denote the set of all critical points of Φ. Let $x_0 \in C(\Phi)$ be a critical point of $\Phi : \mathcal{O}(x) \to \mathbb{R}$. Then the Hessian $D^2\Phi|_{x_0}$ of Φ at x_0 is positive semidefinite and $D^2\Phi|_{x_0}$ degenerates exactly on the tangent space $T_{x_0}C(\Phi)$.

Any function $F : \mathcal{O}(x) \to \mathbb{R}$ which satisfies condition (c) of the above theorem is called a *Morse-Bott* function. We say that Φ is a *perfect Morse-Bott* function.

Example (Similarity Action) Let $\alpha : GL(n) \times \mathbb{R}^{n \times n} \to \mathbb{R}^{n \times n}, (S, A) \mapsto SAS^{-1}$, be the similarity action on $\mathbb{R}^{n \times n}$ and let $||A||^2 = \text{tr}(AA')$ be the Frobenius norm. Then $||A|| = ||\Theta A \Theta'||$ for $\Theta \in O(n)$ is an orthogonal invariant Hermitian norm of $\mathbb{R}^{n \times n}$. It is easy to see and a well known fact from invariant theory that a similarity orbit

$$\mathcal{O}(A) = \{SAS^{-1} \mid S \in GL(n)\} \tag{2.9}$$

is a closed subset of $\mathbb{R}^{n \times n}$ if and only if A is diagonalizable over $\mathbb{C}$. A matrix $B = SAS^{-1} \in \mathcal{O}(A)$ is a critical point for the norm function

$$\Phi : \mathcal{O}(A) \to \mathbb{R}, \quad \Phi(B) = \text{tr}(BB') \tag{2.10}$$

if and only if $BB' = B'B$, i.e. if and only if B is *normal*. Thus the Kempf-Ness theorem implies that, for A diagonalizable over $\mathbb{C}$, the normal matrices $B \in \mathcal{O}(A)$ are the global minima of (2.10) and that there are no other critical points. ∎

Main Points of Section The important notions of group actions and orbits are introduced. An example of a group action is the similarity action on $n \times n$ matrices and the orbits here are the equivalence classes formed by similar matrices. Closed orbits are important since on those there always exists an element with minimal distance to the origin. The Kempf-Ness theorem characterizes the closed orbits of a $GL(n)$ group action and shows that there are no critical points except for global minima of the norm distance function. In the case of the similarity action, the normal matrices are precisely those orbit elements which have minimal distance to the zero matrix.

6.3 Global Analysis of Cost Functions

We now apply the Kempf-Ness theorem to our cost functions $\Phi : \mathcal{O}(X, Y) \to \mathbb{R}$ and $\Phi_N : \mathcal{O}(X, Y) \to \mathbb{R}$ introduced in Section 6.1.

Let $V = \mathbb{R}^{k \times n} \times \mathbb{R}^{n \times \ell}$ denote the vector space of all pairs of $k \times n$ and $n \times \ell$ matrices X and Y. The group $GL(n)$ of invertible $n \times n$ real matrices T acts on V by the linear algebraic group action

$$\alpha : GL(n) \times (\mathbb{R}^{k \times n} \times \mathbb{R}^{n \times \ell}) \to \mathbb{R}^{k \times n} \times \mathbb{R}^{n \times \ell}$$

$$(T, (X, Y)) \mapsto (XT^{-1}, TY). \tag{3.1}$$

The orbits

$$\mathcal{O}(X,Y) = \{(XT^{-1}, TY) \mid T \in GL(n)\}$$

are thus smooth manifolds, Appendix C. We endow V with the Hermitian inner product defined by

$$\langle (X_1, Y_1), (X_2, Y_2) \rangle = \operatorname{tr}(X_1' X_2 + Y_1 Y_2') \tag{3.2}$$

The induced Hermitian norm

$$\|(X,Y)\|^2 = \operatorname{tr}(X'X + YY') \tag{3.3}$$

is orthogonally invariant, that is $\|(X\Theta^{-1}, \Theta Y)\| = \|(X,Y)\|$ for all orthogonal $n \times n$ matrices $\Theta \in O(n)$.

Lemma 3.1 *Let $(X,Y) \in \mathbb{R}^{k \times n} \times \mathbb{R}^{n \times \ell}$. Then*

(a) $\mathcal{O}(X,Y)$ is a smooth submanifold of $\mathbb{R}^{k \times n} \times \mathbb{R}^{n \times \ell}$.

(b) The tangent space of $\mathcal{O} = \mathcal{O}(X,Y)$ at (X,Y) is

$$T_{(X,Y)}\mathcal{O} = \{(-X\Lambda, \Lambda Y) \mid \Lambda \in \mathbb{R}^{n \times n}\} \tag{3.4}$$

Proof $\mathcal{O}(X,Y)$ is an orbit of the smooth algebraic Lie group action (3.1) and thus, by Appendix C, a smooth submanifold of $\mathbb{R}^{k \times n} \times \mathbb{R}^{n \times \ell}$. To prove (b) we consider the smooth map

$$\varphi : GL(n) \to \mathcal{O}(X,Y), \quad T \mapsto (XT^{-1}, TY) \tag{3.5}$$

The derivative of φ at the identity matrix I_n is the surjective linear map $D\varphi|_{I_n} : T_{I_n} GL(n) \to T_{(X,Y)}\mathcal{O}$ defined by

$$D\varphi|_{I_n}(\Lambda) = (-X\Lambda, \Lambda Y)$$

for $\Lambda \in T_{I_n} GL(n) = \mathbb{R}^{n \times n}$. The result follows. ∎

The above characterization of the tangent spaces of course remains in force if the point $(X,Y) \in \mathcal{O}$ is replaced by any other point $(X_1, Y_1) \in \mathcal{O}$. In the sequel we will often denote a general point (X_1, Y_1) of $\mathcal{O}(X,Y)$ simply by (X,Y).

The following lemma is useful in order to characterize the closed orbits $\mathcal{O}(X,Y)$ for the group action (3.1). Let $\bar{A}$ denote the topological closure of subset A of a topological space M.

Lemma 3.2 *(a) $(X_1, Y_1) \in \mathcal{O}(X,Y)$ if and only if $X_1 Y_1 = XY$, $\ker(Y_1) = \ker(Y)$ and $\operatorname{image}(X_1) = \operatorname{image}(X)$.*

6.3. GLOBAL ANALYSIS OF COST FUNCTIONS

(b) $(X_1, Y_1) \in \overline{\mathcal{O}(X, Y)}$ if and only if $X_1 Y_1 = XY$, ker $(Y_1) \supset$ ker (Y) and image $(X_1) \subset$ image (X).

(c) $\mathcal{O}(X, Y)$ is closed if and only if ker $(Y) =$ ker (XY) and image $(X) =$ image (XY).

Proof See Kraft (1984) ∎

Corollary 3.3 $\mathcal{O}(X, Y)$ is closed if and only if $\operatorname{rk}(X) = \operatorname{rk}(Y) = \operatorname{rk}(XY)$.

Proof By Lemma 3.2 the orbit $\mathcal{O}(X, Y)$ is closed if and only if dim ker $(Y) =$ dim ker (XY) and dim image$(X) =$ dim image(XY). But $\operatorname{rk}(Y)$ + dim $\operatorname{ker}(Y) = \ell$ and thus the condition is satisfied if and only if $\operatorname{rk}(X) = \operatorname{rk}(Y) = \operatorname{rk}(XY)$. ∎

Let $\Phi : \mathcal{O}(X, Y) \to \mathbb{R}$ be the smooth function on $\mathcal{O}(X, Y)$ defined by

$$\begin{aligned} \Phi(XT^{-1}, TY) &= \|XT^{-1}\|^2 + \|TY\|^2 \\ &= \operatorname{tr}((T')^{-1} X' X T^{-1} + T Y Y' T') \end{aligned} \quad (3.6)$$

We have

Theorem 3.4 (a) An element $(X_0, Y_0) \in \mathcal{O}(X, Y)$ is a critical point of Φ if and only if $X_0' X_0 = Y_0 Y_0'$.

(b) There exists a minimum of $\Phi : \mathcal{O}(X, Y) \to \mathbb{R}$ in $\mathcal{O}(X, Y)$ if and only if $\operatorname{rk}(X) = \operatorname{rk}(Y) = \operatorname{rk}(XY)$.

(c) All critical points of $\Phi : \mathcal{O}(X, Y) \to \mathbb{R}$ are global minima. If (X_1, Y_1), $(X_2, Y_2) \in \mathcal{O}(X, Y)$ are global minima, then there exists an orthogonal transformation $\Theta \in O(n)$ such that $(X_2, Y_2) = (X_1 \Theta^{-1}, \Theta Y_1)$.

(d) Let $C(\Phi) \subset \mathcal{O}(X, Y)$ denote the set of all critical points of $\Phi : \mathcal{O}(X, Y) \to \mathbb{R}$. Then the Hessian $D^2 \Phi|_{(X_0, Y_0)}$ is positive semidefinite at each critical point (X_0, Y_0) of Φ and $D^2 \Phi|_{(X_0, Y_0)}$ degenerates exactly on the tangent space $T_{(X_0, Y_0)} C(\Phi)$ of $C(\Phi)$.

Proof The derivative $D\Phi|_{(X,Y)} : T_{(X,Y)} \mathcal{O} \to \mathbb{R}$ of Φ at any point $(X, Y) \in \mathcal{O}$ is the linear map defined by

$$D\Phi|_{(X,Y)}(\xi, \eta) = 2\operatorname{tr}(X'\xi + \eta Y').$$

By Lemma 3.1, the tangent vectors of $\mathcal{O}$ at (X, Y) are of the form $(\xi, \eta) = (-X\Lambda, \Lambda Y)$ for an $n \times n$ matrix Λ. Thus

$$D\Phi|_{(X,Y)}(\xi, \eta) = -2\operatorname{tr}[(X'X - YY')\Lambda],$$

which vanishes on $T_{(X,Y)}\mathcal{O}$ if and only if $X'X = YY'$. This proves (a). Parts (b), (c) and (d) are an immediate consequence of the Kempf-Ness Theorem 2.1 together with Corollary 3.3. ∎

Let $N = N' \in \mathbb{R}^{n \times n}$ be an arbitrary symmetric positive definite matrix and let $\Phi_N : \mathcal{O}(X,Y) \to \mathbb{R}$ be the weighted cost function defined by

$$\Phi_N(XT^{-1}, TY) = \operatorname{tr}(N(T')^{-1}X'XT^{-1} + NTYY'T') \qquad (3.7)$$

Note that Φ_N is in general not an orthogonally invariant norm function. Thus the next theorem does not follow immediately from the Kempf-Ness theory.

Theorem 3.5 Let $N = N' > 0$.

(a) (X_0, Y_0) is a critical point of Φ_N if and only if $NX_0'X_0 = Y_0Y_0'N$.

(b) There exists a minimum of $\Phi_N : \mathcal{O}(X,Y) \to \mathbb{R}$ in $\mathcal{O}(X,Y)$ if $\operatorname{rk}(X) = \operatorname{rk}(Y) = \operatorname{rk}(XY)$.

(c) Let $N = \operatorname{diag}(\mu_1, \cdots, \mu_n)$ with $\mu_1 > \cdots > \mu_n > 0$. Then (X_0, Y_0) is a critical point of Φ_N if and only if (X_0, Y_0) is diagonal balanced.

Proof The derivative of $\Phi_N : \mathcal{O}(X,Y) \to \mathbb{R}$ at (X,Y) is $D\Phi_N|_{(X,Y)} : T_{(X,Y)}\mathcal{O} \to \mathbb{R}$ with

$$\begin{aligned} D\Phi_N|_{(X,Y)}(\xi,\eta) &= 2\operatorname{tr}(NX'\xi + N\eta Y') \\ &= -2\operatorname{tr}((NX'X - YY'N)\Lambda) \end{aligned}$$

for $(\xi,\eta) = (-X\Lambda, \Lambda Y)$. Hence $D\Phi_N|_{(X,Y)} = 0$ on $T_{(X,Y)}\mathcal{O}$ if and only if $NX'X = Y'YN$, which proves (a). By Corollary 3.3 the rank equality condition of (b) implies that $\mathcal{O}(X,Y)$ is closed. Consequently, $\Phi_N : \mathcal{O}(X,Y) \to \mathbb{R}_+$ is a proper function and therefore a minimum in $\mathcal{O}(X,Y)$ exists. This proves (b). To prove (c) note that $NX'X = YY'N$ is equivalent to $NX'XN^{-1} = YY'$. By symmetry of YY' thus $N^2X'X = X'XN^2$ and hence

$$(\mu_i^2 - \mu_j^2)x_i'x_j = 0 \text{ for all } i,j = 1, \cdots, n$$

where $X = (x_1, \cdots, x_n)$. Since $\mu_i^2 \neq \mu_j^2$ for $i \neq j$ then $x_i'x_j = 0$ for $i \neq j$ and therefore both $X'X$ and YY' must be diagonal and equal. The result follows. ∎

Finally let us explore what happens if other orthogonal invariant norms are used. Thus let $\Omega_1 \in \mathbb{R}^{k \times k}$, $\Omega_2 \in \mathbb{R}^{\ell \times \ell}$ be symmetric positive definite and let

$$\Psi : \mathcal{O}(X,Y) \to \mathbb{R}, \quad \Psi(X,Y) = \operatorname{tr}(X'\Omega_1 X + Y\Omega_2 Y') \qquad (3.8)$$

The function Ψ defines an orthogonally invariant norm on $\mathcal{O}(X,Y)$. The critical points of $\Psi: \mathcal{O}(X,Y) \to \mathbb{R}$ are easily characterized by

$$X'\Omega_1 X = Y\Omega_2 Y' \tag{3.9}$$

and thus the Kempf-Ness Theorem 2.1 implies the following.

Corollary 3.6 *Let $\Omega_1 \in \mathbb{R}^{k \times k}$ and $\Omega_2 \in \mathbb{R}^{\ell \times \ell}$ be positive definite symmetric matrices. Then:*

(a) The critical points of $\Psi: \mathcal{O}(X,Y) \to \mathbb{R}$ defined by (3.8) are characterized by $X'\Omega_1 X = Y\Omega_2 Y'$.

(b) There exists a minimum of Ψ in $\mathcal{O}(X,Y)$ if and only if $\mathrm{rk}(X) = \mathrm{rk}(Y) = \mathrm{rk}(XY)$.

(c) The critical points of $\Psi: \mathcal{O}(X,Y) \to \mathbb{R}$ are the global minima. If the pairs $(X_1, Y_1), (X_2, Y_2) \in \mathcal{O}(X,Y)$ are global minima, then $(X_2, Y_2) = (X_1 \Theta^{-1}, \Theta Y_1)$ for an orthogonal matrix $\Theta \in O(n)$.

A similar result, analogous to Theorem 3.5, holds for $\Psi_N: \mathcal{O}(X,Y) \to \mathbb{R}$, $\Psi_N(X,Y) = \mathrm{tr}(NX'\Omega_1 X) + \mathrm{tr}(NY\Omega_2 Y')$.

Problem State and prove the analogous result to Theorem 3.5 for the weighted cost function $\Psi_N: \mathcal{O}(X,Y) \to \mathbb{R}$.

Main Points of Section A global analysis of the cost functions for balancing matrix factorizations is developed. The orbits $\mathcal{O}(X,Y)$ are shown to be smooth manifolds and their tangent spaces are computed. The closed orbits are characterized using the Kempf-Ness theorem. The global minima of the cost function are shown to be balanced. Similarly, diagonal balanced factorizations are characterized as the critical points of a weighted cost function. The Kempf-Ness theory does not apply to such weighted cost functions.

6.4 Flows for Balancing Transformations

In this section gradient flows evolving on the Lie group $GL(n)$ of invertible coordinate transformations $T \in GL(n)$ are constructed which converge exponentially fast to the class of all balancing coordinate transformations for a given factorization $H = XY$. Thus let cost functions $\Phi: GL(n) \to \mathbb{R}$ and $\Phi_N: GL(n) \to \mathbb{R}$ be defined by

$$\Phi(T) = \|XT^{-1}\|^2 + \|TY\|^2 \tag{4.1}$$

and

$$\Phi_N(T) = \mathrm{tr}(N(T')^{-1}X'XT^{-1}) + \mathrm{tr}(NTYY'T') \tag{4.2}$$

Given that

$$\Phi(T) = \mathrm{tr}[(T')^{-1}X'XT^{-1} + TYY'T']$$
$$= \mathrm{tr}(X'XP^{-1} + YY'P) \qquad (4.3)$$

for

$$P = T'T \qquad (4.4)$$

it makes sense to consider first the task of minimizing Φ with respect to the positive definite symmetric matrix $P = T'T$.

Gradient flows on positive definite matrices Let $\mathcal{P}(n)$ denote the set of positive definite symmetric $n \times n$ matrices $P = P' > 0$. The function we are going to study is

$$\phi : \mathcal{P}(n) \to \mathbb{R}, \quad \phi(P) = \mathrm{tr}(X'XP^{-1} + YY'P) \qquad (4.5)$$

Lemma 4.1 *Let $(X, Y) \in \mathbb{R}^{k \times n} \times \mathbb{R}^{n \times \ell}$ with $\mathrm{rk}(X) = \mathrm{rk}(Y) = n$. The function $\phi : \mathcal{P}(n) \to \mathbb{R}$ defined by (4.5) has compact sublevel sets, that is $\{P \in \mathcal{P}(n) \mid \mathrm{tr}(X'XP^{-1} + YY'P) \leq a\}$ is a compact subset of $\mathcal{P}(n)$ for all $a \in \mathbb{R}$.*

Proof Let $M_a := \{P \in \mathcal{P}(n) \mid \mathrm{tr}(X'XP^{-1} + YY'P) \leq a\}$ and $N_a := \{T \in GL(n) \mid \|XT^{-1}\|^2 + \|TY\|^2 \leq a\}$. Suppose that X and Y are full rank n. Then

$$\varphi : GL(n) \to \mathcal{O}(X, Y)$$
$$T \mapsto (XT^{-1}, TY)$$

is a homeomorphism which maps N_a homeomorphically onto $\phi(N_a) = \{(X_1, Y_1) \in \mathcal{O}(X, Y) \mid \|X_1\|^2 + \|Y_1\|^2 \leq a\}$. By Corollary 3.3, $\phi(N_a)$ is the intersection of the compact set $\{(X_1, Y_1) \in \mathbb{R}^{k \times n} \times \mathbb{R}^{n \times \ell} \mid \|X_1\|^2 + \|Y_1\|^2 \leq a\}$ with the closed set $\mathcal{O}(X, Y)$ and therefore is compact. Thus N_a must be compact. Consider the continuous map

$$F : M_a \times O(n) \to N_a, \quad F(P, \Theta) = \Theta P^{1/2}.$$

By the polar decomposition of $T \in GL(n)$, F is a bijection and its inverse is given by the continuous map $F^{-1} : N_a \to M_a \times O(n)$, $F^{-1}(T) = (T'T, T(T'T)^{-\frac{1}{2}})$. Thus F is a homeomorphism and the compactness of N_a implies that of M_a. The result follows. ∎

Flanders (1975) seemed to be one of the first who has considered the problem of minimizing the function $\phi : \mathcal{P}(n) \to \mathbb{R}$ defined by (4.5) over the set of positive definite symmetric matrices. His main result is as follows. The proof given here which is a simple consequence of Theorem 3.4 is however much simpler than that of Flanders.

6.4. FLOWS FOR BALANCING TRANSFORMATIONS

Corollary 4.2 *There exists a minimizing positive definite symmetric matrix $P = P' > 0$ for the function $\phi : \mathcal{P}(n) \to \mathbb{R}$, $\phi(P) = \text{tr}(X'XP^{-1} + YY'P)$, if and only if* $\text{rk}(X) = \text{rk}(Y) = \text{rk}(XY)$.

Proof Consider the continuous map $\phi : \mathcal{P}(n) \to \mathcal{O}(X,Y)$ defined by $\phi(P) = (XP^{-\frac{1}{2}}, P^{\frac{1}{2}}Y)$. There exists a minimum for $\phi : \mathcal{P}(n) \to \mathbb{R}$ if and only if there exists a minimum for the function $\mathcal{O}(X,Y) \to \mathbb{R}$, $(X,Y) \mapsto \|X\|^2 + \|Y\|^2$. The result now follows from Theorem 3.4 (b). ∎

We now consider the minimization task for the cost function $\phi : \mathcal{P}(n) \to \mathbb{R}$ from a dynamical systems viewpoint. Certainly $\mathcal{P}(n)$ is a smooth submanifold of $\mathbb{R}^{n \times n}$. For the subsequent results we endow $\mathcal{P}(n)$ with the *induced* Riemannian metric. Thus the inner product of two tangent vectors $\zeta, \eta \in T_P \mathcal{P}(n)$ is $\langle \zeta, \eta \rangle = \text{tr}(\zeta' \eta)$.

Theorem 4.3 *(Linear index gradient flow)* Let $(X,Y) \in \mathbb{R}^{k \times n} \times \mathbb{R}^{n \times \ell}$ with $\text{rk}(X) = \text{rk}(Y) = n$.

(a) *There exists a unique $P_\infty = P'_\infty > 0$ which minimizes $\phi : \mathcal{P}(n) \to \mathbb{R}$, $\phi(P) = \text{tr}(X'XP^{-1} + YY'P)$, and P_∞ is the only critical point of ϕ. This minimum is given by*

$$P_\infty = (YY')^{-\frac{1}{2}}[(YY')^{\frac{1}{2}}(X'X)(YY')^{\frac{1}{2}}]^{\frac{1}{2}}(YY')^{-\frac{1}{2}} \qquad (4.6)$$

and $T_\infty = P_\infty^{\frac{1}{2}}$ is a balancing transformation for (X,Y).

(b) *The gradient flow $\dot{P}(t) = -\nabla \phi(P(t))$ on $\mathcal{P}(n)$ is given by*

$$\boxed{\dot{P} = P^{-1}X'XP^{-1} - YY', \quad P(0) = P_0} \qquad (4.7)$$

For every initial condition $P_0 = P'_0 > 0$, $P(t) \in \mathcal{P}(n)$ exists for all $t \geq 0$ and converges exponentially fast to P_∞ as $t \to \infty$, with a lower bound for the rate of exponential convergence given by

$$\rho \geq 2 \frac{\sigma_{\min}(Y)^3}{\sigma_{\max}(X)} \qquad (4.8)$$

where $\sigma_{\min}(A)$ and $\sigma_{\max}(A)$ are the smallest and largest singular value of a linear operator A respectively.

Proof The existence of P_∞ follows immediately from Lemma 4.1 and Corollary 4.2. Uniqueness of P_∞ follows from Theorem 3.4(c). Similarly by Theorem 3.4 (c) every critical point is a global minimum.

Now simple manipulations using standard results from matrix calculus, reviewed in Appendix A, give

$$\nabla \phi(P) = YY' - P^{-1}X'XP^{-1} \qquad (4.9)$$

for the gradient of ϕ. Thus the critical point of Φ is characterized by

$$\begin{aligned}
\nabla \phi(P_\infty) = 0 &\Leftrightarrow X'X = P_\infty YY' P_\infty \\
&\Leftrightarrow (YY')^{\frac{1}{2}} X'X (YY')^{\frac{1}{2}} = ((YY')^{\frac{1}{2}} P_\infty (YY')^{\frac{1}{2}})^2 \\
&\Leftrightarrow P_\infty = (YY')^{-\frac{1}{2}}[(YY')^{\frac{1}{2}} X'X (YY')^{\frac{1}{2}}]^{\frac{1}{2}} (YY')^{-\frac{1}{2}}
\end{aligned}$$

Also $X'X = P_\infty YY' P_\infty \Leftrightarrow T_\infty^{-1} X'X T_\infty^{-1} = T_\infty YY' T_\infty$ for $T_\infty = P_\infty^{\frac{1}{2}}$. This proves (a) and (4.7). By Lemma 4.1 and standard properties of gradient flows, reviewed in Appendix C, the solution $P(t) \in \mathcal{P}(n)$ exists for all $t \geq 0$ and converges to the unique equilibrium point P_∞. To prove the result about exponential stability, we first review some known results concerning Kronecker products and the vec operation, see Appendix A.10. Recall that the Kronecker product of two matrices A and B is defined by

$$A \otimes B = \begin{bmatrix} a_{11} B & a_{12} B & \cdots & a_{1k} B \\ a_{21} B & a_{22} B & \cdots & a_{2k} B \\ \vdots & \vdots & & \vdots \\ a_{n1} B & a_{n2} B & \cdots & a_{nk} B \end{bmatrix}.$$

The eigenvalues of the Kronecker product of two matrices are given by the product of the eigenvalues of the two matrices, that is

$$\lambda_{ij}(A \otimes B) = \lambda_i(A) \lambda_j(B).$$

Moreover, with $\text{vec}(A)$ defined by

$$\text{vec}(A) = [a_{11} \cdots a_{n1} \, a_{12} \cdots a_{n2} \cdots a_{1m} \cdots a_{nm}]',$$

then

$$\text{vec}(MN) = (I \otimes M)\text{vec}(N) = (N' \otimes I)\text{vec}(M)$$

$$\text{vec}(ABC) = (C' \otimes A)\text{vec}(B).$$

A straightforward computation shows that the linearization of the right hand side of (4.7) at P_∞ is given by the linear operator

$$\xi \mapsto -P_\infty^{-1} \xi P_\infty^{-1} X'X P_\infty^{-1} - P_\infty^{-1} X'X P_\infty^{-1} \xi P_\infty^{-1}.$$

Since $P_\infty^{-1} X'X P_\infty^{-1} = YY'$ it follows that

$$\frac{d}{dt}\text{vec}(P - P_\infty) = (-P_\infty^{-1} \otimes YY' - YY' \otimes P_\infty^{-1})\text{vec}(P - P_\infty) \qquad (4.10)$$

is the linearization of (4.7) around P_∞. The smallest eigenvalue of $J_{SVD} = P_\infty^{-1} \otimes YY' + YY' \otimes P_\infty^{-1}$ can then be estimated as (see Appendix A)

$$\lambda_{\min}(J_{SVD}) \geq 2\lambda_{\min}(P_\infty^{-1})\lambda_{\min}(YY') > 0$$

6.4. FLOWS FOR BALANCING TRANSFORMATIONS

which proves exponential convergence of $P(t)$ to P_∞. To obtain the bound on the rate of convergence let S be an orthogonal transformation that diagonalizes P_∞, so that $SP_\infty S' = \text{diag}(\lambda_1, \cdots, \lambda_n)$. Let $\bar{X} = XS^{-1}$ and $\bar{Y} = SY$. Then $\bar{X}'\bar{X} = SP_\infty S'\bar{Y}\bar{Y}'SP_\infty S'$ and thus by

$$\|\bar{X}e_i\|^2 = e_i'\bar{X}'\bar{X}e_i' = \lambda_i^2\|\bar{Y}'e_i\|^2, \quad i = 1, \cdots, n,$$

we obtain

$$\lambda_{\min}(P_\infty^{-1}) = \min_i \frac{\|\bar{Y}'e_i\|}{\|\bar{X}e_i\|} \geq \frac{\sigma_{\min}(\bar{Y})}{\sigma_{\max}(\bar{X})} = \frac{\sigma_{\min}(Y)}{\sigma_{\max}(X)}$$

and therefore by $\lambda_{\min}(YY') = \sigma_{\min}(Y)^2$ we obtain (4.8). ∎

In the sequel we refer to (4.7) as the *linear index gradient flow*. Instead of minimizing the functional $\phi(P)$ we might as well consider the minimization problem for the quadratic index function

$$\psi(P) = \text{tr}((YY'P)^2 + (X'XP^{-1})^2) \tag{4.11}$$

over all positive definite symmetric matrices $P = P' > 0$.

Since, for $P = T'T$, $\psi(P)$ is equal to $\text{tr}((TYY'T')^2 + ((T')^{-1}X'XT^{-1})^2)$, the minimization of (4.11) is equivalent to the task of minimizing the quartic function $\text{tr}((YY')^2 + (X'X)^2)$ over all full rank factorizations (X, Y) of a given matrix $H = XY$. The cost function ψ has a greater penalty for being away from the minimum than the cost function ϕ, so can be expected to converge more rapidly.

By the formula

$$\begin{aligned} \text{tr}[(YY')^2 + (X'X)^2] &= \|X'X - YY'\|^2 + 2\text{tr}(X'XYY') \\ &= \|X'X - YY'\|^2 + 2\|H\|^2 \end{aligned}$$

we see that the minimization of $\text{tr}[(X'X)^2 + (YY')^2]$ over $\mathcal{F}(H)$ is equivalent to the minimization of the least squares distance $\|X'X - YY'\|^2$ over $\mathcal{F}(H)$.

Theorem 4.4 (*Quadratic index gradient flow*) Let $(X, Y) \in \mathbb{R}^{k \times n} \times \mathbb{R}^{n \times \ell}$ with $\text{rk}(X) = \text{rk}(Y) = n$.

(a) There exists a unique $P_\infty = P_\infty' > 0$ which minimizes $\psi : \mathcal{P}(n) \to \mathbb{R}$, $\psi(P) = \text{tr}((X'XP^{-1})^2 + (YY'P)^2)$, and

$$P_\infty = (YY')^{-\frac{1}{2}}[(YY')^{\frac{1}{2}}(X'X)(YY')^{\frac{1}{2}}]^{\frac{1}{2}}(YY')^{-\frac{1}{2}}$$

is the only critical point of ψ. Also, $T_\infty = P_\infty^{\frac{1}{2}}$ is a balancing coordinate transformation for (X, Y).

(b) The gradient flow $\dot{P}(t) = -\nabla \psi(P(t))$ is

$$\dot{P} = 2P^{-1}X'XP^{-1}X'XP^{-1} - 2YY'PYY', \quad P(0) = P_0 \quad (4.12)$$

For all initial conditions $P_0 = P_0' > 0$, the solution $P(t)$ of (4.12) exists in $\mathcal{P}(n)$ for all $t \geq 0$ and converges exponentially fast to P_∞. A lower bound on the rate of exponential convergence is

$$\rho > 4\sigma_{\min}(Y)^4 \quad (4.13)$$

Proof Again the existence and uniqueness of P_∞ and the critical points of ψ follows as in the proof of Theorem 4.3. Similarly, the expression (4.12) for the gradient flow follows easily from the standard rules of matrix calculus. The only new point we have to check is the bound on the rate of exponential convergence (4.13).

The linearization of the right hand side of (4.12) at the equilibrium point P_∞ is given by the linear map ($\xi = \xi'$)

$$\xi \mapsto -2(P_\infty^{-1}\xi P_\infty^{-1}X'XP_\infty^{-1}X'XP_\infty^{-1} + P_\infty^{-1}X'XP_\infty^{-1}\xi P_\infty^{-1}X'XP_\infty^{-1}$$
$$+ P_\infty^{-1}X'XP_\infty^{-1}X'XP_\infty^{-1}\xi P_\infty^{-1} + YY'\xi YY')$$

That is, using $P_\infty YY' P_\infty = X'X$, by

$$\xi \mapsto -2(P_\infty^{-1}\xi YY' P_\infty YY' + 2YY'\xi YY' + YY' P_\infty YY' \xi P_\infty^{-1}) \quad (4.14)$$

Therefore the linearization of (4.12) at P_∞ is

$$\frac{d}{dt}\text{vec}(P - P_\infty) = -2(2YY' \otimes YY' + P_\infty^{-1} \otimes YY' P_\infty YY'$$
$$+ YY' P_\infty YY' \otimes P_\infty^{-1})\text{vec}(P - P_\infty) \quad (4.15)$$

Therefore, see Appendix A, the minimum eigenvalue of the self adjoint linear operator given by (4.14) is greater than or equal to

$$4\lambda_{\min}(YY')^2 + 4\lambda_{\min}(P_\infty^{-1})\lambda_{\min}(YY' P_\infty YY').$$
$$\geq 4\lambda_{\min}(YY')^2(1 + \lambda_{\min}(P_\infty^{-1})\lambda_{\min}(P_\infty)) > 4\lambda_{\min}(YY')^2 > 0.$$

The result follows. ∎

We refer to (4.12) as the *quadratic index gradient flow*. The above results show that both algorithms converge exponentially fast to P_∞, however their transient behaviour is rather different. In fact, simulation experiment with both gradient flows show that the quadratic index flow seems to behave better than the linear index flow. This is further supported by the following lemma. It compares the rates of exponential convergence of the algorithms and shows that the quadratic index flow is in general faster than the linear index flow.

6.4. FLOWS FOR BALANCING TRANSFORMATIONS

Lemma 4.5 *Let ρ_1 and ρ_2 denote the rates of exponential convergence of (4.7) and (4.12) respectively. Then $\rho_1 < \rho_2$ if $\sigma_{\min}(XY) > \frac{1}{2}$.*

Proof By (4.10), (4.15),

$$\rho_1 = \lambda_{\min}(P_\infty^{-1} \otimes YY' + YY' \otimes P_\infty^{-1}),$$

and

$$\rho_2 = 2\lambda_{\min}(2YY' \otimes YY' + P_\infty^{-1} \otimes YY'P_\infty YY' + YY'P_\infty YY' \otimes P_\infty^{-1}).$$

We need the following lemma; see Appendix A.

Lemma 4.6 *Let $A, B \in \mathbb{R}^{n \times n}$ be symmetric matrices and let $\lambda_1(A) \geq \cdots \geq \lambda_n(A)$ and $\lambda_1(B) \geq \cdots \geq \lambda_n(B)$ be the eigenvalues of A, B, ordered with respect to their magnitude. Then $A - B$ positive definite implies $\lambda_n(A) > \lambda_n(B)$.*

Thus, according to Lemma 4.6, it suffices to prove that

$$4YY' \otimes YY' + 2P_\infty^{-1} \otimes YY'P_\infty YY' +$$
$$2YY'P_\infty YY' \otimes P_\infty^{-1} - P_\infty^{-1} \otimes YY' - YY' \otimes P_\infty^{-1} =$$
$$4YY' \otimes YY' + P_\infty^{-1} \otimes (2YY'P_\infty YY' - YY') + (2YY'P_\infty YY' - YY') \otimes P_\infty^{-1}$$
$$> 0.$$

But this is the case, whenever

$$2YY'P_\infty YY' = 2(YY')^{\frac{1}{2}}[(YY')^{\frac{1}{2}}X'X(YY')^{\frac{1}{2}}]^{\frac{1}{2}}(YY')^{\frac{1}{2}} > YY'$$
$$\Leftrightarrow 2[(YY')^{\frac{1}{2}}X'X(YY')^{\frac{1}{2}}]^{\frac{1}{2}} > I_n$$
$$\Leftrightarrow \lambda_{\min}((YY)^{\frac{1}{2}}X'X(YY')^{\frac{1}{2}}) > \frac{1}{4}$$
$$\Leftrightarrow \lambda_{\min}(X'XYY') > \frac{1}{4}$$

We have used the facts that $\lambda_i(A^2) = \lambda_i(A)^2$ for any operator $A = A'$ and therefore $\lambda_i(A^{\frac{1}{2}})^2 = \lambda_i((A^{\frac{1}{2}})^2) = \lambda_i(A)$ as well as $\lambda_i(AB) = \lambda_i(BA)$. The result follows because

$$\lambda_{\min}(X'XYY') = \lambda_{\min}((XY)(XY)') = \sigma_{\min}(XY)^2.$$

∎

Gradient flows on $GL(n)$ Once the minimization of the cost functions ϕ, ψ on $\mathcal{P}(n)$ has been performed the class of all balancing transformations is $\{\Theta T_\infty \mid \Theta \in O(n)\}$ where $T_\infty = P_\infty^{\frac{1}{2}}$ is the unique symmetric, positive definite balancing transformation for (X, Y). Of course, other balancing transformations than T_∞ might be also of interest to compute in a similar way and therefore one would like to find suitable gradient flows evolving on arbitrary invertible $n \times n$ matrices. See Appendix C for definitions and results on dynamical systems.

Thus for $T \in GL(n)$ we consider

$$\Phi(T) = \text{tr}(TYY'T' + (T')^{-1}X'XT^{-1}) \tag{4.16}$$

and the corresponding gradient flow $\dot T = -\nabla \Phi(T)$ on $GL(n)$. Here and in the sequel we always endow $GL(n)$ with its standard Riemannian metric

$$\langle A, B \rangle = 2\text{tr}(A'B) \tag{4.17}$$

i.e. with the constant Euclidean inner product (4.17) defined on the tangent spaces of $GL(n)$. Here the constant factor 2 is introduced for convenience.

Theorem 4.7 Let $\text{rk}(X) = \text{rk}(Y) = n$.

(a) The gradient flow $\dot T = -\nabla \Phi(T)$ of $\Phi : GL(n) \to \mathbb{R}$ is

$$\boxed{\dot T = (T')^{-1}X'X(T'T)^{-1} - TYY', \quad T(0) = T_0} \tag{4.18}$$

and for any initial condition $T_0 \in GL(n)$ the solution $T(t)$ of (4.18) exists in $GL(n)$ for all $t \geq 0$.

(b) For any initial condition $T_0 \in GL(n)$ the solution $T(t)$ of (4.18) converges to a balancing transformation $T_\infty \in GL(n)$ and all balancing transformations of (X, Y) can be obtained in this way, for suitable initial conditions $T_0 \in GL(n)$. Moreover, $\Phi : GL(n) \to \mathbb{R}$ is a Morse-Bott function.

(c) Let T_∞ be a balancing transformation and let $W^s(T_\infty)$ denote the set of all $T_0 \in GL(n)$ such that the solution $T(t)$ of (4.18) with $T(0) = T_0$ converges to T_∞ as $t \to \infty$. Then $W^s(T_\infty)$ is an immersed invariant submanifold of $GL(n)$ of dimension $\frac{n(n+1)}{2}$ and every solution $T(t) \in W^s(T_\infty)$ converges exponentially fast in $W^s(T_\infty)$ to T_∞.

Proof The derivative of $\Phi : GL(n) \to \mathbb{R}$ at T is the linear operator given by

$$\begin{aligned} D\Phi|_T(\xi) &= 2\text{tr}((YY'T' - T^{-1}(T')^{-1}X'XT^{-1})\xi) \\ &= 2\text{tr}((TYY' - (T')^{-1}X'X(T'T)^{-1})'\xi) \end{aligned}$$

and therefore

$$\nabla \Phi(T) = TYY' - (T')^{-1}X'X(T'T)^{-1}.$$

6.4. FLOWS FOR BALANCING TRANSFORMATIONS

To prove that the gradient flow (4.18) is complete, i.e. that the solutions $T(t)$ exist for all $t \geq 0$, it suffices to show that $\Phi : GL(n) \to \mathbb{R}$ has compact sublevel sets $\{T \in GL(n) \mid \Phi(T) \leq a\}$ for all $a \in \mathbb{R}$. Since $\mathrm{rk}(X) = \mathrm{rk}(Y) = n$ the map $\phi : GL(n) \to \mathcal{O}(X, Y), \phi(T) = (XT^{-1}, TY)$, is a homeomorphism and, using Theorem 3.4, $\{(X_1, Y_1) \in \mathcal{O}(X,Y) \mid \|X_1\|^2 + \|Y_1\|^2 \leq a\}$ is compact for all $a \in \mathbb{R}$. Thus $\{T \in GL(n) \mid \Phi(T) \leq a\} = \phi^{-1}(\{(X_1, Y_1) \in \mathcal{O}(X,Y) \mid \|X_1\|^2 + \|Y_1\|^2 \leq a\})$ is compact. This shows (a).

To prove (b) we note that, by (a) and by the steepest descent property of gradient flows, any solution $T(t)$ converges to the set of equilibrium points $T_\infty \in GL(n)$. The equilibria of (4.18) are characterized by

$$(T'_\infty)^{-1} X' X (T'_\infty T_\infty)^{-1} = T_\infty YY' \Leftrightarrow (T'_\infty)^{-1} X' X T_\infty^{-1} = T_\infty YY' T'_\infty$$

and hence T_∞ is a balancing transformation. This proves (b), except that we haven't yet proved convergence to equilibrium points rather than to the entire set of equilibria; see below.

Let
$$\mathcal{E} := \{T_\infty \in GL(n) \mid (T'_\infty T_\infty) YY'(T'_\infty T_\infty) = X'X\} \quad (4.19)$$

denote the set of equilibria points of (4.18). To prove (c) we need a lemma.

Lemma 4.8 *The tangent space of $\mathcal{E}$ at $T_\infty \in \mathcal{E}$ is*

$$T_{T_\infty} \mathcal{E} = \{S \in \mathbb{R}^{n \times n} \mid S'T_\infty + T'_\infty S = 0\} \quad (4.20)$$

Proof Let P_∞ denote the unique symmetric positive definite solution of $PYY'P = X'X$. Thus $\mathcal{E} = \{T \mid T'T = P_\infty\}$ and therefore $T_{T_\infty}\mathcal{E}$ is the kernel of the derivative of $T \mapsto T'T - P_\infty$ at T_∞, i.e. by $\{S \in \mathbb{R}^{n \times n} \mid S'T_\infty + T'_\infty S = 0\}$. ∎

Let
$$\phi(P) = \mathrm{tr}(YY'P + X'XP^{-1}), \quad \lambda(T) = T'T.$$

Thus $\Phi(T) = \phi(\lambda(T))$. By Theorem 4.3

$$D\phi|_{P_\infty} = 0, \quad D^2\phi|_{P_\infty} > 0.$$

Let L denote the matrix representing the linear operator $D\lambda|_{T_\infty}(S) = T'_\infty S + S'T_\infty$. Using the chain rule we obtain

$$D^2\Phi|_{T_\infty} = L' \cdot D^2\phi|_{P_\infty} \cdot L$$

for all $T_\infty \in \mathcal{E}$. By $D^2\phi|_{P_\infty} > 0$ thus $D^2\Phi|_{T_\infty} \geq 0$ and $D^2\Phi|_{T_\infty}$ degenerates exactly on the kernel of L; i.e. on the tangent space $T_{T_\infty}\mathcal{E}$. Thus Φ is a

Morse-Bott function, as defined in Appendix C. Thus Proposition C.12.3 implies that every solution $T(t)$ of (4.18) converges to an equilibrium point.

From the above we conclude that the set $\mathcal{E}$ of equilibria points is a normally hyperbolic subset for the gradient flow; see Appendix C.11. It follows, see Appendix C, that $W^s(T_\infty)$ the stable manifold of (4.18) at T_∞ is an immersed submanifold of $GL(n)$ of dimension $\dim GL(n) - \dim \mathcal{E} = n^2 - \frac{n(n-1)}{2} = \frac{n(n+1)}{n}$, which is invariant under the gradient flow. Since convergence to an equilibrium is always exponentially fast on stable manifolds this completes the proof of (c). ∎

Now consider the following quadratic version of our objective function $\Phi(T)$. For $T \in GL(n)$, let

$$\Psi(T) := \operatorname{tr}((TYY'T')^2 + ((T')^{-1}X'XT^{-1})^2) \tag{4.21}$$

The proof of the following theorem is completely analogous to the proof of Theorem 4.7 and is therefore omitted.

Theorem 4.9 *(a)* *The gradient flow $\dot{T} = -\nabla \Psi(T)$ of $\Psi(T)$ on $GL(n)$ is*

$$\boxed{\dot{T} = 2[(T')^{-1}X'X(T'T)^{-1}X'X(T'T)^{-1} - TYY'T'TYY']} \tag{4.22}$$

and for all initial conditions $T_0 \in GL(n)$, the solution $T(t)$ of exist in $GL(n)$ for all $t \geq 0$.

(b) For all initial conditions $T_0 \in GL(n)$, every solution $T(t)$ of (4.20) converges to a balancing transformation and all balancing transformations are obtained in this way, for suitable initial conditions $T_0 \in GL(n)$. Moreover, $\Psi : GL(n) \to \mathbb{R}$ is a Morse-Bott function.

(c) For any balancing transformation $T_\infty \in GL(n)$ let $W^s(T_\infty) \subset GL(n)$ denote the set of all $T_0 \in GL(n)$, such that the solution $T(t)$ of (4.20) with initial condition T_0 converges to T_∞ as $t \to \infty$. Then $W^s(T_\infty)$ is an immersed submanifold of $GL(n)$ of dimension $\frac{n(n+1)}{2}$ and is invariant under the flow of (4.20). Every solution $T(t) \in W^s(T_\infty)$ converges exponentially to T_∞.

Diagonal balancing transformations The previous results were concerned with the question of finding balancing transformations via gradient flows; here we address the similar issue of computing diagonal balancing transformations.

We have already shown in Theorem 3.5 that diagonal balancing factorizations of a $k \times \ell$ matrix H can be characterized as the critical points of the

6.4. FLOWS FOR BALANCING TRANSFORMATIONS

weighted cost function

$$\Phi_N : \mathcal{O}(X,Y) \to \mathbb{R}, \quad \Phi_N(X,Y) = \operatorname{tr}(NX'X + NYY')$$

for a fixed diagonal matrix $N = \operatorname{diag}(\mu_1, \cdots, \mu_n)$ with distinct eigenvalues $\mu_1 > \cdots > \mu_n$. Of course a similar result holds for diagonal balancing transformations. Thus let $\Phi_N : GL(n) \to \mathbb{R}$ be defined by

$$\Phi_N(T) = \operatorname{tr}(N(TYY'T' + (T')^{-1}X'XT^{-1})).$$

The proof of the following result parallels that of Theorem 3.5 and is left as an exercise for the reader. Note that for $X'X = YY'$ and T orthogonal, the function $\Phi_N : O(n) \to \mathbb{R}$, restricted to the orthogonal group $O(n)$, coincides with the trace function studied in Chapter 2. Here we are interested in minimizing the function Φ_N rather than maximizing it. We will therefore expect an order reversal relative to that of N occurring for the diagonal entries of the equilibria points.

Lemma 4.10 Let $\operatorname{rk}(X) = \operatorname{rk}(Y) = n$ and $N = \operatorname{diag}(\mu_1, \cdots, \mu_n)$ with $\mu_1 > \cdots > \mu_n > 0$. Then

(a) $T \in GL(n)$ is a critical point for $\Phi_N : GL(n) \to \mathbb{R}$ if and only if T is a diagonal balancing transformation.

(b) The global minimum $T_{\min} \in GL(n)$ has the property $T_{\min} YY' T'_{\min} = \operatorname{diag}(d_1, \cdots, d_n)$ with $d_1 \leq d_2 \leq \cdots \leq d_n$.

Theorem 4.11 Let $\operatorname{rk}(X) = \operatorname{rk}(Y) = n$ and $N = \operatorname{diag}(\mu_1, \cdots, \mu_n)$ with $\mu_1 > \cdots > \mu_n > 0$.

(a) The gradient flow $\dot{T} = -\nabla \Phi_N(T)$ of $\Phi_N : GL(n) \to \mathbb{R}$ with respect to the Riemannian metric (4.17) is

$$\boxed{\dot{T} = (T')^{-1} X'XT^{-1} N(T')^{-1} - NTYY', \quad T(0) = T_0} \quad (4.23)$$

and for all initial conditions $T_0 \in GL(n)$ the solution $T(t)$ of (4.23) exists in $GL(n)$ for all $t \geq 0$.

(b) For any initial condition $T_0 \in GL(n)$ the solution $T(t)$ of (4.23) converges to a diagonal balancing transformation $T_\infty \in GL(n)$. Moreover, all diagonal balancing transformations can be obtained in this way.

(c) Suppose the singular values of $H = XY$ are distinct. Then (4.23) has exactly $2^n n!$ equilibrium points. These are characterized by $(T'_\infty)^{-1} X'X T_\infty^{-1} = T_\infty YY' T'_\infty = D$ where D is a diagonal matrix. There are exactly 2^n stable equilibrium points of (4.23) which are characterized by $(T'_\infty)^{-1} X'X T_\infty^{-1} = T_\infty YY' T'_\infty = D$, where $D = \operatorname{diag}(d_1, \cdots, d_n)$ is diagonal with $d_1 < \cdots < d_n$.

(d) There exists an open and dense subset $\Omega \subset GL(n)$ such that for all $T_0 \in \Omega$ the solution of (4.23) converges exponentially fast to a stable equilibrium point T_∞. The rate of exponential convergence is bounded below by $\lambda_{\min}((T_\infty T'_\infty)^{-1}) \min_{i<j}((d_i - d_j)(\mu_j - \mu_i), 4d_i\mu_i)$. All other equilibria are unstable.

Proof Using Lemma 4.10, the proof of (a) and (b) parallels that of Theorem 4.7 and is left as an exercise for the reader. To prove (c) consider the linearization of (4.23) around an equilibrium T_∞, given by

$$\dot{\xi} = -N\xi T_\infty^{-1} D(T'_\infty)^{-1} - D\xi T_\infty^{-1} N(T'_\infty)^{-1} - (T'_\infty)^{-1}\xi' DN(T'_\infty)^{-1}$$
$$- DN(T'_\infty)^{-1}\xi'(T'_\infty)^{-1},$$

using $(T'_\infty)^{-1} X'X T_\infty = T_\infty Y Y' T'_\infty = D$. Using the change of variables $\eta = \xi T_\infty^{-1}$ thus

$$\dot{\eta}(T_\infty T'_\infty) = -N\eta D - D\eta N - \eta' DN - DN\eta'$$

Thus, using Kronecker products and the vec notation, Appendix A, we obtain

$$(T_\infty T'_\infty \otimes I_n)\text{vec}(\dot{\eta}) = -(D \otimes N + N \otimes D)\text{vec}(\eta)$$
$$- (DN \otimes I_n + I_n \otimes DN)\text{vec}(\eta')$$

Consider first the special case when $T_\infty T'_\infty = I$, and η is denoted η^*

$$\text{vec}(\dot{\eta}^*) = -(D \otimes N + N \otimes D)\text{vec}(\eta^*) - (DN \otimes I + I \otimes DN)\text{vec}(\eta^{*'}). \quad (4.24)$$

Then for $i < j$,

$$\begin{bmatrix} \dot{\eta}^*_{ij} \\ \dot{\eta}^*_{ji} \end{bmatrix} = -\begin{bmatrix} d_i\mu_j + \mu_i d_j & d_j\mu_j + d_i\mu_i \\ d_i\mu_i + d_j\mu_j & d_i\mu_j + \mu_i d_j \end{bmatrix} \begin{bmatrix} \eta^*_{ij} \\ \eta^*_{ji} \end{bmatrix}$$

and for all i,

$$\dot{\eta}^*_{ii} = -4d_i\mu_i\eta^*_{ii}.$$

By assumption, $\mu_i > 0$, and $d_i > 0$ for all i. Thus (4.24) is exponentially stable if and only if $(d_i - d_j)(\mu_j - \mu_i) > 0$ for all $i, j, i < j$, or equivalently, if and only if the diagonal entries of D are distinct and in reverse ordering to those of N. In this case, (4.24) is equivalent to (4.25)

$$\text{vec}(\dot{\eta}^*) = -\mathcal{F}\text{vec}(\eta^*). \quad (4.25)$$

with a symmetric positive definite matrix $\mathcal{F} = \mathcal{F}' > 0$.

6.4. FLOWS FOR BALANCING TRANSFORMATIONS

Consequently, exponential convergence of (4.25) is established with a rate given by $\lambda_{\min}(\mathcal{F})$ as follows:

$$\lambda_{\min}(\mathcal{F}) = \min_{i<j}(\lambda_{\min}(\begin{bmatrix} d_i\mu_j + \mu_i d_j & d_j\mu_j + d_i\mu_i \\ d_i\mu_i + d_j\mu_j & d_i\mu_j + \mu_i d_j \end{bmatrix}), 4d_i\mu_i)$$

$$= \min_{i<j}(d_i - d_j)(\mu_j - \mu_i), (4d_i\mu_i)$$

Since $(T_\infty T'_\infty \otimes I)$ is positive definite, exponential stability of (4.25) implies exponential stability of

$$((T_\infty T'_\infty) \otimes I)\text{vec}(\dot\eta) = -\mathcal{F}\text{vec}(\eta).$$

The rate of exponential convergence is given by $\lambda_{\min}(((T_\infty T'_\infty)^{-1} \otimes I)\mathcal{F})$. Now since $A = A' > 0, B = B' > 0$ implies $\lambda_{\min}(AB) \geq \lambda_{\min}(A)\lambda_{\min}(B)$, a lower bound on the convergence rate is given from

$$\lambda_{\min}(((T_\infty T'_\infty)^{-1} \otimes I)\mathcal{F}) \geq \lambda_{\min}((T_\infty T'_\infty)^{-1} \otimes I)\lambda_{\min}(\mathcal{F})$$
$$= \lambda_{\min}((T_\infty T'_\infty)^{-1}) \min_{i<j}(d_i - d_j)(\mu_j - \mu_i), 4d_i\mu_i)$$

as claimed. Since there are only a finite number of equilibria, the union of the stable manifolds of the unstable equilibria points is a closed subset of $\mathcal{O}(X,Y)$ of codimension at least one. Thus its complement Ω in $\mathcal{O}(X,Y)$ is open and dense and coincides with the union of the stable manifolds of the stable equilibria. This completes the proof. ■

Problems   **1.** Show that $\langle\langle \zeta, \nu \rangle\rangle := \text{tr}(P^{-1}\zeta P^{-1}\eta)$ defines an inner product on the tangent spaces $T_P\mathcal{P}(n)$ for $P \in \mathcal{P}(n)$.

2. Prove that $\langle\langle\,,\,\rangle\rangle$ defines a Riemannian metric on $\mathcal{P}(n)$. We refer to this as the *intrinsic Riemannian metric* on $\mathcal{P}(n)$.

3. Prove that the gradient flow of (6.4.5) with respect to the intrinsic Riemannian metric is the Riccati equation

$$\dot P = -\text{grad } \phi(P) = X'X - PYY'P.$$

4. Prove the analogous result to Theorem 4.3 for this Riccati equation.

5. Concrete proofs of Theorem 4.9, and Lemma 4.10.

Main Points of Section   Gradient flows can be developed for matrix factorizations which converge exponentially fast to balancing factorizations, including to diagonal balanced factorizations. Such flows side step the usual requirement to find the balancing transformations.

6.5 Flows on the Factors X and Y

In the previous section gradient flows for balancing or diagonal balancing coordinate transformations T_∞ and $P_\infty = T'_\infty T_\infty$ were analyzed. Here we address the related issue of finding gradient flows for the cost functions $\Phi : \mathcal{O}(X,Y) \to \mathbb{R}$ and $\Phi_N : \mathcal{O}(X,Y) \to \mathbb{R}$ on the manifold $\mathcal{O}(X,Y)$ of factorizations of a given matrix $H = XY$.

Thus for arbitrary integers k, ℓ and n let

$$\mathcal{O}(X,Y) = \{(XT^{-1}, TY) \in \mathbb{R}^{k \times n} \times \mathbb{R}^{n \times \ell} \mid T \in GL(n)\}$$

denote the $GL(n)$-orbit of (X,Y). We endow the vector space $\mathbb{R}^{k \times n} \times \mathbb{R}^{n \times \ell}$ with its standard inner product (3.2) repeated here as

$$\langle (X_1, Y_1), (X_2, Y_2) \rangle = \operatorname{tr}(X'_1 X_2 + Y_1 Y'_2) \tag{5.1}$$

By Lemma 3.1, $\mathcal{O}(X,Y)$ is a submanifold of $\mathbb{R}^{k \times n} \times \mathbb{R}^{n \times \ell}$ and thus the inner product (5.1) on $\mathbb{R}^{k \times n} \times \mathbb{R}^{n \times \ell}$ induces an inner product on each tangent space $T_{(X,Y)}\mathcal{O}$ of $\mathcal{O}$ by

$$\langle (-X\Lambda_1, \Lambda_1 Y), (-X\Lambda_2, \Lambda_2 Y) \rangle = \operatorname{tr}(\Lambda'_2 X' X \Lambda_1 + \Lambda_2 Y Y' \Lambda_1) \tag{5.2}$$

and therefore a Riemannian metric on $\mathcal{O}(X,Y)$; see Appendix C and (3.4). We refer to this Riemannian metric as the *induced Riemannian metric* on $\mathcal{O}(X,Y)$. It turns out that the gradient flow of the above cost functions with respect to this Riemannian metric has a rather complicated form; see Problems.

A second, different, Riemannian metric on $\mathcal{O}(X,Y)$ is constructed as follows. Here we assume that $\operatorname{rk}(X) = \operatorname{rk}(Y) = n$. Instead of defining the inner product of tangent vectors $(-X\Lambda_1, \Lambda_1 Y), (-X\Lambda_2, \Lambda_2 Y) \in T_{(X,Y)}\mathcal{O}$ as in (5.2) we set

$$\langle\langle (-X\Lambda_1, \Lambda_1 Y), (-X\Lambda_2, \Lambda_2 Y) \rangle\rangle := 2\operatorname{tr}(\Lambda'_1 \Lambda_2) \tag{5.3}$$

It is easily seen (using $\operatorname{rk}(X) = \operatorname{rk}(Y) = n$) that this defines an inner product on each tangent space $T_{(X,Y)}\mathcal{O}$ and in fact a Riemannian metric on $\mathcal{O} = \mathcal{O}(X,Y)$. We refer to this as the *normal Riemannian metric* on $\mathcal{O}(X,Y)$. This is in fact a particularly convenient Riemannian metric with which to work. In this, the associated gradient takes a particularly simple form.

Theorem 5.1 *Let* $\operatorname{rk}(X) = \operatorname{rk}(Y) = n$. *Consider the cost function* $\Phi : \mathcal{O}(X,Y) \to \mathbb{R}$, $\Phi(X,Y) = \operatorname{tr}(X'X + YY')$.

(a) The gradient flow $(\dot{X}, \dot{Y}) = -\operatorname{grad} \Phi(X,Y)$ *for the normal Riemannian metric is:*

$$\begin{aligned} \dot{X} &= -X(X'X - YY'), & X(0) &= X_0 \\ \dot{Y} &= (X'X - YY')Y, & Y(0) &= Y_0 \end{aligned} \tag{5.4}$$

6.5. FLOWS ON THE FACTORS X AND Y

(b) For every initial condition $(X_0, Y_0) \in \mathcal{O}(X,Y)$ the solutions $(X(t), Y(t))$ of (5.4) exist for all $t \geq 0$ and $(X(t), Y(t)) \in \mathcal{O}(X,Y)$ for all $t \geq 0$.

(c) For any initial condition $(X_0, Y_0) \in \mathcal{O}(X,Y)$ the solutions $(X(t), Y(t))$ of (5.4) converge to a balanced factorization (X_∞, Y_∞) of $H = XY$. Moreover the convergence to the set of all balanced factorizations of H is exponentially fast.

Proof Let grad $\Phi = (\text{grad}\Phi_1, \text{grad}\Phi_2)$ denote the gradient of $\Phi : \mathcal{O}(X,Y) \to \mathbb{R}$ with respect to the normal Riemannian metric. The derivative of Φ at $(X, Y) \in \mathcal{O}$ is the linear map $D\Phi|_{(X,Y)} : T_{(X,Y)}\mathcal{O} \to \mathbb{R}$ defined by

$$D\Phi|_{(X,Y)}(-X\Lambda, \Lambda Y) = 2\text{tr}((YY' - X'X)\Lambda) \tag{5.5}$$

By definition of the gradient of a function, see Appendix C, $\text{grad}\Phi(X, Y)$ is characterized by

$$\text{grad}\Phi(X, Y) \in T_{(X,Y)}\mathcal{O} \tag{5.6}$$

and

$$D\Phi|_{(X,Y)}(-X\Lambda, \Lambda Y) = \langle\langle (\text{grad}\Phi_1, \text{grad}\Phi_2), (-X\Lambda, \Lambda Y)\rangle\rangle \tag{5.7}$$

for all $\Lambda \in \mathbb{R}^{n \times n}$. By Lemma 3.1, (5.6) is equivalent to

$$\text{grad}\Phi(X, Y) = (-X\Lambda_1, \Lambda_1 Y) \tag{5.8}$$

for an $n \times n$ matrix Λ_1. Thus, using (5.5), (5.7) is equivalent to

$$\begin{aligned} 2\text{tr}((YY' - X'X)\Lambda) &= \langle\langle (-X\Lambda_1, \Lambda_1 Y), (-X\Lambda, \Lambda Y)\rangle\rangle \\ &= 2\text{tr}(\Lambda_1' \Lambda) \end{aligned}$$

for all $\Lambda \in \mathbb{R}^{n \times n}$. Thus

$$YY' - X'X = \Lambda_1 \tag{5.9}$$

Therefore $\text{grad}\Phi(X, Y) = (-X(YY' - X'X), (YY' - X'X)Y)$. This proves (a).

To prove (b) note that $\Phi(X(t), Y(t))$ decreases along every solution of (5.4). By Corollary 3.3 $\{(X, Y) \in \mathcal{O} \mid \Phi(X, Y) \leq \Phi(X_0, Y_0)\}$ is a compact set. Therefore $(X(t), Y(t))$ stays in that compact subset of $\mathcal{O}(X, Y)$ and therefore exists for all $t \geq 0$. This proves (b). Since (5.4) is a gradient flow of $\Phi : \mathcal{O}(X, Y) \to \mathbb{R}$ and since $\Phi : \mathcal{O}(X, Y) \to \mathbb{R}$ has compact sublevel sets the solutions $(X(t), Y(t))$ all converge to the set of equilibria of (5.4). But (X_∞, Y_∞) is an equilibrium point of (5.4) if and only if $X'_\infty X_\infty = Y_\infty Y'_\infty$. Thus the equilibria are in both cases the balanced factorizations of $H = XY$.

By Theorem 3.4 the Hessian of $\Phi : \mathcal{O}(X, Y) \to \mathbb{R}$ is positive semidefinite at each critical point and degenerates exactly on the tangent spaces at the set of critical points. This implies that the linearization of (5.4) is exponentially stable in directions transverse to the tangent spaces of the set of equilibria. Thus Proposition C.12.3 implies that any solutions $(X(t), Y(t))$ of (5.4) converges to an equilibrium point. Exponential convergence follows from the stable manifold theory, as summarized in Appendix C.11. ∎

A similar approach also works for the weighted cost functions

$$\Phi_N : \mathcal{O}(X,Y) \to \mathbb{R}, \quad \Phi_N(X,Y) = \operatorname{tr}(N(X'X + YY')).$$

We have the following result.

Theorem 5.2 *Consider the weighted cost function $\Phi_N : \mathcal{O}(X,Y) \to \mathbb{R}$, $\Phi_N(X,Y) = \operatorname{tr}(N(X'X + YY'))$, with $N = \operatorname{diag}(\mu_1, \cdots, \mu_n), \mu_1 > \cdots > \mu_n > 0$. Let $\operatorname{rk}(X) = \operatorname{rk}(Y) = n$.*

(a) The gradient flow $(\dot{X}, \dot{Y}) = -\operatorname{grad} \Phi_N(X,Y)$ for the normal Riemannian metric is

$$\begin{array}{rll} \dot{X} & = -X(X'XN - NYY'), & X(0) = X_0 \\ \dot{Y} & = (X'XN - NYY')Y, & Y(0) = Y_0 \end{array} \quad (5.10)$$

(b) For any initial condition $(X_0, Y_0) \in \mathcal{O}(X,Y)$ the solutions $(X(t), Y(t))$ of (5.10) exist for all $t \geq 0$ and $(X(t), Y(t)) \in \mathcal{O}(X,Y)$ for all $t \geq 0$.

(c) For any initial condition $(X_0, Y_0) \in \mathcal{O}(X,Y)$ the solutions $(X(t), Y(t))$ of (5.10) converge to a diagonal balanced factorization (X_∞, Y_∞) of $H = XY$.

(d) Suppose that the singular values of $H = X_0 Y_0$ are distinct. There are exactly $2^n n!$ equilibrium points of (5.10) The set of such asymptotically stable equilibria is characterized by $X'_\infty X_\infty = Y_\infty Y'_\infty = D$, where $D = \operatorname{diag}(d_1, \cdots, d_n)$ satisfies $d_1 < \cdots < d_n$. In particular there are exactly 2^n asymptotically stable equilibria, all yielding the same value of Φ_N. The diagonal entries d_i are the singular values of $H = X_0 Y_0$ arranged in increasing order. All other equilibria are unstable.

(e) There exists an open and dense subset $\Omega \subset \mathcal{O}(X,Y)$ such that for all $(X_0, Y_0) \in \Omega$ the solutions of (5.10) converge exponentially fast to a stable equilibrium point (X_∞, Y_∞).

Proof The proof of (a), (b) goes *mutatis mutandis* as for (a) and (b) in Theorem 5.1. Similarly for (c), except that we have yet to check that the equilibria of (5.10) are the diagonal balanced factorizations. The equilibria of (5.10) are just the critical points of the function $\Phi_N : \mathcal{O}(X,Y) \to \mathbb{R}$ and the desired result follows from Theorem 3.5 (c).

6.5. FLOWS ON THE FACTORS X AND Y

To prove (d), first note that we cannot apply, as for Theorem 5.1, the Kempf-Ness theory. Therefore we give a direct argument based on linearizations at the equilibria. The equilibrium points of (5.10) are characterized by $X'_\infty X_\infty = Y_\infty Y'_\infty = D$, where $D = \text{diag}(d_1, \cdots, d_n)$. By linearizing (5.10) around an equilibrium point (X_∞, Y_∞) we obtain

$$\begin{aligned}\dot\xi &= -X_\infty(\xi'X_\infty N + X'_\infty \xi N - N\eta Y'_\infty - NY_\infty \eta') \\ &= (\xi'X_\infty N + X'_\infty \xi N - N\eta Y'_\infty - NY_\infty \eta')Y_\infty\end{aligned} \quad (5.11)$$

where $(\xi, \eta) = (X_\infty \Lambda, -\Lambda Y_\infty) \in T_{(X_\infty, Y_\infty)}\mathcal{O}(X,Y)$ denotes a tangent vector. Since (X_∞, Y_∞) are full rank matrices (5.11) is equivalent to the linear ODE on $\mathbb{R}^{n\times n}$

$$\begin{aligned}\dot\Lambda &= -(\Lambda'X'_\infty X_\infty N + X'_\infty X_\infty \Lambda N + N\Lambda Y_\infty Y'_\infty + NY_\infty Y'_\infty \Lambda') \\ &= -(\Lambda' DN + D\Lambda N + N\Lambda D + ND\Lambda')\end{aligned} \quad (5.12)$$

The remainder of the proof follows the pattern of the proof for Theorem 4.11. ∎

An alternative way to derive differential equations for balancing is by the following augmented systems of differential equations. They are of a somewhat simpler form, being quadratic matrix differential equations.

Theorem 5.3 Let $(X,Y) \in \mathbb{R}^{k\times n} \times \mathbb{R}^{n\times \ell}$, $\Lambda \in \mathbb{R}^{n\times n}$, $N = \text{diag}(\mu_1, \cdots, \mu_n)$ with $\mu_1 > \cdots > \mu_n > 0$ and $\kappa > 0$ a scaling factor. Assume $\text{rk}(X) = \text{rk}(Y) = n$.

(a) The solutions $(X(t), Y(t))$ of

$$\boxed{\begin{aligned}\dot X &= -X\Lambda, \quad X(0) = X_0 \\ \dot Y &= \Lambda Y, \quad Y(0) = Y_0 \\ \dot\Lambda &= -\kappa \Lambda + X'X - YY', \quad \Lambda(0) = \Lambda_0\end{aligned}} \quad (5.13)$$

and of

$$\boxed{\begin{aligned}\dot X &= -X\Lambda_N, \quad X(0) = X_0 \\ \dot Y &= \Lambda_N Y, \quad Y(0) = Y_0 \\ \dot\Lambda_N &= -\kappa \Lambda_N + X'XN - NYY', \quad \Lambda_N(0) = \Lambda_0\end{aligned}} \quad (5.14)$$

exist for all $t \geq 0$ and satisfy $X(t)Y(t) = X_0 Y_0$.

(b) Every solution $(X(t), Y(t), \Lambda_N(t))$ of (5.14) converges to an equilibrium point $(X_\infty, Y_\infty, \Lambda_\infty)$, characterized by $\Lambda_\infty = 0$ and (X_∞, Y_∞) is a diagonal balanced factorization.

(c) Every solution $(X(t), Y(t), \Lambda(t))$ of (5.13) converges to the set of equilibria $(X_\infty, Y_\infty, 0)$. The equilibrium points of (5.13) are characterized by $\Lambda_\infty = 0$ and (X_∞, Y_∞) is a balanced factorization of (X_0, Y_0).

Proof For any solution $(X(t), Y(t), \Lambda(t))$ of (5.13), we have

$$\frac{d}{dt} X(t) Y(t) = \dot{X} Y + X \dot{Y}$$
$$= -X \Lambda Y + X \Lambda Y \equiv 0$$

and thus $X(t) Y(t) = X(0) Y(0)$ for all t. Similarly for (5.14).

For any symmetric $n \times n$ matrix N let $\Omega_N : \mathcal{O}(X_0, Y_0) \times \mathbb{R}^{n \times n} \to \mathbb{R}$ be defined by

$$\Omega_N(X, Y, \Lambda) = \Phi_N(X, Y) + \frac{1}{2} \text{tr}(\Lambda \Lambda') \tag{5.15}$$

By Corollary 3.3 the function $\Omega_N : \mathcal{O}(X_0, Y_0) \times \mathbb{R}^{n \times n} \to \mathbb{R}$ has compact sublevel sets and $\Omega_N(X, Y, \Lambda) \geq 0$ for all (X, Y, Λ). For any solution $(X(t), Y(t), \Lambda(t))$ of (5.14) we have

$$\frac{d}{dt} \Omega_N(X(t), Y(t), \Lambda(t)) = \text{tr}(N X' \dot{X} + N \dot{Y} Y' + \dot{\Lambda} \Lambda')$$
$$= \text{tr}(-N X' X \Lambda + N \Lambda Y Y' + (-\kappa \Lambda + X' X N - N Y Y') \Lambda')$$
$$= -\kappa \text{tr}(\Lambda \Lambda') \leq 0.$$

Thus Ω_N decreases along the solutions of (5.14) and $\dot{\Omega}_N = 0$ if and only if $\Lambda = 0$. This shows that Ω_N is a weak Lyapunov function for (5.14) and La Salle's invariance principle (Appendix B) asserts that the solutions of (5.14) converge to a connected invariant subset of $\mathcal{O}(X_0, Y_0) \times \{0\}$. But a subset of $\mathcal{O}(X_0, Y_0) \times \{0\}$ is invariant under the flow of (5.14) if and only if the elements of the subset satisfy $X'XN = NYY'$. Specializing now to the case where $N = \text{diag}(\mu_1, \cdots, \mu_n)$ with $\mu_1 > \ldots > \mu_n > 0$ we see that the set of equilibria points of (5.14) is finite. Thus the result follows from Proposition C.12.2. If $N = I_n$, then we conclude from Proposition C.12.2 (b) that the solutions of (5.13) converge to a connected component of the set of equilibria. This completes the proof. ∎

As a generalization of the task of minimizing the cost function $\Phi : \mathcal{O}(X, Y) \to \mathbb{R}$ let us consider the minimization problem for the functions

$$\Phi_{(X_0, Y_0)} : \mathcal{O}(X, Y) \to \mathbb{R}, \quad \Phi_{(X_0, Y_0)} = \|X - X_0\|^2 + \|Y - Y_0\|^2 \tag{5.16}$$

for arbitrary $(X_0, Y_0) \in \mathbb{R}^{k \times n} \times \mathbb{R}^{n \times \ell}$.

6.5. FLOWS ON THE FACTORS X AND Y

Theorem 5.4 *Suppose* $\mathrm{rk}(X) = \mathrm{rk}(Y) = n$. *The gradient flow of* $\Phi_{(X_0,Y_0)} : \mathcal{O}(X,Y) \to \mathbb{R}$ *with respect to the normal Riemannian metric is*

$$\boxed{\begin{aligned} \dot{X} &= X((Y-Y_0)Y' - X'(X-X_0)), & X(0) &= X_0 \\ \dot{Y} &= ((Y-Y_0)Y' - X'(X-X_0))Y, & Y(0) &= Y_0 \end{aligned}} \quad (5.17)$$

The solutions $(X(t), Y(t))$ *exist for all* $t \geq 0$ *and converge to a connected component of the set of equilibrium points, characterized by*

$$Y_\infty(Y'_\infty - Y'_0) = (X'_\infty - X'_0)X_\infty \quad (5.18)$$

Proof The derivative of $\Phi_{(X_0,Y_0)}$ at (X,Y) is

$$\begin{aligned} D\Phi|_{(X_0,Y_0)}(-X\Lambda, \Lambda Y) &= 2\mathrm{tr}(-(X-X_0)'X\Lambda + \Lambda Y(Y-Y_0)') \\ &= 2\mathrm{tr}(((Y-Y_0)Y' - X'(X-X_0))'\Lambda)) \end{aligned}$$

The gradient of $\Phi_{(X_0,Y_0)}$ at (X,Y) is $\mathrm{grad}\Phi_{(X_0,Y_0)}(X,Y) = (-X\Lambda_1, \Lambda_1 Y)$ with

$$\begin{aligned} D\Phi|_{(X_0,Y_0)}(-X\Lambda, \Lambda Y) &= \langle\langle \mathrm{grad}\Phi|_{(X_0,Y_0)}(X,Y), (-X\Lambda, \Lambda Y)\rangle\rangle \\ &= 2\mathrm{tr}(\Lambda'_1 \Lambda) \end{aligned}$$

and hence
$$\Lambda_1 = (Y-Y_0)Y' - X'(X-X_0).$$

Therefore

$$\begin{aligned} \mathrm{grad}\Phi_{(X_0,Y_0)}(X,Y) = &(-X((Y-Y_0)Y' - X'(X-X_0)), \\ &((Y-Y_0)Y' - X'(X-X_0))Y) \end{aligned}$$

which proves (5.17). By assumption, $\mathcal{O}(X,Y)$ is closed and therefore function $\Phi_{(X_0,Y_0)} : \mathcal{O}(X,Y) \to \mathbb{R}$ is proper. Thus the solutions of (5.17) for all $t \geq 0$ and, using Proposition C.12.2, converge to a connected component of the set of equilibria. This completes the proof. ∎

For generic choices of (X_0, Y_0) there are only a finite numbers of equilibria points of (5.17) but an explicit characterization seems hard to obtain. Also the number of local minima of $\Phi_{(X_0,Y_0)}$ on $\mathcal{O}(X,Y)$ is unknown and the dynamical classification of the critical points of $\Phi_{(X_0,Y_0)}$, i.e. a local phase portrait analysis of the gradient flow (5.17), is an unsolved problem.

Problems 1. Let $\mathrm{rk}X = \mathrm{rk}Y = n$ and let $N = N' \in \mathbb{R}^{n \times n}$ be an arbitrary symmetric matrix. Prove that the gradient flow $\dot{X} = -\nabla_X \Phi_N(X,Y), \dot{Y} = -\nabla_Y \Phi_N(X,Y)$, of $\Phi_N : \mathcal{O}(X,Y) \to \mathbb{R}$, $\Phi_N(X,Y) = \frac{1}{2}\mathrm{tr}(N(X'X + YY'))'$ with respect to the induced Riemannian metric (5.2) is

$$\begin{aligned} \dot{X} &= -X\Lambda_N(X,Y) \\ \dot{Y} &= \Lambda_N(X,Y)Y \end{aligned}$$

where
$$\Lambda_N(X,Y) = \int_0^\infty e^{-sX'X}(X'XN - NYY')e^{-sYY'}\,ds$$
is the uniquely determined solution of the Lyapunov equation
$$X'X\Lambda_N + \Lambda_N YY' = X'XN - NYY'.$$

2. For matrix triples $(X,Y,Z) \in \mathbb{R}^{k\times n} \times \mathbb{R}^{n\times m} \times \mathbb{R}^{m\times \ell}$ let
$$\mathcal{O}(X,Y,Z) = \{(XS^{-1}, SYT^{-1}, TZ) \mid S \in GL(n), T \in GL(m)\}.$$
Prove that $\mathcal{O} = \mathcal{O}(X,Y,Z)$ is a smooth manifold with tangent spaces
$$T_{(X,Y,Z)}\mathcal{O} = \{(-X\Lambda, \Lambda Y - YL, LZ) \mid \Lambda \in \mathbb{R}^{n\times n}, L \in \mathbb{R}^{m\times m}\}.$$
Show that
$$\boxed{\begin{aligned} \dot X &= -X(X'X - YY') \\ \dot Y &= (X'X - YY')Y - Y(Y'Y - ZZ') \\ \dot Z &= (Y'Y - ZZ')Z \end{aligned}} \qquad (5.19)$$
is a gradient flow ($\dot X = -\operatorname{grad}_X \phi(X,Y,Z)$, $\dot Y = -\operatorname{grad}_Y \phi(X,Y,Z)$, $\dot Z = -\operatorname{grad}_Z \phi(X,Y,Z)$) of $\phi: \mathcal{O}(X,Y,Z) \mapsto \mathbb{R}$, $\phi(X,Y,Z) = \|X\|^2 + \|Y\|^2 + \|Z\|^2$.

(Hint: Consider the natural generalization of the normal Riemannian metric (5.3) as a Riemannian metric on $\mathcal{O}(X,Y,Z)$!).

Main Points of Section Gradient flows on positive definite matrices and balancing transformations, including diagonal balancing transformations, can be used to construct balancing matrix factorizations. Under reasonable (generic) conditions, the convergence is exponentially fast.

6.6 Recursive Balancing Matrix Factorizations

In order to illustrate, rather than fully develop, discretization possibilities for the matrix factorization flows of this chapter, we first focus on a discretization of the balancing flow (5.4) on matrix factors X, Y, and then discretize the diagonal balancing flow (5.10).

Balancing Flow Following the discretization strategy for the double bracket flow of Chapter 2 and certain subsequent flows, we are led to consider the recursion, for all $k \in \mathbb{N}$,
$$\boxed{\begin{aligned} X_{k+1} &= X_k e^{-\alpha_k(X'_k X_k - Y_k Y'_k)} \\ Y_{k+1} &= e^{\alpha_k(X'_k X_k - Y_k Y'_k)} Y_k \end{aligned}}, \qquad (6.1)$$

6.6. RECURSIVE BALANCING MATRIX FACTORIZATIONS

with step-size selection

$$\alpha_k = \frac{1}{2\lambda_{\max}(X_k'X_k + Y_kY_k')} \qquad (6.2)$$

initialized by X_0, Y_0 such that $X_0Y_0 = H$. Here $X_k \in \mathbb{R}^{m \times n}, Y_k \in \mathbb{R}^{n \times \ell}$ for all k and $H \in \mathbb{R}^{m \times \ell}$

Theorem 6.1 *(Balancing flow on factors) Consider the recursion (6.1), (6.2) initialized by X_0, Y_0 such that $X_0Y_0 = H$ and $\text{rk} X_0 = \text{rk} Y_0 = n$. Then*

(a) for all $k \in \mathbb{N}$

$$X_k Y_k = H . \qquad (6.3)$$

(b) The fixed points (X_∞, Y_∞) of the recursion are the balanced matrix factorizations, satisfying

$$X_\infty' X_\infty = Y_\infty Y_\infty' , \quad H = X_\infty Y_\infty . \qquad (6.4)$$

(c) Every solution (X_k, Y_k), $k \in \mathbb{N}$, of (6.1) converges to the class of balanced matrix factorizations (X_∞, Y_∞) of H, satisfying (6.4).

(d) The linearization of the flow at the equilibria, in transversal directions to the set of equilibria, is exponentially stable with eigenvalues satisfying, for all i, j

$$0 \leq 1 - \frac{\lambda_i(X_\infty' X_\infty) + \lambda_j(X_\infty' X_\infty)}{2\lambda_{\max}(X_\infty' X_\infty)} < 1 . \qquad (6.5)$$

Proof Part (a) is obvious. For part (b), first observe that for any specific sequence of positive real numbers α_k, the fixed points (X_∞, Y_∞) of the flow (6.1) satisfy (6.4). To proceed, let us introduce the notation

$$\begin{aligned} X(\alpha) &= Xe^{-\alpha\Delta} , \ Y(\alpha) = e^{\alpha\Delta}Y , \ \Delta = X'X - YY' , \\ \Phi(\alpha) : &= \Phi(X(\alpha), Y(\alpha)) = \text{tr}(X'(\alpha)X(\alpha) + Y(\alpha)Y'(\alpha)) , \end{aligned}$$

and study the effects of the step size selection α on the potential function $\Phi(\alpha)$. Thus observe that

$$\begin{aligned} \Delta\Phi(\alpha) : &= \Phi(\alpha) - \Phi(0) \\ &= \text{tr}\left(X'X(e^{-2\alpha\Delta} - I) + YY'(e^{2\alpha\Delta} - I)\right) \end{aligned}$$

$$
\begin{aligned}
&= \sum_{j=1}^{\infty} \frac{(2\alpha)^j}{j!} \operatorname{tr}\left((X'X + (-1)^j YY')(-\Delta)^j\right) \\
&= \sum_{j=1}^{\infty} \left\{ \frac{(2\alpha)^{2j-1}}{(2j-1)!} \operatorname{tr}\left((X'X - YY')\Delta^{2j-1}\right) \right. \\
&\qquad \left. + \frac{(2\alpha)^{2j}}{(2j)!} \operatorname{tr}\left((X'X + YY')\Delta^{2j}\right) \right\} \\
&= \sum_{j=1}^{\infty} \operatorname{tr}\left((X'X + YY' - \frac{2j}{2\alpha}I)\frac{(2\alpha\Delta)^{2j}}{(2j)!}\right) \\
&\leq \sum_{j=1}^{\infty} \operatorname{tr}(X'X + YY' - \frac{1}{\alpha}I)\frac{(2\alpha\Delta)^{2j}}{(2j)!} .
\end{aligned}
$$

With the selection α^* of (6.2), deleting subscript k, then $\Delta\Phi(\alpha^*) < 0$. Consequently, under (6.1) (6.2), for all k

$$\Phi(X_{k+1}, Y_{k+1}) < \Phi(X_k, Y_k) .$$

Moreover, $\Phi(X_{k+1}, Y_{k+1}) = \Phi(X_k, Y_k)$ if and only if

$$\sum_{j=1}^{\infty} \frac{1}{(2j)!} \operatorname{tr}\left(2\alpha_k(X'_k X_k - Y_k Y'_k)\right)^{2j} = 0,$$

or equivalently, since $\alpha_k > 0$ for all k, $X'_k X_k = Y_k Y'_k$. Thus the fixed points (X_∞, Y_∞) of the flow (6.1) satisfy (6.4), as claimed, and result (b) of the theorem is established. Moreover, the ω-limit set of a solution (X_k, Y_k) is contained in the compact set $\mathcal{O}(X_0, Y_0) \cap \{(X, Y) \mid \Phi(X, Y) \leq \Phi(X_0, Y_0)\}$, and thus is a nonempty compact subset of $\mathcal{O}(X_0, Y_0)$. This, together with the property that Φ decreases along solutions, establishes (c).

For part (d), note that the linearization of the flow at (X_∞, Y_∞) with $X'_\infty X_\infty = Y_\infty Y'_\infty$ is

$$\Lambda_{k+1} = \Lambda_k - \alpha_\infty (\Lambda_k Y_\infty Y'_\infty + Y_\infty Y'_\infty \Lambda'_k + X'_\infty X_\infty \Lambda_k + \Lambda'_k X'_\infty X_\infty)$$

where $\Lambda \in \mathbb{R}^{n \times n}$ parametrizes uniquely the tangent space $T_{(X_\infty, Y_\infty)}\mathcal{O}(X_0, Y_0)$. The skew-symmetric part of Λ_k corresponds to the linearization in directions tangent to the set of equilibria. Similarly the symmetric part of Λ_k corresponds to the linearization in directions transverse to the set of equilibria. Actually, the skew-symmetric part of the Λ_k remains constant. Only the dynamics of the symmetric part of Λ_k is of interest here. Thus, the linearization with $\Lambda = \Lambda'$ is,

$$
\begin{aligned}
\Lambda_{k+1} &= \Lambda_k - 2\alpha_\infty (\Lambda_k X'_\infty X_\infty + X'_\infty X_\infty \Lambda) , \\
\operatorname{vec}(\Lambda_{k+1}) &= \left(I - 2\alpha_\infty \left((X'_\infty X_\infty) \otimes I + I \otimes (X'_\infty X_\infty)\right)\right)(\operatorname{vec}\Lambda_k) .
\end{aligned}
$$

6.6. RECURSIVE BALANCING MATRIX FACTORIZATIONS

Since $X'_\infty X_\infty = Y_\infty Y'_\infty$ is square, the set of eigenvalues of $((X'_\infty X_\infty) \otimes I + I \otimes (X'_\infty X_\infty))$ is given by the set of eigenvalues $[\lambda_i(X'_\infty X_\infty) + \lambda_j(X'_\infty X_\infty)]$ for all i,j. The result (d) follows. ∎

Remark Other alternative step-size selections are

(i) $\quad \hat{\alpha}_k = \frac{1}{2}\left(\lambda_{\max}(X'_k X_k) + \lambda_{\max}(Y_k Y'_k)\right)^{-1}$

$\quad\quad\quad = \frac{1}{2}\left(\sigma_{\max}(X_k)^2 + \sigma_{\max}(Y_k)^2\right)^{-1}.$

(ii) $\quad \tilde{\alpha}_k = \frac{1}{2}\left(\|X_k\|^2 + \|Y_k\|^2\right)$

Diagonal Balancing Flows The natural generalization of the balancing flow (6.1), (6.2) to diagonal balancing is a discrete-time version of (5.10). Thus consider

$$\begin{aligned} X_{k+1} &= X_k e^{-\alpha_k(X'_k X_k N - N Y_k Y'_k)} \\ Y_{k+1} &= e^{\alpha_k(X'_k X_k N - N Y_k Y'_k)} Y_k \end{aligned} \quad (6.6)$$

where

$$\alpha_k = \frac{1}{2\|X'_k X_k N - N Y_k Y'_k\|} \log\left(\frac{\|(X'_k X_k - Y_k Y'_k)N + N(X'_k X_k - Y_k Y'_k)\|^2}{4\|N\|\,\|X'_k X_k N - N Y_k Y'_k\|(\|X_k\|^2 + \|Y_k\|^2)} + 1\right), \quad (6.7)$$

initialized by full rank matrices X_0, Y_0 such that $X_0 Y_0 = H$. Again $X_k \in \mathbb{R}^{m \times n}, Y_k \in \mathbb{R}^{n \times \ell}$ for all k and $H \in \mathbb{R}^{m \times \ell}$. We shall consider $N = \text{diag}(\mu_1, \cdots \mu_n)$ with $\mu_1 > \mu_2 \cdots > \mu_n$.

Theorem 6.2 *Consider the recursion (6.6) (6.7) initialized by (X_0, Y_0) such that $X_0 Y_0 = H$ and $\text{rk} X_0 = \text{rk} Y_0 = n$. Assume that $N = \text{diag}(\mu_1, \ldots, \mu_n)$ with $\mu_1 > \ldots > \mu_n$. Then*

(a) for all $k \in \mathbb{N}$

$$X_k Y_k = H. \quad (6.8)$$

(b) The fixed points X_∞, Y_∞ of the recursion are the diagonal balanced matrix factorizations satisfying

$$X'_\infty X_\infty = Y_\infty Y'_\infty = D, \quad (6.9)$$

with $D = \text{diag}(d_1, \cdots, d_n)$.

(c) Every solution (X_k, Y_k), $k \in \mathbb{N}$, of (6.6), (6.7) converges to the class of diagonal balanced matrix factorizations (X_∞, Y_∞) of H. The singular values of H are $d_1, \ldots, d_n$.

(d) Suppose that the singular values of H are distinct. There are exactly $2^n n!$ fixed points of (6.6). The set of asymptotically stable fixed points is characterized by (6.9) with $d_1 < \ldots < d_n$. The linearization of the flow at the fixed points is exponentially stable.

Proof The proof follows that of Theorem 5.2, and Theorem 6.1. Only its new aspects are presented here. The main point is to derive the step-size selection (6.7) from an upper bound on $\Delta \Phi_N(\alpha)$. Consider the linear operator $\mathcal{A}_F: \mathbb{R}^{n \times n} \to \mathbb{R}^{n \times n}$ defined by $\mathcal{A}_F(B) = FB + B'F'$. Let $\mathcal{A}_F^m(B) = \mathcal{A}_F(\mathcal{A}_F^{m-1}(B))$, $\mathcal{A}_F^0(B) = B$, be defined recursively for $m \in \mathbb{N}$. In the sequel we only interested in the case where $B = B'$ is symmetric, so that $\mathcal{A}_F$ maps the set of symmetric matrices to itself. The following identity is easily verified by differentiating both sides

$$e^{\alpha F} B e^{\alpha F'} = \sum_{m=0}^{\infty} \frac{\alpha^m}{m!} \mathcal{A}_F^m(B).$$

Then

$$\begin{aligned}
\|e^{\alpha F} B e^{\alpha F'} - \alpha \mathcal{A}_F(B) - B\| &= \|\sum_{m=2}^{\infty} \frac{\alpha^m}{m!} \mathcal{A}_F^m(B)\| \\
&\leq \sum_{m=2}^{\infty} \frac{|\alpha|^m}{m!} \|\mathcal{A}_F^m(B)\| \\
&\leq \sum_{m=2}^{\infty} \frac{|\alpha|^m}{m!} 2^m \|F\|^m \|B\| \\
&= (e^{2|\alpha| \|F\|} - 2|\alpha| \|F\| - 1) \|B\|.
\end{aligned}$$

Setting $B = X'X$ and $F = X'XN - NYY'$, then simple manipulations give that

$$\begin{aligned}
\Delta \Phi_N(\alpha) &= \operatorname{tr}\left(N(e^{-\alpha F'} X'X e^{-\alpha F} + \alpha \mathcal{A}_F(X'X) - X'X) \right. \\
&\quad + \left. N(e^{\alpha F'} YY' e^{\alpha F} - \alpha \mathcal{A}_F(YY') - YY')\right) \\
&\quad - \alpha \operatorname{tr}(F(F+F')).
\end{aligned}$$

Applying the exponential bound above, then

$$\begin{aligned}
\Delta \Phi_N(\alpha) &\leq -\frac{\alpha}{2} \|F + F'\|^2 \\
&\quad + \left(e^{2|\alpha| \|F\|} - 2|\alpha| \|F\| + 1\right) \|N\| (\|X\|^2 + \|Y\|^2) \\
&=: \Delta \Phi_N^U(\alpha).
\end{aligned}$$

Since $\frac{d^2}{d\alpha^2} \Delta \Phi_N^U(\alpha) > 0$ for $\alpha \geq 0$, the upper bound $\Delta \Phi_N^U(\alpha)$ is a strictly convex function of $\alpha \geq 0$. Thus there is a unique minimum $\alpha^* \geq 0$ of

6.6. RECURSIVE BALANCING MATRIX FACTORIZATIONS

$\triangle \Phi_N^U(\alpha)$ obtained by setting $\frac{d}{d\alpha}\Phi_N^U(\alpha) = 0$. This leads to

$$\alpha_* = \frac{1}{2||F||} \log\left(\frac{||F+F'||^2}{4||F||\,||N||(||X||^2+||Y||^2)} + 1\right)$$

and justifies the variable step-size selection (6.7).

A somewhat tedious argument shows that – under the hypothesis of the theorem – the linearization of (6.6) at an equilibrium point (X_∞, Y_∞) satisfying (6.9) is exponentially stable. We omit these details. ∎

Remarks 1. The discrete-time flows of this section on matrix factorization for balancing inherit the essential properties of the continuous-time flows, namely exponential convergence rates.

2. Actually, the flows of this section can be viewed as hybrid flows in that a linear system continuous-time flow can be used to calculate the matrix exponential and at discrete-time instants a recursive update of the linear system parameters is calculated. Alternatively, matrix Padé approximations could perhaps be most in lieu of matrix exponentials. ∎

Main Points of Section There are discretizations of the continuous-time flows on matrix factors for balancing, including diagonal balancing. These exhibit the same convergence properties as the continuous-time flows, including exponential convergence rates.

Notes for Chapter 6

Matrix factorizations (1.1) are of interest in many areas. We have already mentioned an application in system theory, where factorizations of Hankel matrices by the observability and controllability matrices are crucial for realization theory. As a reference to linear system theory we mention Kailath (1980). Factorizations (1.1) also arise in neural network theory; see e.g. Baldi and Hornik (1989).

In this chapter techniques from invariant theory are used in a substantial way. For references on invariant theory and algebraic group actions we refer to Weyl (1946), Dieudonné and Carrell (1971), Kraft (1984) and Mumford and Fogarty (1982). It is a classical result from invariant theory and linear algebra that any two full rank factorizations $(X,Y), (X_1,Y_1) \in \mathbb{R}^{n\times r} \times \mathbb{R}^{r\times m}$ of a $n \times m$-matrix $A = XY = X_1Y_1$ satisfy $(X_1,Y_1) = (XT^{-1}, TY)$ for a unique invertible $r \times r$-matrix T. In fact, the first fundamental theorem of invariant theory states that any polynomial in the coefficients of X and Y which is invariant under the $GL(r,\mathbb{R})$ group action $(X,Y) \mapsto (XT^{-1}, TY)$ can be written as a polynomial in the coefficient of XY. See Weyl (1946), and Gardner (1975) for proofs.

The geometry of orbits of a group action plays a central role in invariant theory. Of special significance are the closed orbits, as these can be separated by polynomial invariants. For a thorough study of the geometry of the orbits $\mathcal{O}(X, Y)$ including a proof of Lemma 3.2 we refer to Kraft (1984).

The standard notions from topology are used throughout this chapter. An algebraic group such as $GL(n, \mathbb{C})$ however also has a different, coarser topology: the Zariski topology.

It can be shown that an orbit of an algebraic group action $GL(n, \mathbb{C}) \times \mathbb{C}^N \to \mathbb{C}^n$ is a Zariski-closed subset of $\mathbb{C}^N$ if and only if it is a closed subset of $\mathbb{C}^N$ in the usual sense of Euclidean topology. Similar results hold for real algebraic group actions; see Slodowy (1989).

The Hilbert-Mumford criterion, Mumford and Fogarty (1982), Kraft (1984), yields an effective test for checking whether or not an orbit is closed.

The Kempf-Ness theorem can be generalized to algebraic group actions of reductive groups. See Kempf and Ness (1979) for proofs in the complex case and Slodowy (1989) for a proof in the real case. Although this has not been made explicitly in the chapter, the theory of Kempf and Ness can be linked to the concept of the moment map arising in symplectic geometry and Hamiltonian mechanics. See Guillemin and Sternberg (1984). The book of Arnold (1989) is an excellent source for classical mechanics.

The problem of minimizing the trace function $\varphi : P(n) \to \mathbb{R}$ defined by (4.5) over the set of positive definite symmetric matrices has been considered by Flanders (1975) and Anderson and Olkin (1978). For related work we refer to Wimmer (1988). A proof of Corollary 4.2 is due to Flanders (1975).

The simple proof of Corollary 4.2, which is based on the Kempf-Ness theorem is due to Helmke (1993).

References

Anderson, T.W. and I. Olkin (1978) An extremal problem for positive definite matrices, *Linear and Multilinear Algebra* 6, pp. 257-262.

Arnold, V.I. (1989) *Mathematical Methods of Classical Mechanics* Second Edition, Graduate Texts in Mathematics 60, Springer, New York.

Azad, J. and J.J. Loeb (1990) On a theorem of Kempf and Ness, *Indiana Univ. Math. J.* 39, pp. 61-65,

Baldi, P. and K. Hornik (1989) Neural networks and principal component analysis: learning from examples without local minima, *Neural Networks* 2, 53-58.

6.6. RECURSIVE BALANCING MATRIX FACTORIZATIONS

Dieudonné, J. and Carrell, J.B. (1971) *Invariant theory, old and new*, Academic Press, New York.

Flanders, H. (1975) An extremal problem on the space of positive definite matrices, *Linear and Multilinear Algebra* 3, pp. 33-39.

Guillemin, V. and S. Sternberg (1984) *Symplectic Techniques in Physics*, Cambridge University Press, Cambridge.

Helmke, U. (1993) Balanced realizations for linear systems: a variational approach, *SIAM J. Control and Opt.* 31, 1-15.

Helmke, U., J.B. Moore and J.E. Perkins (1993) Dynamical systems that compute balanced realizations and the singular value decomposition, to appear in *SIAM J. Matrix Analysis*.

Kailath, T. (1980) *Linear Systems*, Prentice Hall, Englewood Cliffs.

Kempf, G. and L. Ness (1979) The length of vectors in representation spaces, in *Algebraic Geometry* (K. Lonsted, ed.) *Lecture Notes in Math.* 732, pp. 233-244.

Kraft, H. (1984) *Geometrische Methoden in der Invariantentheorie*, Aspects of Mathematics, D1, Vieweg, Braunschweig.

Mumford, D. and J. Fogarty (1982) *Geometry Invariant Theory*, Ergebnisse der Mathematik und ihrer Grenzgebiete 124, Springer, Berlin.

Slodowy, P. (1989) Zur Geometrie der Bahnen reell reduktiver Gruppen, *Algebraic Transformation Groups and Invariant Theory* (H. Kraft, P. Slodowy and T. Springer, eds.), Birkhäuser, Boston, pp. 133-144.

Weyl, H. (1946) *The Classical Groups*, Princeton University Press, N.J.

Wimmer, K. (1988) Extremal problems for Hölder norms of matrices and realizations of linear systems, *SIAM J. Matrix Anal. Appl.* 9, 314-322.

Chapter 7

Invariant Theory and System Balancing

7.1 Introduction

In this chapter balanced realizations arising in linear system theory are discussed from a matrix least squares viewpoint. The optimization problems we consider are those of minimizing the distance $||(A_0, B_0, C_0) - (A, B, C)||^2$ of an initial system (A_0, B_0, C_0) from a manifold M of realizations (A, B, C). Usually M is the set of realizations of a given transfer function but other situations are also of interest. If B_0, C_0 and B, C are all zero then we obtain the least squares matrix estimation problems studied in Chapters 1-4. Thus the present chapter extends the results studied in earlier chapters on numerical linear algebra to the system theory context. The results in this chapter have been obtained by Helmke (1993).

In the sequel, we will focus on the special situation where (A_0, B_0, C_0) = $(0, 0, 0)$, in which case it turns out that we can replace the norm functions by a more general class of norm functions, including p-norms. This is mainly for technical reasons. A more important reason is that we can apply the relevant results from invariant theory without the need for a generalization of the theory.

Specifically, we consider continuous-time linear dynamical systems

$$\begin{aligned} \dot{x}(t) &= Ax(t) + Bu(t) \\ y(t) &= Cx(t) \ , \ t \in \mathbb{R}, \end{aligned} \qquad (1.1)$$

or discrete-time linear dynamical systems of the form

$$x_{k+1} = Ax_k + Bu_k$$

$$y_k = Cx_k, \quad k \in \mathbb{N}_0 \tag{1.2}$$

where $(A, B, C) \in \mathbb{R}^{n \times n} \times \mathbb{R}^{n \times m} \times \mathbb{R}^{p \times n}$ and x, u, y are vectors of suitable dimensions. The systems (1.1) or (1.2) are called asymptotically stable if the eigenvalues of A are in the open left half plane $\mathbb{C}^- = \{z \in \mathbb{C} \mid \operatorname{Re} z < 0\}$ or in the unit disc $D = \{z \in \mathbb{C} \mid |z| < 1\}$, respectively.

Given any asymptotically stable system (A, B, C), the controllability Gramian W_c and observability Gramian W_o are defined in the discrete and continuous-time case, respectively, by

$$W_c = \sum_{k=0}^{\infty} A^k BB'(A')^k, \quad W_o = \sum_{k=0}^{\infty} (A')^k C'CA^k$$

$$W_o = \int_0^{\infty} e^{tA} BB' e^{tA'} dt, \quad W_o = \int_0^{\infty} e^{tA'} C'C e^{tA} dt \tag{1.3}$$

For unstable systems, finite Gramians are defined, respectively, by

$$W_c^{(N)} = \sum_{k=0}^{N} A^k BB'(A')^k, \quad W_o^{(N)} = \sum_{k=0}^{N} (A')^k C'CA^k$$

$$W_c(T) = \int_0^T e^{tA} BB' e^{tA'} dt, \quad W_o(T) = \int_0^T e^{tA'} C'C e^{tA} dt \tag{1.4}$$

for $N \in \mathbb{N}$ and $T > 0$ a real number. Thus (A, B, C) is controllable or observable if and only $W_c > 0$ or $W_o > 0$. In particular, the 'sizes' of W_c and W_o as expressed e.g. by the norms $\|W_c\|, \|W_o\|$ or by the eigenvalues of W_c, W_o measure the controllability and observability properties of (A, B, C). Functions such as

$$f(A, B, C) = \operatorname{tr}(W_c) + \operatorname{tr}(W_o)$$

or, more generally

$$f_p(A, B, C) = \operatorname{tr}(W_c^p) + \operatorname{tr}(W_o^p), \quad p \in \mathbb{N},$$

measure the controllability and observability properties of (A, B, C).

In the sequel we concentrate on asymptotically stable systems and the associated infinite Gramians (1.3), although many results also carry over to finite Gramians (1.4).

A realization (A, B, C) is called *balanced* if the controllability and observability Gramians are equal

$$W_c = W_o \tag{1.5}$$

and is called *diagonal balanced* if

$$W_c = W_o = D$$

7.1. INTRODUCTION

For a stable system (A, B, C), a realization in which the Gramians are equal and diagonal as

$$W_c = W_o = \text{diag}(\sigma_1, \cdots, \sigma_n)$$

is termed a *diagonally balanced* realization. For a minimal, that is controllable and observable, realization (A, B, C), the *singular values* σ_i are all positive. For a non minimal realization of McMillan degree $m < n$, then $\sigma_{m+i} = 0$ for $i > 0$. Corresponding definitions and results apply for Gramians defined on finite intervals T. Also, when the controllability and observability Gramians are equal but not necessarily diagonal the realizations are termed *balanced*. Such realizations are unique only to within orthogonal basis transformations.

Balanced truncation is where a system (A, B, C) with $A \in \mathbb{R}^{n \times n}, B \in \mathbb{R}^{n \times m}, C \in \mathbb{R}^{p \times n}$ is approximated by an rth order system with $r < n$ and $\sigma_r > \sigma_{r+1}$, the last $(n-r)$ rows of (A, B) and last $(n-r)$ columns of $\begin{bmatrix} A \\ C \end{bmatrix}$ of a balanced realization are set to zero to form a reduced rth order realization $(A_r, B_r, C_r) \in \mathbb{R}^{r \times r} \times \mathbb{R}^{r \times m} \times \mathbb{R}^{p \times r}$. A theorem of Pernebo and Silverman (1982) states that if (A, B, C) is balanced and minimal, and $\sigma_r > \sigma_{r+1}$, then the reduced r-th order realization (A_r, B_r, C_r) is also balanced and minimal.

Diagonal balanced realizations of asymptotically stable transfer functions were introduced by B. C. Moore (1981) and have quickly found widespread use in model reduction theory and system approximations. In such diagonal balanced realizations, the controllability and observability properties are reflected in a symmetric way thus, as we have seen, allowing the possibility of model truncation. However, model reduction theory is not the only reason why one is interested in balanced realizations. For example, in digital control and signal processing an important issue is that of optimal finite-wordlength representations of linear systems. Balanced realizations or other related classes of realizations (but not necessarily diagonal balanced realizations!) play an important role in these issues; see Mullis and Roberts (1988), Hwang (1977).

To see the connection with matrix least squares problems we consider the manifolds

$$\mathcal{O}_{\mathbb{C}}(A, B, C) = \{(TAT^{-1}, TB, CT^{-1}) \in \mathbb{R}^{n \times n} \times \mathbb{R}^{n \times m} \times \mathbb{R}^{p \times n} | T \in GL(n)\}$$

for arbitrary n, m, p. If (A, B, C) is controllable and observable then $\mathcal{O}_{\mathbb{C}}(A, B, C)$ is the set of *all* realizations of a transfer function

$$G(z) = C(zI - A)^{-1}B \in \mathbb{R}^{p \times m}(z), \qquad (1.6)$$

see Kalman's realization theorem; Appendix B.5.

Given any $p \times m$ transfer function $G(z) = C(zI-A)^{-1}B$ we are interested in the minima, maxima and other critical points of smooth functions $f : \mathcal{O}_{\mathbb{C}}(A,B,C) \to \mathbb{R}$ defined on the set $\mathcal{O}_{\mathbb{C}}(A,B,C)$ of all realizations of $G(z)$. Balanced and diagonal balanced realizations are shown to be characterized as the critical points of the respective cost functions

$$f : \mathcal{O}_{\mathbb{C}}(A,B,C) \to \mathbb{R}, \quad f(A,B,C) = \text{tr}(W_c) + \text{tr}(W_o) \tag{1.7}$$

and

$$f_N : \mathcal{O}_{\mathbb{C}}(A,B,C) \to \mathbb{R}, \quad f_N(A,B,C) = \text{tr}(NW_c) + \text{tr}(NW_o) \tag{1.8}$$

for a symmetric matrix $N = N'$. Suitable tools for analyzing the critical point structure of such cost functions come from both invariant theory and several complex variable theory. It has been developed by Kempf and Ness (1979) and, more recently, by Azad and Loeb (1990).

For technical reasons the results in this chapter have to be developed over the field of complex numbers. Subsequently, we are able to deduce the corresponding results over the field of real numbers, this being of main interest. Later chapters develop results building on the real domain theory in this chapter.

7.2 Plurisubharmonic Functions

In this section we recall some basic facts and definitions from several complex variable theory concerning plurisubharmonic functions. One reason why plurisubharmonic functions are of interest is that they provide a coordinate free generalization of convex functions. They also play an important role in several complex variable theory, where they are used as a tool in the solution of Levi's problem concerning the characterization of holomorphy domains in $\mathbb{C}^n$. For textbooks on several complex variable theory we refer to Krantz (1982) and Vladimirov (1966).

Let D be an open, connected subset of $\mathbb{C} = \mathbb{R}^2$. An upper semicontinuous function $f : D \to \mathbb{R} \cup \{-\infty\}$ is called *subharmonic* if the mean value inequality

$$f(a) \leq \frac{1}{2\pi} \int_0^{2\pi} f(a + re^{i\theta}) d\theta \tag{2.1}$$

holds for every $a \in D$ and each open disc $B(a,r) \subset \overline{B(a,r)} \subset D$; $B(a,r) = \{z \in \mathbb{C} \mid |z-a| < r\}$. More generally, subharmonic functions can be defined for any open subset D of $\mathbb{R}^n$. If $f : D \to \mathbb{R}$ is twice continuously differentiable then f is subharmonic if and only if

$$\Delta f = \sum_{i=1}^n \frac{\partial^2 f}{\partial x_i^2} \geq 0 \tag{2.2}$$

7.2. PLURISUBHARMONIC FUNCTIONS

on D. For $n = 1$, condition (2.2) is just the usual condition for convexity of a function. Thus, for $n = 1$, the class of subharmonic functions on $\mathbb{R}$ coincides with the class of convex functions.

Let $X \subset \mathbb{C}^n$ be an open and connected subset of $\mathbb{C}^n$. An upper semicontinuous function $f : X \to \mathbb{R} \cup \{-\infty\}$ is called *plurisubharmonic (plush)* if the restriction of f to any one-dimensional complex disc is subharmonic, i.e. if for all $a, b \in \mathbb{C}^n$ and $z \in \mathbb{C}$ with $a + bz \in X$ the function

$$z \longmapsto f(a + bz) \tag{2.3}$$

is subharmonic. The class of plurisubharmonic functions constitutes a natural extension of the class of convex functions: Any convex function on X is plurisubharmonic. We list a number of further properties of plurisubharmonic functions:

Properties of Plurisubharmonic Functions

(i) Let $f : X \to \mathbb{C}$ be holomorphic. Then the functions $\log |f|$ and $|f|^p, p > 0$ real, are plurisubharmonic (plush).

(ii) Let $f : X \to \mathbb{R}$ be twice continuously differentiable. The f is plush if and only if the *Levi form* of f

$$L(f) := \left(\frac{\partial^2 f}{\partial z_i \partial \bar{z}_j}\right) \tag{2.4}$$

is positive semidefinite on X. We say that $f : X \to \mathbb{R}$ is *strictly plush* if the Levi form $L(f)$ is positive definite, i.e. $L(f) > 0$, on X.

(iii) Let $f, g : X \to \mathbb{R}$ be plush, $a \geq 0$ real. Then the functions $f + g$ and $a \cdot f$ are plush.

(iv) Let $f : X \to \mathbb{R}$ be plush and let $\varphi : \mathbb{R} \to \mathbb{R}$ be a convex and monotonically increasing function. Then the composed map $\varphi \circ f : X \to \mathbb{R}$ is plush.

(v) Let $\varphi : X \to Y$ be holomorphic and $f : Y \to \mathbb{R}$ be plush. Then $f \circ \varphi : X \to \mathbb{R}$ is plush. By property (v), the notion of plurisubharmonic functions can be extended to functions on any complex manifold. We then have the following important property:

(vi) Let $M \subset X$ be a complex submanifold and $f : X \to \mathbb{R}$ (strictly) plush. Then the restriction $f|_M : M \to \mathbb{R}$ of f to M is strictly plush.

(vii) Any norm on $\mathbb{C}^n$ is certainly a convex function of its arguments and therefore plush. More generally we have by property (vi).

(viii) Let $||\cdot||$ be any norm on $\mathbb{C}^n$ and let X be a complex submanifold of $\mathbb{C}^n$. Then for every $a \in \mathbb{C}^n$ the distance function $\phi_a : X \to \mathbb{R}$, $\phi_a(x) = ||x - a||$, is plurisubharmonic.

Let $\langle \ , \ \rangle$ be any (positive definite) Hermitian inner product on $\mathbb{C}^n$. Thus for all $z = (z_1, \cdots, z_n)$ and $w = (w_1, \cdots, w_n) \in \mathbb{C}^n$ we have

$$\langle z, w \rangle = \sum_{i,j=1}^{n} a_{ij} z_i \bar{w}_j \qquad (2.5)$$

where $A = (a_{ij}) \in \mathbb{C}^{n \times n}$ a uniquely determined positive definite complex Hermitian $n \times n$-matrix. If $||z|| := \langle z, z \rangle^{\frac{1}{2}}$ is the induced norm on $\mathbb{C}^n$ then the Levi form of $f : \mathbb{C}^n \to \mathbb{R}$, $f(z) = ||z||^2$, is $L(f) = A > 0$ and therefore $f : \mathbb{C}^n \to \mathbb{R}$ is strictly plurisubharmonic. More generally, if $||\cdot||$ is the induced norm of positive definite Hermitian inner product on $\mathbb{C}^n$ and $X \subset \mathbb{C}^n$ is a complex analytic submanifold, then the distance functions $\phi_a : X \to \mathbb{R}$, $\phi_a(x) = ||x - a||^2$, are strictly plurisubharmonic, $a \notin X$.

Problems 1. For $A \in \mathbb{C}^{n \times n}$ let $||A||_F := [\text{tr}(AA^*)]^{\frac{1}{2}}$ be the Frobenius norm. Show that $f : \mathbb{C}^{n \times n} \to \mathbb{R}, f(A) = ||A||_F^2$, is a strictly plurisubharmonic function.

2. Show that the condition number, in terms of Frobenius norms,

$$K(A) = ||A||_F \ ||A^{-1}||_F$$

defines a strictly plurisubharmonic function $K^2 : GL(n, \mathbb{C}) \to \mathbb{R}, K^2(A) = ||A||_F^2 \ ||A^{-1}||_F^2$, on the open subset of invertible complex $n \times n$ - matrices.

Main Points of Section Plurisubharmonic functions extend the class of subharmonic functions for a single complex variable to several complex variables. They are also a natural coordinate-free extension of the class of convex functions and have better structural properties. Least squares distance functions from a point to a complex analytic submanifold are always strictly plurisubharmonic. Strictly plurisubharmonic functions are characterized by the Levi form being positive definite.

7.3 The Azad-Loeb Theorem

We derive a generalization of the Kempf-Ness theorem, employed in the previous chapter, by Azad and Loeb (1990). The result plays a central role in our approach to balanced realizations.

Let $GL(n, \mathbb{C})$ denote the group of all invertible complex $n \times n$ matrices and let $U(n, \mathbb{C}) \subset GL(n, \mathbb{C})$ denote the subgroup consisting of all complex unitary matrices Θ characterized by $\Theta\Theta^* = \Theta^*\Theta = I_n$.

7.3. THE AZAD-LOEB THEOREM

A *holomorphic group action* of $GL(n, \mathbb{C})$ on a finite dimensional complex vector space V is a holomorphic map

$$\alpha : GL(n, \mathbb{C}) \times V \to V \ , \ (g, x) \mapsto g \cdot x \qquad (3.1)$$

such that for all $x \in V$ and $g, h \in GL(n, \mathbb{C})$

$$g \cdot (h \cdot x) = (gh) \cdot x \ , \ e \cdot x = x$$

where $e = I_n$ denotes the $n \times n$ identity matrix. Given an element $x \in V$, the subset of V

$$GL(n, \mathbb{C}) \cdot x = \{g \cdot x \mid g \in GL(n, \mathbb{C})\} \qquad (3.2)$$

is called an *orbit* of α. Since $GL(n, \mathbb{C})$ is a complex manifold, each orbit $GL(n, \mathbb{C}) \cdot x$ is a complex manifold which is biholomorphic to the complex homogeneous space $GL(n, \mathbb{C})/H$, where $H = \{g \in GL(n, \mathbb{C}) \mid g \cdot x = x\}$ is the stabilizer subgroup of x, Appendix C.

A function $\varphi : GL(n, \mathbb{C}) \cdot x \to \mathbb{R}$ on a $GL(n, \mathbb{C})$ - orbit $GL(n, \mathbb{C}) \cdot x$ is called *unitarily invariant* if for all $g \in GL(n, \mathbb{C})$ and all unitary matrices $\Theta \in U(n, \mathbb{C})$

$$\varphi(\Theta g \cdot x) = \varphi(g \cdot x) \qquad (3.3)$$

holds. We are interested in the critical points of unitarily invariant plurisubharmonic functions, defined on $GL(n, \mathbb{C})$ - orbits of a holomorphic group action α. The following result (except for part (b)) is a special case of a more general result due to Azad and Loeb (1990).

Theorem 3.1 *Let $\varphi : GL(n, \mathbb{C}) \cdot x \to \mathbb{R}$ be a unitarily invariant plurisubharmonic function defined on a $GL(n, \mathbb{C})$-orbit $GL(n, \mathbb{C}) \cdot x$ of a holomorphic group action α. Suppose that a global minimum of φ exists on $GL(n, \mathbb{C}) \cdot x$. Then*

(a)   The local minima of $\varphi : GL(n, \mathbb{C}) \cdot x \to \mathbb{R}$ coincide with the global minima. If φ is smooth then all its critical points are global minima.

(b)   The set of global minima is connected.

(c)   If φ is a smooth strictly plurisubharmonic function on $GL(n, \mathbb{C}) \cdot x$, then any critical point of φ is a point where φ assumes its global minimum. The set of global minima is a single $U(n, \mathbb{C})$ − orbit.

Proof   Let $GL(n, \mathbb{C}) = U(n, \mathbb{C}) \cdot \mathcal{P}(n)$ denote the polar decomposition. Here $\mathcal{P}(n)$ denotes the set of positive definite Hermitian matrices and $U(n, \mathbb{C}) = \{e^{i\Omega} \mid \Omega^* = -\Omega\}$ is the group of unitary matrices. Suppose that

$x_0 \in GL(n, \mathbb{C}) \cdot x$ is a local minimum (a critical point, respectively) of φ. By property (v) of plurisubharmonic functions, the scalar function

$$\phi(z) = \varphi(e^{iz\Omega} \cdot x_0), \ z \in \mathbb{C} \tag{3.4}$$

is (pluri)subharmonic for all $\Omega^* = -\Omega$. Since $e^{iz\Omega}$ is a unitary matrix for purely imaginary z, the invariance property of φ implies that $\phi(z)$ depends only on the real part of $Re(z)$ of z. Thus for t real, $t \mapsto \phi(t)$ is a convex function. Thus $\varphi(e^{it\Omega} \cdot x_0) \geq \varphi(x_0)$ for all $t \in \mathbb{R}$ and all $\Omega^* = -\Omega$. By the unitary invariance of φ it follows that $\varphi(g \cdot x_0) \geq \varphi(x_0)$ for all $g \in GL(n, \mathbb{C})$. This proves (a).

To prove (b) suppose that $x_0, x_1 \in GL(n, \mathbb{C}) \cdot x$ are global minima of φ. Thus for $x_1 = ue^{i\Omega} \cdot x_0$ with $\Omega^* = -\Omega, u \in U(n, \mathbb{C})$, we have

$$\varphi(ue^{it\Omega} \cdot x_0) = \varphi(x_0) \tag{3.5}$$

for $t = 0, 1$. By the above argument, $t \mapsto \varphi(ue^{it\Omega} \cdot x_0)$ is convex and therefore $\varphi(t) = \varphi(0)$ for all $0 \leq t \leq 1$. Since $U(n, \mathbb{C})$ is connected there exists a continuous path $[0, 1] \to U(n, \mathbb{C}), t \mapsto u_t$, with $u_0 = I_n, u_1 = u$. Thus $t \mapsto u_t e^{it\Omega} \cdot x_0$ is a continuous path connecting x_0 with x_1 and $\varphi(u_t e^{it\Omega} \cdot x_0) = \varphi(x_0)$ for all $0 \leq t \leq 1$. The result follows.

For (c) note that everything has been proved except for the last statement. But this follows immediately from Lemma 2 in Azad and Loeb (1990). ∎

Since any norm function induced by an unitarily invariant positive definite Hermitian inner product on V is strictly plurisubharmonic on any $GL(n, \mathbb{C})$-orbit of a holomorphic group action x, part (ii) of the Kempf-Ness Theorem 6.2.1 (in the complex case) follows immediately from the Azad-Loeb result.

Main Points of Section The theorem of Azad and Loeb generalizes the scope of the Kempf-Ness theorem from unitarily invariant Hermitian norms to arbitrary unitarily invariant strictly plurisubharmonic functions. This extension is crucial for the subsequent application to balancing.

7.4 Application to Balancing

We now show how the above result can be used to study balanced realizations. The work is based on Helmke (1993).

Consider the complex vector space of triples (A, B, C)

$$L_{\mathbb{C}}(n, m, p) := \{(A, B, C) \in \mathbb{C}^{n \times n} \times \mathbb{C}^{n \times m} \times \mathbb{C}^{p \times n}\}. \tag{4.1}$$

7.4. APPLICATION TO BALANCING

The complex Lie group $GL(n, \mathbb{C})$ of complex invertible $n \times n$ matrices acts on $L_{\mathbb{C}}(n, m, p)$ via the holomorphic group action

$$\sigma : GL(n, \mathbb{C}) \times L_{\mathbb{C}}(n, m, p) \to L_{\mathbb{C}}(n, m, p)$$
$$(S, (A, B, C)) \mapsto (SAS^{-1}, SB, CS^{-1}). \quad (4.2)$$

The orbits of σ

$$\mathcal{O}_{\mathbb{C}}(A, B, C) = \{(SAS^{-1}, SB, CS^{-1}) \mid S \in GL(n, \mathbb{C})\} \quad (4.3)$$

are complex homogenous spaces and thus complex submanifolds of the space $L_{\mathbb{C}}(n, m, p)$.

Of course, by a fundamental theorem in linear systems theory (Kalman's realization theorem; Appendix B) the orbits (4.3) of controllable and observable triples (A, B, C) are in one-to-one correspondence with strictly proper complex rational transfer functions $G(z) \in \mathbb{C}(z)^{p \times m}$ via

$$\mathcal{O}_{\mathbb{C}}(A, B, C) \leftrightarrow G(z) = C(zI - A)^{-1}B. \quad (4.4)$$

A function $f : \mathcal{O}_{\mathbb{C}}(A, B, C) \to \mathbb{R}$ is called *unitarily invariant* if for all unitary matrices $S \in U(n, \mathbb{C})$, $SS^* = I_n$,

$$f(SAS^{-1}, SB, CS^{-1}) = f(A, B, C) \quad (4.5)$$

holds. We are interested in the critical point structure of smooth, unitary invariant plurisubharmonic functions $f : \mathcal{O}_{\mathbb{C}}(A, B, C) \to \mathbb{R}$ on $GL(n, \mathbb{C})$-orbits $\mathcal{O}_{\mathbb{C}}(A, B, C)$. A particular case of interest is where the function $f : \mathcal{O}_{\mathbb{C}}(A, B, C) \to \mathbb{R}$ is induced from a suitable norm on $L_{\mathbb{C}}(n, m, p)$.

Thus let $\langle \, , \, \rangle$ denote a positive definite Hermitian inner product on the $\mathbb{C}$-vector space $L_{\mathbb{C}}(n, m, p)$. The induced Hermitian norm of (A, B, C) is defined by

$$\|(A, B, C)\|^2 = \langle (A, B, C), (A, B, C) \rangle \quad (4.6)$$

A Hermitian norm (4.6) is called unitarily invariant, if

$$\|(SAS^{-1}, SB, CS^{-1})\| = \|(A, B, C)\| \quad (4.7)$$

holds for all unitary transformations $S, with SS^* = I_n$, and $(A, B, C) \in L_{\mathbb{C}}(n, m, p)$. Any Hermitian norm (4.6) defines a smooth strictly plurisubharmonic function

$$\phi : \mathcal{O}_{\mathbb{C}}(A, B, C) \to \mathbb{R}$$
$$(SAS^{-1}, SB, CS^{-1}) \mapsto \|(SAS^{-1}, SB, CS^{-1})\|^2 \quad (4.8)$$

on $\mathcal{O}_{\mathbb{C}}(A, B, C)$.

In the sequel we fix a strictly proper transfer function $G(z) \in \mathbb{C}^{p \times m}(z)$ of McMillan degree n and an initial controllable and observable realization $(A, B, C) \in L_\mathbb{C}(n, m, p)$ of $G(z)$:

$$G(z) = C(zI - A)^{-1}B. \tag{4.9}$$

Thus the $GL(n, \mathbb{C})$-orbit $\mathcal{O}_\mathbb{C}(A, B, C)$ parametrizes the set of all (minimal) realizations of $G(z)$.

Our goal is to study the variation of the norm $\|(SAS^{-1}, SB, CS^{-1})\|^2$ as S varies in $GL(n, \mathbb{C})$. More generally, we want to obtain answers to the questions:

1) Given a function $f : \mathcal{O}_\mathbb{C}(A, B, C) \to \mathbb{R}$, does there exists a realization of $G(z)$ which minimizes f?

2) How can one characterize the set of realizations of a transfer function which minimize $f : \mathcal{O}_\mathbb{C}(A, B, C) \to \mathbb{R}$?

As we will see, the theorems of Kempf-Ness and Azad-Loeb give a rather general solution to these questions. Let $f : \mathcal{O}_\mathbb{C}(A, B, C) \to \mathbb{R}$ be a smooth function on $\mathcal{O}_\mathbb{C}(A, B, C)$ and let $\|\cdot\|$ denote a Hermitian norm defined on $L_\mathbb{C}(n, m, p)$.

A realization

$$(F, G, H) = (S_0 A S_0^{-1}, S_0 B, C S_0^{-1}) \tag{4.10}$$

of a transfer function $G(s) = C(sI - A)^{-1}B$ is called *norm balanced* and *function balanced* respectively, if the function

$$\phi : GL(n, \mathbb{C}) \longrightarrow \mathbb{R}$$
$$S \longmapsto \|(SAS^{-1}, SB, CS^{-1})\|^2 \tag{4.11}$$

or the function

$$\mathcal{F} : GL(n, \mathbb{C}) \longrightarrow \mathbb{R}$$
$$S \longmapsto f(SAS^{-1}, SB, CS^{-1}) \tag{4.12}$$

respectively, has a critical point at $S = S_0$; i.e. the Fréchet derivative

$$D\phi|_{S_0} = 0 \tag{4.13}$$

respectively,

$$D\mathcal{F}|_{S_0} = 0 \tag{4.14}$$

vanishes. (F, G, H) is called *norm minimal* or *function minimizing*, if $\phi(S_0)$, respectively, $\mathcal{F}(S_0)$ is a global minimum for the function (4.11) or (4.12) on $GL(n, \mathbb{C})$, see Figure 7.4.1

7.4. APPLICATION TO BALANCING

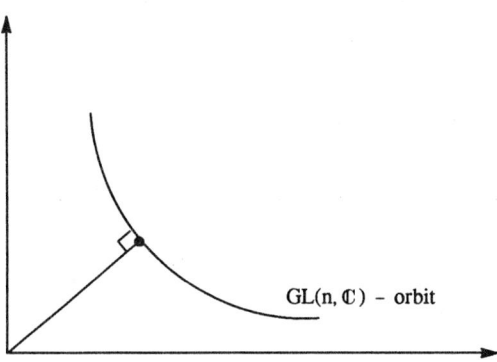

Figure 7.4.1: Norm minimality

We need the following characterization of controllable and observable realizations. Using the terminology of geometric invariant theory, these are shown to be the $GL(n,\mathbb{C})$-stable points for the similarity action $(A,B,C) \mapsto (SAS^{-1}, SB, CS^{-1})$.

Lemma 4.1 $(A,B,C) \in L_{\mathbb{C}}(n,m,p)$ is controllable and observable if and only if the following conditions are satisfied:

(a) The similarity orbit $\mathcal{O}_{\mathbb{C}}(A,B,C) := \{(SAS^{-1}, SB, CS^{-1}) \mid S \in GL(n,\mathbb{C})\}$ is a closed subset of $L_{\mathbb{C}}(n,m,p)$.

(b) $\dim_{\mathbb{C}} \mathcal{O}_{\mathbb{C}}(A,B,C) = n^2$.

Proof The necessity of (b) is obvious, since $GL(n,\mathbb{C})$ acts freely on controllable and observable triples via similarity. The necessity of (a) follows from realization theory, since $\mathcal{O}_{\mathbb{C}}(A,B,C)$ is a fibre of the continuous map

$$\mathcal{H}: L_{\mathbb{C}}(n,m,p) \rightarrow \prod_{i=1}^{\infty} \mathbb{C}^{p \times m}$$

$$(F,G,H) \mapsto (HF^iG \mid i \in \mathbb{N}_0) \qquad (4.15)$$

where $\prod_{i=1}^{\infty} \mathbb{C}^{p \times m}$ is endowed with the product topology.

To prove the sufficiency, let us assume that, e.g., (A,B) is not controllable while conditions (a), (b) are satisfied. Without loss of generality

$$A = \begin{bmatrix} A_{11} & A_{12} \\ 0 & A_{22} \end{bmatrix}, \quad B = \begin{bmatrix} B_1 \\ 0 \end{bmatrix}, \quad C = [C_1, C_2]$$

and (A_{11}, B_1) controllable, $A_{ii} \in \mathbb{C}^{n_i \times n_i}$ for $i = 1, 2$. Consider the one-parameter group of transformations

$$S_t := \begin{bmatrix} I_{n_1} & 0 \\ 0 & t^{-1} I_{n_2} \end{bmatrix} \in GL(n, \mathbb{C})$$

for $t \neq 0$. Then $(A_t, B_t, C_t) := (S_t A S_t^{-1}, S_t B, C S_t^{-1}) \in \mathcal{O}_{\mathbb{C}}(A, B, C)$ with

$$A_t = \begin{bmatrix} A_{11} & t A_{22} \\ 0 & A_{22} \end{bmatrix}, \quad B_t := \begin{bmatrix} B_1 \\ 0 \end{bmatrix}, \quad C_t := [C_1, t C_2]$$

Since $\mathcal{O}_{\mathbb{C}}(A, B, C)$ is closed

$$(A_0, B_0, C_0) = \left(\begin{bmatrix} A_{11} & 0 \\ 0 & A_{22} \end{bmatrix}, \begin{bmatrix} B_1 \\ 0 \end{bmatrix}, [C_1, 0] \right) \in \mathcal{O}_{\mathbb{C}}(A, B, C)$$

the stabilizer of which is $S_t, t \in \mathbb{C}^*$, there is a contradiction to condition (b). Thus (A, B) must be controllable, and similarly (A, C) must be observable. This proves the lemma. ∎

The above lemma shows that the orbits $\mathcal{O}_{\mathbb{C}}(A, B, C)$ of controllable and observable realizations (A, B, C) are *closed* subsets of $L_{\mathbb{C}}(n, m, p)$. It is still possible for other realizations to generate closed orbits with complex dimension strictly less than n^2. The following result characterizes completely which orbits $\mathcal{O}_{\mathbb{C}}(A, B, C)$ are closed subsets of $L_{\mathbb{C}}(n, m, p)$.

Let

$$R(A, B) = (B, AB, \cdots, A^n B) \tag{4.16}$$

$$O(A, C) = \begin{bmatrix} C \\ CA \\ \vdots \\ CA^n \end{bmatrix} \tag{4.17}$$

denote the controllability and observability matrices of (A, B, C) respectively and let

$$\mathcal{H}(A, B, C) = O(A, C) R(A, B)$$

$$= \begin{bmatrix} CB & CAB & \cdots & CA^n B \\ CAB & & & \vdots \\ \vdots & & \ddots & \\ CA^n B & & \cdots & CA^{2n} B \end{bmatrix} \tag{4.18}$$

denote the corresponding $(n+1) \times (n+1)$ block Hankel matrix.

7.4. APPLICATION TO BALANCING

Lemma 4.2 *A similarity orbit $\mathcal{O}_{\mathbb{C}}(A,B,C)$ is a closed subset of $L_{\mathbb{C}}(n, m, p)$ if and only if there exists $(F, G, H) \in \mathcal{O}_{\mathbb{C}}(A, B, C)$ of the form*

$$F = \begin{bmatrix} F_{11} & 0 \\ 0 & F_{22} \end{bmatrix}, \quad G = \begin{bmatrix} G_1 \\ 0 \end{bmatrix}, \quad H = [H_1, 0]$$

such that (F_{11}, G_1, H_1) is controllable and observable and F_{22} is diagonalizable. In particular, a necessary condition for $\mathcal{O}_{\mathbb{C}}(A, B, C)$ to be closed is

$$\text{rk}R(A,B) = \text{rk}O(A,C) = \text{rk}\mathcal{H}(A,B,C) \tag{4.19}$$

Proof Suppose $\mathcal{O}_{\mathbb{C}}(A,B,C)$ is a closed subset of $L_{\mathbb{C}}(n,m,p)$ and assume, for example, that (A, B) is not controllable. Then there exists $(F, G, H) \in \mathcal{O}_{\mathbb{C}}(A, B, C)$ with

$$\begin{bmatrix} F_{11} & F_{12} \\ 0 & F_{22} \end{bmatrix}, \quad G = \begin{bmatrix} G_1 \\ 0 \end{bmatrix}, \quad H = [H_1, H_2]$$

and (F_{11}, G_1) controllable. The same argument as for the proof of Lemma 4.1 shows that $F_{12} = 0, H_2 = 0$. Moreover, arguing similarly for (A, C) we may assume that (F_{11}, H_1) is observable. Suppose F_{22} is not diagonalizable. Then there exists a convergent sequence of matrices $(F_{22}^{(n)} | n \in \mathbb{N})$ with $F_{22}^{\infty} = \lim_{n \to \infty} F_{22}^{(n)}$, such that $F_{22}^{(n)}$ is similar to F_{22} for all $n \in \mathbb{N}$ but F_{22}^{∞} is not. As $\mathcal{O}_{\mathbb{C}}(A, B, C)$ is closed, it follows that for

$$F^{(n)} = \begin{bmatrix} F_{11} & 0 \\ 0 & F_{22}^{(n)} \end{bmatrix}$$

one has $(F^{(n)}, G, H) \in \mathcal{O}_{\mathbb{C}}(A, B, C)$ for all $n \in \mathbb{N}$. Moreover, $(F^{(\infty)}, G, H) = \lim_{n \to \infty}(F^{(n)}, G, H) \in \mathcal{O}_{\mathbb{C}}(A, B, C)$. But this implies that F_{22}^{∞} is similar to F_{22}, therefore leading to a contradiction. This proves the necessity part of the theorem.

Conversely, suppose that (A, B, C) is given as

$$A = \begin{bmatrix} A_{11} & 0 \\ 0 & A_{22} \end{bmatrix}, \quad B = \begin{bmatrix} B_1 \\ 0 \end{bmatrix}, \quad C = [C_1, 0]$$

with (A_{11}, B_1, C_1) controllable and observable, and A_{22} diagonalizable. Then suppose $\mathcal{O}_{\mathbb{C}}(A, B, C)$ is not a closed subset of $L_{\mathbb{C}}(n, m, p)$. By the closed orbit lemma, Kraft (1984), there exists a closed orbit $\mathcal{O}_{\mathbb{C}}(F, G, H) \subset \overline{\mathcal{O}_{\mathbb{C}}(A, B, C)}$. Using the same arguments as above for the necessary part of the lemma we may assume without loss of generality that

$$F = \begin{bmatrix} F_{11} & 0 \\ 0 & F_{22} \end{bmatrix}, \quad G = \begin{bmatrix} G_1 \\ 0 \end{bmatrix}, \quad H = [H_1, 0]$$

with (F_{11}, G_1, H_1) controllable and observable and F_{22} diagonalizable. As the entries of the Hankel matrix $\mathcal{H}(A, B, C)$ depend continuously on A, B, C, we obtain

$$\mathcal{H}(F_{11}, G_1, H_1) = \mathcal{H}(F, G, H)$$
$$\|$$
$$\mathcal{H}(A_{11}, B_1, C_1) = \mathcal{H}(A, B, C)$$

Thus the Hankel matrices of (F_{11}, G_1, H_1) and (A_{11}, B_1, C_1) coincide and therefore, by Kalman's realization theorem,

$$(F_{11}, G_1, H_1) = (S_1 A_{11} S_1^{-1}, S_1 B_1, C_1 S_1^{-1})$$

by an invertible matrix S_1. Since F is in the closure of the similarity orbit of A, their characteristic polynomials coincide: $\det(sI - F) = \det(sI - A)$. Thus

$$\det(sI - F_{22}) = \frac{\det(sI - F)}{\det(sI - F_{11})}$$
$$= \frac{\det(sI - A)}{\det(sI - A_{11})}$$
$$= \det(sI - A_{22})$$

and F_{22}, A_{22} must have the same eigenvalues. Both matrices are diagonalizable and therefore F_{22} is similar to A_{22}. Therefore $(F, G, H) \in \mathcal{O}_{\mathbb{C}}(A, B, C)$ and $\mathcal{O}_{\mathbb{C}}(A, B, C) = \mathcal{O}_{\mathbb{C}}(F, G, H)$ is closed. This contradiction completes the proof. ∎

The following theorems are our main existence and uniqueness results for function minimizing realizations. They are immediate consequences of Lemma 4.1 and the Azad-Loeb Theorem 3.1 and the Kempf-Ness Theorem 6.2.1, respectively.

Recall that a continuous function $f : X \to \mathbb{R}$ on a topological space X is called *proper* if the inverse image $f^{-1}([a, b])$ of any compact interval $[a, b] \subset \mathbb{R}$ is a compact subset of X. For every proper function $f : X \to \mathbb{R}$ the image $f(X)$ is a closed subset of $\mathbb{R}$.

Theorem 4.3 Let $(A, B, C) \in L_{\mathbb{C}}(n, m, p)$ with $G(z) = C(zI - A)^{-1} B$. Let $f : \mathcal{O}_{\mathbb{C}}(A, B, C) \to \mathbb{R}_+$ with $\mathbb{R}_+ = [0, \infty)$ be a smooth unitarily invariant, strictly plurisubharmonic function on $\mathcal{O}_{\mathbb{C}}(A, B, C)$ which is proper. Then:

(a) There exists a global minimum of f in $\mathcal{O}_{\mathbb{C}}(A, B, C)$.

(b) A realization $(F, G, H) \in \mathcal{O}_{\mathbb{C}}(A, B, C)$ is a critical point of f if and only if it minimizes f.

7.4. APPLICATION TO BALANCING

(c)   If $(A_1, B_1, C_1), (A_2, B_2, C_2) \in \mathcal{O}_{\mathbb{C}}(A, B, C)$ are minima of f, then there exists a unitary transformation $S \in U(n, \mathbb{C})$ such that

$$(A_2, B_2, C_2) = (SA_1S^{-1}, SB_1, C_1S^{-1}).$$

S is uniquely determined if (A, B, C) is controllable and observable.

Proof   Part (a) follows because any proper function $f : \mathcal{O}_{\mathbb{C}}(A, B, C) \to \mathbb{R}_+$ possesses a minimum. Parts (b) and (c) are immediate consequences of the Azad-Loeb Theorem.   ∎

A direct application of Theorem 4.3 to the question of norm balancing yields the following result.

Theorem 4.4   *Let $\|\cdot\|^2$ be a norm on $L_{\mathbb{C}}(n, m, p)$ which is induced from an invariant unitarily Hermitian inner product and let $G(z) \in \mathbb{C}^{p \times m}(z)$ denote a strictly proper rational transfer function of McMillan degree n.*

(a)   There exists a norm minimal realization (A, B, C) of $G(z)$.

(b)   A controllable and observable realization $(A, B, C) \in L_{\mathbb{C}}(n, m, p)$ of $G(z)$ is norm minimal if and only if it is norm balanced.

(c)   If $(A_1, B_1, C_1), (A_2, B_2, C_2) \in L_{\mathbb{C}}(n, m, p)$ are controllable and observable norm minimal realizations of $G(z)$, then there exists a uniquely determined unitary transformation $S \in U(n, \mathbb{C})$ such that

$$(A_2, B_2, C_2) = (SA_1S^{-1}, SB_1, C_1S^{-1}).$$

Proof   Let (A, B, C) be a controllable and observable realization of $G(z)$. By Lemmma 4.1, the similarity orbit $\mathcal{O}_{\mathbb{C}}(A, B, C)$ is a closed complex submanifold of $L_{\mathbb{C}}(n, m, p)$. Thus the norm function $(A, B, C) \mapsto \|(A, B, C)\|^2$ defines a proper strictly plurisubharmonic function on $\mathcal{O}_{\mathbb{C}}(A, B, C)$. The result now follows immediately from Theorem 4.3   ∎

Corollary 4.5   *Let $G(z) \in \mathbb{R}^{p \times m}(z)$ denote a strictly proper real rational transfer function of McMillan degree n and let $\|\cdot\|^2$ be a unitarily invariant Hermitian norm on the complex vector space $L_{\mathbb{C}}(n, m, p)$. Similarly, let $f : \mathcal{O}_{\mathbb{C}}(A, B, C) \to \mathbb{R}_+$ be a smooth unitarily invariant strictly plurisubharmonic function on the complex similarity orbit which is proper and invariant under complex conjugation. That is, $f(\bar{F}, \bar{G}, \bar{H}) = f(F, G, H)$ for all $(F, G, H) \in \mathcal{O}_{\mathbb{C}}(A, B, C)$. Then:*

(a*)   There exists a real norm (function) minimal realization (A, B, C) of $G(z)$.

(b*) A real controllable and observable realization $(A, B, C) \in L_{\mathbb{C}}(n, m, p)$ of $G(z)$ is norm (function) minimal if and only if it is norm (function) balanced.

(c*) If $(A_1, B_1, C_1), (A_2, B_2, C_2) \in L_{\mathbb{C}}(n, m, p)$ are real controllable and observable norm (function) minimal realizations of $G(z)$, then there exists a uniquely determined real orthogonal transformation $S \in O(n, \mathbb{R})$ such that (A_1, B_1, C_1) transforms into a real realization $(A_2, B_2, C_2) = (SA_1S^{-1}, SB_1, C_1S^{-1})$.

Proof The following observation is crucial for establishing the above real versions of Theorems 4.3, 4.4. If (A_1, B_1, C_1) and (A_2, B_2, C_2) are real controllable and observable realizations of a transfer function $G(z)$, then any co-ordinate transformation S with $(A_2, B_2, C_2) = (SA_1S^{-1}, SB_1, C_1S^{-1})$ is necessarily real, i.e. $S \in GL(n, \mathbb{R})$. This follows immediately from Kalman's realization theorem, see Appendix (B.5). With this observation, then the proof follows easily from Theorem 4.3. ∎

Remarks 1. The norm balanced realizations were also considered by Verriest (1988) where they are referred to as "optimally clustered". Verriest points out the invariance of the norm under orthogonal transformations but does not show global minimality of the norm balanced realizations. An example is given which shows that not every similarity orbit $\mathcal{O}_{\mathbb{C}}(A, B, C)$ allows a norm balanced realization. In fact, by the Kempf-Ness Theorem 6.2.1 (i), there exists a norm balanced realization in $\mathcal{O}_{\mathbb{C}}(A, B, C)$ if and only if $\mathcal{O}_{\mathbb{C}}(A, B, C)$ is a closed subset of $L_{\mathbb{C}}(n, m, p)$. Therefore Lemma 4.2 characterizes all orbits $\mathcal{O}_{\mathbb{C}}(A, B, C)$ which contain a norm balanced realization.

2. In the above theorems we use the trivial fact that $L_{\mathbb{C}}(n, m, p)$ is a complex manifold, however, the vector space structure of $L_{\mathbb{C}}(n, m, p)$ is not essential. ∎

Balanced realizations for the class of asymptotically stable linear systems were first introduced by B.C. Moore (1981) and are defined by the condition that the controllability and observability Gramians are equal and diagonal. We will now show that these balanced realizations can be treated as a special case of the theorems above. For simplicity we consider only the discrete-time case and complex systems (A, B, C).

A complex realization (A, B, C) is called N-balanced if and only if

$$\sum_{k=0}^{N} A^k BB^*(A^*)^k = \sum_{k=0}^{N} (A^*)^k C^* C A^k \qquad (4.20)$$

An asymptotically stable realization (A, B, C) (i.e. $\sigma(A) < 1$) is said to be

7.4. APPLICATION TO BALANCING

balanced, or ∞-*balanced*, if and only if

$$\sum_{k=0}^{\infty} A^k BB^*(A^*)^k = \sum_{k=0}^{\infty} (A^*)^k C^* C A^k \qquad (4.21)$$

We say that (A, B, C) is *diagonally N-balanced*, or *diagonally balanced* respectively, if the Hermitian matrix (4.20), or (4.21), respectively, is diagonal.

Note that the above terminology differs slightly from the common terminology in the sense that for balanced realizations controllability, respectively observability Gramians, are not required to be diagonal. Of course, this can always be achieved by an orthogonal change of basis in the state space. In Gray and Verriest (1987) realizations (A, B, C) satisfying (4.20) or (4.21) are called essentially balanced.

In order to prove the existence of N-balanced realizations we consider the minimization problem for the *N-Gramian norm* function

$$f_N(A, B, C) := \operatorname{tr}\left(\sum_{k=0}^{N} (A^k BB^*(A^*)^k + (A^*)^k C^* C A^k) \right) \qquad (4.22)$$

for $N \in \mathbb{N} \cup \{\infty\}$. Let $(A, B, C) \in L_{\mathbb{C}}(n, m, p)$ be a controllable and observable realization with $N \geq n$. Consider the smooth unitarily invariant function on the $GL(n, \mathbb{C})$-orbit

$$f_N : \mathcal{O}_{\mathbb{C}}(A, B, C) \longrightarrow \mathbb{R}_+$$

$$(SAS^{-1}, SB, CS^{-1}) \longmapsto f_N(SAS^{-1}, SB, CS^{-1}) \qquad (4.23)$$

where $f_N(SAS^{-1}, SB, CS^{-1})$ is defined by (4.22) for any $N \in \mathbb{N} \cup \{\infty\}$. In order to apply Theorem 4.3 to the cost function $f_N(A, B, C)$, we need the following lemma.

Lemma 4.6 *Let $(A, B, C) \in L_{\mathbb{C}}(n, m, p)$ be a controllable and observable realization with $N \geq n, N \in \mathbb{N} \cup \{\infty\}$. For $N = \infty$ assume, in addition, that A has all its eigenvalues in the open unit disc. Then the function $f_N : \mathcal{O}_{\mathbb{C}}(A, B, C) \to \mathbb{R}$ defined by (4.23) is a proper, unitarily invariant strictly plurisubharmonic function.*

Proof: Obviously $f_N: \mathcal{O}_{\mathbb{C}}(A, B, C) \to \mathbb{R}_+$ is a smooth, unitarily invariant function for all $N \in \mathbb{N} \cup \{\infty\}$. For simplicity we restrict attention to the case where N is finite. The case $N = \infty$ then follows by a simple limiting argument. Let $R_N(A, B) = (B, \ldots, A^N B)$ and $O_N(A, C) = (C', A'C', \ldots, (A')^N C')'$ denote the controllability and observability matrices of length $N+1$, respectively. Let

$$\mathcal{O}_{\mathbb{C}}(O_N, R_N) = \{(O_N(A, C)T^{-1}, T R_N(A, B) \mid T \in GL(n, \mathbb{C})\}$$

denote the complex orbit of $(O_N(A,C), R_N(A,B))$ — see Chapter 6. By controllability and observability of (A,B,C) and using $N \geq n$, both $\mathcal{O}_{\mathbb{C}}(A,B,C)$ and $\mathcal{O}_{\mathbb{C}}(O_N, R_N)$ are closed complex submanifolds. Moreover, for $N \geq n$, the map $\rho_N: \mathcal{O}_{\mathbb{C}}(A,B,C) \to \mathcal{O}_{\mathbb{C}}(O_N, R_N)$ defined by $(SAS^{-1}, SB, CS^{-1}) \mapsto (O_N(A,C)S^{-1}, SR_N(A,B))$ is seen to be biholomorphic. The Frobenius norm defines a strictly plurisubharmonic function

$$\varphi: \mathcal{O}_{\mathbb{C}}(O_N, R_N) \to \mathbb{R},$$
$$\varphi(O_N(A,C)S^{-1}, SR_N(A,B)) = \|O_N(A,C)S^{-1}\|^2 + \|SR_N(A,B)\|^2.$$

Thus $f_N = \varphi \circ \rho_N: \mathcal{O}_{\mathbb{C}}(A,B,C) \to \mathbb{R}$ with

$$f_N(SAS^{-1}, SB, CS^{-1}) = \|O_N(A,C)S^{-1}\|^2 + \|SR_N(A,B)\|^2$$

is strictly plurisubharmonic, too. Similarly, as $\mathcal{O}_{\mathbb{C}}(O_N, R_N)$ is closed, the Frobenius norm $\varphi: \mathcal{O}_{\mathbb{C}}(O_N, R_N) \to \mathbb{R}_+$ is a proper function. Since $\rho_N: \mathcal{O}_{\mathbb{C}}(A,B,C) \to \mathcal{O}_{\mathbb{C}}(O, R)$ is a homeomorphism, then also $F_N = \varphi \circ \rho_N$ is proper. This completes the proof of the lemma. ∎

Using this lemma we can now apply Theorem 4.3. To compute the critical points of $f_N : \mathcal{O}_{\mathbb{C}}(A,B,C) \longrightarrow \mathbb{R}_+$, we consider the induced function on $GL(n, \mathbb{C})$:

$$F_N : GL(n, \mathbb{C}) \longrightarrow \mathbb{R}_+$$
$$S \longmapsto f_N(SAS^{-1}, SB, CS^{-1}) \qquad (4.24)$$

for any $N \in \mathbb{N} \cup \{\infty\}$. A simple calculation of the gradient vector ∇F_N at $S = I_n$ shows that

$$\nabla F_N(I_n) = 2 \sum_{k=0}^{N} (A^k BB^*(A^*)^k - (A^*)^k C^* C A^k) \qquad (4.25)$$

for any $N \in \mathbb{N} \cup \{\infty\}$. We conclude

Corollary 4.7 *Given a complex rational strictly proper transfer function $G(z)$ of McMillan degree n, then for all finite $N \geq n$ there exists a realization (A_*, B_*, C_*) of $G(z)$ that is N-balanced. If $(A_1, B_1, C_1), (A_2, B_2, C_2,)$ are N-balanced realizations of $G(z)$ of order n, $N \geq n$, then*

$$(A_2, B_2, C_2) = (SA_1S^{-1}, SB_1, C_1S^{-1})$$

for a uniquely determined unitary transformation $S \in U(n, \mathbb{C})$. A realization (A_1, B_1, C_1) is N-balanced if and only if it minimizes the N-Gramian norm taken over all realizations of $G(z)$ of order n.

For asymptotically stable linear systems we obtain the following amplifications of Moore's fundamental existence and uniqueness theorem for balanced realizations, see Moore (1981).

Theorem 4.8 *Suppose there is given a complex, rational, strictly-proper transfer function $G(z)$ of McMillan degree n and with all poles in the open unit disc. Then there exists a balanced realization (A_1, B_1, C_1) of $G(z)$ of order n. If $(A_1, B_1, C_1), (A_2, B_2, C_2)$ are two balanced realizations of $G(z)$ of order n, then*

$$(A_2, B_2, C_2) = (SA_1S^{-1}, SB_1, C_1S^{-1})$$

for a uniquely determined unitary transformation $S \in U(n, \mathbb{C})$. An n-dimensional realization (A_1, B_1, C_1) of $G(z)$ is balanced if and only if (A_1, B_1, C_1) minimizes the Gramian norm (4.23) (for $N = \infty$), taken over all realizations of $G(z)$ of order n.

Problems 1. Deduce Corollary 4.7 from Theorem 3.1. Hint: Work with the factorization of the $(N+1) \times (N+1)$ block Hankel $\mathcal{H}_N = O_N \cdot R_N$, where $R_N = (B, AB, \cdots, A^N B)$, $O_N := (C', A'C', \cdots, (A')^N C')'$.

2. Prove that the critical points of the cost function $f_p : \mathcal{O}(A, B, C) \to \mathbb{R}$, $f_p(A, B, C) = \text{tr}(W_c(A, B)^p + W_o(A, C)^p)$, $p \in \mathbb{N}$, are the balanced realizations (A_0, B_0, C_0) satisfying $W_c(A_0, B_0) = W_o(A_0, C_0)$.

Main Points of Section Function balanced realizations are the critical points of a function defined on the manifold of all realizations of a fixed transfer function. Function minimizing realizations are those realizations of given transfer function which minimize a given function. In the case of strictly plurisubharmonic functions, a general existence and uniqueness result on such function balanced realizations is obtained using the Azad-Loeb theorem. Standard balanced realizations are shown to be function minimizing, where the function is the sum of the eigenvalues of the controllability and observability Gramians. This function is unitarily invariant strictly plurisubharmonic.

7.5 Euclidean Norm Balancing

The simplest candidate of a unitarily invariant Hermitian norm on the space $L_{\mathbb{C}}(n, m, p)$ is the *standard Euclidean norm*, defined by

$$\|(A, B, C)\|^2 := \text{tr}(AA^*) + \text{tr}(BB^*) + \text{tr}(C^*C) \qquad (5.1)$$

where $X^* = \bar{X}'$ denotes Hermitian transpose. Applying Theorem 4.4, or more precisely its Corollary 4.5, to this norm yields the following result, which describes a new class of norm minimal realizations.

Theorem 5.1 *Let $G(z)$ be a real, rational, strictly-proper transfer function of McMillan degree n. Then:*

(a) There exists a (real) controllable and observable realization (A, B, C) of $G(z)$ with
$$AA' + BB' = A'A + C'C \qquad (5.2)$$

(b) If $(A_1, B_1, C_1), (A_2, B_2, C_2)$ are (real) controllable and observable realizations of $G(z)$ satisfying (5.2). Then there exists a unique orthogonal transformation $S \in O(n, \mathbb{R})$ with
$$(A_2, B_2, C_2) = (SA_1S^{-1}, SB_1, C_1S^{-1})$$

(c) An n-dimensional realization (A, B, C) of $G(z)$ satisfies (5.2) if and only if it minimizes the Euclidean norm (5.1) taken over all possible n-dimensional realizations of $G(z)$.

Proof For any controllable and observable realization (A, B, C) of $G(z)$, consider the function $\phi : GL(n, \mathbb{R}) \to \mathbb{R}$ defined by the Euclidean norm
$$\phi(S) = \|(SAS^{-1}, SB, CS^{-1})\|^2$$
Thus
$$\phi(S) = \operatorname{tr}(SAS^{-1}(S')^{-1}A'S') + \operatorname{tr}(SBB'S') + \operatorname{tr}((S')^{-1}C'CS^{-1})$$
The gradient vector of ϕ at $S = I_n$ is
$$\nabla \phi(I_n) = 2(AA' - A'A + BB' - C'C) \qquad (5.3)$$
and thus (A, B, C) is norm balanced for the Euclidean norm (5.1) if and only if
$$AA' - A'A + BB' - C'C = 0$$
which is equivalent to (5.2). The result now follows immediately from Theorem 4.4 and its Corollary 4.5. ∎

Similar results hold for symmetric or Hamiltonian transfer functions. Recall that a real rational $m \times m$-transfer function $G(z)$ is called a *symmetric realization* if for all $z \in \mathbb{C}$
$$G(z) = G(z)' \qquad (5.4)$$
and a *Hamiltonian realization*, if for all $z \in \mathbb{C}$
$$G(z) = G(-z)' \qquad (5.5)$$

Every strictly proper symmetric transfer function $G(z)$ of McMillan degree n has a minimal *signature symmetric* realization (A, B, C) satisfying
$$(AI_{pq})' = AI_{pq}, C' = I_{pq}B \qquad (5.6)$$

7.5. EUCLIDEAN NORM BALANCING

where $p + q = n$ and $I_{pq} = \text{diag}(\varepsilon_1, \cdots, \varepsilon_n)$ with

$$\varepsilon_i = \begin{cases} 1 & i = 1, \cdots, p \\ -1 & i = p+1, \cdots, n \end{cases}$$

Here $p - q$ is the Cauchy-Maslov index of $G(z)$; cf. Anderson and Bitmead (1977), Byrnes and Duncan (1982). Similarly, every strictly proper Hamiltonian transfer function $G(z) \in \mathbb{R}(z)^{m \times m}$ of McMillan degree $2n$ has a minimal Hamiltonian realization (A, B, C) satisfying

$$(AJ)' = AJ \quad , \quad C' = JB \tag{5.7}$$

where

$$J = \begin{bmatrix} 0 & -I_n \\ I_n & 0 \end{bmatrix} \tag{5.8}$$

is the standard complex structure. Let $O(p, q)$, respectively, $Sp(n, \mathbb{R})$ denote the (real) stabilizer groups of I_{pq} respectively, J, i.e.

$$\begin{aligned} T \in O(p, q) &\iff T' I_{pq} T = I_{pq} \, , \, T \in GL(n, \mathbb{R}) \\ T \in Sp(n, \mathbb{R}) &\iff T' JT = J \, , \, T \in GL(2n, \mathbb{R}) \end{aligned} \tag{5.9}$$

Analogues of Theorem 5.1 are now presented for the symmetric and Hamiltonian transfer functions, The proof of Theorem 5.2 requires a more subtle argument than merely an application of the Corollary 4.5.

Theorem 5.2 *(a) Every strictly proper symmetric transfer function $G(z) \subset \mathbb{R}(z)^{m \times m}$ with McMillan degree n and Cauchy-Maslov index $p - q$ has a controllable and observable signature symmetric realization (A, B, C) satisfying*

$$\begin{aligned} (AI_{pq})' &= AI_{pq} \, , \, C' = I_{pq} B \tag{5.10} \\ AA' + BB' &= A'A + C'C \tag{5.11} \end{aligned}$$

(b) If $(A_1, B_1, C_1), (A_2, B_2, C_2)$ are two minimal realizations of $G(z)$ satisfying (5.10),(5.11), then there exists a unique orthogonal transformation $S = \text{diag}(S_1, S_2) \in O(p) \times O(q) \subset O(n)$ with

$$(A_2, B_2, C_2) = (SA_1 S^{-1}, SB_1, C_1 S^{-1})$$

Proof We first prove (a). By Theorem 5.1 there exists a minimal realization (A, B, C) of the symmetric transfer function $G(s)$ which satisfies (5.11) and (A, B, C) is unique up to an orthogonal change of basis $S \in O(n, \mathbb{R})$. By the symmetry of $G(s)$, with (A, B, C) also (A', C', B') is a realization of

$G(s)$. Note that if (A, B, C) satisfies (5.11), also (A', C', B') satisfies (5.11). Thus, by the above uniqueness property (ii) of Theorem 5.1 there exists a unique $S \in O(n, \mathbb{R})$ with

$$(A', C', B') = (SAS', SB, CS') \tag{5.12}$$

By transposing (5.12) we also obtain

$$(A, B, C) = (SA'S', SC', B'S') \tag{5.13}$$

or equivalently

$$(A', C', B') = (S'AS, S'B, CS) \tag{5.14}$$

Thus (by minimality of (A, B, C)) $S = S'$ is symmetric orthogonal. A straightforward argument, see Byrnes and Duncan (1982) for details, shows that the signature of S is equal to the Cauchy-Maslov index $p - q$ of $G(s) = C(sI - A)^{-1}B$. Thus $S = T'I_{pq}T$ for $T \in O(n, \mathbb{R})$ and the new realization

$$(F, G, H) := (TAT', TB, CT')$$

satisfies (5.11).

To prove part (b), let

$$(A_2, B_2, C_2) = (SA_1 S^{-1}, SB_1, C_1 S^{-1})$$

be two realizations of $G(s)$ which satisfy (5.10), (5.11). Byrnes and Duncan (1982), have shown that (5.10) implies $S \in O(p, q)$. By Theorem 5.1 (b) also $S \in O(n, \mathbb{R})$. Since $O(p) \times O(q) = O(p, q) \cap O(n, \mathbb{R})$ the result follows. ∎

A similar result holds for Hamiltonian transfer functions.

Theorem 5.3 (a) *Every strictly proper Hamiltonian transfer function $G(s) \in \mathbb{R}(s)^{m \times m}$ with McMillan degree $2n$ has a controllable and observable Hamiltonian realization (A, B, C) satisfying*

$$(AJ)' = AJ, \quad C' = JB \tag{5.15}$$
$$AA' + BB' = A'A + C'C \tag{5.16}$$

(b) *If $(A_1, B_1, C_1), (A_2, B_2, C_2)$ are two minimal realizations of $G(s)$ satisfying (5.15), (5.16), then there exists a unique symplectic transformation $S \in Sp(n, \mathbb{R}) \cap O(2n, \mathbb{R})$ with*

$$(A_2, B_2, C_2) = (SA_1 S^{-1}, SB_1, C_1 S^{-1})$$

7.5. EUCLIDEAN NORM BALANCING

Proof This result can be deduced directly from Theorem 4.4, generalized to the complex similarity orbits for the reductive group $Sp(n, \mathbb{C})$. Alternatively, a proof can be constructed along the lines of that for Theorem 5.2.

Problems 1. Prove that every strictly proper, asymptotically stable symmetric transfer function $G(z) \in \mathbb{R}(z)^{m \times m}$ with McMillan degree n and Cauchy-Maslov index $p - q$ has a controllable and observable signature-symmetric balanced realization satisfying

$$(AI_{pq})' = AI_{pq}, C' = I_{pq}B, W_c(A, B) = W_o(A, C) = \text{diagonal}.$$

2. Prove as follows an extension of Theorem 5.1 to bilinear systems. Let $W(n) = \{I = (i_1, \ldots, i_k) \mid i_j \in \{1, 2\}, 1 \leq k \leq n\} \cup \{\emptyset\}$ be ordered lexicographically. A bilinear system $(A_1, A_2, B, C) \in \mathbb{R}^{n \times n} \times \mathbb{R}^{n \times n} \times \mathbb{R}^{n \times m} \times \mathbb{R}^{p \times n}$ is called *span controllable* if and only if $R(A_1, A_2, B) = (A_I B \mid I \in W(n)) \in \mathbb{R}^{n \times N}$, $N = m(2^{n+1} - 1)$, has full rank n. Here $A_\emptyset := I_n$. Similarly (A_1, A_2, B, C) is called *span observable* if and only if $O(A_1, A_2, C) = (CA_I \mid I \in W(n)) \in \mathbb{R}^{M \times n}$, $M = p(2^{n+1} - 1)$, has full rank n.

(a) Prove that, for (A_1, A_2, B, C) span controllable and observable, the similarity orbit

$$\mathcal{O} = \mathcal{O}(A_1, A_2, B, C) = \{(SA_1S^{-1}, SA_2S^{-1}, SB, CS^{-1}) \mid S \in GL(n)\}$$

is a closed submanifold of $\mathbb{R}^{2n^2 + n(m+p)}$ with tangent space

$$T_{(A_1, A_2, B, C)}\mathcal{O} = \{([X, A_1], [X, A_2], XB, -CX) \mid X \in \mathbb{R}^{n \times n}\}.$$

(b) Show that the critical points $(F_1, F_2, G, H) \in \mathcal{O}(A_1, A_2, B, C)$ of $\Phi: \mathcal{O}(A_1, A_2, B, C) \to \mathbb{R}$,

$$\Phi(F_1, F_2, G, H) = ||F_1||^2 + ||F_2||^2 + ||G||^2 + ||H||^2,$$

are characterized by

$$F_1 F_1' + F_2 F_2' + GG' = F_1' F_1 + F_2' F_2 + H'H.$$

(c) Prove that, for (A_1, A_2, B, C) span controllable and observable, the critical points of $\Phi: \mathcal{O}(A_1, A_2, B, C) \to \mathbb{R}$ form a uniquely determined $O(n)$-orbit

$$\{(SF_1S^{-1}, SF_2S^{-1}, SG, HS^{-1}) \mid S \in O(n)\}.$$

∎

Main Points of Section Function minimizing realization of a transfer function, where the distance function is the matrix Euclidean norm, are characterized. These realizations minimize the Euclidean norm distance to the zero realization $(0,0,0)$. The existence and characterization of signature symmetric norm minimal realizations for symmetric or Hamiltonian transfer functions is shown.

Notes for Chapter 7

For results on linear system theory we refer to Kailath (1980) and Sontag (1990). Balanced realizations for asymptotically stable linear systems were introduced by B.C. Moore (1981) as a tool for model reduction and system approximation theory. For important results on balanced realizations see Moore (1981), Pernebo and Silverman (1982), Jonkheere and Silverman (1983), Verriest (1983), Ober (1987). In signal processing and sensitivity analysis, balanced realizations were introduced in the pioneering work of Mullis and Roberts (1976); see also Hwang (1977). In these papers sensitivity measures such as weighted trace functions of controllability and observability Gramians are considered and the global minima are characterized as balanced, input balanced or sensitivity optimal realizations.

For further results in this direction see Williamson (1986), Williamson and Skelton (1989). A systematic optimization approach for defining and studying various classes of balanced realizations has been developed by Helmke (1993). The further development of this variational approach to balancing has been the initial motivation of writing this book. We also mention the closely related work of Gray and Verriest (1987, 1988), Verriest (1986) and Verriest (1988) who apply differential geometric methods to characterize balanced realizations. For further applications of differential geometry to define balanced realizations for multi-mode systems and sensitivity optimal singular systems we refer to Verriest (1988) and Gray and Verriest (1989).

Techniques from geometric invariant theory and complex analysis, such as the Kempf-Ness or the Azad-Loeb theorems, for the study of system balancing and sensitivity optimization tasks have been introduced by Helmke (1992, 1993). Both the Kempf-Ness theorem and the theorem by Azad and Loeb are initially stated over the field of complex numbers. In fact, for applications of these tools to systems balancing it becomes crucial to work over the field of complex numbers, even if one is ultimately interested in real data. Therefore we first focus in this chapter on the complex case and then deduce the real case from the complex one. In the later chapters we mainly concentrate on the real case.

For textbooks on several complex variable theory we refer to Krantz (1982) and Vladimirov (1966).

Preliminary results on norm balanced realizations were obtained by Verriest (1988), where they are referred to as "optimally clustered". The existence and uniqueness results Theorems 5.1 – 5.3 were obtained by Helmke (1993). In contrast to balanced realizations, no simple algebraic methods such as the SVD are available to compute norm balanced realizations. To this date, the dynamical systems methods for finding norm balanced realizations are the only available tools to compute such realizations. For an application of these ideas to solve an outstanding problem in sensitivity optimization of linear systems see Chapter 9.

References

Anderson, B.D.O. and R.R. Bitmead (1977) The matrix Cauchy index: Properties and applications, *SIAM J. Appl. Math.* 33, 655-672.

Azad, H. and J.J. Loeb (1990) On a theorem of Kempf and Ness, *Indiana Univ. J.*, 39, 61-65.

Byrnes, C.I. and T.W. Duncan (1989) On certain topological invariants arising in system theory, in *New Directions in Applied Mathematics*, (P.J. Hilton and G.S. Young, eds.), Springer Verlag, pp. 29-72.

Gray, W.S. and E.I. Verriest (1989) On the sensitivity of generalized state-space systems, *Proc. of the 28th Conference on Decision and Control*, Tampa, 1337-1342.

Gray, W.S. and E.I. Verriest (1987) Optimality properties of balanced realizations: minimum sensitivity, *Proc. of the 26th Conference on Decision and Control*, Los Angeles, 124-128.

Helmke, U. (1992) A several complex variables approach to sensitivity analysis and structured singular values, *Journal of Mathematical Systems, Estimation, and Control*, 2, 1-13.

Helmke, U. (1993) Balanced realizations for linear systems: a variational approach, *SIAM J. Control and Optimization* 31, 1-15.

Hwang, S.Y. (1977) Minimum uncorrelated unit noise in state space digital filtering, *IEEE Trans. Acoust. Speech Signal Process.*, ASSP-25, 273-281.

Jonckheere, E. and L.M. Silverman (1983) A new set of invariants for linear systems: Applications to reduced order compensator design, *IEEE Transactions on Automatic Control* 28, 953-964.

Kailath, T. (1980) *Linear Systems*, Prentice Hall, Englewood Cliffs.

Kempf, G. and L. Ness (1979) The length of vectors in representation spaces, *Algebraic Geometry*, (K. Lonsted, ed.), *Lecture Notes in Math.*, 732, Springer Verlag, 233-244.

Krantz, S.G. (1982) *Function Theory of Several Complex Variables*, John Wiley, New York.

Moore, B.C. (1981) Principal component analysis in linear systems: Controllability, observability and model reduction, *IEEE Transactions on Automatic Control*, AC-26, 17-32.

Mullis, C.T. and R.A. Roberts (1976) Synthesis of minimum roundoff noise fixed point digital filters, *IEEE Trans. Circuits Sys.*, C AS-23, 551-562.

Ober, R. (1987) Balanced realizations: Canonical form, parametrization, model reduction, *International Journal of Control* 46, 643-670.

Pernebo, L. and L.M. Silverman (1982) Model reduction via balanced state space representations, *IEEE Transactions on Automatic Control* 27, 282-287.

Sontag, E.D. (1990) *Mathematical Control Systems: Deterministic Finite Dimensional Systems*, Texts in Applied Mathematics, Springer Verlag, New York.

Verriest, E.I. (1983) On generalized balanced realizations, *IEEE Transaction on Automatic Control* AC-28, 833-844.

Verriest, E.I. (1986) Model reduction via balancing, and connections with other methods, *Modelling and Approximation of Stochastic Processes* (U.B. Desai, Ed.), Kluwer Academic Publ., 123-154.

Verriest, E.I. and W.S. Gray (1988) Robust design problems: A geometric approach, in *Linear Circuits, Systems and Signal Processing: Theory and Applications* (Byrnes, Martin and Saeks, eds.), North-Holland, Amsterdam, 321-328.

Verriest, E.I. (1988) Minimum sensitivity implementation for multi-mode systems, *Proc. IEEE Conf. on Decision and Control*, Austin, TX, 2165-2170.

Vladimirov, V.S. (1966) *Methods of the Theory of Functions of Many Complex Variables*, Cambridge, MA: MIT Press.

Williamson, D. (1986) A property of internally balanced and low noise structures, *IEEE Transactions on Automatic Control*, AC-31, 633-634.

Williamson, D. and R.E. Skelton (1989) Optimal q-Markov COVER for finite wordlength implementation, *Math. Systems Theory* 22, 253-273.

Chapter 8

Balancing via Gradient Flows

8.1 Introduction

In the previous chapter we investigate balanced realizations from a variational viewpoint, i.e. as the critical points of objective functions defined on the manifold of realizations of a given transfer function. This leads us naturally to the computation of balanced realizations using steepest descent methods. In this chapter, gradient flows for the balancing cost functions are constructed which evolve on the class of positive definite matrices and converge exponentially fast to the class of balanced realizations. Also gradient flows are considered which evolve on the Lie group of invertible coordinate transformations. Again there is exponential convergence to the class of balancing transformations. Of course, explicit algebraic methods are available to compute balanced realizations which are reliable and comparatively easy to implement on a digital computer, see Laub, Heath, Paige and Ward(1987), Safanov and Chiang (1989).

So why should one consider continuous-time gradient flows for balancing, an approach which seems much more complicated and involved? This question of course brings us back to our former goal of seeking to replace algebraic methods of computing by analytic or geometric ones, i.e. via the analysis of certain well designed differential equations whose limiting solutions solve the specific computational task. We believe that the flexibility one gains while working with various discretizations of continuous-time systems, opens the possibility of new algorithms to solve a given problem than a purely algebraic approach would allow. We see the possibility of adaptive schemes, and perhaps even faster algorithms based on this approach. It is for this reason that we proceed with the analysis of this chapter.

For any asymptotically stable minimal realization (A, B, C) the controllability Gramian W_c and observability Gramian W_o are defined in discrete time and continuous time, respectively, by

$$W_c = \sum_{k=0}^{\infty} A^k BB' A'^k, \quad W_o = \sum_{k=0}^{\infty} A'^k C'C A^k \qquad (1.1)$$

$$W_c = \int_0^{\infty} e^{At} BB' e^{A't} dt, \quad W_o = \int_0^{\infty} e^{A't} C'C e^{At} dt. \qquad (1.2)$$

For an unstable system the controllability and observability Gramians are defined by finite sums or integrals, rather than the above infinite sums or integrals. In the following, we assume asymptotic stability and deal with infinite sums, however, all results hold in the unstable case for finite sum Gramians such as in (7.1.4).

Any linear change of coordinates in the state space $\mathbb{R}^n$ by an invertible transformation $T \in GL(n, \mathbb{R})$ changes the realization by $(A, B, C) \mapsto (TAT^{-1}, TB, CT^{-1})$ and thus transforms the Gramians via

$$W_c \mapsto TW_c T', \quad W_o \mapsto (T')^{-1} W_o T^{-1} \qquad (1.3)$$

We call a state space representation (A, B, C) of the transfer function $G(s)$ *balanced* if $W_c = W_o$. This is more general than the usual definition of balanced realizations, Moore (1981), which requires that $W_c = W_o = diagonal$, which is one particular realization of our class of balanced realizations. In this latter case we refer to (A, B, C) as a *diagonal balanced realization*.

In the sequel we consider a given, fixed asymptotically stable minimal realization (A, B, C) of a transfer function $G(s) = C(sI - A)^{-1}B$.

To get a quantitative measure of how the Gramians change we consider the function

$$\phi(T) = \text{tr}(TW_c T' + (T')^{-1} W_o T^{-1}) = \text{tr}(W_c T'T + W_o (T'T)^{-1})$$
$$= \text{tr}(W_c P + W_o P^{-1}) \qquad (1.4)$$

with

$$P = T'T \qquad (1.5)$$

Note that $\phi(T)$ is the sum of the eigenvalues of the controllability and observability Gramians of (TAT^{-1}, TB, CT^{-1}) and thus is a crude numerical measure of the controllability and observability properties of (TAT^{-1}, TB, CT^{-1}).

8.2. FLOWS ON POSITIVE DEFINITE MATRICES

Moreover, as we have seen in Chapter 7, the critical points of the potential $\phi(\cdot)$ are balanced realizations.

Main Points of Section In seeking a balanced realization of a state space linear system representation, it makes sense to work with a quantitative measure of how the controllability and observability Gramians change with the co-ordinate basis transformation. We choose the sum of the eigenvalues of these two Gramians.

8.2 Flows on Positive Definite Matrices

In this and the next section, we consider a variety of gradient flows for balancing. The intention is that the reader browse through these to grasp the range of possibilities of the gradient flow approach.

Let $\mathcal{P}(n)$ denote the set of positive definite symmetric $n \times n$ matrices $P = P' > 0$. The function we are going to study is

$$\phi : \mathcal{P}(n) \to \mathbb{R}, \quad \phi(P) = \text{tr}(W_c P + W_o P^{-1}) \tag{2.1}$$

Let $X, Y \in \mathbb{R}^{n \times n}$ be invertible square root factors of the positive definite Gramians W_c, W_o so that

$$X'X = W_o, \quad YY' = W_c \tag{2.2}$$

The following results are then immediate consequences of the more general results from Chapter 6, Section 4.

Lemma 2.1 *Let W_c, W_o be defined by (1.1) for an asymptotically stable controllable and observable realization (A, B, C). Then the function $\phi : \mathcal{P}(n) \to \mathbb{R}, \phi(P) = \text{tr}(W_c P + W_o P^{-1})$ has compact sublevel sets, i.e. for all $a \in \mathbb{R}$, $\{P \in \mathcal{P}(n) \mid \text{tr}(W_c P + W_o P^{-1}) \leq a\}$ is a compact subset of $\mathcal{P}(n)$. There exists a minimizing positive definite symmetric matrix $P = P' > 0$ for the function $\phi : \mathcal{P}(n) \to \mathbb{R}$ defined by (2.1).*

Theorem 2.2 *(Linear index gradient flow)*

(a) There exists a unique $P_\infty = P'_\infty > 0$ which minimizes $\phi : \mathcal{P}(n) \to \mathbb{R}, \phi(P) = \text{tr}(W_c P + W_o P^{-1})$, and P_∞ is the only critical point of ϕ. This minimum is given by

$$P_\infty = W_c^{-\frac{1}{2}} (W_c^{\frac{1}{2}} W_o W_c^{\frac{1}{2}})^{\frac{1}{2}} W_c^{-\frac{1}{2}} \tag{2.3}$$

$T_\infty = P_\infty^{\frac{1}{2}}$ *is a balancing transformation for (A, B, C).*

(b) The gradient flow $\dot P(t) = -\nabla\phi(P(t))$ on $\mathcal{P}(n)$ is given by

$$\dot P = P^{-1}W_o P^{-1} - W_c, \quad P(0) = P_0.\quad (2.4)$$

For every initial condition $P_0 = P_0' > 0$, $P(t)$ exists for all $t \geq 0$ and converges exponentially fast to P_∞ as $t \to \infty$, with a lower bound for the rate of exponential convergence given by

$$\rho \geq 2\frac{\lambda_{\min}(W_c)^{3/2}}{\lambda_{\max}(W_o)^{1/2}} \quad (2.5)$$

where $\lambda_{\min}(A)$, $\lambda_{\max}(A)$, denote the smallest and largest eigenvalue of A, respectively.

In the sequel we refer to (2.4) as the *linear index gradient flow*. Instead of minimizing $\phi(P)$ we can just as well consider the minimization problem for the quadratic index function

$$\psi : \mathcal{P}(n) \to \mathbb{R}, \quad \psi(P) = \operatorname{tr}((W_c P)^2 + (W_o P^{-1})^2) \quad (2.6)$$

over all positive symmetric matrices $P = P' > 0$. Since, for $P = T'T$, $\psi(P)$ is equal to $\operatorname{tr}((TW_c T')^2 + ((T')^{-1}W_o T^{-1})^2)$, the minimization problem for (2.6) is equivalent to the task of minimizing $\operatorname{tr}(W_c^2 + W_o^2) = \|W_c\|^2 + \|W_o\|^2$ over all realizations of a given transfer function $G(s) = C(sI - A)^{-1}B$ where $\|X\|$ denotes the Frobenius norm. Thus $\psi(P)$ is the sum of the squared eigenvalues of the controllability and observability Gramians of (TAT^{-1}, TB, CT^{-1}). The quadratic index minimization task above can also be reformulated as minimizing $\|TW_c T' - (T')^{-1}W_o T^{-1}\|^2$. A theorem corresponding to Theorem 6.4.2 now applies.

Theorem 2.3 *(Quadratic index gradient flow)*

(a) There exists a unique $P_\infty = P_\infty' > 0$ which minimizes $\psi : \mathcal{P}(n) \to \mathbb{R}$, $\psi(P) = \operatorname{tr}((W_c P)^2 + (W_o P^{-1})^2)$, and $P_\infty = W_o^{-\frac{1}{2}}(W_c^{\frac{1}{2}} W_o W_c^{\frac{1}{2}})^{\frac{1}{2}} W_c^{-\frac{1}{2}}$ is the only critical point of ψ. Also, $T_\infty = P_\infty^{\frac{1}{2}}$ is a balancing coordinate transformation for (A, B, C).

(b) The gradient flow $\dot P(t) = -\nabla \psi(P(t))$ on $\mathcal{P}(n)$ is

$$\dot P = 2P^{-1}W_o P^{-1} W_o P^{-1} - 2W_c P W_c.\quad (2.7)$$

For all initial conditions $P_o = P_0' > 0$, the solution $P(t)$ of (2.7) exists for all $t \geq 0$ and converges exponentially fast to P_∞. A lower bound on the rate of exponential convergence is

$$\rho > 4\lambda_{\min}(W_c)^2 \quad (2.8)$$

8.2. FLOWS ON POSITIVE DEFINITE MATRICES

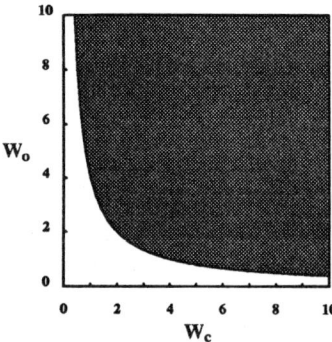

Figure 8.2.1: The quadratic index flow is faster than the linear index flow in the shaded region

We refer to (2.7) as *quadratic index gradient flow*. Both algorithms converge exponentially fast to P_∞, although the rate of convergence is rather slow if the smallest singular value of W_c is near to zero. The convergence rate of the linear index flow, however, depends inversely on the maximal eigenvalue of the observability Gramian. In contrast, the convergence of the quadratic index flow is robust with respect to the observability properties of the system.

In general, the quadratic index flow seems to behave better than the linear index flow. The following lemma, which is a special case of Lemma 6.4.5, compares the rates of exponential convergence of the algorithms and shows that the quadratic index flow is in general faster than the linear index flow.

Lemma 2.4 *Let ρ_1 and ρ_2 respectively denote the rates of exponential convergence of (2.5) and (2.8) respectively. Then $\rho_1 < \rho_2$ if the smallest singular value of the Hankel operator of (A,B,C) is $> \frac{1}{2}$, or equivalently, if $\lambda_{\min}(W_o W_c) > \frac{1}{4}$.*

Simulations: The following simulations show the exponential convergence of the diagonal elements of P towards the solution matrix P_∞ and illustrate what might effect the convergence rate. In Figure 8.2.2a-c we have

$$W_o = \begin{bmatrix} 7 & 4 & 4 & 3 \\ 4 & 4 & 2 & 2 \\ 4 & 2 & 4 & 1 \\ 3 & 2 & 1 & 5 \end{bmatrix} \text{ and } W_c = \begin{bmatrix} 5 & 2 & 0 & 3 \\ 2 & 7 & -1 & -1 \\ 0 & -1 & 5 & 2 \\ 3 & -1 & 2 & 6 \end{bmatrix}$$

so that $\lambda_{\min}(W_o W_c) \approx 1.7142 > \frac{1}{4}$. Figure 8.2.2a concerns the linear index flow while Figure 8.2.2b shows the evolution of the quadratic index flow, both using $P(0) = P_1$ where

$$P(0) = P_1 = \begin{bmatrix} 1 & 0 & 0 & 0 \\ 0 & 1 & 0 & 0 \\ 0 & 0 & 1 & 0 \\ 0 & 0 & 0 & 1 \end{bmatrix} \quad P(0) = P_2 = \begin{bmatrix} 2 & 1 & 0 & 0 \\ 1 & 2 & 1 & 0 \\ 0 & 1 & 2 & 1 \\ 0 & 0 & 1 & 2 \end{bmatrix}.$$

Figure 8.2.2c shows the evolution of both algorithms with a starting value of $P(0) = P_2$. These three simulations demonstrate that the quadratic algorithm converges more rapidly than the linear algorithm when $\lambda_{\min}(W_o W_c) > \frac{1}{4}$. However, the rapid convergence rate is achieved at the expense of approximately twice the computational complexity, and consequently the computing time on a serial machine may increase.

In Figure 8.2.2d

$$W_o = \begin{bmatrix} 7 & 4 & 4 & 3 \\ 4 & 4 & 2 & 2 \\ 4 & 2 & 4 & 1 \\ 3 & 2 & 1 & 3 \end{bmatrix} \text{ and } W_c = \begin{bmatrix} 5 & 4 & 0 & 3 \\ 4 & 7 & -1 & -1 \\ 0 & -1 & 5 & 2 \\ 3 & -1 & 2 & 6 \end{bmatrix}$$

so that $\lambda_{\min}(W_o W_c) \approx 0.207 < \frac{1}{4}$. Figure 8.2.2d compares the linear index flow behaviour with that of the quadratic index flow for $P(0) = P_1$. This simulation demonstrates that the linear algorithm does not necessarily converge more rapidly than the quadratic algorithm when $\lambda_{\min}(W_o W_c) < \frac{1}{4}$, because the bounds on convergence rates are conservative.

Riccati Equations A different approach to determine a positive definite matrix $P = P' > 0$ with $W_o = PW_cP$ is via Riccati equations. Instead of solving the gradient flow (2.4) we consider the Riccati equation

$$\boxed{\dot{P} = W_o - PW_cP, \quad P(0) = P_0.} \tag{2.9}$$

It follows from the general theory of Riccati equations that for every positive definite symmetric matrix P_0, (2.9) has a unique solution as a positive definite symmetric $P(t)$ defined for all $t \geq 0$, see for example Anderson and Moore (1990). Moreover, $P(t)$ converges exponentially fast to P_∞ with $P_\infty W_c P_\infty = W_o$.

The Riccati equation (2.9) can also be interpreted as a gradient flow for the cost function $\phi : \mathcal{P}(n) \to \mathbb{R}$, $\phi(P) = \text{tr}(W_c P + W_o P^{-1})$ by considering a different Riemannian matrix to that used above. Consider the Riemannian metric on $\mathcal{P}(n)$ defined by

$$\langle\langle \xi, \eta \rangle\rangle := \text{tr}(P^{-1}\xi P^{-1}\eta) \tag{2.10}$$

for tangent vectors $\xi, \eta \in T_P(\mathcal{P}(n))$. This is easily seen to define a symmetric positive definite inner product on $T_P \mathcal{P}(n)$. The Fréchet derivative of ϕ :

8.2. FLOWS ON POSITIVE DEFINITE MATRICES

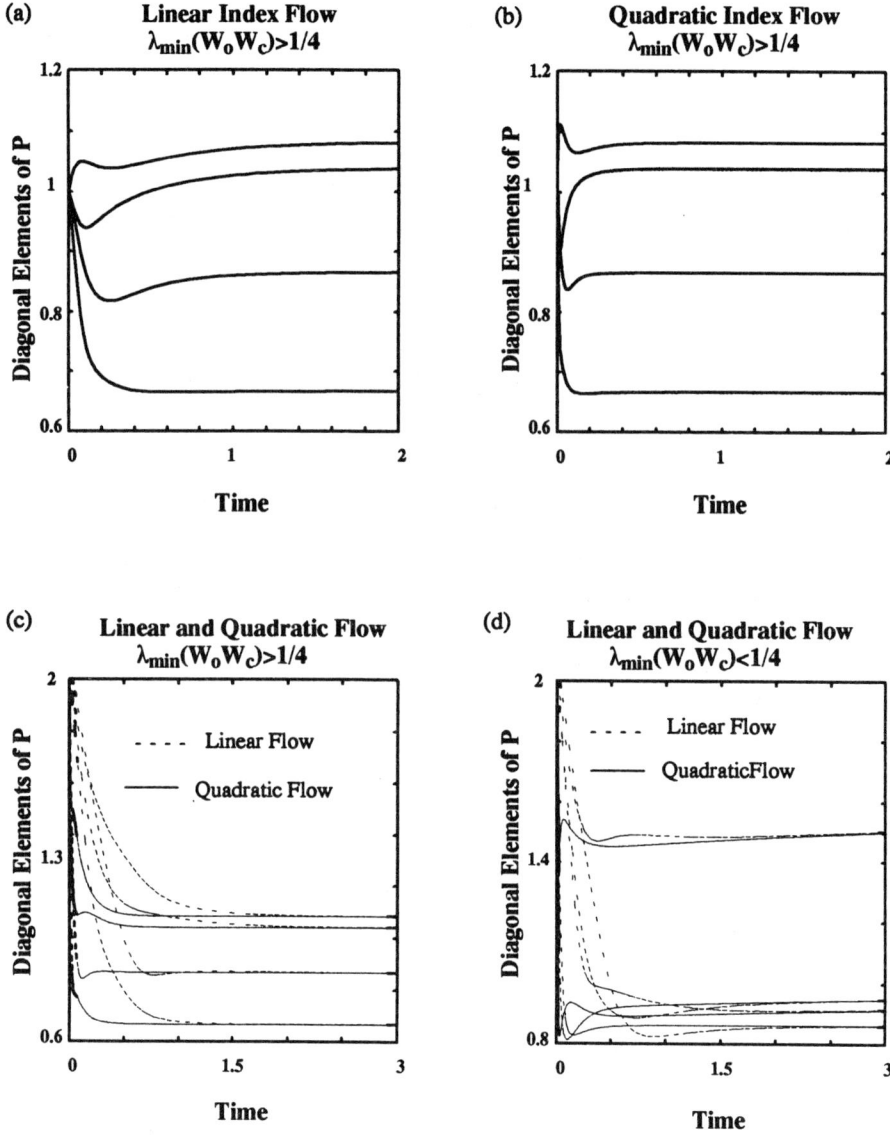

Figure 8.2.2: The linear and quadratic flows

$\mathcal{P}(n) \to \mathbb{R}$ at P is the linear map $D\phi|_P : T_P\mathcal{P}(n) \to \mathbb{R}$ on the tangent space defined by

$$D\phi|_P(\xi) = \text{tr}((W_c - P^{-1}W_o P^{-1})\xi), \; \xi \in T_P(\mathcal{P}(n))$$

Thus the gradient $\text{grad}\phi(P)$ of ϕ with respect to the Riemannian metric (2.10) on $P(n)$ is characterized by

$$\begin{aligned} D\phi|_P(\xi) &= \langle\langle \text{grad}\phi(P), \xi \rangle\rangle \\ &= \text{tr}(P^{-1}\text{grad}\phi(P)P^{-1}\xi) \end{aligned}$$

for all $\xi \in T_P\mathcal{P}(n)$. Thus $P^{-1}\text{grad}\phi(P)P^{-1} = W_c - P^{-1}W_o P^{-1}$, or equivalently

$$\text{grad}\phi(P) = PW_c P - W_o \qquad (2.11)$$

Therefore (2.9) is seen as the gradient flow $\dot{P} = -\text{grad}\phi(P)$ of ϕ on $\mathcal{P}(n)$ and thus has equivalent convergence properties to (2.4). In particular, the solution $P(t), t \geq 0$ converges exponentially to P_∞.

This Riccati equation approach is particularly to balanced realizations suitable if one is dealing with time-varying matrices W_c and W_o.

As a further illustration of the above approach we consider the following optimization problem over a convex set of positive semidefinite matrices. Let $A \in \mathbb{R}^{m \times n}$, $\text{rk}A = m < n$, $B \in \mathbb{R}^{m \times n}$ and consider the convex subset of positive semidefinite matrices

$$\mathcal{C} = \{P \in \mathbb{R}^{n \times n} \mid P = P' \geq 0,\; AP = B\}$$

with nonempty interior

$$\overset{\circ}{\mathcal{C}} = \{P \in \mathbb{R}^{n \times n} \mid P = P' > 0,\; AP = B\}.$$

A Riemannian metric on $\overset{\circ}{\mathcal{C}}$ is

$$\langle\langle \xi, \eta \rangle\rangle = \text{tr}(P^{-1}\xi' P^{-1}\eta) \quad \forall\, \xi, \eta \in T_P(\overset{\circ}{\mathcal{C}})$$

with

$$T_P(\overset{\circ}{\mathcal{C}}) = \{\xi \in \mathbb{R}^{n \times n} \mid A\xi = 0\}.$$

Similarly, we consider for $A \in \mathbb{R}^{m \times n}$, $\text{rk}A = m < n$, and $B \in \mathbb{R}^{m \times m}$

$$\begin{aligned} \mathcal{D} &= \{P \in \mathbb{R}^{n \times n} \mid P = P' \geq 0,\; APA' = B\} \\ \overset{\circ}{\mathcal{D}} &= \{P \in \mathbb{R}^{n \times n} \mid P = P' > 0,\; APA' = B\} \end{aligned}$$

Here the Riemannian metric is the same as above and

$$T_P(\overset{\circ}{\mathcal{D}}) = \{\xi \in \mathbb{R}^{n \times n} \mid A\xi A' = 0\}.$$

8.2. FLOWS ON POSITIVE DEFINITE MATRICES

We now consider the cost function

$$\phi : \overset{\circ}{\mathcal{C}} \to \mathbb{R} \quad \text{respectively} \quad \phi : \overset{\circ}{\mathcal{D}} \to \mathbb{R}$$

defined by

$$\phi(P) = \text{tr}(W_1 P + W_2 P^{-1})$$

for symmetric matrices $W_1, W_2 \in \mathbb{R}^{n \times n}$.

The gradient of $\phi : \overset{\circ}{\mathcal{C}} \to \mathbb{R}$ is characterized by

(i) $\text{grad}\phi(P) \in T_P(\overset{\circ}{\mathcal{C}})$

(ii) $\langle\langle \text{grad}\phi(P), \xi \rangle\rangle = D\phi|_P(\xi) \ \forall \ \xi \in T_P(\overset{\circ}{\mathcal{C}})$

Now (i) $\iff A \cdot \text{grad}\phi(P) = 0$
and (ii) $\iff \text{tr}((W_1 - P^{-1}W_c P^{-1})\xi)$
$= \text{tr}((P^{-1}\text{grad}\phi(P)P^{-1})^{-1}\xi) \ \forall \ \xi \in T_P(\overset{\circ}{\mathcal{C}})$
$\iff W_1 - P^{-1}W_2 P^{-1} - P^{-1}\text{grad}\phi(P)P^{-1} \in T_P(\overset{\circ}{\mathcal{C}})^{\perp}$
$\iff W_1 - P^{-1}W_2 P^{-1} - P^{-1}\text{grad}\phi(P)P^{-1} = A'\Lambda$

From this it follows, as above [1] that

$$\text{grad}\phi(P) = (I_n - PA'(APA')^{-1}A) \underbrace{P(W_1 - P^{-1}W_2 P^{-1})P}_{PW_1 P - W_2}$$

Thus the gradient flow on $\overset{\circ}{\mathcal{C}}$ is $(\dot{P} = \text{grad}\phi(P))$

$$\dot{P} = (I_n - PA'(APA')^{-1}A)(PW_1 P - W_2)$$

For $\overset{\circ}{\mathcal{D}}$ the gradient $\text{grad}\phi(P)$ of $\phi : \overset{\circ}{\mathcal{D}} \to \mathbb{R}$, is characterized by

(i) $A \cdot \text{grad}\phi(P) A' = 0$

(ii) $\text{tr}((P^{-1}\text{grad}\phi(P)P^{-1})'\xi) = \text{tr}((W_1 - P^{-1}W_2 P^{-1})\xi) \ \forall \ \xi : A\xi A' = 0$.

Hence

(ii) $\iff P^{-1}\text{grad}\phi(P)P^{-1} - (W_1 - P^{-1}W_2 P^{-1}) \in \{\xi \mid A\xi A' = 0\}^{\perp}$

Lemma 2.5 Let $A \in \mathbb{R}^{m \times n}$, $AA' > 0$.

$$\{\xi \in \mathbb{R}^{n \times n} \mid A\xi A' = 0\}^{\perp} = \{A'\Lambda A \mid \Lambda \in \mathbb{R}^{m \times m}\}$$

---
[1] $P \nabla \phi(P) P - PA'\Lambda P = \text{grad}\phi$ and $\Lambda = (APA')^{-1}AP \nabla \phi(P)$

Proof As

$$\text{tr}(A'\Lambda' A\xi) = \text{tr}(\Lambda' A\xi A') = 0 \ \forall\, \xi \text{ with } A\xi A' = 0,$$

the right hand side set in the lemma is contained in the left hand side set. Both spaces have the same dimension. Thus the result follows. ∎

Hence with this lemma

$$(\text{ii}) \iff P^{-1}\text{grad}\phi(P)P^{-1} - \nabla\phi(P) = A'\Lambda A$$

and therefore

$$\text{grad}\phi(P) = P\nabla\phi(P)P + PA'\Lambda AP.$$

Thus

$$AP\nabla\phi(P)PA' = -APA'\Lambda APA',$$

i.e.

$$\Lambda = -(APA')^{-1}AP\nabla\phi(P)PA'(APA')^{-1}$$

and therefore

$$\text{grad}\phi(P) = P\nabla\phi(P)P - PA'(APA')^{-1}AP\nabla\phi(P)PA'(APA')^{-1}AP$$

We conclude

Theorem 2.6 *The gradient flow* $\dot{P} = \text{grad}\phi(P)$ *of the cost function* $\phi(P) = \text{tr}(W_1 P + W_2 P^{-1})$ *on the constraint sets* $\overset{\circ}{\mathcal{C}}$ *and* $\overset{\circ}{\mathcal{D}}$, *respectively, is*

(a) $\boxed{\dot{P} = (I_n - PA'(APA')^{-1}A)(PW_1 P - W_2)}$

for $P \in \overset{\circ}{\mathcal{C}}$.

(b) $\boxed{\begin{aligned}\dot{P} &= PW_1 P - W_2 \\ &\quad - PA'(APA')^{-1}A(PW_1 P - W_2)A'(APA')^{-1}AP\end{aligned}}$

for $P \in \overset{\circ}{\mathcal{D}}$.

In both cases the underlying Riemannian metric is defined as above.

Dual Flows We also note the following transformed versions of the linear and quadratic gradient flow. Let $Q = P^{-1}$, then (2.4) is equivalent to

$$\boxed{\dot{Q} = QW_c Q - Q^2 W_o Q^2} \tag{2.12}$$

and (2.7) is equivalent to

$$\boxed{\dot{Q} = 2QW_c Q^{-1} W_c Q - 2Q^2 W_o QW_o Q^2\,.} \tag{2.13}$$

8.2. FLOWS ON POSITIVE DEFINITE MATRICES

We refer to (2.12) and (2.13) as the transformed linear and quadratic index gradient algorithms respectively. Clearly the analogue statements of Theorems 2.2, 2.3 remain valid. We state for simplicity only the result for the transformed gradient algorithm (2.12).

Theorem 2.7 *The transformed gradient flow (2.12) converges exponentially from every initial condition $Q_o = Q'_0 > 0$ to $Q_\infty = P_\infty^{-1}$. The rate of exponential convergence is the same as for the Linear Index Gradient Flow.*

Proof It remains to prove the last statement. This is true since $P \mapsto P^{-1} = Q$ is a diffeomorphism of the set of positive definite symmetric matrices onto itself. Therefore the matrix

$$J = -P_\infty^{-1} \otimes W_c - W_c \otimes P_\infty^{-1}$$

of the linearization $\dot\xi = J \cdot \xi$ of (2.4) at P_∞ is similar to the matrix $\hat J$ of the linearization

$$\dot\eta = -Q_\infty \eta Q_\infty^{-1} W_c Q_\infty - Q_\infty W_c Q_\infty^{-1} \eta Q_\infty$$

of (2.12) via the invertible linear map $\xi \mapsto \eta = -Q_\infty^{-1} \xi Q_\infty$. It follows that J and $\hat J$ have the same eigenvalues which completes the proof. ∎

The *dual* versions of the linear and quadratic gradient flows are defined as follows. Consider the objective functions

$$\phi^d(P) = \operatorname{tr}(W_c P^{-1} + W_o P)$$

and

$$\psi^d(P) = \operatorname{tr}((W_c P^{-1})^2 + (W_o P)^2)$$

Thus $\phi^d(P) = \phi(P^{-1}), \psi^d(P) = \psi(P^{-1})$. The gradient flows of ϕ^d and ψ^d respectively are the *dual linear flow*

$$\boxed{\dot P = -W_o + P^{-1} W_c P^{-1}} \qquad (2.14)$$

and the *dual quadratic flow*

$$\boxed{\dot P = 2 P^{-1} W_c P^{-1} W_c P^{-1} - 2 W_o P W_o.} \qquad (2.15)$$

Thus the dual gradient flows are obtained from the gradient flows (2.4), (2.7) by interchanging the matrices W_c and W_o.

In particular, (2.14) converges exponentially to P_∞^{-1}, with a lower bound on the rate of convergence given by

$$\rho^d \geq 2 \frac{\lambda_{\min}(W_o)^{3/2}}{\lambda_{\max}(W_c)^{1/2}} \qquad (2.16)$$

while (2.15) converges exponentially to P_∞^{-1}, with a lower bound on the rate of convergence given by

$$\rho^d \geq 4\lambda_{\min}(W_o)^2 . \tag{2.17}$$

Thus in those cases where W_c is ill-conditioned, i.e. the norm of W_c is small, we may prefer to use (2.16).

In this way we can obtain a relatively fast algorithm for computing P_∞^{-1} and thus, by inverting P_∞^{-1}, we obtain the desired solution P_∞. By transforming the gradient flows (2.12), respectively, (2.13) with the transformation $P \to P^{-1} = Q$ we obtain the following results.

Theorem 2.8 *For every initial condition $P_o = P_o' > 0$, the solution $P(t)$ of the ordinary differential equation*

$$\boxed{\dot{P} = PW_oP - P^2W_cP^2} \tag{2.18}$$

exists within $\mathcal{P}(n)$ for all $t \geq 0$ and converges exponentially fast to P_∞, with a lower bound on the rate of convergence

$$\rho \geq 2\frac{\lambda_{\min}(W_o)^{3/2}}{\lambda_{\max}(W_c)^{1/2}} \tag{2.19}$$

Theorem 2.9 *For every initial condition $P_o = P_o' > 0$, the solution $P(t)$ of the ordinary differential equation*

$$\boxed{\dot{P} = 2PW_oP^{-1}W_oP - 2P^2W_cPW_cP^2} \tag{2.20}$$

exists in $\mathcal{P}(n)$ for all $t \geq 0$ and converges exponentially fast to P_∞. A lower bound on the rate of convergence is

$$\rho \geq 4\lambda_{\min}(W_o)^2 \tag{2.21}$$

Proofs Apply the transformation $P \to P^{-1} = Q$ to (2.14), respectively, (2.15) to achieve (2.19), respectively (2.21). The same bounds on the rate of convergence as (2.16), (2.17) carry over to (2.18) ,(2.20). ∎

Thus, in cases where $\|W_c\|$ (respectively $\lambda_{\min}(W_c)$) is small, one may prefer to use (2.18) (respectively (2.20)) as a fast method to compute P_∞.

Simulations: These observations are also illustrated by the following figures, which show the evolution of the diagonal elements of P towards the solution matrix P_∞. In Figure 8.2.3a-b, W_o and W_c are the same as used in generating Figure 8.2.2d simulations so that $\lambda_{\min}(W_oW_c) \approx 0.207 < \frac{1}{4}$. Figure 8.2.3a uses (2.14) while Figure 8.2.3b shows the evolution of

8.2. FLOWS ON POSITIVE DEFINITE MATRICES

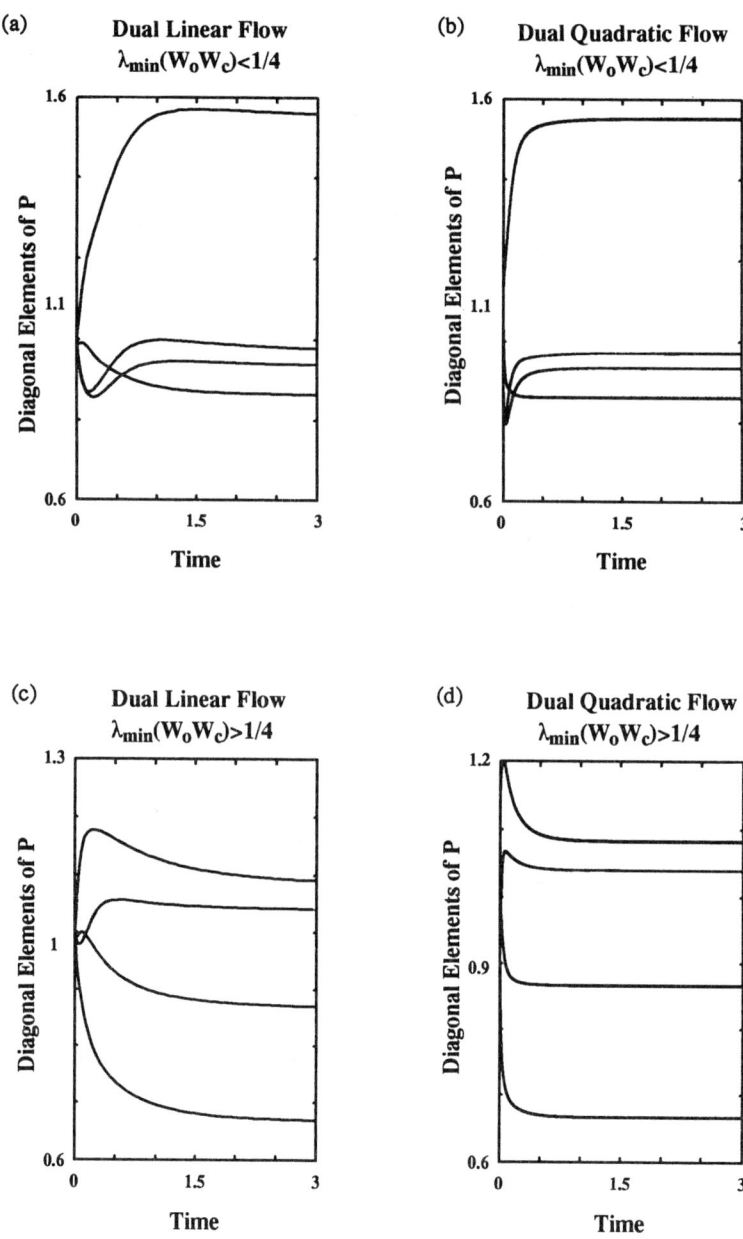

Figure 8.2.3: The linear and quadratic dual flows

(2.15), both using $P(0) = P_1$. These simulations demonstrate that the quadratic algorithm converges more rapidly than the linear algorithm in the case when $\lambda_{\min}(W_o W_c), \frac{1}{4}$. In Figure 8.2.3c-d W_o and W_c are those used in Figure 8.2.2a-e simulations with $\lambda_{\min}(W_o W_c) \approx 1.7142 > \frac{1}{4}$. Figure 8.2.3c uses (2.14) while Figure 8.2.3d shows the evolution of (2.15), both using $P(0) = P_1$. These simulations demonstrate that the linear algorithm does not necessarily converge more rapidly than the quadratic algorithm in this case when $\lambda_{\min}(W_o W_c) > \frac{1}{4}$.

Newton-Raphson Gradient Flows The following modification of the gradient algorithms (2.4), (2.7) is a continuous-time version of the (discrete-time) *Newton-Raphson method* to minimize $\phi(P)$, respectively $\psi(P)$.

The Hessians of ϕ and ψ are given by:

$$H_\phi(P) = -P^{-1} \otimes P^{-1} W_o P^{-1} - P^{-1} W_o P^{-1} \otimes P^{-1} \quad (2.22)$$

and

$$\begin{aligned} H_\psi(P) &= -2P^{-1} \otimes P^{-1} W_o P^{-1} W_o P^{-1} - 2P^{-1} W_o P^{-1} \otimes P^{-1} W_o P^{-1} \\ &\quad -2P^{-1} W_o P^{-1} W_o P^{-1} \otimes P^{-1} - 2W_c \otimes W_c \end{aligned} \quad (2.23)$$

Thus

$$H_\phi(P)^{-1} = -(P \otimes P)[P \otimes W_o + W_o \otimes P]^{-1}(P \otimes P) \quad (2.24)$$

and

$$\begin{aligned} H_\psi(P)^{-1} &= -(P \otimes P)[P \otimes W_o P^{-1} W_o + W_o \otimes W_o + W_o P^{-1} W_o \otimes P \\ &\quad + P W_c P \otimes P W_c P]^{-1}(P \otimes P) \end{aligned} \quad (2.25)$$

We now consider the *Newton-Raphson gradient* flows

$$\boxed{\operatorname{vec}(\dot{P}) = -H_\phi(P)^{-1}\operatorname{vec}(\nabla \phi(P))} \quad (2.26)$$

respectively,

$$\boxed{\operatorname{vec}(\dot{P}) = -H_\psi(P)^{-1}\operatorname{vec}(\nabla \psi(P))} \quad (2.27)$$

where ϕ and ψ are defined by (2.4), (2.7). The advantage of these flows is that the linearization of (2.26), (2.27) at the equilibrium P_∞ has all eigenvalues equal to -1 and hence each component of P converges to P_∞ at the same rate.

Theorem 2.10 *Let $\kappa > 0$ be given. The Newton-Raphson gradient flow (2.26) is*

$$\boxed{\frac{d}{dt}\operatorname{vec}(P) = \kappa(P \otimes P)[P \otimes W_o + W_o \otimes P]^{-1}(P \otimes P)\operatorname{vec}(P^{-1}W_o P^{-1} - W_c)}$$

$$(2.28)$$

8.2. FLOWS ON POSITIVE DEFINITE MATRICES

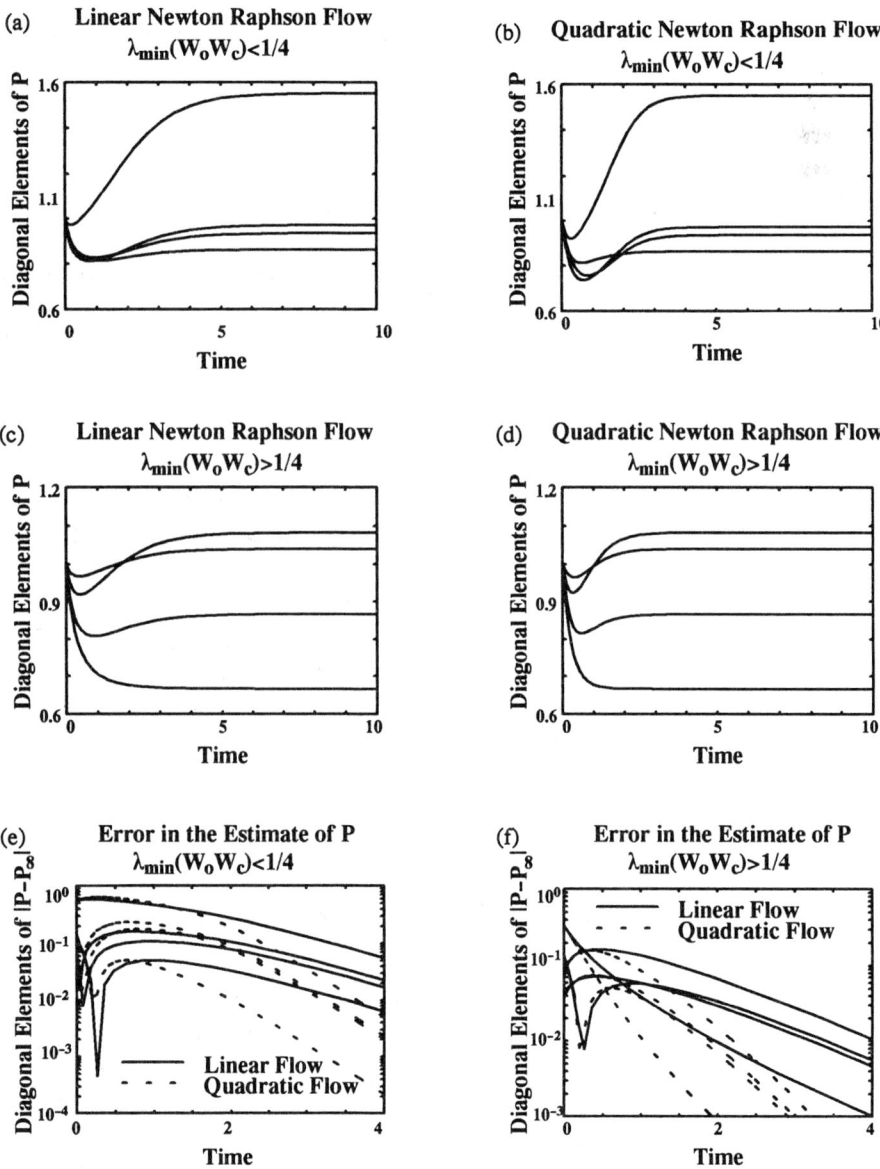

Figure 8.2.4: The linear and quadratic Newton Raphson flows

which converges exponentially fast from every initial condition $P_0 = P_0' > 0$ to P_∞. The eigenvalues of the linearization of (2.28) at P_∞ are all equal to $-\kappa$.

Simulations: Figure 8.2.4a-d shows the evolution of these Newton Raphson equations. It can be observed in Figure 8.2.4e-f that all of the elements of P, in one equation evolution, converge to P_∞ at the same exponential rate. In Figure 8.2.4a,b,e, $W_o = W_1, W_c = W_2$ with $\lambda_{\min}(W_o W_c) \approx 0.207 < \frac{1}{4}$. Figure 8.2.4a uses (2.27) while Figure 8.2.4b shows the evolution of (2.26) both using $P(0) = P_1$. These simulations demonstrate that the linear algorithm does not necessarily converge more rapidly than the quadratic algorithm when $\lambda_{\min}(W_o W_c) \approx 1.7142 > \frac{1}{4}$. Figure 8.2.4c uses (2.27) while Figure 8.2.4d shows the evolution of (2.26), both using $P(0) = P_1$. These simulations demonstrate that the quadratic algorithm converges more rapidly than the linear algorithm in this case when $\lambda_{\min}(W_o W_c) > \frac{1}{4}$.

Main Points of Section With T denoting the co-ordinate basis transformation of a linear system state space realization, then gradient flows on the class of positive definite matrices $P = TT'$ are developed which lead to equality of the controllability and observability Gramians. A different choice of the Riemannian metric leads to the Riccati Equation. Constrained optimization tasks on positive definite matrices can also be solved along similar lines. A Newton-Raphson flow involving second derivatives of the cost function is developed, and flows on $Q = P^{-1}$ are also studied. A variation involving a quadratic index is also introduced. Each flow converges exponentially fast to the balanced realization, but the rates depend on the eigenvalues of the Gramians, so that at various extremes of conditioning, one flow converges more rapidly than the others.

8.3 Flows for Balancing Transformations

In the previous section we studied gradient flows which converged to $P_\infty = T_\infty^2$, where T_∞ is the unique symmetric positive definite balancing transformation for a given asymptotically stably system (A, B, C). Thus T_∞ is obtained as the unique symmetric positive definite square root of P_∞. In this section we address the general problem of determining *all* balancing transformations $T \in GL(n, \mathbb{R})$ for a given asymptotically stable system (A, B, C), using a suitable gradient flow on the set $GL(n, \mathbb{R})$ of all invertible $n \times n$ matrices.

Thus for $T \in GL(n)$ we consider the cost function defined by

$$\Phi(T) = \text{tr}(TW_c T' + (T')^{-1} W_o T^{-1}) \tag{3.1}$$

8.3. FLOWS FOR BALANCING TRANSFORMATIONS

and the associated gradient flow $\dot{T} = -\nabla \Phi(T)$ on $GL(n)$. Of course, in order to define the gradient of a function we have to specify a Riemannian metric. Here and in the sequel we always endow $GL(n, \mathbb{R})$ with its standard Riemannian metric

$$\langle A, B \rangle = 2\mathrm{tr}(A'B) \tag{3.2}$$

i.e. with the constant Euclidean inner product (3.2) defined on the tangent spaces of $GL(n, \mathbb{R})$. For the geometric terminology used in part (c) of the following theorem we refer to Appendix B; cf. also the digression on dynamical systems in Chapter 6.

Theorem 3.1 (a) *The gradient flow $\dot{T} = -\nabla \Phi(T)$ of the cost function Φ on $GL(n, \mathbb{R})$ is*

$$\dot{T} = (T')^{-1}W_o(T'T)^{-1} - TW_c \tag{3.3}$$

and for any initial condition $T_0 \in GL(n, \mathbb{R})$ the solution $T(t)$ of (3.3), $T(0) = T_0$, exists, and is invertible, for all $t \geq 0$.

(b) *For any initial condition $T_0 \in GL(n, \mathbb{R})$, the solution $T(t)$ of (3.3) converges to a balancing transformation $T_\infty \in GL(n, \mathbb{R})$ and all balancing transformations can be obtained in this way, for suitable initial conditions $T_0 \in GL(n, \mathbb{R})$.*

(c) *Let T_∞ be a balancing transformation and let $\mathrm{In}(T_\infty)$ denote the set of all $T_0 \in GL(n, \mathbb{R})$, such that the solution $T(t)$ of (3.3) with $T(0) = T_0$ converges to T_∞ at $t \to \infty$. Then the stable manifold $W^s(T_\infty)$ is an immersed invariant submanifold of $GL(n, \mathbb{R})$ of dimension $\frac{n(n+1)}{2}$ and every solution $T(t) \in W^s(T_\infty)$ converges exponentially fast in $W^s(T_\infty)$ to T_∞.*

Proof Follows immediately from Theorem 6.4.7 with the substitution $W_c = YY', W_o = X'X$. ∎

In the same way the gradient flow of the quadratic version of the objective function $\Phi(T)$ is derived. For $T \in GL(n, \mathbb{R})$, let

$$\Psi(T) = \mathrm{tr}((TW_cT')^2 + ((T')^{-1}W_oT^{-1})^2) \tag{3.4}$$

The next theorem is an immediate consequence of Theorem 6.4.9, using the substitutions $W_c = YY', W_o = X'X$.

Theorem 3.2 (a) *The gradient flow $\dot{T} = -\nabla \Psi(T)$ of the objective function Ψ on $GL(n, \mathbb{R})$ is*

$$\boxed{\dot{T} = 2((T')^{-1}W_o(T'T)^{-1}W_o(T'T)^{-1} - TW_cT'TW_c)} \tag{3.5}$$

and for all initial conditions $T_0 \in GL(n, \mathbb{R})$, the solutions $T(t) \in GL(n, \mathbb{R})$ of (3.5) exist for all $t \geq 0$.

(b) For all initial conditions $T_0 \in GL(n, \mathbb{R})$, every solution $T(t)$ of (3.5) converges to a balancing transformation and all balancing transformations are obtained in this way, for a suitable initial conditions $T_0 \in GL(n, \mathbb{R})$.

(c) For any balancing transformation $T_\infty \in GL(n, \mathbb{R})$ let $W^s(T_\infty) \subset GL(n, \mathbb{R})$ denote the set of all $T_0 \in GL(n, \mathbb{R})$, such that the solution $T(t)$ of (3.5) with initial condition T_0 converges to T_∞ as $t \to \infty$. Then $W^s(T_\infty)$ is an immersed submanifold of $GL(n, \mathbb{R})$ of dimension $\frac{n(n+1)}{2}$ and is invariant under the flow of (3.5). Every solution $T(t) \in W^s(T_\infty)$ converges exponentially to T_∞.

Diagonal balancing transformations Here we address the related issue of computing diagonal balancing transformations for a given asymptotically stable minimal realization (A, B, C).

Again the results are special cases of the more general results derived in Chapter 6, Section 4. Let us consider a fixed diagonal matrix $N = \text{diag}(\mu_1, \cdots, \mu_n)$ with distinct eigenvalues $\mu_1 > \cdots > \mu_n$. Using N, a weighted cost function for balancing is defined by

$$\Phi_N(T) = \text{tr}(NTW_cT' + N(T')^{-1}W_oT^{-1}) \tag{3.6}$$

We have the following immediate corollaries of Lemma 6.4.10 and Theorem 6.4.11.

Lemma 3.3 Let $N = \text{diag}(\mu_1, \cdots, \mu_n)$ with $\mu_1 > \cdots > \mu_n$. Then

(a) $T \in GL(n, \mathbb{R})$ is a critical point of $\Phi_N : GL(n, \mathbb{R}) \to \mathbb{R}$ if and only if T is a diagonal balancing transformation, i.e.

$$TW_cT' = (T')^{-1}W_oT^{-1} = \text{diagonal}$$

(b) A global minimum $T_{\min} \in GL(n, \mathbb{R})$ of Φ_N exists, if (A, B, C) is controllable and observable.

Theorem 3.4 Let (A, B, C) be asymptotically stable, controllable and observable and let $N = \text{diag}(\mu_1, \cdots, \mu_n)$ with $\mu_1 > \cdots > \mu_n$.

(a) The gradient flow $\dot{T} = -\nabla \Phi_N(T)$ of the weighted cost function $\Phi_N : GL(n) \to \mathbb{R}$ is

$$\boxed{\dot{T} = (T')^{-1}W_oT^{-1}N(T')^{-1} - NTW_c, \quad T(0) = T_0} \tag{3.7}$$

and for all initial conditions $T_0 \in GL(n)$ the solution of (3.7) exists and is invertible for all $t \geq 0$.

(b) For any initial condition $T_0 \in GL(n)$ the solution $T(t)$ of (3.7) converges to a diagonal balancing coordinate transformation $T_\infty \in GL(n)$ and all diagonal balancing transformations can be obtained in this way, for a suitable initial condition $T_0 \in GL(n)$.

(c) Suppose the Hankel singular values d_i, $i = 1, \ldots, n$, of (A, B, C) are distinct. Then (3.7) has exactly $2^n n!$ equilibrium points T_∞. These are characterized by $(T'_\infty)^{-1} W_o T_\infty^{-1} = T_\infty W_c T'_\infty = D$ where D is a diagonal matrix. There are exactly 2^n stable equilibrium points of (3.7), where now $D = \text{diag}(d_1, \ldots, d_n)$ is diagonal with $d_1 < \ldots < d_n$. There exists an open and dense subset $\Omega \subset GL(n)$ such that for all $T_0 \in \Omega$ the solution of (3.7) converges exponentially fast to a stable equilibrium point T_∞. The rate of exponential convergence is bounded below by

$$\lambda_{\min}(T_\infty T'_\infty)^{-1} \cdot \min_{i<j}((d_i - d_j)(\mu_j - \mu_i), 4d_i\mu_i).$$

All other equilibria are unstable.

Main Points of Section Gradient flows on the co-ordinate basis transformations are studied for both balancing and diagonal balancing. Again, exponential convergence rates are achieved.

8.4 Balancing via Isodynamical Flows

In this section we construct ordinary differential equations

$$\begin{aligned} \dot{A} &= f(A, B, C) \\ \dot{B} &= g(A, B, C) \\ \dot{C} &= h(A, B, C) \end{aligned}$$

evolving on the space of all realizations (A, B, C) of a given transfer function $G(s)$, with the property that their solutions $(A(t), B(t), C(t))$ all converge for $t \to \infty$ to balanced realizations $(\bar{A}, \bar{B}, \bar{C})$ of $G(s)$. These equations generalize the class of isospectral flows, defined by $B = 0$, $C = 0$, and are thus of theoretical interest.

A differential equation

$$\begin{aligned} \dot{A}(t) &= f(t, A(t), B(t), C(t)) \\ \dot{B}(t) &= g(t, A(t), B(t), C(t)) \\ \dot{C}(t) &= h(t, A(t), B(t), C(t)) \end{aligned} \qquad (4.1)$$

defined on the vector space of all triples $(A, B, C) \in \mathbb{R}^{n \times n} \times \mathbb{R}^{n \times m} \times \mathbb{R}^{p \times n}$ is called *isodynamical* if every solution $(A(t), B(t), C(t))$ of (4.1) is of the form

$$(A(t), B(t), C(t)) = (S(t)A(0)S(t)^{-1}, S(t)B(0), C(0)S(t)^{-1}) \quad (4.2)$$

with $S(t) \in GL(n)$, $S(0) = I_n$. Condition (4.2) implies that the transfer function

$$G_t(s) = C(t)(sI - A(t))^{-1}B(t) = C(0)(sI - A(0))^{-1}B(0) \quad (4.3)$$

is independent of t. Conversely if (4.3) holds with $(A(0), B(0), C(0))$ controllable and observable, then $(A(t), B(t), C(t))$ is of the form (4.2). This is the content of Kalman's realization theorem, see Appendix B.

There is a simple characterization of isodynamical flows which extends the characterization of isospectral flows given in Chapter 1.

Lemma 4.1 *Let $I \subset \mathbb{R}$ be an interval and let $\Lambda(t) \in \mathbb{R}^{n \times n}, t \in I$, be a continuous time-varying family of matrices. Then*

$$\boxed{\begin{aligned} \dot{A}(t) &= \Lambda(t)A(t) - A(t)\Lambda(t) \\ \dot{B}(t) &= \Lambda(t)B(t) \\ \dot{C}(t) &= -C(t)\Lambda(t) \end{aligned}} \quad (4.4)$$

is isodynamical. Conversely, every isodynamical differential equation (4.1) on $\mathbb{R}^{n \times n} \times \mathbb{R}^{n \times m} \times \mathbb{R}^{p \times n}$ is of the form (4.4).

Proof Let $T(t)$ denote the unique solution of the linear differential equation

$$\dot{T}(t) = \Lambda(t)T(t), \quad T(0) = I_n,$$

and let $(\hat{A}(t), \hat{B}(t), \hat{C}(t)) = (T(t)A(0)T(t)^{-1}, T(t)B(0), C(0)T(t)^{-1})$. Then $(\hat{A}(0), \hat{B}(0), \hat{C}(0)) = (A(0), B(0), C(0))$ and

$$\begin{aligned} \frac{d}{dt}\hat{A}(t) &= \dot{T}(t)A(0)T(t)^{-1} - T(t)A(0)T(t)^{-1}\dot{T}(t)T(t)^{-1} \\ &= \Lambda(t)\hat{A}(t) - \hat{A}(t)\Lambda(t) \\ \frac{d}{dt}\hat{B}(t) &= \dot{T}(t)B(0) = \Lambda(t)\hat{B}(t) \\ \frac{d}{dt}\hat{C}(t) &= C(0) - T(t)^{-1}\dot{T}(t)T(t)^{-1} = -\hat{C}(t)\Lambda(t). \end{aligned}$$

Thus every solution $(\hat{A}(t), \hat{B}(t), \hat{C}(t))$ of an isodynamical flow satisfies (4.4). On the other hand, for every solution $(A(t), B(t), C(t))$ of (4.4), $(\hat{A}(t), \hat{B}(t), \hat{C}(t))$ is of the form (4.2). By the uniqueness of the solutions of (4.4), $(\hat{A}(t), \hat{B}(t), \hat{C}(t)) = (A(t), B(t), C(t))$, $t \in I$, and the result follows. ∎

8.4. BALANCING VIA ISODYNAMICAL FLOWS

In the sequel let (A, B, C) denote a fixed asymptotically stable realization. At this point we do not necessarily assume that (A, B, C) is controllable or observable. Let

$$\mathcal{O}(A, B, C) = \{(SAS^{-1}, SB, CS^{-1}) \mid S \in GL(n, \mathbb{R})\} \quad (4.5)$$

Since $\mathcal{O}(A, B, C)$ is an orbit for the similarity action on $\mathbb{R}^{n \times (n+m+p)}$, see Chapter 7, it is a smooth submanifold of $\mathbb{R}^{n \times (n+m+p)}$.

Let $\Phi : \mathcal{O}(A, B, C) \to \mathbb{R}$ denote the cost function defined by

$$\Phi(SAS^{-1}, SB, CS^{-1}) = \text{tr}(W_c(SAS^{-1}, SB) + W_o(SAS^{-1}, CS^{-1})) \quad (4.6)$$

where $W_c(F, G)$ and $W_o(F, H)$ denote the controllability and observability Gramians of (F, G, H) respectively. The following proposition summarizes some important properties of $\mathcal{O}(A, B, C)$ and $\Phi : \mathcal{O}(A, B, C) \to \mathbb{R}$.

Proposition 4.2 Let $(A, B, C) \in \mathbb{R}^{n \times (n+m+p)}$.

(a) $\mathcal{O}(A, B, C)$ is a smooth submanifold of $\mathbb{R}^{n \times (n+m+p)}$. The tangent space of $\mathcal{O}(A, B, C)$ at $(F, G, H) \in \mathcal{O}(A, B, C)$ is

$$T_{(F,G,H)}\mathcal{O}(A, B, C) = \{([X, F], XG, -HX) \in \mathbb{R}^{n \times (m+m+p)} | X \in \mathbb{R}^{n \times n}\} \quad (4.7)$$

Let (A, B, C) be asymptotically stable, controllable and observable. Then

(b) $\mathcal{O}(A, B, C)$ is a closed subset of $\mathbb{R}^{n \times (n+m+p)}$.

(c) The function $\Phi : \mathcal{O}(A, B, C) \to \mathbb{R}$ defined by (4.6) is smooth and has compact sublevel sets.

Proof $\mathcal{O}(A, B, C)$ is an orbit of the $GL(n, \mathbb{R})$ similarity action

$$\sigma : GL(n) \times \mathbb{R}^{n \times (n+m+p)} \to \mathbb{R}^{n \times (n+m+p)}$$

$$(S, (A, B, C)) \to (SAS^{-1}, SB, CS^{-1})$$

and thus a smooth submanifold of $\mathbb{R}^{n \times (n+m+p)}$; see Appendix C. This proves (a). Now assume (A, B, C) is asymptotically stable, controllable and observable. Then (b), (c) follow immediately from Lemma 6.4.1, Lemma 6.4.10. ∎

We now address the issue of finding gradient flows for the objective function $\Phi : \mathcal{O}(A, B, C) \to \mathbb{R}$, relative to some Riemannian metric on $\mathcal{O}(A, B, C)$. We endow the vector space $\mathbb{R}^{n \times (n+m+p)}$ of all triples (A, B, C) with its standard Euclidean inner product $\langle \, , \, \rangle$ defined by

$$\langle (A_1, B_1, C_1), (A_2, B_2, C_2) \rangle = \text{tr}(A_1 A_2' + B_1 B_2' + C_1 C_2') \quad (4.8)$$

Since the orbit $\mathcal{O}(A,B,C)$ is a submanifold of $\mathbb{R}^{n\times(n+m+p)}$, the inner product $\langle\ ,\ \rangle$ on $\mathbb{R}^{n\times(n+m+p)}$ induces an inner product on each tangent space $T_{(F,G,H)}\mathcal{O}(A,B,C)$ by

$$\langle([X_1,F], X_1G, -HX_1), ([X_2,F], X_2G, -HX_2)\rangle$$
$$= \text{tr}([X_1,F]\cdot [X_2,F]' + X_1GG'X_2' + HX_1X_2'H') \quad (4.9)$$

and therefore defines a Riemannian metric on $\mathcal{O}(A,B,C)$; see Appendix C and (4.7). We refer to this Riemannian metric as the *induced Riemannian metric* on $\mathcal{O}(A,B,C)$

A second, and for the subsequent development, a more important Riemannian metric on $\mathcal{O}(A,B,C)$ is defined as follows. Here we assume that (A,B,C) is controllable and observable. Instead of defining the inner product of tangent vectors $([X_1,F], X_1G, -HX_1), ([X_2,F], X_2G, -HX_2) \in T_{(F,G,H)}\mathcal{O}(A,B,C)$ as in (4.9) we set

$$\langle\!\langle\, ([X_1,F], X_1G, -HX_1), ([X_2,F], X_2G, -HX_2)\,\rangle\!\rangle := 2\text{tr}(X_1'X_2) \quad (4.10)$$

The next lemma shows that (4.10) defines an inner product on the tangent space.

Lemma 4.3 *Let (A,B,C) be controllable or observable. Then $([X,A], XB, -CX) = (0,0,0)$ implies $X = 0$.*

Proof If $XB = 0$ and $AX = XA$ then $X(B, AB, \cdots, A^{n-1}B) = 0$. Thus by controllability $X = 0$. Similarly for observability of (A,B,C). ∎

It is easily verified, using controllability and observability of (F,G,H), that (4.10) defines a Riemannian metric on $\mathcal{O}(A,B,C)$. We refer to this as the *normal Riemannian metric* on $\mathcal{O}(A,B,C)$. In order to determine the gradient flow of $\Phi: \mathcal{O}(A,B,C) \to \mathbb{R}$ we need the following lemma.

Lemma 4.4 *Let $\Phi: \mathcal{O}(A,B,C) \to \mathbb{R}$ be defined by $\Phi(F,G,H) = \text{tr}(W_c(F,G) + W_o(F,H))$ for all $(F,G,H) \in \mathcal{O}(A,B,C)$. Then the derivative of Φ at $(F,G,H) \in \mathcal{O}(A,B,C)$ is the linear map on $T_{(F,G,H)}\mathcal{O}(A,B,C)$ defined by*

$$D\Phi_{(F,G,H)}([X,F], XG, -HX) = 2\text{tr}((W_c(F,G) - W_o(F,H))X) \quad (4.11)$$

Proof Consider the smooth map $\sigma: GL(n) \to \mathcal{O}(A,B,C)$ defined by $\sigma(S) = (SFS^{-1}, SG, HS^{-1})$ for $S \in GL(n)$. The composed map of Φ with σ is $\phi = \Phi \cdot \sigma$ with

$$\phi(S) = \text{tr}(SW_c(F,G)S' + (S')^{-1}W_o(F,H)S^{-1})$$

8.4. BALANCING VIA ISODYNAMICAL FLOWS

By the chain rule we have $D\phi|_I(X) = D\Phi|_{(F,G,H)}([X,F], XG, -HX)$ and

$$D\phi|_I(X) = 2\mathrm{tr}((W_c(F,G) - W_o(F,H))X)$$

This proves the result. ∎

Theorem 4.5 *Let (A, B, C) be asymptotically stable, controllable and observable. Consider the cost function $\Phi : \mathcal{O}(A, B, C) \to \mathbb{R}, \Phi(F, G, H) = \mathrm{tr}(W_c(F,G) + W_o(F,H))$. (a) The gradient flow $\dot{A} = -\nabla_A \Phi(A,B,C), \dot{B} = -\nabla_B \Phi(A,B,C), \dot{C} = -\nabla_C \Phi(A,B,C)$ for the normal Riemannian metric on $\mathcal{O}(A,B,C)$ is*

$$
\begin{aligned}
\dot{A}(t) &= -[A(t), W_c(A(t), B(t)) - W_o(A(t), C(t))] \\
\dot{B}(t) &= -(W_c(A(t), B(t)) - W_o(A(t), C(t)))B(t) \\
\dot{C}(t) &= C(t)(W_c(A(t), B(t)) - W_o(A(t), C(t))).
\end{aligned}
\quad (4.12)
$$

(b) For every initial condition $(A(0), B(0), C(0)) \in \mathcal{O}(A,B,C)$ the solution $(A(t), B(t), C(t))$ of (4.12) exist for all $t \geq 0$ and (4.12) is an isodynamical flow on the open set of asymptotically stable controllable and observable systems (A, B, C).

(c) For any initial condition $(A(0), B(0), C(0)) \in \mathcal{O}(A,B,C)$ the solution $(A(t), B(t), C(t))$ of (4.12) converges to a balanced realization $(A_\infty, B_\infty, C_\infty)$ of the transfer function $C(sI - A)^{-1}B$. Moreover the convergence to the set of all balanced realizations of $C(sI - A)^{-1}B$ is exponentially fast.

Proof The proof is similar to that of Theorem 6.5.1. Let $\mathrm{grad}\Phi = (\mathrm{grad}\Phi_1, \mathrm{grad}\Phi_2, \mathrm{grad}\Phi_3)$ denote the three components of the gradient of Φ with respect to the normal Riemannian metric. The derivative, i.e. tangent map, of Φ at $(F, G, H) \in \mathcal{O}(A, B, C)$ is the linear map $D\Phi|_{(F,G,H)} : T_{(F,G,H)}\mathcal{O}(A,B,C) \to \mathbb{R}$ defined by

$$D\Phi|_{(F,G,H)}([X,F], XG, -HX) = 2\mathrm{tr}(X(W_c(F,G) - W_o(F,H))), \quad (4.13)$$

see Lemma 5.1. By definition of the gradient of a function, see Appendix C, $\mathrm{grad}\Phi(F, G, H)$ is characterized by the conditions

$$\mathrm{grad}\Phi(F,G,H) \in T_{(F,G,H)}\mathcal{O}(A,B,C) \quad (4.14)$$

and

$$D\Phi|_{(F,G,H)}([X,F], XG, -HX)$$
$$= \langle\langle (\mathrm{grad}\Phi_1, \mathrm{grad}\Phi_2, \mathrm{grad}\Phi_3), ([X,F], XG, -HX)\rangle\rangle \quad (4.15)$$

for all $X \in \mathbb{R}^{n \times n}$. By Proposition 4.2, (4.14) is equivalent to

$$\mathrm{grad}\Phi(F, G, H) = ([X_1, F], X_1 G, -H X_1) \qquad (4.16)$$

for $X_1 \in \mathbb{R}^{n \times n}$. Note that by Lemma 4.4, the matrix X_1 is uniquely determined. Thus (4.15) is equivalent to

$$2\mathrm{tr}((W_c(F, G) - W_o(F, H))X)$$

$$= \langle\langle([X_1, F], X_1 G, -H X_1), ([X, F], XG, -HX)\rangle\rangle$$
$$= 2\mathrm{tr}(X_1' X)$$

for all $X \in \mathbb{R}^{n \times n}$. Thus

$$X_1 = W_c(F, G) - W_o(F, H)$$

and therefore $\mathrm{grad}\Phi(F, G, H) = ([W_c(F, G) - W_o(F, H), F], (W_c(F, G) - W_o(F, H))G, -H(W_c(F, G) - W_o(F, H)))$. This proves (a). The rest of the argument goes as in the proof of Theorem 6.5.1. Explicitly, for (b) note that $\Phi(A(t), B(t), C(t))$ decreases along every solution of (4.12). By Proposition 4.2, $\{(F, G, H) \in \mathcal{O}(A, B, C) \mid \Phi(F, G, H) \leq \Phi(F_o, G_o, H_o)\}$ is a compact set. Therefore $(A(t), B(t), C(t))$ stays in that compact subset (for $(F_o, G_o, H_o) = (A(0), B(0), C(0))$) and thus exists for all $t \geq 0$. By Lemma 4.1 the flows are isodynamical. This proves (b). Since (4.12) is a gradient flow of $\Phi : \mathcal{O}(A, B, C) \to \mathbb{R}$ and since $\Phi : \mathcal{O}(A, B, C) \to \mathbb{R}$ has compact sublevel sets the solutions $(A(t), B(t), C(t))$ all converge to the equilibria points of (4.12), i.e. to the critical points of $\Phi : \mathcal{O}(A, B, C) \to \mathbb{R}$. But the critical points of $\Phi : \mathcal{O}(A, B, C) \to \mathbb{R}$ are just the balanced realizations $(F, G, H) \in \mathcal{O}(A, B, C)$, being characterized by $W_c(F, G) = W_o(F, H)$; see Lemma 4.4. Exponential convergence follows since the Hessian of Φ at a critical point is positive definite, cf. Lemma 6.4.10. ■

A similar ODE approach also works for diagonal balanced realizations. Here we consider the weighted cost function

$$\Phi_N : \mathcal{O}(A, B, C) \to \mathbb{R},$$

$$\Phi_N(F, G, H) = \mathrm{tr}(N(W_c(F, G) + W_o(F, H))) \qquad (4.17)$$

for a real diagonal matrix $N = \mathrm{diag}(\mu_1, \cdots, \mu_n)$, $\mu_1 > \cdots > \mu_n$. We have the following result.

Theorem 4.6 *Let (A, B, C) be asymptotically stable, controllable and observable. Consider the weighted cost function $\Phi_N : \mathcal{O}(A, B, C) \to \mathbb{R}$, $\Phi_N(F, G, H) = \mathrm{tr}(N(W_c(F, G) + W_o(F, H)))$, with $N = \mathrm{diag}(\mu_1, \cdots, \mu_n)$, $\mu_1 > \cdots > \mu_n$.*

8.4. BALANCING VIA ISODYNAMICAL FLOWS

(a) The gradient flow $(\dot{A} = -\nabla_A \Phi_N, \dot{B} = -\nabla_B \Phi_N, \dot{C} = -\nabla_C \Phi_N)$ for the normal Riemannian metric is

$$\boxed{\begin{aligned} \dot{A} &= -[A, W_c(A,B)N - NW_o(A,C)] \\ \dot{B} &= -(W_c(A,B)N - NW_o(A,C))B \\ \dot{C} &= C(W_c(A,B)N - NW_o(A,C)) \end{aligned}} \quad (4.18)$$

(b) For any initial condition $(A(0), B(0), C(0)) \in \mathcal{O}(A,B,C)$ the solution $(A(t), B(t), C(t))$ of (4.18) exists for all $t \geq 0$ and the flow (4.18) is isodynamical.

(c) For any initial condition $((A(0), B(0), C(0)) \in \mathcal{O}(A,B,C)$ the solution matrices $(A(t), B(t), C(t))$ of (4.18) converges to a diagonal balanced realization $(A_\infty, B_\infty, C_\infty)$ of $C_0(sI - A_0)^{-1}B_0$.

Proof The proof of (a), (b) goes – mutatis mutandis – as for (a), (b) in Theorem 4.5. Similarly for (c) once we have checked that the critical point of $\Phi_N : \mathcal{O}(A,B,C) \to \mathbb{R}$ are the diagonal balanced realizations. But this follows using the same arguments given in Theorem 6.3.5. ∎

We emphasize that the above theorems give a direct method to compute balanced or diagonal balanced realizations, without computing any balancing coordinate transformations.

Discrete-time Balancing Flows Based on earlier observations in this chapter, it is no surprise that discrete-time flows to achieve balanced realizations and diagonal balanced realizations follow directly from the balanced matrix factorization algorithms of Chapter 6 and inherit their convergence properties. Here we restrict attention to isodynamical balancing flows derived by taking the matrix factors X, Y to be the m-th controllability and observability matrices as follows:

$$Y = R_m(A,B) = (B, AB, \cdots, A^{m-1}B) \quad (4.19)$$

$$X = O_m(A,C) = \begin{bmatrix} C \\ CA \\ \vdots \\ CA^{m-1} \end{bmatrix}. \quad (4.20)$$

Now the observability and controllability Gramians are

$$W_o^{(m)}(A,C) = X'X, \; W_c^{(m)}(A,B) = YY' \quad (4.21)$$

From now on let us assume that $m \geq n+1$. The case $m = \infty$ is allowed if A is allowed to be discrete-time stable. We claim that the matrix factorization balancing flow (6.6.1) (6.6.2) specialized to this case gives by simple manipulations the discrete-time isodynamical flow

$$\begin{aligned}
A_{k+1} &= e^{-\alpha_k(W_c^{(m)}(A_k,B_k)-W_o^{(m)}(A_k,C_k))}A_k e^{\alpha_k(W_c^{(m)}(A_k,B_k)-W_o^{(m)}(A_k,C_k))} \\
B_{k+1} &= e^{-\alpha_k(W_c^{(m)}(A_k,B_k)-W_o^{(m)}(A_k,C_k))}B_k \\
C_{k+1} &= C_k e^{\alpha_k(W_c^{(m)}(A_k,B_k)-W_o^{(m)}(A_k,C_k))}
\end{aligned} \quad (4.22)$$

$$\alpha_k = \frac{1}{2\lambda_{max}(W_c^{(m)}(A_k,B_k)+W_o^{(m)}(A_k,C_k))} \quad (4.23)$$

In fact from (6.6.1), (6.6.2) we obtain for $T_k = e^{-\alpha_k(Y_k Y_k' - X_k' X_k)} \in GL(n)$ that

$$\begin{aligned}
X_{k+1} &= X_k T_k^{-1}, \quad Y_{k+1} = T_k Y_k; \quad k \in \mathbb{N}, \\
X_0 &= O_m(A_0,C_0), \quad Y_0 = R_m(A_0,B_0)
\end{aligned}$$

A simple induction argument on k then establishes the existence of (A_k, B_k, C_k) with

$$X_k = O_m(A_k,C_k), \quad Y_k = O_m(A_k,B_k)$$

for all $k \in \mathbb{N}_0$. The above recursion on (X_k, Y_k) is now equivalent to

$$\begin{aligned}
O_m(A_{k+1},C_{k+1}) &= O_m(A_k,C_k)T_k^{-1} = O_m(T_k A_k T_k^{-1}, C_k T_k^{-1}) \\
R_m(A_{k+1},B_{k+1}) &= T_k R_m(A_k,B_k) = R_m(T_k A_k T_k^{-1}, T_k B_k)
\end{aligned}$$

for $k \in \mathbb{N}$. Thus, using $m \geq n+1$, we obtain

$$A_{k+1} = T_k A_k T_k^{-1}, \quad B_{k+1} = T_k B_k, \quad C_{k+1} = C_k T_k^{-1},$$

which is equivalent to (4.22). Likewise for the *isodynamical diagonal balancing* flows, we have

$$\begin{aligned}
A_{k+1} &= e^{-\alpha_k(W_c^{(m)}(A_k,B_k)N - N W_o^{(m)}(A_k,C_k))} \cdot \\
&\quad A_k e^{\alpha_k(W_c^{(m)}(A_k,B_k)N - N W_o^{(m)}(A_k,C_k))} \\
B_{k+1} &= e^{-\alpha_k(W_c^{(m)}(A_k,B_k)N - N W_o^{(m)}(A_k,C_k))} B_k \\
C_{k+1} &= C_k e^{\alpha_k(W_c^{(m)}(A_k,B_k)N - N W_o^{(m)}(A_k,C_k))}
\end{aligned} \quad (4.24)$$

$$\alpha_k = \frac{1}{2\|W_o^{(m)}N - N W_c^{(m)}\|} \log\left(\frac{\|(W_o^{(m)} - W_c^{(m)})N + N(W_o^{(m)} - W_c^{(m)})\|}{4\|N\| \cdot \|(W_o^{(m)}N - N W_c^{(m)}\| \operatorname{tr}(W_c^{(m)} + W_o^{(m)})} + 1\right) \quad (4.25)$$

where the A_k, B_k, C_k dependance of α_k has been omitted for simplicity. Also, as before N is taken for most applications as $N = \text{diag}(\mu_1, \cdots, \mu_n)$, $\mu_1 > \cdots > \mu_n$.

By applying Theorem 6.6.2 we conclude

Theorem 4.7 *Let $(A_0, B_0, C_0) \in L(n, m, p)$ be controllable and observable and let $N = \text{diag}(\mu_1, \ldots, \mu_n)$ with $\mu_1 > \ldots > \mu_n$. Then:*

(a) The recursion (4.24) is a discrete-time isodynamical flow.

(b) Every solution (A_k, B_k, C_k) of (4.24) converges to the set of all diagonal balanced realizations of the transfer function $C_0(sI - A_0)^{-1} B_0$.

(c) Suppose that the singular values of the Hankel matrix $\mathcal{H}_m = O_m(A_0, C_0) R_m(A_0, B_0)$ are distinct. There are exactly $2^n n!$ fixed points of (4.24), corresponding to the diagonal balanced realizations of $C_0(sI - A_0)^{-1} B_0$. The set of asymptotically stable fixed points correspond to those diagonal balanced realizations $(A_\infty, B_\infty, C_\infty)$ with $W_c(A_\infty, B_\infty) = W_o(A_\infty, C_\infty) = \text{diag}(\sigma_1, \ldots, \sigma_n)$, where $\sigma_1 < \ldots < \sigma_n$ are the singular values of $\mathcal{H}_m$.

Main Points of Section Ordinary differential equations on the linear system matrices A, B, C are achieved which converge exponentially to balancing, or diagonal balancing matrices. This avoids the need to work with co-ordinate basis transformations. Discrete-time isodynamical flows for balancing and diagonal balancing are in essence matrix factorization flows for balancing and inherit the same geometric and dynamical system properties.

8.5 Euclidean Norm Optimal Realizations

We now consider another type of balancing that can be applied to systems regardless of their stability properties. This form of balancing was introduced in Verriest (1988), Helmke (1993) and interesting connections exist with least squares matching problems arising in computer graphics, Brockett (1989). The techniques in this section are crucial to our analysis of L^2-sensitivity optimization problems as studied in the next chapter.

The limiting solution of a gradient algorithm appears to be the only possible way of finding such least squares optimal realizations and direct algebraic algorithms for the solutions are unknown.

Rather than minimizing the sum of the traces of the controllability and observability Gramians (which may not exist) the task here is to minimize the *least squares* or *Euclidean norm* of the realization, that is

$$\|(TAT^{-1}, TB, CT^{-1})\|^2 = \mathrm{tr}(TAT^{-1}T'^{-1}A'T' + TBB'T' + CT^{-1}T'^{-1}C') \tag{5.1}$$

Observe that (5.1) measures the distance of (TAT^{-1}, TB, CT^{-1}) to the zero realization $(0,0,0)$. Eising has considered the problem of finding the best approximations of a given realization (A, B, C) by uncontrollable or unobservable systems with respect to the Euclidean norm (5.1). Thus this norm is sometimes called the *Eising distance*.

Using the substitution $P = T'T$ the function to be minimized, over the class of symmetric positive definite matrices P, is $\Gamma : \mathcal{P}(n) \to \mathbb{R}$,

$$\Gamma(P) = \mathrm{tr}(AP^{-1}A'P + BB'P + P^{-1}C'C). \tag{5.2}$$

The gradient $\nabla \Gamma(P)$ is then

$$\nabla \Gamma(P) = AP^{-1}A' - P^{-1}A'PAP^{-1} + BB' - P^{-1}C'CP^{-1} \tag{5.3}$$

and the associated gradient flow $\dot{P} = -\nabla \Gamma(P)$ is

$$\boxed{\dot{P} = -AP^{-1}A' - BB' + P^{-1}(A'PA + C'C)P^{-1}.} \tag{5.4}$$

Any equilibrium P_∞ of such a gradient flow will be characterized by

$$AP_\infty^{-1}A' + BB' = P_\infty^{-1}(A'P_\infty A + C'C)P_\infty^{-1}. \tag{5.5}$$

Any $P_\infty = P_\infty' > 0$ satisfying (5.5) is called *Euclidean norm optimal*.

Let $P_\infty = P_\infty' > 0$ satisfy (5.5) and $T_\infty = P_\infty^{\frac{1}{2}}$ be the positive definite symmetric square root. Then $(F, G, H) = (T_\infty A T_\infty^{-1}, T_\infty B, C T_\infty^{-1})$ satisfies

$$FF' + GG' = F'F + H'H. \tag{5.6}$$

Any realization $(F, G, H) \in \mathcal{O}(A, B, C)$ which satisfies (5.6) is called *Euclidean norm balanced*. A realization $(F, G, H) \in \mathcal{O}(A, B, C)$ is called Euclidean *diagonal norm balanced* if

$$FF' + GG' = F'F + H'H = \text{diagonal}. \tag{5.7}$$

The diagonal entries $\sigma_1 \geq \cdots \geq \sigma_n > 0$ of (5.7) are called the *generalized (norm) singular values* of (F, G, H) and $FF' + GG'$, $F'F + H'H$ are called the *Euclidean norm controllability Gramian* and *Euclidean norm observability Gramian*, respectively.

For the subsequent stability of the gradient flow we need the following technical lemma.

8.5. EUCLIDEAN NORM OPTIMAL REALIZATIONS

Lemma 5.1 *Let (A, B, C) be controllable or observable (but not necessarily asymptotically stable!). Then the linear operator*

$$I_n \otimes AA' + A'A \otimes I_n - A \otimes A - A' \otimes A' + I_n \otimes BB' + C'C \otimes I_n \quad (5.8)$$

has all its eigenvalues in $\mathbb{C}^+ = \{\lambda \in \mathbb{C} \mid \operatorname{Re}\lambda > 0\}$.

Proof Suppose there exists $\lambda \in \mathbb{C}$ and $X \in \mathbb{C}^{n \times n}$ such that

$$XAA' + A'AX - A'XA - AXA' + XBB' + C'CX = \lambda \cdot X$$

Let $X^* = \bar{X}'$ denote the Hermitian transpose. Then

$$\operatorname{tr}(XAA'X^* + AXX^*A' - AXA'X^* - A'XAX^* + XBB'X^* + CXX^*C')$$
$$= (\operatorname{Re}\lambda) \cdot \|X\|^2$$

A straightforward manipulation shows that the left hand side is equal to

$$\|AX - XA\|^2 + \|XB\|^2 + \|CX\|^2 \geq 0$$

and therefore $\operatorname{Re}\lambda \geq 0$.

Suppose $AX = XA, XB = 0 = CX$. Then by Lemma 4.4 $X = 0$. Thus taking vec operations and assuming $X \neq 0$ implies the result. ∎

Lemma 5.2 *Given a controllable and observable realization (A, B, C), the linearization of the flow (5.4) at any equilibrium point $P_\infty > 0$ is exponentially stable.*

Proof The linearization (5.4) about P_∞ is given by $\frac{d}{dt}\operatorname{vec}(\xi) = J \cdot \operatorname{vec}(\xi)$ where

$$\begin{aligned}J &= AP_\infty^{-1} \otimes AP_\infty^{-1} - P_\infty^{-1} \otimes P_\infty^{-1}(A'P_\infty A + C'C)P_\infty^{-1} \\ &\quad - P_\infty^{-1}(A'P_\infty A + C'C)P_\infty^{-1} \otimes P_\infty^{-1} + P_\infty^{-1}A' \otimes P_\infty^{-1}A'\end{aligned}$$

Let $F = P_\infty^{\frac{1}{2}}AP_\infty^{-\frac{1}{2}}$, $G = P_\infty^{\frac{1}{2}}B$, $H = CP_\infty^{-\frac{1}{2}}$. Then with

$$\bar{J} = (P_\infty^{\frac{1}{2}} \otimes P_\infty^{\frac{1}{2}})J(P_\infty^{\frac{1}{2}} \otimes P_\infty^{\frac{1}{2}})$$

$$\bar{J} = F \otimes F - I \otimes F'F + F' \otimes F' - F'F \otimes I - I \otimes H'H - H'H \otimes I \quad (5.9)$$

and using (5.6) gives

$$\bar{J} = F \otimes F + F' \otimes F' - I \otimes (FF' + GG') - (F'F + H'H) \otimes I \quad (5.10)$$

By Lemma 5.1 the matrix $\bar{J}$ has only negative eigenvalues. The symmetric matrices $\bar{J}$ and J are congruent, i.e. $\bar{J} = X'JX$ for an invertible $n \times n$

matrix X. By the inertia theorem, see Appendix A, J and $\bar{J}$ have the same rank and signatures and thus the numbers of positive respectively negative eigenvalues coincide. Therefore J has only negative eigenvalues which proves the result. ∎

The uniqueness of P_∞ is a consequence of the Kempf-Ness theorem and follows from Theorem 7.5.1. Using Lemma 5.2 we can now give an alternative uniqueness proof which uses Morse theory. It is intuitively clear that on a mountain with more than one maximum or minimum there should also be a saddle point. This is the content of the celebrated Birkhoff minimax theorem and Morse theory offers a systematic generalization of such results. Thus, while the intuitive basis of the uniqueness of P_∞ should be obvious from Theorem 7.4.4, we show how the result follows by the formal machinery of Morse theory. See Milnor (1963) for a thorough account on Morse theory.

Proposition 5.3 *Given any controllable and observable system (A, B, C), there exists a unique $P_\infty = P'_\infty > 0$ which minimizes $\Gamma : \mathcal{P}(n) \to \mathbb{R}$ and $T_\infty = P_\infty^{\frac{1}{2}}$ is a least squares optimal coordinate transformation. Also, $\Gamma : \mathcal{P}(n) \to \mathbb{R}$ has compact sublevel sets.*

Proof Consider the continuous map

$$\tau : \mathcal{P}(n) \to \mathbb{R}^{n \times (n+m+p)}$$

defined by

$$\tau(P) = (P^{\frac{1}{2}} A P^{-\frac{1}{2}}, P^{\frac{1}{2}} B, C P^{-\frac{1}{2}}).$$

By Lemma 7.4.1 the similarity orbit $\mathcal{O}(A, B, C)$ is a closed subset of $\mathbb{R}^{n \times (n+m+p)}$ and therefore $M_a := \{(F, G, H) \in \mathcal{O}(A, B, C) \mid \operatorname{tr}(FF' + GG' + H'H) \leq a\}$ is compact for all $a \in \mathbb{R}$. The function $\tau : \mathcal{P}(n) \to \mathcal{O}(A, B, C)$ is a homeomorphism and therefore $\tau^{-1}(M_a) = \{P \in \mathcal{P}(n) \mid \Gamma(P) \leq a\}$ is compact. This proves that $\Gamma : \mathcal{P}(n) \to \mathbb{R}$ has compact sublevel sets. By Lemma 5.2, the Hessian of Γ at each critical point is positive definite. A smooth function $f : \mathcal{P}(n) \to \mathbb{R}$ is called a *Morse function* if it has compact sublevel sets and if the Hessian at each critical point of f is invertible. Morse functions have isolated critical points, let $c_i(f)$ denote the number of critical points where the Hessian has exactly i negative eigenvalues, counted with multiplicities. Thus $c_0(f)$ is the number of local minima of f. The Morse inequalities bound the numbers $c_i(f), i \in \mathbb{N}_o$ for a Morse function f in terms of topological invariants of the space $\mathcal{P}(n)$, which are thus independent of f. These are the so-called *Betti numbers* of $\mathcal{P}(n)$.

The i-th Betti number $b_i(\mathcal{P}(n))$ is defined as the rank of the i-th (singular) homology group $H_i(\mathcal{P}(n))$ of $\mathcal{P}(n)$. Also, $b_0(\mathcal{P}(n))$ is the number of

8.5. EUCLIDEAN NORM OPTIMAL REALIZATIONS

connected components of $\mathcal{P}(n)$. Since $\mathcal{P}(n)$ is homeomorphic to the Euclidean space $\mathbb{R}^{\frac{n(n+1)}{2}}$ we have

$$b_i(\mathcal{P}(n)) = \begin{cases} 1 & i = 0 \\ 0 & i \geq 1. \end{cases} \quad (5.11)$$

From the above, $\Gamma : \mathcal{P}(n) \to \mathbb{R}$ is a Morse function on $\mathcal{P}(n)$. The Morse inequalities for Γ are

$$\begin{aligned} c_0(\Gamma) &\geq b_0(\mathcal{P}(n)) \\ c_0(\Gamma) - c_1(\Gamma) &\leq b_0(\mathcal{P}(n)) - b_1(\mathcal{P}(n)) \\ c_0(\Gamma) - c_1(\Gamma) + c_2(\Gamma) &\geq b_0(\mathcal{P}(n)) - b_1(\mathcal{P}(n)) + b_2(\mathcal{P}(n)) \\ &\cdots \\ \sum_{i=0}^{\dim \mathcal{P}(n)} (-1)^i c_i(\Gamma) &= \sum_{i=0}^{\dim \mathcal{P}(n)} (-1)^i b_i(\mathcal{P}(n)). \end{aligned}$$

Since $c_i(\Gamma) = 0$ for $i \geq 1$ the Morse inequalities imply, using (5.11),

$$\begin{aligned} c_0(\Gamma) &\geq 1 \\ c_0(\Gamma) - c_1(\Gamma) &= c_0(\Gamma) \leq b_0(\mathcal{P}(n)) = 1. \end{aligned}$$

Hence $c_0(\Gamma) = 1$, and $c_i(\Gamma) = 0$ for $i \geq 1$, i.e. Γ has a unique local = global minimum. This completes the proof. ∎

We summarize the above results in a theorem.

Theorem 5.4 *Let (A, B, C) be controllable and observable and let $\Gamma : \mathcal{P}(n) \to \mathbb{R}$ be the smooth function defined by $\Gamma(P) = \operatorname{tr}(AP^{-1}A'P + BB'P + C'CP^{-1})$. The gradient flow $\dot{P}(t) = -\nabla \Gamma(P(t))$ on $\mathcal{P}(n)$ is given by*

$$\boxed{\dot{P} = -AP^{-1}A' - BB' + P^{-1}(A'PA + C'C)P^{-1}.} \quad (5.12)$$

For every initial condition $P_0 = P_0' > 0$, the solution $P(t) \in \mathcal{P}(n)$ of the gradient flow exists for all $t \geq 0$ and $P(t)$ converges to the uniquely determined positive definite matrix P_∞ satisfying (5.5).

As before we can find equations that evolve on the space of realizations rather than on the set of positive definite squared transformation matrices.

Theorem 5.5 *Let the matrices (A, B, C) be controllable and observable and let $\Gamma : \mathcal{O}(A, B, C) \to \mathbb{R}$ be the least squares cost function defined by $\Gamma(F, G, H) = \operatorname{tr}(FF' + GG' + H'H)$ for $(F, G, H) \in \mathcal{O}(A, B, C)$.*

(a) The gradient flow ($\dot{A} = -\nabla_A \Gamma$, $\dot{B} = -\nabla_B \Gamma$, $\dot{C} = -\nabla_C \Gamma$) for the normal Riemannian metric on $\mathcal{O}(A,B,C)$ given by (4.10) is

$$\boxed{\begin{aligned} \dot{A} &= [A, [A, A'] + BB' - C'C] \\ \dot{B} &= -([A, A'] + BB' - C'C)B \\ \dot{C} &= C([A, A'] + BB' - C'C) \end{aligned}} \quad (5.13)$$

(b) For all initial conditions $(A(0), B(0), C(0)) \in \mathcal{O}(A,B,C)$, the solution $(A(t), B(t), C(t))$ of (5.13), exist for all $t \geq 0$ and (5.13), is an isodynamical flow on the set of all controllable and observable triples (A, B, C).

(c) For any initial condition, the solution $(A(t), B(t), C(t))$ of (5.13) converges to a Euclidean norm optimal realization, characterized by

$$AA' + BB' = A'A + C'C \quad (5.14)$$

(d) Convergence to the class of Euclidean norm optimal realizations (5.14) is exponentially fast.

Proof Let $\text{grad}\Gamma = (\text{grad}\Gamma_1, \text{grad}\Gamma_2, \text{grad}\Gamma_3)$ denote the gradient of $\Gamma : \mathcal{O}(A,B,C) \to \mathbb{R}$ with respect to the normal Riemannian metric. Thus for each $(F, G, H) \in \mathcal{O}(A,B,C)$, then $\text{grad}\Gamma(F, G, H)$ is characterized by the condition

$$\text{grad}\Gamma(F, G, H) \in T_{(F,G,H)}\mathcal{O}(A,B,C) \quad (5.15)$$

and

$$D\Gamma|_{(F,G,H)}([X, F], XG, -HX) =$$

$$\langle\langle (\text{grad}\Gamma_1, \text{grad}\Gamma_2, \text{grad}\Gamma_3), ([X, F], XG, -HX) \rangle\rangle \quad (5.16)$$

for all $X \in \mathbb{R}^{n \times n}$. The derivative of Γ at (F, G, H) is the linear map defined on the tangent space $T_{(F,G,H)}\mathcal{O}(A,B,C)$ by

$$D\Gamma|_{(F,G,H)}([X, F], XG, -HX) = 2\text{tr}[([F, F'] + GG' - H'H)X] \quad (5.17)$$

for all $X \in \mathbb{R}^{n \times n}$. By (4.15)

$$\text{grad}\Gamma(F, G, H) = ([X_1, F], X_1 G, -H X_1)$$

for a uniquely determined $n \times n$ matrix X_1. Thus with (5.17), then (5.16) is equivalent to

$$X_1 = [F, F'] + GG' - H'H \quad (5.18)$$

This proves (a). The proofs of (b), (c) are similar to the proofs of Theorem 4.5. We only have to note that $\Gamma : \mathcal{O}(A, B, C) \to \mathbb{R}$ has compact sublevel sets (this follows immediately from the closedness of $\mathcal{O}(A, B, C)$ in $\mathbb{R}^{n \times (n+m+p)}$)

8.5. EUCLIDEAN NORM OPTIMAL REALIZATIONS

and that the Hessian of Γ on the normal bundle of the set of equilibria is positive definite. ∎

The above theorem is a natural generalization of Theorem 6.5.1. In fact, for $A = 0$, (5.13) is equivalent to the gradient flow (6.5.4). Similarly for $B = 0, C = 0$, (5.13) is the double bracket flow $\dot{A} = [A, [A, A']]$.

It is inconvenient to work with the gradient flow (5.13) as it involves solving a cubic matrix differential equation. A more convenient form leading to a quadratic differential equation is to augment the system with a suitable differential equation for a matrix parameter Λ.

Problem Show that the system of differential equations

$$\begin{aligned}
\dot{A} &= A\Lambda - \Lambda A \\
\dot{B} &= -\Lambda B \\
\dot{C} &= C\Lambda \\
\dot{\Lambda} &= -\Lambda + [A, A'] + BB' - C'C
\end{aligned} \qquad (5.19)$$

is Lyapunov stable, where $\Lambda = \Lambda'$ is symmetric. Moreover, every solution $(A(t), B(t), C(t), \Lambda(t))$ of (5.19) exists for all $t \geq 0$ and converges to $(A_\infty, B_\infty, C_\infty, 0)$ where $(A_\infty, B_\infty, C_\infty)$ is Euclidean norm balanced.

Main Points of Section A form of balancing involving a least squares (Euclidean) norm on the system matrices is achieved via gradient flow techniques. There does not appear to be an explicit algebraic solution to such balancing. At this stage, the range of application areas of the results of this section is not clear, but there are some applications in system identification under study, and there is a generalization which is developed in Chapter 9 to sensitivity minimizations.

Notes for Chapter 8

Numerical algorithms for computing balanced realizations are described in Laub et al. (1987), Safanov and Chiang (1989). These are based on standard numerical software. Least squares optimal realizations have been considered by Verriest (1988) and Helmke (1993). Numerical linear algebra methods for computing such realizations are unknown and the gradient flow techniques developed in this chapter are the only available computational tools.

An interesting feature of the dynamical systems for balancing as described in this chapter is their robustness with respect to losses of controllability or observability properties. For example, the linear and quadratic gradient flows on positive definite matrices have robust or high rates of exponential convergence, respectively, if the observability properties of the system are poor. Thus for certain applications such robustness properties may prove useful.

Discrete-time versions of the gradient flows evolving on positive definite matrices are described in Yan, Moore and Helmke (1993). Such recursive algorithms for balancing are studied following similar ideas for solving Riccati equations as described by Hitz and Anderson (1972).

Isodynamical flows are a natural generalization of isospectral matrix flows (where $B = C = 0$). Conversely, isodynamical flows may also be viewed as special cases of isospectral flows (evolving on matrices $\begin{bmatrix} A & B \\ C & 0 \end{bmatrix}$). Moreover, if $A = 0$, then the class of isodynamical flows is equivalent those evolving on orbits $\mathcal{O}(B, C)$, studied in Chapter 6. For some early ideas concerning isodynamical flows see Hermann (1979). Flows on spaces of linear systems are discussed in Krishnaprasad (1979) and Brockett and Faybusovich (1991).

A number of open research problems concerning the material of Chapter 8 are listed below

1) Is there a cost function whose critical points are the Euclidean norm diagonal balanced realizations? N.B.: The trace function $tr(N(AA' + BB' + C'C))$ does not seem to work!

2) Pernebo and Silverman (1982) have shown that truncated subsystems of diagonal balanced realizations of asymptotically stable systems are also diagonal balanced. Is there a "cost function" proof of this? What about other types of balancing? The truncation property does not hold for Euclidean norm balancing.

3) Develop a *Singular Perturbation Approach* to balanced realization model reduction, i.e. find a class of singular perturbation isodynamical flows which converge to reduced order systems.

4) Study measures for the degree of balancedness! What is the minimal distance of a realization to the class of balanced ones? There are two cases of interest: Distance to the entire set of *all* balanced realizations or to the subset of all balanced realizations *of a given transfer function*.

References

Brockett, R.W. (1989) Least squares matching problems, *Linear Algebra Appl.*, 122-127, 701-777.

Brockett, R.W. and L.E. Faybusovich (1991) Toda flows, inverse spectral transform and realization theory, *Systems and Control Letters* 16, 79-88.

Helmke, U. (1993) Balanced realizations for linear systems: a variational approach, *SIAM J. Control and Optimization* 31, 1-15.

Hermann, R. (1979) *Cartanian geometry, nonlinear waves, and control theory, Part A*, Interdisciplinary Mathematics, Vol. 20, Math. Sci. Press, Brookline MA 02146.

Hitz, K.L. and B.D.O. Anderson (1972) Iterative method of computing the limiting solution of the matrix Riccati differential equation, *Proc. IEE* 119, 1402-1406.

Krishnaprasad, P.S. (1979) Symplectic mechanics and rational functions, *Richerche di Automatica* 10, 107-135.

Laub, A.J., M.T. Heath, C.C. Paige and R.C. Ward (1987) Computation of system balancing transformations and other applicaitons of simultaneous diagonalization algorithms, *IEEE Trans. Automatic Control* AC-32, 115-121.

Milnor, J. (1963) *Morse Theory*, Ann. of Math. Studies, 51, Princeton University Press, Princeton, N.J.

Moore, B.C. (1981) Principal component analysis in linear systems: controllability, observability, and model reduction, *IEEE Transactions on Automatic Control* AC-26, 17-32.

Pernebo, L. and L.M. Silverman (1982) Model reduction via balanced state space representations, *IEEE Transactions on Automatic Control* AC-27, 282-287.

Safonov, M.G. and R.Y. Chiang (1989) A Schur method for balanced-truncation model reduction, *IEEE Transactions on Automatic Control* July, 729-733.

Verriest, E.I. (1988) Minimum sensitivity implementation for multi-mode systems, *Proc. IEEE Conference on Decision and Control*, Austin, 2165-2170.

Yan, W., Moore, J.B. and U. Helmke (1993) Recursive algorithms for solving a class of nonlinear matrix equations with applications to certain sensitivity optimization problems, *SIAM J. of Control and Optimization*, to appear.

Chapter 9

Sensitivity Optimization

9.1 A Sensitivity Minimizing Gradient Flow

In this chapter, the further development and application of matrix least squares optimization techniques is made to minimizing linear system parameter sensitivity measures. Even for scalar linear systems, the sensitivity of the system transfer function with respect to the system state space realization parameters is expressed in terms of norms of matrices. Thus a matrix least squares approach is a very natural choice for sensitivity minimization. See also Helmke and Moore (1993), Yan and Moore (1992), Yan, Moore and Helmke (1993).

For practical implementations of linear systems in signal processing and control, an important issue is that of sensitivity of the input/output behaviour with respect to the internal parameters. Such sensitivity is dependent on the co-ordinate basis of the state space realization of the parametrized system. In this section we tackle, via gradient flow techniques, the task of L^2-sensitivity minimization over the class of all co-ordinate basis. The existence of L^2-sensitivity optimal realizations is shown and it is proved that the class of all optimal realizations is obtained by orthogonal transformations from a single optimal one. Gradient flows are constructed whose solutions converge exponentially fast to the class of sensitivity optimal realizations. The results parallel the norm-balancing theory developed in Chapters 7 and 8. Since the sensitivity optimum co-ordinate basis selections are unique to within arbitrary orthogonal transformations, there remains the possibility of 'super' optimal designs over this optimum class to achieve scaling and sparseness constraints, with the view to good finite-word-length filter design. Alternatively, one could incorporate such constraints in a more complex 'sensitivity' measure, or as side constraints in the optimization, as developed in the last section.

As far as we know, the matrix least squares L^2-sensitivity minimization problem studied here can not be solved by algebraic means. For the authors, this chapter is a clear and convincing demonstration of the power of the gradient flow techniques of this book. It sets the stage for future gradient flow studies to solve optimization problems in signal processing and control theory not readily amenable to algebraic solution.

The classic problem of robust continuous-time filter design insensitive to component value uncertainty is these days often replaced by a corresponding problem in digital filter design. Finite-word-length constraints for the variables and coefficients challenge the designer to work with realizations with input/output properties relatively insensitive to coefficient values, as discussed in Williamson (1991), and Roberts and Mullis (1987). Input/output shift operator representations involving regression state vectors are notoriously sensitive to their coefficients, particularly as the system order increases. The so called delta operator representations are clearly superior for fast sampled continuous-time systems, as are the more general delay representation direct forms, as discussed in Williamson (1991) and Middleton and Goodwin (1990). The challenge is to select appropriate sensitivity measures for which optimal co-ordinate basis selections can be calculated, and which translate to practical robust designs.

A natural sensitivity measure to use for robust filter design is an L^2-index as in Thiele (1986). Thus consider a discrete-time filter transfer function $H(z)$ with the (minimal) state space representation

$$H(z) = c(zI - A)^{-1}b \qquad (1.1)$$

where $(A, b, c) \in \mathbb{R}^{n \times n} \times \mathbb{R}^{n \times 1} \times \mathbb{R}^{1 \times n}$. The L^2-sensitivity measure is

$$\mathcal{S}(A, b, c) = \|\frac{\partial H}{\partial A}(z)\|_2^2 + \|\frac{\partial H}{\partial b}(z)\|_2^2 + \|\frac{\partial H}{\partial c}(z)\|_2^2 \qquad (1.2)$$

where

$$\|X(z)\|_2^2 = \frac{1}{2\pi i} \oint_{|z|=1} \text{tr}(X(z)X(z)^*) \frac{dz}{z},$$

and $X(z)^* = X'(z^{-1})$. The associated optimization task is to select a co-ordinate basis transformation T such that $\mathcal{S}(TAT^{-1}, Tb, cT^{-1})$ is minimized. Incidentally, adding a direct feedthrough constant d in (1.1) changes the index (1.2) by a constant and does not change the analysis. Also for $p \times m$ transfer function matrices $H(z) = (H_{ij}(z))$

$$H(z) = C(zI - A)^{-1}B, \quad \text{with} \quad H_{ij}(z) = c_i(zI - A)^{-1}b_j \qquad (1.3)$$

9.1. A SENSITIVITY MINIMIZING GRADIENT FLOW

where $B = (b_1, b_2, \cdots, b_m)$, $C' = (c'_1, c'_2, \cdots, c'_p)$, the corresponding sensitivity measure is

$$\mathcal{S}(A,B,C) = \|\frac{\partial H}{\partial A}\|_2^2 + \|\frac{\partial H}{\partial B}\|_2^2 + \|\frac{\partial H}{\partial C}\|_2^2 = \sum_{i=1}^{p}\sum_{j=1}^{m} \mathcal{S}(A, b_j, c_i) \quad (1.4)$$

or more generally $\mathcal{S}(TAT^{-1}, TB, CT^{-1})$.

In tackling such sensitivity problems, the key observation of Thiele (1986) is that although the term $\|\frac{\partial H}{\partial A}\|_2^2$ appears difficult to work with technically, such difficulties can be circumvented by working instead with a term $\|\frac{\partial H}{\partial A}\|_1^2$, or rather an upper bound on this. There results a mixed L^2/L^1 sensitivity bound optimization which is mainly motivated because it allows explicit solution of the optimization problem. Moreover, the optimal realization turns out to be, conveniently, a balanced realization which is also the optimal solution for the case when the term $\|\frac{\partial H}{\partial A}\|_2^2$ is deleted from the sensitivity measure. More recently in Li, Anderson and Gevers (1992), the theory for frequency shaped designs within this L^2/L^1 framework as first proposed by Thiele (1986) has been developed further. It is shown that, in general, the sensitivity term involving $\frac{\partial H}{\partial A}$ does then affect the optimal solution. Also the fact that the L^2/L^1 bound optimal T is not unique is exploited. In fact the 'optimal' T is only unique to within arbitrary orthogonal transformations. This allows then further selections, such as the Schur form or Hessenberg forms, which achieve sparse matrices. One obvious advantage of working with sparse matrices is that zero elements can be implemented with effectively infinite precision.

In this chapter, we achieve a complete theory for L^2 sensitivity realization optimization both for discrete-time and continuous-time, linear, time-invariant systems. Constrained sensitivity optimization is studied in the last section.

Our aim in this section is to show that an optimal coordinate basis transformation $\bar{T}$ exists and that every other optimal coordinate transformation differs from $\bar{T}$ by an arbitrary orthogonal left factor. Thus $\bar{P} = \bar{T}'\bar{T}$ is uniquely determined. While explicit algebraic constructions for $\bar{T}$ or $\bar{P}$ are unknown and appear hard to obtain we propose to use steepest descent methods in order to find the optimal solution. Thus for example, with the sensitivities $\mathcal{S}(TAT^{-1}, Tb, cT^{-1})$ expressed as a function of $P = T'T$, denoted $\bar{\mathcal{S}}(P)$, the gradient flow on the set of positive definite symmetric matrices P is constructed as

$$\dot{P} = -\nabla \bar{\mathcal{S}}(P) \quad (1.5)$$

with

$$\bar{P} = \lim_{t \to \infty} P(t) \quad (1.6)$$

The calculation of the sensitivities $\mathcal{S}(P)$ and the gradient $\nabla \mathcal{S}(P)$, in the first instance, requires unit circle integration of matrix transfer functions expressed in terms of P, A, b, c. For corresponding continuous-time results, the integrations would be along the imaginary axis in the complex plane. Such contour integrations can be circumvented by solving appropriate Lyapunov equations.

Optimal L^2-sensitivity Measure We consider the task of optimizing a total L^2-sensitivity function over the class of all minimal realizations (A, b, c) of single-input, single-output, discrete-time, asymptotically stable, rational transfer functions

$$H(z) = c(zI - A)^{-1}b. \tag{1.7}$$

Continuous-time, asymptotically stable, transfer functions can be treated in a similar way and present no additional difficulty. Here asymptotic stability for discrete-time systems means that A has its eigenvalues all in the open unit disc $\{z \in \mathbb{C} \mid |z| < 1\}$. Likewise, continuous time systems (A, b, c) are called asymptotically stable if the eigenvalues of A all have real parts less than zero. In the sequel $(A, b, c) \in \mathbb{R}^{n \times n} \times \mathbb{R}^{n \times 1} \times \mathbb{R}^{1 \times n}$ always denotes an asymptotically stable controllable and observable realization of $H(z)$. Given any such initial minimal realization (A, b, c) of $H(z)$, the similarity orbit

$$\mathcal{R}_H := \{(TAT^{-1}, Tb, cT^{-1}) \mid T \in GL(n)\} \tag{1.8}$$

of all minimal realizations of $H(z)$ is a smooth closed submanifold of $\mathbb{R}^{n \times n} \times \mathbb{R}^{n \times 1} \times \mathbb{R}^{1 \times n}$, see Lemma 7.4.1. Note that we here depart from our previous notation $\mathcal{O}(A, b, c)$ for (1.8).

To define a sensitivity measure on $\mathcal{R}_H$ we first have to determine the partial derivatives of the transfer function $H(z)$ with respect to the variables A, b and c. Of course, in the course of such computations we have to regard the complex variable $z \in \mathbb{C}$ as a fixed but arbitrary constant. Let us first introduce the definitions

$$\begin{aligned} \mathcal{B}(z) &:= (zI - A)^{-1}b \\ \mathcal{C}(z) &:= c(zI - A)^{-1} \\ \mathcal{A}(z) &:= \mathcal{B}(z) \cdot \mathcal{C}(z) \end{aligned} \tag{1.9}$$

By applying the standard rules of matrix calculus, see Appendix A, we obtain well known formulas for the partial derivatives as follows:

$$\frac{\partial H(z)}{\partial A} = \mathcal{A}(z)', \quad \frac{\partial H(z)}{\partial b} = \mathcal{C}(z)', \quad \frac{\partial H(z)}{\partial c} = \mathcal{B}(z)' \tag{1.10}$$

We see that the derivatives $\frac{\partial H}{\partial A}(z), \frac{\partial H}{\partial b}(z), \frac{\partial H}{\partial c}(z)$ are stable rational matrix valued functions of z and therefore their L^2-norms exist as the unit circle

9.1. A SENSITIVITY MINIMIZING GRADIENT FLOW

contour integrals:

$$\|\frac{\partial H}{\partial A}\|_2^2 = \|\mathcal{A}(z)\|_2^2 = \frac{1}{2\pi i}\oint \text{tr}(\mathcal{A}(z)\mathcal{A}(z)^*)\frac{dz}{z}$$

$$\|\frac{\partial H}{\partial b}\|_2^2 = \|\mathcal{C}(z)\|_2^2 = \frac{1}{2\pi i}\oint \text{tr}(\mathcal{C}(z)^*\mathcal{C}(z))\frac{dz}{z}$$

$$\|\frac{\partial H}{\partial c}\|_2^2 = \|\mathcal{B}(z)\|_2^2 = \frac{1}{2\pi i}\oint \text{tr}(\mathcal{B}(z)\mathcal{B}(z)^*)\frac{dz}{z} \quad (1.11)$$

The *total L^2-sensitivity function* $S: \mathcal{R}_H \to \mathbb{R}$ is defined by (1.2), which can be reformulated using the following identities for the controllability Gramian and observability Gramian obtained by expanding $\mathcal{B}(z)$ and $\mathcal{C}(z)$ in a Laurent series.

$$W_c = \sum_{k=0}^{\infty} A^k bb'(A')^k = \frac{1}{2\pi i}\oint \mathcal{B}(z)\mathcal{B}(z)^* \frac{dz}{z}$$

$$W_o = \sum_{k=0}^{\infty} (A')^k c'cA^k = \frac{1}{2\pi i}\oint \mathcal{C}(z)^*\mathcal{C}(z)\frac{dz}{z} \quad (1.12)$$

as

$$\boxed{S(A,b,c) = \frac{1}{2\pi i}\oint \text{tr}(\mathcal{A}(z)\mathcal{A}(z)^*)\frac{dz}{z} + \text{tr}W_c + \text{tr}W_o} \quad (1.13)$$

The first term involving contour integration can be expressed in explicit form as

$$\|\frac{\partial H}{\partial A}\|_2^2 = \frac{1}{2\pi i}\oint \text{tr}(\mathcal{A}\mathcal{A}^*)\frac{dz}{z}$$

$$= \sum_{\substack{k,l,r,s \\ r+l=k+s}} \text{tr}((A')^k c'b'(A')^l A^s bcA^r)$$

$$= \sum_{\substack{k,l,r,s \\ r+l=k+s}} cA^r(A')^k c' \cdot b'(A')^l A^s b \quad (1.14)$$

Consider now the effect on $S(A,b,c)$ of a transformation to the state space $(A,b,c) \mapsto (TAT^{-1}, Tb, cT^{-1})$, $T \in GL(n,\mathbb{R})$. It is easily seen that

$$S(TAT^{-1}, Tb, cT^{-1}) = \frac{1}{2\pi i}\oint \text{tr}(T\mathcal{A}(z)T^{-1}(T')^{-1}\mathcal{A}(z)^*T')\frac{dz}{z}$$

$$+\frac{1}{2\pi i}\oint \text{tr}(T\mathcal{B}(z)\mathcal{B}(z)^*T')\frac{dz}{z}$$

$$+\frac{1}{2\pi i}\oint \text{tr}((T')^{-1}\mathcal{C}(z)^*\mathcal{C}(z)T^{-1})\frac{dz}{z}$$

Setting $P = T'T$, we can reformulate $S(TAT^{-1}, Tb, cT^{-1})$ as a cost function on $\mathcal{P}(n)$, the set of positive definite matrices in $\mathbb{R}^{n \times n}$, as

$$\bar{S}(P) = \frac{1}{2\pi i}\text{tr}\oint (\mathcal{A}(z)P^{-1}\mathcal{A}(z)^*P + \mathcal{B}(z)\mathcal{B}(z)^*P + \mathcal{C}(z)^*\mathcal{C}(z)P^{-1})\frac{dz}{z} \quad (1.15)$$

Lemma 1.1 *The smooth sensitivity function $\bar{S} : \mathcal{P}(n) \to \mathbb{R}$ defined by (1.15) has compact sublevel sets.*

Proof Obviously $\bar{S}(P) \geq 0$ for all $P \in \mathcal{P}(n)$. For any $P \in \mathcal{P}(n)$ the inequality
$$\bar{S}(P) \geq \text{tr}(W_c P) + \text{tr}(W_o P^{-1})$$
holds by (1.13). By Lemma 6.4.1, $\{P \in \mathcal{P}(n) \mid \text{tr}(W_c P + W_o P^{-1}) \leq a\}$ is a compact subset of $\mathcal{P}(n)$. Thus
$$\{P \in \mathcal{P}(n) \mid \bar{S}(P) \leq a\} \subset \{P \in \mathcal{P}(n) \mid \text{tr}(W_c P + W_o P^{-1}) \leq a\}$$
is a closed subset of a compact set and therefore also compact. ∎

Any continuous function $f : \mathcal{P}(n) \to \mathbb{R}$ with compact sublevel sets is proper and thus possesses a minimum, see Appendix C.1. Thus Lemma 1.1 immediately implies the following corollary.

Corollary 1.2 *The sensitivity function $\bar{S} : \mathcal{P}(n) \to \mathbb{R}$ (and $S : \mathcal{R}_H \to \mathbb{R}$) defined by (1.15) ((respectively (1.13)) assumes its global minimum, i.e. there exists $P_{\min} \in \mathcal{P}(n)$ (and $(A_{\min}, b_{\min}, c_{\min}) \in \mathcal{R}_H$) such that*
$$\bar{S}(P_{\min}) = \inf_{P \in \mathcal{P}(n)} \bar{S}(P)$$
$$S(A_{\min}, b_{\min}, c_{\min}) = \inf_{(F,g,h) \in \mathcal{R}_H} S(F, g, h)$$

Following the analysis developed in Chapters 6 and 8 for Euclidean norm balancing, we now derive the gradient flow $\dot{P} = -\nabla \bar{S}(P)$. Although the equations look forbidding at first, they are the only available reasons for solving the L^2-sensitivity optimal problem. Subsequently, we show how to work with Lyapunov equations rather than contour integrations. Using (1.15), we obtain for the total derivative of $\bar{S}$ at P

$$D\bar{S}|_P(\xi) = \frac{1}{2\pi i} \oint \text{tr}(\mathcal{A}(z)P^{-1}\mathcal{A}(z)^*\xi - \mathcal{A}(z)P^{-1}\xi P^{-1}\mathcal{A}(z)^* P$$
$$+ \mathcal{B}(z)\mathcal{B}(z)^*\xi - \mathcal{C}(z)^*\mathcal{C}(z)P^{-1}\xi P^{-1})\frac{dz}{z}$$
$$= \frac{1}{2\pi i} \oint \text{tr}(\mathcal{A}(z)P^{-1}\mathcal{A}(z)^* - P^{-1}\mathcal{A}(z)^*P\mathcal{A}(z)P^{-1} + \mathcal{B}(z)\mathcal{B}(z)^*$$
$$- P^{-1}\mathcal{C}(z)^*\mathcal{C}(z)P^{-1})\xi \frac{dz}{z}$$

Therefore the gradient for $\bar{S} : \mathcal{P}(n) \to \mathbb{R}$ with respect to the induced Riemannian metric on $\mathcal{P}(n)$ is given by

$$\nabla \bar{S}(P) = \frac{1}{2\pi i} \oint (\mathcal{A}(z)P^{-1}\mathcal{A}(z)^* - P^{-1}\mathcal{A}(z)^*P\mathcal{A}(z)P^{-1} + \mathcal{B}(z)\mathcal{B}(z)^*$$
$$- P^{-1}\mathcal{C}(z)^*\mathcal{C}(z)P^{-1})\frac{dz}{z}$$

9.1. A SENSITIVITY MINIMIZING GRADIENT FLOW

The gradient flow $\dot P = -\nabla \bar S(P)$ of $\bar S$ is thus seen to be

$$\dot P = \frac{1}{2\pi i}\oint (P^{-1}\mathcal{A}(z)^*P\mathcal{A}(z)P^{-1} - \mathcal{A}(z)P^{-1}\mathcal{A}(z)^* - \mathcal{B}(z)\mathcal{B}(z)^* \\ + P^{-1}\mathcal{C}(z)^*\mathcal{C}(z)P^{-1})\frac{dz}{z}$$

(1.16)

The equilibrium points P_∞ of this gradient flow are characterized by $\dot P = 0$, and consequently

$$\frac{1}{2\pi i}\oint (\mathcal{A}(z)P_\infty^{-1}\mathcal{A}(z)^* + \mathcal{B}(z)\mathcal{B}(z)^*)\frac{dz}{z}$$
$$= P_\infty^{-1}\frac{1}{2\pi i}\oint (\mathcal{A}(z)^*P_\infty \mathcal{A}(z) + \mathcal{C}(z)^*\mathcal{C}(z))\frac{dz}{z}P_\infty^{-1} \quad (1.17)$$

Any co-ordinate transformation $T_\infty \in GL(n)$ such that $P_\infty = T'_\infty T_\infty$ satisfies (1.17) is called an L^2-*sensitivity optimal co-ordinate transformation*. Moreover, any realization $(A_{\min}, b_{\min}, c_{\min})$ which minimizes the L^2-sensitivity index $S: \mathcal{R}_H \to \mathbb{R}$ is called L^2-*sensitivity optimal*. Having established the existence of the L^2-sensitivity optimal transformations, and realizations, respectively, we now turn to develop a gradient flow approach to achieve the L^2-sensitivity optimization.

The following results are straightforward modifications of analogous results developed in Chapters 7 and 8 for the case of Euclidean norm balanced realizations. The only significant modification is the introduction of contour integrations. We need the following technical lemma.

Lemma 1.3 *Let (A, b, c) be a controllable or observable asymptotically stable realization. Let $(\mathcal{A}(z), \mathcal{B}(z), \mathcal{C}(z))$ be defined by (1.9), (1.10). Then the linear operator*

$$\frac{1}{2\pi i}\oint (I \otimes (\mathcal{A}(z)\mathcal{A}(z)^* + \mathcal{B}(z)\mathcal{B}(z)^*) + (\mathcal{A}(z)^*\mathcal{A}(z) + \mathcal{C}(z)^*\mathcal{C}(z)) \otimes I$$
$$- \mathcal{A}(z) \otimes \mathcal{A}(z) - \mathcal{A}(z)^* \otimes \mathcal{A}(z)^*)\frac{dz}{z}$$

has all its eigenvalues in $\mathbb{C}^+ = \{\lambda \in \mathbb{C} \mid \operatorname{Re}\lambda > 0\}$.

Proof Suppose there exists $\lambda \in \mathbb{C}$ and a non zero matrix $X \in \mathbb{C}^{n\times n}$ such that (writing $\mathcal{A}$ instead of $\mathcal{A}(z)$, etc)

$$\frac{1}{2\pi i}\oint (X\mathcal{A}\mathcal{A}^* + \mathcal{A}^*\mathcal{A}X - \mathcal{A}^*X\mathcal{A} - \mathcal{A}X\mathcal{A}^* + X\mathcal{B}\mathcal{B}^* + \mathcal{C}^*\mathcal{C}X)\frac{dz}{z} = \lambda X$$

Let $X^* = \bar X'$ denote the Hermitian transpose. Then

$$\frac{1}{2\pi i}\oint \operatorname{tr}(X\mathcal{A}\mathcal{A}^*X^* + \mathcal{A}XX^*\mathcal{A}^* - \mathcal{A}X\mathcal{A}^*X^* - \mathcal{A}^*X\mathcal{A}X^*$$
$$+ X\mathcal{B}\mathcal{B}^*X^* + \mathcal{C}XX^*\mathcal{C}^*)\frac{dz}{z}$$

$$= \mathrm{Re}\lambda \|X\|^2$$

A straightforward manipulation shows that left hand side is equal to

$$\frac{1}{2\pi i}\oint (\|\mathcal{A}(z)X - X\mathcal{A}(z)\|^2 + \|X\mathcal{B}(z)\|^2 + \|\mathcal{C}(z)X\|^2)\frac{dz}{z} \geq 0$$

Thus $\mathrm{Re}\lambda \geq 0$. The integral on the left is zero if and only if

$$\mathcal{A}(z)X = X\mathcal{A}(z), \quad X\mathcal{B}(z) = 0, \quad \mathcal{C}(z)X = 0$$

for all $|z| = 1$. Now suppose that $\mathrm{Re}\lambda = 0$. Then any X satisfying these equations satisfies

$$X\mathcal{B}(z) = X(zI - A)^{-1}b = 0 \tag{1.18}$$

for all $|z| = 1$ and hence, by the identity theorem for analytic functions, for all $z \in \mathbb{C}$. Thus controllability of (A,b) implies $X = 0$ and similarly for (A,c) observable. Thus $\mathrm{Re}\lambda \neq 0$ and the proof is complete. ∎

Lemma 1.4 *Given any initial controllable and observable asymptotically stable realization (A, b, c), then the linearization of the gradient flow (1.16) at any equilibrium point $P_\infty > 0$ is exponentially stable.*

Proof The linearization of (1.16) at P_∞ is given by

$$\bar{J} = \frac{1}{2\pi i}\oint \left[\begin{array}{c} AP_\infty^{-1} \otimes AP_\infty^{-1} - P_\infty^{-1} \otimes P_\infty^{-1}[A^* P_\infty A + C^* C] P_\infty^{-1} \\ -P_\infty^{-1}(A^* P_\infty A + C^* C) P_\infty^{-1} \otimes P_\infty^{-1} + P_\infty^{-1} A^* \otimes P_\infty^{-1} A^* \end{array} \right] \frac{dz}{z}$$

Let $\mathcal{F} := P_\infty^{\frac{1}{2}} A P_\infty^{\frac{1}{2}}, \mathcal{G} = P_\infty^{\frac{1}{2}} B, \mathcal{H} = C P_\infty^{-\frac{1}{2}}$. Then with

$$\hat{J} = (P_\infty^{\frac{1}{2}} \otimes P_\infty^{\frac{1}{2}}) \bar{J} (P_\infty^{\frac{1}{2}} \otimes P_\infty^{\frac{1}{2}})$$

we have

$$\hat{J} = \frac{1}{2\pi i}\oint (\mathcal{F} \otimes \mathcal{F} + \mathcal{F}^* \otimes \mathcal{F}^* - I \otimes (\mathcal{F}^*\mathcal{F} + \mathcal{H}^*\mathcal{H}) - (\mathcal{F}^*\mathcal{F} + \mathcal{H}^*\mathcal{H}) \otimes I)\frac{dz}{z}$$

Using (1.17) gives

$$\hat{J} = \frac{1}{2\pi i}\oint (\mathcal{F} \otimes \mathcal{F} + \mathcal{F}^* \otimes \mathcal{F}^* - I \otimes (\mathcal{F}\mathcal{F}^* + \mathcal{G}\mathcal{G}^*) - (\mathcal{F}^*\mathcal{F} + \mathcal{H}^*\mathcal{H}) \otimes I)\frac{dz}{z}$$

By Lemma 1.4, $\hat{J}$ has only negative eigenvalues. Therefore, by Sylvester's inertia theorem, see Appendix A, $\bar{J}$ has only negative eigenvalues. The result follows. ∎

We can now state and prove one of the main results of this section.

9.1. A SENSITIVITY MINIMIZING GRADIENT FLOW

Theorem 1.5 *Consider any minimal asymptotically stable realization (A, b, c) of the transfer function $H(z)$. Let $(\mathcal{A}(z), \mathcal{B}(z), \mathcal{C}(z))$ be defined by (1.9), (1.10).*

(a) There exists a unique $P_\infty = P'_\infty > 0$ which minimizes $\bar{S}(P)$ and $T_\infty = P_\infty^{1/2}$ is an L^2-sensitivity optimal transformation. Also, P_∞ is the uniquely determined critical point of the sensitivity function $\bar{S} : \mathcal{P}(n) \to \mathbb{R}$, characterized by (1.17).

(b) The gradient flow $\dot{P}(t) = -\nabla \bar{S}(P(t))$ is given by (1.16) and for every initial condition $P_0 = P'_0 > 0$, the solution $P(t)$ exists for all $t \geq 0$ and converges exponentially fast to P_∞.

(c) A realization $(A, b, c) \in \mathcal{R}_H$ is a critical point for the total L^2-sensitivity function $S : \mathcal{R}_H \to \mathbb{R}$ if and only if (A, b, c) is a global minimum for $S : \mathcal{R}_H \to \mathbb{R}$. The set of global minima of $S : \mathcal{R}_H \to \mathbb{R}$ is a single $O(n)$-orbit $\{(SAS^{-1}, Sb, cS^{-1}) \mid S'S = SS' = I\}$.

Proof We give two proofs. The first proof runs along similar lines to the proof of Proposition 8.5.3. The existence of P_∞ is shown in Corollary 1.2. By Lemma 1.1 the function $\bar{S} : \mathcal{P}(n) \to \mathbb{R}$ has compact sublevel sets and therefore the solutions of the gradient flow (1.16) exist for all $t \geq 0$ and converges to a critical point. By Lemma 1.4 the linearization of the gradient flow at each critical point is exponentially stable, with all its eigenvalues on the negative real axis. As the Hessian of $\bar{S}$ at each critical point is congruent to $\hat{\mathcal{J}}$, it is positive definite. Thus $\bar{S}$ has only isolated minima as critical points and $\bar{S}$ is a Morse function. The set $\mathcal{P}(n)$ of symmetric positive definite matrices is connected and homeomorphic to Euclidean space $\mathbb{R}^{\binom{n+1}{2}}$. From the above, all critical points have index $\binom{n+1}{2}$. Thus by the Morse inequalities, as in the proof of Proposition 8.5.3, Milnor (1963),

$$c_0(\bar{S}) = b_0(\mathcal{P}(n)) = 1$$

and therefore there is only one critical point, which is a local and indeed global minimum. This completes the proof of (a), (b). Part (c) follows since every L^2-sensitivity optimal realization is of the form $(ST_\infty^{\frac{1}{2}} AT_\infty^{-\frac{1}{2}} S^{-1}, ST_\infty^{\frac{1}{2}} b, CT_\infty^{-\frac{1}{2}} S^{-1})$ for a unique orthogonal matrix $S \in O(n)$. The result follows.

For the second proof we apply the Azad-Loeb theorem. It is easy to see, using elementary properties for plurisubharmonic functions that the function $(A, b, c) \mapsto \frac{1}{2\pi i} \oint ||\mathcal{A}(z)||^2 \frac{dz}{z}$ defines a plurisubharmonic function on the complex similarity orbit $\mathcal{O}_\mathbb{C}(A, b, c) = \{(TAT^{-1}, Tb, cT^{-1}) \mid T \in GL(n, \mathbb{C})\}$. Similarly, as in Chapter 8, the sum of the controllability and observability Gramians defines a strictly plurisubharmonic function $(A, b, c) \mapsto \text{tr } W_c +$

tr W_o on $\mathcal{O}_{\mathbb{C}}(A,b,c)$. As the sum of a plurisubharmonic and a strictly plush function is strictly plush, we see by formula (1.13) that the sensitivity function $\mathcal{S}(A,b,c)$ is a strictly plush function on $\mathcal{O}_{\mathbb{C}}(A,b,c)$.

As in Chapter 7, the real case follows from the complex case. Thus part (c) of Theorem 1.5 follows immediately from the Azad-Loeb theorem. ∎

Remark By Theorem 1.5 part(c) any two optimal realizations which minimize the L^2-sensitivity function (1.2) are related by an orthogonal similarity transformation. One may thus use this freedom to transform any optimal realization into a canonical form for orthogonal similarity transformations, such as e.g. into the Hessenberg form or the Schur form. In particular the above theorem implies that there exists *a unique sensitivity optimal realization* (A_H, b_H, c_H) which is in *Hessenberg form*

$$A_H = \begin{bmatrix} * & \cdots & \cdots & * \\ \otimes & \ddots & & \vdots \\ & \ddots & \ddots & \vdots \\ 0 & & \otimes & * \end{bmatrix}, \quad b_H = \begin{bmatrix} \otimes \\ 0 \\ \vdots \\ 0 \end{bmatrix}, \quad c_H = [* \cdots *]$$

where the entries denoted by $\otimes$ are positive. These condensed forms are useful since they reduce the complexity of the realization. ∎

From an examination of the L^2 sensitivity optimal condition (1.17), it makes sense to introduce the two modified Gramian matrices, termed L^2-*sensitivity Gramians* of $H(z)$ as follows

$$\begin{aligned} \tilde{W}_c &\triangleq \frac{1}{2\pi i} \oint (\mathcal{A}(z)\mathcal{A}(z)^* + \mathcal{B}(z)\mathcal{B}(z)^*) \frac{dz}{z} \\ \tilde{W}_o &\triangleq \frac{1}{2\pi i} \oint (\mathcal{A}(z)^*\mathcal{A}(z) + \mathcal{C}(z)^*\mathcal{C}(z)) \frac{dz}{z} \end{aligned} \quad (1.19)$$

These Gramians are clearly both a generalization of the standard Gramians W_c, W_o of (1.12) and of the *Euclidean norm* Gramians appearing in (8.5.6).

Now the above theorem implies the corollary.

Corollary 1.6 *With the L^2-sensitivity Gramian definitions (1.19), the necessary and sufficient condition for a realization (A,b,c) to be L^2-sensitivity optimal is the L^2-sensitivity balancing property*

$$\tilde{W}_c = \tilde{W}_o . \quad (1.20)$$

Any controllable and observable realization of $H(z)$ with the above property is said to be L^2-*sensitivity balanced*.

9.1. A SENSITIVITY MINIMIZING GRADIENT FLOW

Let us also denote the singular values of $\tilde{W}_o$, when $\tilde{W}_o = \tilde{W}_c$, as the Hankel singular values of an L^2-sensitivity optimal realization of $H(z)$.

Moreover, the following properties regarding the L^2-sensitivity Gramians apply

(i) If the two L^2-sensitivity optimal realizations (A_1, b_1, c_1, d_1) and (A_2, b_2, c_2, d_2) are related by a similarity transformation T, i.e.

$$\begin{bmatrix} A_2 & b_2 \\ c_2 & d_2 \end{bmatrix} = \begin{bmatrix} T & 0 \\ 0 & I \end{bmatrix} \begin{bmatrix} A_1 & b_1 \\ c_1 & d_1 \end{bmatrix} \begin{bmatrix} T & 0 \\ 0 & I \end{bmatrix}^{-1}$$

then $P = T'T = I_n$ and T is orthogonal. Moreover, in obvious notation

$$\tilde{W}_c^{(2)} = T\tilde{W}_c^{(1)}T'$$

so that the L^2-sensitivity Hankel singular values are invariant of orthogonal changes of co-ordinates.

(ii) There exists an L^2-sensitivity optimal realization such that its L^2-sensitivity Gramians are diagonal with diagonal elements in descending order.

(iii) The eigenvalues of $\tilde{W}_c \tilde{W}_o$ are invariant under orthogonal state space transformations.

(iv) If $(\tilde{W}_c, \tilde{W}_o)$ are the L^2-sensitivity Gramians of (A, b, c), then so are $(\tilde{W}_o', \tilde{W}_c')$ of (A', c', b').

Contour integrations by Lyapunov equations The next result is motivated by the desire to avoid contour integrations in calculating the L^2-sensitivity gradient flows, and L^2-sensitivity Gramians. Note the emergence of coupled Riccati-like equations with Lyapunov equations.

Proposition 1.7 *Given a minimal realization (A, b, c) of $H(z)$. Let*

$$R = \begin{bmatrix} R_{11} & R_{12} \\ R_{21} & R_{22} \end{bmatrix} \quad \text{and} \quad Q = \begin{bmatrix} Q_{11} & Q_{12} \\ Q_{21} & Q_{22} \end{bmatrix}$$

be the solutions to the following two Lyapunov equations, respectively,

$$\begin{bmatrix} A & bc \\ 0 & A \end{bmatrix} \begin{bmatrix} R_{11} & R_{12} \\ R_{21} & R_{22} \end{bmatrix} \begin{bmatrix} A' & 0 \\ c'b' & A' \end{bmatrix} - \begin{bmatrix} R_{11} & R_{12} \\ R_{21} & R_{22} \end{bmatrix}$$
$$= -\begin{bmatrix} bb' & 0 \\ 0 & I \end{bmatrix}$$

$$\begin{bmatrix} A' & c'b' \\ 0 & A' \end{bmatrix} \begin{bmatrix} Q_{11} & Q_{12} \\ Q_{21} & Q_{22} \end{bmatrix} \begin{bmatrix} A & 0 \\ bc & A \end{bmatrix} - \begin{bmatrix} Q_{11} & Q_{12} \\ Q_{21} & Q_{22} \end{bmatrix}$$

$$= -\begin{bmatrix} c'c & 0 \\ 0 & I \end{bmatrix} \quad (1.21)$$

Then the L^2-sensitivity Gramian pair $(\tilde{W}_c, \tilde{W}_o)$ of (A, b, c) equals (R_{11}, Q_{11}). Moreover, the differential equation (1.16) can be written in the equivalent form

$$\boxed{\dot{P} = P^{-1} Q_{11}(P) P^{-1} - R_{11}(P)} \quad (1.22)$$

where

$$\begin{bmatrix} A & bc \\ 0 & A \end{bmatrix} \begin{bmatrix} R_{11}(P) & R_{12}(P) \\ R_{21}(P) & R_{22}(P) \end{bmatrix} \begin{bmatrix} A' & 0 \\ c'b' & A' \end{bmatrix} - \begin{bmatrix} R_{11}(P) & R_{12}(P) \\ R_{21}(P) & R_{22}(P) \end{bmatrix}$$

$$= -\begin{bmatrix} bb' & 0 \\ 0 & P^{-1} \end{bmatrix} \quad (1.23)$$

$$\begin{bmatrix} A' & c'b' \\ 0 & A' \end{bmatrix} \begin{bmatrix} Q_{11}(P) & Q_{12}(P) \\ Q_{21}(P) & Q_{22}(P) \end{bmatrix} \begin{bmatrix} A & 0 \\ bc & A \end{bmatrix} - \begin{bmatrix} Q_{11}(P) & Q_{12}(P) \\ Q_{21}(P) & Q_{22}(P) \end{bmatrix}$$

$$= -\begin{bmatrix} c'c & 0 \\ 0 & P \end{bmatrix} \quad (1.24)$$

Proof It is routine to compute the augmented state equations and thereby the transfer function of the concatenation of two linear systems to yield

$$(\mathcal{A}(z), \mathcal{B}(z)) = \mathcal{B}(z)(\mathcal{C}(z), I) = C_a(zI - A_a)^{-1} B_a$$

where

$$A_a = \begin{bmatrix} A & bc \\ 0 & A \end{bmatrix}, \quad B_a = \begin{bmatrix} 0 & b \\ I & 0 \end{bmatrix}, \quad C_a = [I \ 0].$$

Thus we have

$$\tilde{W}_c = \frac{1}{2\pi i} \oint (\mathcal{A}(z), \mathcal{B}(z))(\mathcal{A}(z), \mathcal{B}(z))^* \frac{dz}{z}$$

$$= C_a \left\{ \frac{1}{2\pi i} \oint \left((zI - A_a)^{-1} B_a B'_a (\bar{z}I - A'_a)^{-1} \right)^* \frac{dz}{z} \right\} C'_a$$

$$= C_a R C'_a = R_{11}$$

Similarly, it can be proved that $\tilde{W}_o = Q_{11}$. As a consequence, (1.22)-(1.24) follow. ∎

Remark Given $P > 0$, $R_{11}(P)$ and $Q_{11}(P)$ can be computed either indirectly by solving the Lyapunov equations (2.7)-(2.7) or directly by using an iterative algorithm which needs n iterations and can give an exact value, where n is the order of $H(z)$. Actually, a simpler recursion for L^2-sensitivity minimization based on these equations is studied in Yan, Moore and Helmke (1993), and summarized in Section 9.3. ∎

Problems 1. Let $H(z)$ be a scalar transfer function. Show, using the symmetry of $H(z) = H(z)'$, that if $(A_{\min}, b_{\min}, c_{\min}) \in \mathcal{R}_H$ is an L^2-sensitivity optimal realization then also $(A'_{\min}, b'_{\min}, c'_{\min})$ is.

2. Show that there exists a unique symmetric matrix $\Theta \in O(n)$ such that $(A'_{\min}, b'_{\min}, c'_{\min}) = (\Theta A_{\min} \Theta', \Theta b_{\min}, c_{\min} \Theta')$.

Main Points of Section The theory of norm-balancing via gradient flows completely solves the L^2 sensitivity minimization for linear time-invariant systems. Since the optimal co-ordinate basis selections are optimal to within a class of orthogonal matrices, there is the possibility for 'super' optimal designs based on other considerations, such as sparseness of matrices and scaling/overflow for state variables. Furthermore, the results open up the possibility for tackling other sensitivity measures such as those with frequency shaping and including other constraints in the optimization such as scaling and sparseness constraints. Certain of these issues are addressed in later sections.

9.2 Related L^2-Sensitivity Minimization Flows

In this section, we first explore isodynamical flows to find optimal L^2-sensitivity realizations. Differential equations are constructed for the matrices A, b, c, so that these converge to the optimal ones. Next, the results of the previous section are generalized to cope with frequency shaped indices, so that emphasis can be placed in certain frequency bands to the sensitivity of $H(z)$ to its realizations. Then, the matrix transfer function case is considered, and the case of sensitivity to the first N Markov parameters of $\mathcal{H}(z)$.

Isodynamical Flows Following the approach to find balanced realizations via differential equations as developed in Chapter 7, we construct certain ordinary differential equations

$$\dot{A} = f(A, b, c)$$
$$\dot{b} = g(A, b, c)$$
$$\dot{c} = h(A, b, c)$$

which evolve on the space $\mathcal{R}_H$ of all realizations of the transfer function $H(z)$, with the property that the solutions $(A(t), b(t), c(t))$ all converge for $t \to \infty$ to a sensitivity optimal realization of $H(z)$. This approach avoids computing any sensitivity optimizing coordinate transformation matrices T. We have already considered in Chapter 8 the task of minimizing certain cost functions defined on $\mathcal{R}_H$ via gradient flows. Here we consider the related task of minimizing the L^2-sensitivity index via gradient flows. Following the approach developed in Chapter 8 we endow $\mathcal{R}_H$ with its normal Riemannian metric. Thus for tangent vectors $([X_1, A], X_1 b, -cX_1)$, $([X_2, A], X_2 b, -cX_2)$ $\in T_{(A,b,c)} \mathcal{R}_H$ we work with the inner product

$$\langle\langle ([X_1, A], X_1 b, -cX_1), ([X_2, A], X_2 b, -cX_2) \rangle\rangle := 2\mathrm{tr}(X_1' X_2), \qquad (2.1)$$

which defines the normal Riemannian metric, see (8.4.1).

Consider the L^2-sensitivity function

$$\begin{aligned} \mathcal{S} : \mathcal{R}_H &\to \mathbb{R} \\ \mathcal{S}(A, b, c) &= \frac{1}{2\pi i} \oint \mathrm{tr}(\mathcal{A}(z)\mathcal{A}(z)^*) \frac{dz}{z} + \mathrm{tr}(W_c) + \mathrm{tr}(W_o), \end{aligned} \qquad (2.2)$$

This has critical points which correspond to L^2-sensitivity optimal realizations.

Theorem 2.1 *Let $(A, b, c) \in \mathcal{R}_H$ be a stable controllable and observable realization of the transfer function $H(z)$ and let $\Lambda(A, b, c)$ be defined by*

$$\boxed{\Lambda(A, b, c) = \frac{1}{2\pi i} \oint (\mathcal{A}(z)\mathcal{A}(z)^* - \mathcal{A}(z)^* \mathcal{A}(z)) \frac{dz}{z} + (W_c - W_o)} \qquad (2.3)$$

(a) Then the gradient flow of the L^2-sensitivity function $\mathcal{S} : \mathcal{R}_H \to \mathbb{R}$ on $\mathcal{R}_H (\dot{A} = -\mathrm{grad}_A \mathcal{S}, \dot{b} = -\mathrm{grad}_b \mathcal{S}, \dot{c} = -\mathrm{grad}_c \mathcal{S})$ with respect to the normal Riemannian metric is

$$\boxed{\begin{aligned} \dot{A} &= A\Lambda(A, b, c) - \Lambda(A, b, c) A \\ \dot{b} &= -\Lambda(A, b, c) b \\ \dot{c} &= c\Lambda(A, b, c) \end{aligned}} \qquad (2.4)$$

For all initial conditions $(A_0, b_0, c_0) \in \mathcal{R}_H$ the solution $(A(t), b(t), c(t))$ of (2.4) exists for all $t \geq 0$ and converges for $t \to \infty$ to an L^2-sensitivity optimal realization $(\bar{A}, \bar{b}, \bar{c})$ of $H(z)$.

(b) Convergence to the class of sensitivity optimal realizations is exponentially fast.

(c) The transfer function of any solution $(A(t), b(t), c(t))$ is independent of t, i.e. the flow is isodynamical.

Proof By Proposition 8.4.2 and Lemma 8.4.3, the gradient vector field $\mathcal{S} := (\mathrm{grad}_A \mathcal{S}, \mathrm{grad}_b \mathcal{S}, \mathrm{grad}_c \mathcal{S})$ is of the form $\mathrm{grad}_A \mathcal{S} = \Lambda A - A\Lambda$, $\mathrm{grad}_b \mathcal{S} = \Lambda b$, $\mathrm{grad}_c \mathcal{S} = -c\Lambda$, for a uniquely determined $n \times n$ matrix Λ. Recall that a general element of the tangent space $T_{(A,b,c)} \mathcal{R}_H$ is $([X, A], Xb, -cX)$ for

9.2. RELATED L^2-SENSITIVITY MINIMIZATION FLOWS

$X \in \mathbb{R}^{n \times n}$, so that the gradient for the normal Riemannian metric satisfies

$$DS|_{(A,b,c)}([X,A], Xb, -cX) =$$
$$= \langle\langle([\Lambda, A], \Lambda b, -c\Lambda) , ([X,A], Xb, -cX)\rangle\rangle$$
$$= 2\mathrm{tr}(\Lambda' X)$$

at $(A, b, c) \in \mathcal{R}_H$.

Let $\bar{S} : GL(n) \to \mathbb{R}$ be defined by $\bar{S}(T) = S(TAT^{-1}, Tb, cT^{-1})$. A straightforward computation using the chain rule shows that the total derivative of the function $S(A, b, c)$ is given by

$$DS|_{(A,b,c)}([X,A], Xb, -cX) = D\bar{S}|_{I_n}(X)$$

$$= \frac{1}{\pi i} \oint \mathrm{tr}(X(\mathcal{A}(z)\mathcal{A}(z)^* - \mathcal{A}(z)^*\mathcal{A}(z) + \mathcal{B}(z)\mathcal{B}(z)^* - \mathcal{C}(z)^*\mathcal{C}(z)))\frac{dz}{z}$$

$$= \frac{1}{\pi i}\mathrm{tr}(X \oint (\mathcal{A}(z)\mathcal{A}(z)^* - \mathcal{A}(z)^*\mathcal{A}(z)))\frac{dz}{z} + 2\mathrm{tr}(X(W_c - W_o))$$

for all $X \in \mathbb{R}^{n \times n}$. This shows that (2.4) is the gradient flow ($\dot{A} = -\mathrm{grad}_A S$, $\dot{b} = -\mathrm{grad}_b S, \dot{c} = -\mathrm{grad}_c S$) of S. The other statements in Theorem 2.1 follow easily from Theorem 8.4.5. ∎

Remark Similar results to the previous ones hold for parameter dependent sensitivity function such as

$$S_\varepsilon(A, b, c) = \varepsilon\|\frac{\partial H}{\partial A}\|_2^2 + \|\frac{\partial H}{\partial b}\|_2^2 + \|\frac{\partial H}{\partial c}\|_2^2$$

for $\varepsilon \geq 0$. Note that for $\varepsilon = 0$ the function $S_0(A, b, c)$ is equal to the sum of the traces of the controllability and observability Gramians. The optimization of the function $S_0 : \mathcal{R}_H \to \mathbb{R}$ is well understood, and in this case the class of sensitivity optimal realizations consists of the balanced realizations, as studied in Chapters 7, 8. This opens perhaps the possibility of using continuation type of methods in order to compute L^2-sensitivity optimal realization for $S : \mathcal{R}_H \to \mathbb{R}$, where $\varepsilon = 1$. ∎

Frequency Shaped Sensitivity Minimization In practical applications of sensitivity minimization for filters, it makes sense to introduce frequency domain weightings. It may for example be that insensitivity in the pass band is more important that in the stop band. It is perhaps worth mentioning that one case where frequency shaping is crucial is to the case of cascade filters $H = H_1 H_2$ where $H_i(z) = c_i(zI - A_i)^{-1}b_i$. Then $\frac{\partial H}{\partial A_1} = \frac{\partial H_1}{\partial A_1}H_2$ and $\frac{\partial H}{\partial A_2} = H_1\frac{\partial H_2}{\partial A_2}$ and H_2, H_1 can be viewed as the frequency shaping filters. It could be that each filter H_i is implemented with a different degree of precision. Other frequency shaping filter selections to minimize pole and zero sensitivities are discussed in Thiele (1986).

Let us consider a generalization of the index $S(A, b, c)$ of (1.2) as

$$S_w(A, b, c) = \|w_A \frac{\partial H}{\partial A}\|_2^2 + \|w_b \frac{\partial H}{\partial b}\|_2^2 + \|w_c \frac{\partial H}{\partial c}\|_2^2 \qquad (2.5)$$

where weighting w_A etc denote the scalar transfer functions $w_A(z)$ etc. Thus

$$\begin{aligned} w_c(z) \frac{\partial H}{\partial c}(z) &= w_c(z) \mathcal{B}(z)' =: \mathcal{B}_w(z)' \\ w_b(z) \frac{\partial H}{\partial b}(z) &= w_b(z) \mathcal{C}(z)' =: \mathcal{C}_w(z)' \\ w_A(z) \frac{\partial H}{\partial A}(z) &= w_A(z) \mathcal{C}(z) \mathcal{B}(z)' =: \mathcal{A}_w(z)' \end{aligned} \qquad (2.6)$$

The important point to note is that the derivations in this more general frequency shaped case are now a simple generalization of earlier ones. Thus the weighted index S_w expressed as a function of positive definite matrices $P = T'T$ is now given from a generalization of (1.15) as

$$\begin{aligned} \bar{S}_w(P) &= \frac{1}{2\pi i} \mathrm{tr} \oint (\mathcal{A}_w(z) P^{-1} \mathcal{A}_w(z)^* P \\ &\quad + \mathcal{B}_w(z) \mathcal{B}_w(z)^* P + \mathcal{C}_w(z)^* \mathcal{C}_w(z) P^{-1}) \frac{dz}{z}. \end{aligned} \qquad (2.7)$$

The gradient flow equations are corresponding generalizations of (1.16):

$$\begin{aligned} \dot{P} &= \frac{1}{2\pi i} \oint (P^{-1} \mathcal{A}_w(z)^* P \mathcal{A}_w(z) P^{-1} \\ &\quad - \mathcal{A}_w(z) P^{-1} \mathcal{A}_w(z)^* - \mathcal{B}_w(z) \mathcal{B}_w(z)^* \\ &\quad + P^{-1} \mathcal{C}_w(z)^* \mathcal{C}_w(z) P^{-1}) \frac{dz}{z}. \end{aligned} \qquad (2.8)$$

With arbitrary initial conditions satisfying $P_0 = P_0' > 0$, the unique limiting equilibrium point is $P_\infty = P_\infty' > 0$, characterized by

$$\begin{aligned} &\frac{1}{2\pi i} \oint (\mathcal{A}_w(z) P_\infty^{-1} \mathcal{A}_w(z)^* + \mathcal{B}_w(z) \mathcal{B}_w(z)^*) \frac{dz}{z} \\ &= P_\infty^{-1} \frac{1}{2\pi i} \oint (\mathcal{A}_w(z)^* P_\infty \mathcal{A}_w(z) + \mathcal{C}_w(z)^* \mathcal{C}_w(z)) \frac{dz}{z} P_\infty^{-1} \end{aligned} \qquad (2.9)$$

To complete the full analysis, it is necessary to introduce the genericity conditions on w_b, w_c:

$$\text{No zeros of } w_b(z), w_c(z) \text{ are poles of } H(z). \qquad (2.10)$$

Under (2.10), and minimality of (A, b, c), the weighted Gramians are positive definite:

$$\begin{aligned} W_{w_c} &= \frac{1}{2\pi i} \oint \mathcal{B}_w(z) \mathcal{B}_w(z)^* \frac{dz}{z} > 0 \\ W_{w_o} &= \frac{1}{2\pi i} \oint \mathcal{C}_w(z)^* \mathcal{C}_w(z) \frac{dz}{z} > 0 \end{aligned} \qquad (2.11)$$

9.2. RELATED L^2-SENSITIVITY MINIMIZATION FLOWS 283

Moreover, in generalizing (1.18) and its dual, likewise, then

$$XB_w(z) = X(zI - A)^{-1}bw_b(z) = 0$$
$$C_w(z)X = w_c(z)c(zI - A)^{-1}X = 0$$

for all $z \in \mathbb{C}$ implies $X = 0$ as required. The frequency shaped generalizations are summarized as a theorem.

Theorem 2.2 *Consider any controllable and observable asymptotically stable realization (A, b, c) of the transfer function $H(z)$. Consider also the weighted index (2.5), (2.7) under definition (2.7).*

(a) There exists a unique $P_\infty = P'_\infty > 0$ which minimizes $\bar{S}_w(P), P \in \mathcal{P}(n)$, and $T_\infty = P_\infty^{\frac{1}{2}}$ is an L^2-frequency shaped sensitivity optimal transformation. Also, P_∞ is the uniquely determined critical point of $\bar{S}_w : \mathcal{P}(n) \to \mathbb{R}$.

(b) The gradient flow $\dot{P}(t) = -\nabla \bar{S}_w(P(t))$ on $\mathcal{P}(n)$ is given by (2.8). Moreover, its solution $P(t)$ of (2.8) exists for all $t \geq 0$ and converges exponentially fast to P_∞ satisfying (2.9).

Sensitivity of Markov Parameters   It may be that a filter $H_N(z)$ has a finite Markov expansion

$$H_N(z) = \sum_{k=0}^{N} cA^k b z^{-k-1}.$$

Then definitions (1.9) specialize as

$$\mathcal{B}_N(z) = \sum_{k=0}^{N} A^k b z^{-k-1}, \quad \mathcal{C}_N(z) = \sum_{k=0}^{N} cA^k z^{-k-1},$$

and the associated Gramians are finite sums

$$W_c^{(N)} = \sum_{k=0}^{N} A^k bb'(A')^k, \quad W_o^{(N)} = \sum_{k=0}^{N} (A')^k c'cA^k$$

In this case the sensitivity function for H_N is with $\mathcal{A}_N(z) = \mathcal{B}_N(z)\mathcal{C}_N(z)$,

$$S_N(A, b, c) = \|\frac{\partial H_N}{\partial A}\|_2^2 + \|\frac{\partial H_N}{\partial b}\|_2^2 + \|\frac{\partial H_N}{\partial c}\|_2^2$$

$$= \frac{1}{2\pi i} \oint \operatorname{tr}(\mathcal{A}_N(z)\mathcal{A}_N(z)^*)\frac{dz}{z} + \operatorname{tr}(W_c^{(N)}) + \operatorname{tr}(W_o^{(N)})$$

$$= \boxed{\sum_{\substack{0 \leq k, l, r, s \leq N \\ r+l = k+s}} b'(A')^r A^k bc A^s (A')^l c' + \operatorname{tr}(W_c^N) + \operatorname{tr}(W_o^N)}$$

This is just the L^2-sensitivity function for the first $N+1$ Markov parameters. Of course, $\mathcal{S}_N$ is defined for an arbitrary, even unstable, realization. If (A, b, c) were stable, then the functions $\mathcal{S}_N : \mathcal{R}_H \to \mathbb{R}$ converge uniformly on compact subsets of $\mathcal{R}_H$ to the L^2-sensitivity function (1.2) as $N \to \infty$. Thus one may use the function $\mathcal{S}_N$ in order to approximate the sensitivity function (1.2). If N is greater or equal to the McMillan degree of (A, b, c), then earlier results all remain in force if $\mathcal{S}$ is replaced by $\mathcal{S}_N$ and likewise $(zI - A)^{-1}$ by the finite sum $\sum_{k=0}^{N} A^k z^{-k-1}$.

The Matrix Transfer Function Case Once the multivariable sensitivity norm is formulated as the sum of scalar variable sensitivity norms as in (1.3), (1.4), the generalization of the scalar variable results to the multivariable case is straightforward.

Thus let $H(z) = (H_{ij}(z)) = C(zI - A)^{-1}B$ be a $p \times m$ strictly proper transfer function with a controllable and observable realization $(A, B, C) \in \mathbb{R}^{n \times n} \times \mathbb{R}^{n \times m} \times \mathbb{R}^{p \times n}$, so that $H_{ij}(z) = c_i(zI - A)^{-1}b_j$ for $i = 1, \ldots, p$, $j = 1, \ldots, m$. The set of all controllable and observable realizations of $H(z)$ then is the similarity orbit

$$\mathcal{R}_H = \{(TAT^{-1}, TB, CT^{-1}) \mid T \in GL(n)\}.$$

By Lemma 7.4.1, $\mathcal{R}_H$ is a closed submanifold of $\mathbb{R}^{n \times n} \times \mathbb{R}^{n \times m} \times \mathbb{R}^{p \times n}$. Here, again, $\mathcal{R}_H$ is endowed with the normal Riemannian metric defined by (2.1).

The multivariable sensitivity norm $\mathcal{S}(A, B, C)$ is then seen to be the sum of scalar variable sensitivities as

$$\mathcal{S}(A, B, C) = \sum_{i=1}^{p}\sum_{j=1}^{m} \mathcal{S}(A, b_j, c_i).$$

Similarly $\mathcal{S} : \mathcal{P}(n) \to \mathbb{R}$, $\mathcal{S}(P) = \mathcal{S}(TAT^{-1}, TB, CT^{-1})$, is

$$\bar{\mathcal{S}}(P) = \sum_{i=1}^{p}\sum_{j=1}^{m} \bar{\mathcal{S}}_{ij}(P)$$

$$\bar{\mathcal{S}}_{ij}(P) = \frac{1}{2\pi i}\mathrm{tr}\oint (\mathcal{A}_{ij}(z)P^{-1}\mathcal{A}_{ij}(z)^*P$$
$$+ \mathcal{B}_{ij}(z)\mathcal{B}_{ij}(z)^*P + \mathcal{C}_{ij}(z)^*\mathcal{C}_{ij}(z)P^{-1})\frac{dz}{z} \qquad (2.12)$$

with, following (1.9), (1.10)

$$\mathcal{C}_{ij}(z) = \left(\frac{\partial H_{ij}(z)}{\partial b}\right)' = c_i(zI - A)^{-1}$$

9.2. RELATED L^2-SENSITIVITY MINIMIZATION FLOWS

$$\mathcal{B}_{ij}(z) = \left(\frac{\partial H_{ij}(z)}{\partial c}\right)' = (zI - A)^{-1}b_j$$

$$\mathcal{A}_{ij}(z) = \left(\frac{\partial H_{ij}(z)}{\partial A}\right)' = \mathcal{B}_{ij}(z)\mathcal{C}_{ij}(z) \qquad (2.13)$$

The gradient flow (1.5) is generalized as

$$\dot{P} = -\nabla \bar{\mathcal{S}}(P) = -\sum_{i=1}^{p}\sum_{j=1}^{m}\nabla \bar{\mathcal{S}}_{ij}(P). \qquad (2.14)$$

Therefore

$$\bar{\mathcal{S}}(P) = \frac{1}{2\pi i}\mathrm{tr}\oint\left(\sum_{i=1}^{p}\sum_{j=1}^{m}\mathcal{A}_{ij}(z)P^{-1}\mathcal{A}_{ij}(z)^*P\right)\frac{dz}{z} +$$
$$p\,\mathrm{tr}(W_c P) + m\,\mathrm{tr}(W_o P^{-1}). \qquad (2.15)$$

The equilibrium point P_∞ of the gradient flow is characterized by

$$\nabla \bar{\mathcal{S}}(P_\infty) = \sum_{i=1}^{p}\sum_{j=1}^{m}\nabla \bar{\mathcal{S}}_{ij}(P_\infty) = 0 \qquad (2.16)$$

which, following (1.17), is now equivalent to

$$\frac{1}{2\pi i}\oint\sum_{i=1}^{p}\sum_{j=1}^{m}(\mathcal{A}_{ij}(z)P_\infty^{-1}\mathcal{A}_{ij}(z)^* + \mathcal{B}_{ij}(z)\mathcal{B}_{ij}(z)^*)\frac{dz}{z}$$
$$= P_\infty^{-1}\frac{1}{2\pi i}\oint\sum_{i=1}^{p}\sum_{j=1}^{m}(\mathcal{A}_{ij}(z)^*P_\infty \mathcal{A}_{ij}(z) + \mathcal{C}_{ij}(z)^*\mathcal{C}_{ij}(z))\frac{dz}{z}P_\infty^{-1} \qquad (2.17)$$

The introduction of the summation terms is clearly the only effect of generalization to the multivariable case, and changes none of the arguments in the proofs. A point at which care must be taken is in generalizing (1.18). Now $X\mathcal{B}_{ij}(z) = 0$ for all i,j, and thus $X(zI - A)^{-1}B = 0$, requiring (A, B) controllable to imply $X = 0$. Likewise, the condition (A, c) observable generalizes as (A, C) observable.

We remark that certain notational elegance can be achieved using vec and Kronecker product notation in the multivariable case, but this can also obscure the essential simplicity of the generalization as in the above discussion.

We summarize the first main multivariable results as a theorem, whose proof follows Theorem 1.5 as indicated above.

Theorem 2.3 *Consider any controllable and observable asymptotically stable realization (A, B, C) of the transfer function $H(z) = C(zI - A)^{-1}B$, $H(z) = (H_{ij}(z))$ with $H_{ij}(z) = c_i(zI - A)^{-1}b_j$. Then:*

(a) *There exists a unique $P_\infty = P_\infty' > 0$ which minimizes the sensitivity index (1.3), also written as $\bar{S}(P) = \sum_{i=1}^{p} \sum_{j=1}^{m} \bar{S}_{ij}(P)$, and $T_\infty = P_\infty^{\frac{1}{2}}$ is an L^2-sensitivity optimal transformation. Also, P_∞ is the uniquely determined critical point of the sensitivity function $\bar{S} : \mathcal{P}(n) \to \mathbb{R}$.*

(b) *The gradient flow $\dot{P} = -\nabla \bar{S}(P)$ of $\bar{S} : \mathcal{P}(n) \to \mathbb{R}$ is given by (2.14)-(2.13). The solution exists for arbitrary $P_0 = P_0' > 0$ and for all $t \geq 0$, and converges exponentially fast to P_∞ satisfying (2.16)-(2.17).*

(c) *A realization $(A, B, C) \in \mathcal{R}_H$ is a critical point for the total L^2-sensitivity function $S : \mathcal{R}_H \to \mathbb{R}$ if and only if (A, B, C) is a global minimum for $S : \mathcal{R}_H \to \mathbb{R}$. The set of global minima of $S : \mathcal{R}_H \to \mathbb{R}$ is a single $O(n)$ orbit $\{(SAS^{-1}, SB, CS^{-1}) \mid S'S = SS' = I\}$.*

The second main multivariable result generalizes Theorem 2.1 as follows.

Theorem 2.4 *Let $(A, B, C) \in \mathcal{R}_H$ be a stable controllable and observable realization of the transfer function $H(z)$, and let $\Lambda = \Lambda(A, B, C)$ be defined by*

$$\Lambda(A, B, C) = \sum_{i=1}^{p} \sum_{j=1}^{m} \frac{1}{2\pi i} \oint (\mathcal{A}_{ij}(z) \mathcal{A}_{ij}(z)^* - \mathcal{A}_{ij}(z)^* \mathcal{A}_{ij}(z)) \frac{dz}{z}$$
$$+ (pW_c - mW_o)$$

Then the gradient flow of the L^2-sensitivity function $S : \mathcal{R}_H \to \mathbb{R}$ on $\mathcal{R}_H (\dot{A} = -\nabla_A S, \dot{B} = -\nabla_B S, \dot{C} = -\nabla_C S)$ with respect to the induced Riemannian metric is

$$\boxed{\begin{aligned} \dot{A} &= A\Lambda(A, B, C) - \Lambda(A, B, C)A \\ \dot{B} &= -\Lambda(A, B, C)B \\ \dot{C} &= C\Lambda(A, B, C). \end{aligned}}$$

For all initial conditions $(A_0, B_0, C_0) \in \mathcal{R}_H$ the solution $(A(t), B(t), C(t))$ exists for all $t \geq 0$ and converges for $t \to \infty$ exponentially fast to an L^2-sensitivity optimal realization $(\bar{A}, \bar{B}, \bar{C})$ of $H(z)$, with the transfer function of any solution $(A(t), B(t), C(t))$ independent of t.

Proof Follows that of Theorem 2.1 with the multivariable generalizations. ∎

9.3 Recursive L^2-Sensitivity Balancing

Main Points of Section Differential equations on the system matrices A, b, c which converge exponentially fast to optimal L^2-sensitivity realizations are available. Also generalizations of the gradient flows to cope with frequency shaping and the matrix transfer function case are seen to be straightforward, as is the case of L^2-sensitivity minimization for just the first N Markov parameters of the transfer function.

9.3 Recursive L^2-Sensitivity Balancing

To achieve L^2-sensitivity optimal realizations on a digital computer, it makes sense to seek for recursive algorithms which converge to the same solution set as that for the differential equations of the previous sections.

In this section, we present such a recursive scheme based on our work in Yan, Moore and Helmke (1993). This work in turn can be viewed as a generalization of the work of Hitz and Anderson (1972) on Riccati equations. Speed improvements of two orders of magnitude of the recursive algorithm over Runge-Kutta implementations of the differential equations are typical. Full proofs of the results are omitted here since they differ in spirit from the other analysis in the book. The results are included because of their practical significance.

For the L^2-sensitivity minimization problem, we have the following result.

We concentrate on the task of finding the unique minimum $P_\infty \in \mathcal{P}(n)$ of the L^2-sensitivity cost $\bar{S}: \mathcal{P}(n) \to \mathbb{R}$. The set of all sensitivity optimal coordinate transformations T_∞ is then given as $T'_\infty T_\infty = P_\infty$.

Theorem 3.1 *Given an initial stable, controllable and observable realization (A, b, c) of $H(z) = c(zI - A)^{-1}b$. The solution of the difference equation*

$$\boxed{\begin{aligned} P_{k+1} &= P_k - 2[P_k + \tilde{W}_o(P_k)/\alpha] \\ &\quad \times [2P_k + \tilde{W}_o(P_k)/\alpha + \alpha \tilde{W}_c(P_k)^{-1}]^{-1}[P_k + \tilde{W}_o(P_k)/\alpha] \\ &\quad + 2\tilde{W}_o(P_k)/\alpha \end{aligned}} \quad (3.1)$$

converges exponentially to $P_\infty \in \mathcal{P}(n)$ from any initial condition $P_0 \in \mathcal{P}(n)$ and P_∞ minimizes the L^2-sensitivity function $\bar{S}: \mathcal{P}(n) \to \mathbb{R}$. Here α is any positive constant. $\tilde{W}_c(P)$ and $\tilde{W}_o(P)$ are defined by

$$\tilde{W}_c(P) \triangleq \frac{1}{2\pi i} \oint (\mathcal{A}(z)P^{-1}\mathcal{A}(z)^* + \mathcal{B}(z)\mathcal{B}(z)^*) \frac{dz}{z} \quad (3.2)$$

$$\tilde{W}_o(P) \triangleq \frac{1}{2\pi i} \oint (\mathcal{A}(z)^* P \mathcal{A}(z) + \mathcal{C}(z)^* \mathcal{C}(z)) \frac{dz}{z}. \quad (3.3)$$

Proof See Yan, Moore and Helmke (1993) for full details. Suffice it to say here, the proof technique works with an upper bound P_k^U for P_k which is monotonically decreasing, and a lower bound P_k^L for P_k which is monotonically increasing. In fact, P_k^U is a solution of (3.1) with initial condition P_0^U suitably "large" satisfying $P_0^U > P_0 > 0$. Also, P_k^L is a solution of (3.1) with initial condition P_0^L suitably "small" satisfying $0 < P_0^L < P_0$. Since it can be shown that $\lim_{k\to\infty} P_k^U = P_k^L$, then $\lim_{k\to\infty} P_k = P_\infty$ is claimed. ∎

Remarks 1. Of course, the calculation of the Gramians $\tilde{W}_c(P), \tilde{W}_o(P)$ on (3.2) (3.3) can be achieved by the contour integrations as indicated, or by solving Lyapunov equations. The latter approach leads then to the more convenient recursive implementation of the next theorem.

2. When $\tilde{W}_c(P)$ and $\tilde{W}_o(P)$ are replaced by constant matrices, then the algorithm specializes to the Riccati equation studied by Hitz and Anderson (1972). This earlier result stimulated us to conjecture the above theorem. The proof of the theorem is not in any sense a straightforward consequence of the Riccati theory as developed in Hitz and Anderson (1972). ∎

Theorem 3.2 *With the same hypotheses as in Theorem 3.1, let $U(P)$ and $V(Q)$ be the solutions of the Lyapunov equations*

$$U(P) = \begin{bmatrix} A' & c'b' \\ 0 & A' \end{bmatrix} U(P) \begin{bmatrix} A & 0 \\ bc & A \end{bmatrix} + \begin{bmatrix} c'c & 0 \\ 0 & P \end{bmatrix} \quad (3.4)$$

$$V(Q) = \begin{bmatrix} A & bc \\ 0 & A \end{bmatrix} V(Q) \begin{bmatrix} A' & 0 \\ c'b' & A' \end{bmatrix} + \begin{bmatrix} bb' & 0 \\ 0 & Q \end{bmatrix} \quad (3.5)$$

Then for any $\alpha > 0$ and any initial condition $(P_0, U_0, V_0) \in \mathcal{P}(n) \times \mathcal{P}(2n) \times \mathcal{P}(2n)$, the solution (P_k, U_k, V_k) of the system of difference equations

$$\boxed{\begin{aligned} P_{k+1} = P_k & \\ -2(P_k + U_k^{11}/\alpha)[2P_k + U_k^{11}/\alpha + \alpha(V_k^{11})^{-1}]^{-1}(P_k + U_k^{11}/\alpha) + 2U_k^{11}/\alpha & \end{aligned}} \quad (3.6)$$

$$U_{k+1} \triangleq \begin{bmatrix} U_{k+1}^{11} & U_{k+1}^{12} \\ U_{k+1}^{21} & U_{k+1}^{22} \end{bmatrix}$$

$$= \begin{bmatrix} A' & c'b' \\ 0 & A' \end{bmatrix} \begin{bmatrix} U_k^{11} & U_k^{12} \\ U_k^{21} & U_k^{22} \end{bmatrix} \begin{bmatrix} A & 0 \\ bc & A \end{bmatrix} + \begin{bmatrix} c'c & 0 \\ 0 & P_k \end{bmatrix} \quad (3.7)$$

$$V_{k+1} \triangleq \begin{bmatrix} V_{k+1}^{11} & V_{k+1}^{12} \\ V_{k+1}^{21} & V_{k+1}^{22} \end{bmatrix}$$

$$= \begin{bmatrix} A & bc \\ 0 & A \end{bmatrix} \begin{bmatrix} V_k^{11} & V_k^{12} \\ V_k^{21} & V_k^{22} \end{bmatrix} \begin{bmatrix} A' & 0 \\ c'b' & A' \end{bmatrix} + \begin{bmatrix} bb' & 0 \\ 0 & P_k^{-1} \end{bmatrix} \quad (3.8)$$

9.3. RECURSIVE L^2-SENSITIVITY BALANCING

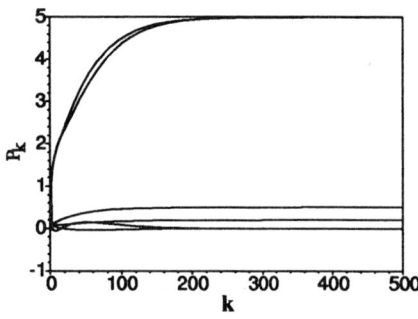
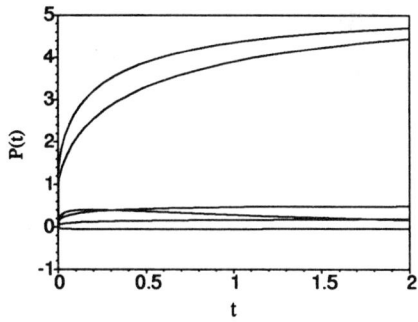

Figure 9.3.1: The trajectory of P_k of (3.1) with $\alpha = 300$

Figure 9.3.2: The trajectory of $P(t)$ of (1.16)

converges to $(P_\infty, U(P_\infty), V(P_\infty^{-1})) \in \mathcal{P}(n) \times \mathcal{P}(2n) \times \mathcal{P}(2n)$.

Here P_∞ minimizes the sensitivity function $\bar{S}: \mathcal{P}(n) \to \mathbb{R}$ and $U_\infty = U(P_\infty), V_\infty = V(P_\infty^{-1})$ are the L^2-sensitivity Gramians $\tilde{W}_c(P_\infty), \tilde{W}_o(P_\infty)$ respectively.

Proof See Yan, Moore and Helmke (1993) for details. ∎

Remark There does not appear to be any guarantee of exponential convergences of these difference equations, at least with the proof techniques as in Theorem 3.1. ∎

Example To demonstrate the effectiveness of the proposed algorithms, consider a specific minimal state-space realization (A, b, c) with

$$A = \begin{pmatrix} 0.5 & 0 & 1.0 \\ 0 & -0.25 & 0 \\ 0 & 0 & 0.1 \end{pmatrix}, \quad b = \begin{pmatrix} 0 \\ 1 \\ 2 \end{pmatrix}, \quad c = \begin{pmatrix} 1 & 5 & 10 \end{pmatrix}$$

Recall that there exists a unique positive definite matrix P_∞ such that the realization (TAT^{-1}, Tb, cT^{-1}) is L^2-sensitivity optimal for any similarity transformation T with $T'_\infty T_\infty = P_\infty$. It turns out that P_∞ is exactly given by

$$P_\infty = \begin{pmatrix} 0.2 & 0 & 0.5 \\ 0 & 5.0 & 0 \\ 0.5 & 0 & 5.0 \end{pmatrix} \tag{3.9}$$

which indeed satisfies (1.17).

We first take the algorithm (3.1) with $\alpha = 300$ and implement it starting from the identity matrix. The resulting trajectory P_k during the first 500 iterations is shown in Figure 9.3.1 and is clearly seen to converge very rapidly to P_∞. In contrast, referring to Figure 9.3.2, using Runge-Kutta integration

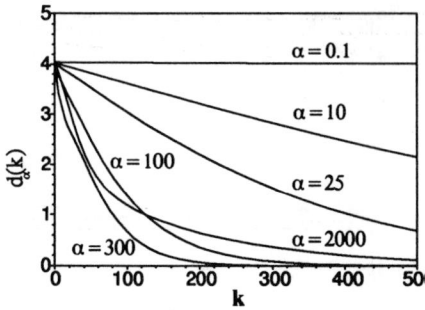 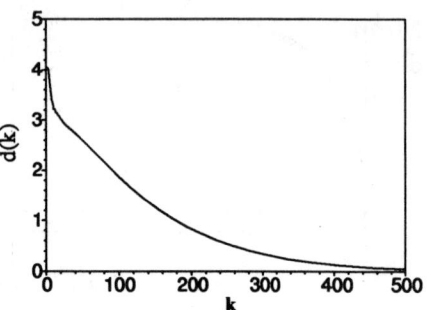

Figure 9.3.3: Effect of different α on the convergence rate of (3.1)

Figure 9.3.4: Convergence of algorithm (3.6)-(3.8) with $\alpha = 300$

methods to solve the differential equation (1.16) with the same initial condition, even after 2400 iterations the solution does not appear to be close to P_∞ and it is converging very slowly to P_∞. Adding a scalar factor to the differential equation (1.16) does not reduce significantly the computer time required for solving it on a digital computer.

Next we examine the effect of α on the convergence rate of the algorithm (3.1). For this purpose, define the deviation between P_k and the true solution P_∞ in (3.9) as

$$d_\alpha(k) = ||P_k - P_\infty||_2$$

where $||\cdot||_2$ denotes the spectral norm of a matrix. Implement (3.1) with

$$\alpha = 0.1,\ 10,\ 25,\ 100,\ 300,\ 2000,$$

respectively, and depict the evolution of the associated deviation $d_\alpha(k)$ for each α in Figure 9.3.3. Then one can see that $\alpha = 300$ is the best choice. In addition, as long as $\alpha \leq 300$, the larger α, the faster the convergence of the algorithm. On the other hand, it should be observed that a larger α is not always better than a smaller α and that too small α can make the convergence extremely slow.

Finally, let us turn to the algorithm (3.6)-(3.8) with $\alpha = 300$, where all the initial matrices required for the implementation are set to identity matrices of appropriate dimension. Define

$$d(k) = ||P_k - P_\infty||_2$$

as the deviation between the first component P_k of the solution and P_∞. Figure 9.3.4 gives its evolution and manifestly exhibits the convergence of the algorithm. Indeed, the algorithm (3.6)-(3.8) is faster compared to (3.1) in terms of the execution time; but with the same number of iterations the former does not produce a solution as satisfactory as the latter. ∎

9.4 L^2-Sensitivity Model Reduction

Main Points of Section Recursive L^2-sensitivity balancing schemes are devised which achieve the same convergence points as the continuous-time gradient flows of the previous sections. The algorithms are considerably faster than those obtained by applying standard numerical software for integrating the gradient flows.

9.4 L^2-Sensitivity Model Reduction

We consider an application of L^2-sensitivity minimization to model reduction, first developed in Yan and Moore (1992). Recall that a L^2-sensitivity optimal realization (A, b, c) of $H(z)$ can be found so that

$$\tilde{W}_c = \tilde{W}_o = \text{diag}(\sigma_1, \sigma_2, \cdots, \sigma_n) = \begin{bmatrix} \Sigma_1 & 0 \\ 0 & \Sigma_2 \end{bmatrix} \quad (4.1)$$

where $\sigma_1 \geq \cdots \geq \sigma_{n_1} > \sigma_{n_1+1} \geq \cdots \geq \sigma_n$ and Σ_i is $n_i \times n_i$, $i = 1, 2$. Partition compatibly (A, b, c) as

$$A = \begin{bmatrix} A_{11} & A_{12} \\ A_{21} & A_{22} \end{bmatrix}, \quad b = \begin{bmatrix} b_1 \\ b_2 \end{bmatrix}, \quad c = \begin{bmatrix} c_1 & c_2 \end{bmatrix} \quad (4.2)$$

Then it is not hard to prove that

$$\|\frac{\partial H}{\partial A_{11}}\|_2^2 + \|\frac{\partial H}{\partial b_1}\|_2^2 \geq \|\frac{\partial H}{\partial A_{22}}\|_2^2 + \|\frac{\partial H}{\partial c_2}\|_2^2 + \text{trace}(\Sigma_1 - \Sigma_2) \quad (4.3)$$

$$\|\frac{\partial H}{\partial A_{11}}\|_2^2 + \|\frac{\partial H}{\partial c_1}\|_2^2 \geq \|\frac{\partial H}{\partial A_{22}}\|_2^2 + \|\frac{\partial H}{\partial b_2}\|_2^2 + \text{trace}(\Sigma_1 - \Sigma_2) \quad (4.4)$$

which suggests that the system is generally less sensitive with respect to (A_{22}, b_2, c_2) than to (A_{11}, b_1, c_1). In this way, the n_1^{th} order model (A_{11}, b_1, c_1) may be used as an approximation to the full order model. However, it should be pointed out that in general the realization (A_{11}, b_1, c_1) is no longer sensitivity optimal.

Let us present a simple example to illustrate the procedure of performing model reduction based on L^2-sensitivity balanced truncation.

Example Consider the discrete-time state transfer function $H(z) = c(zI - A)^{-1}b$ with

$$A = \begin{bmatrix} 0 & 1 \\ -0.25 & -1 \end{bmatrix}, \quad b = \begin{bmatrix} 0 \\ 1 \end{bmatrix}, \quad c = \begin{bmatrix} 1 & 5 \end{bmatrix} \quad (4.5)$$

This is a standard Lyapunov balanced realization with discrete-time controllability and observability Gramians

$$W_c = W_o = \text{diag}(10.0415, 0.7082) \quad (4.6)$$

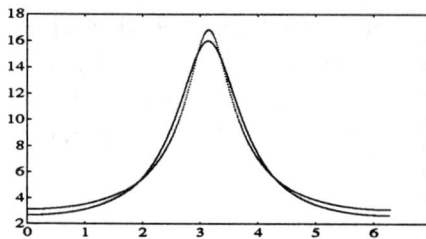

Figure 9.4.1: Frequency responses of $H(z)$, $H_1(z)$

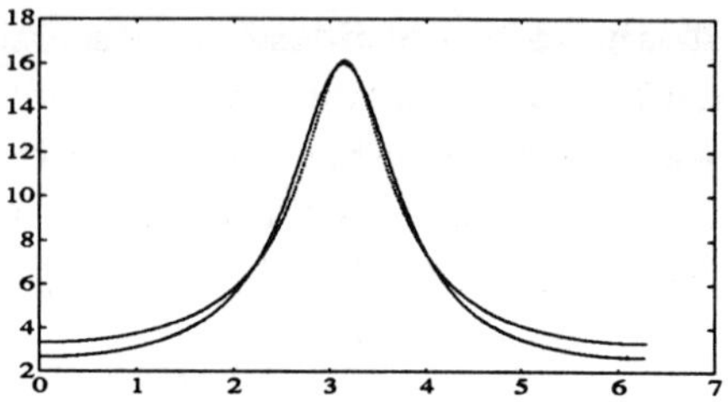

Figure 9.4.2: Frequency responses of $H(z)$, $H_2(z)$

The first order model resulting from direct truncation is

$$H_1(z) = \frac{5.2909}{z + 0.6861}.$$

The magnitude plot of the reduced order model $H_1(z)$ is shown in Fig. 9.4.1 with point symbol and compared to that of the full order model $H(z)$. An L^2-sensitivity balanced realization $(\tilde{A}, \tilde{b}, \tilde{c})$ of $H(z)$ satisfying (4.1) is found to be

$$\tilde{A} = \begin{bmatrix} -0.6564 & 0.1564 \\ -0.1564 & -0.3436 \end{bmatrix}, \quad \tilde{b} = \begin{bmatrix} -2.3558 \\ 0.7415 \end{bmatrix}, \quad \tilde{c} = \begin{bmatrix} -2.3558 & -0.7415 \end{bmatrix}$$

with the associated L^2-sensitivity Gramians being

$$\tilde{W}_c = \tilde{W}_o = \text{diag}(269.0124, 9.6175)$$

Thus by directly truncating this realization, there results a first order model

$$H_2(z) = \frac{5.5498}{z + 0.6564} \qquad (4.7)$$

which is stable with a dotted magnitude plot as depicted in Figure 9.4.2. Comparing Figure 9.4.1 and Figure 9.4.2, one can see that $H_2(z)$ and $H_1(z)$ have their respective strengths in approximating $H(z)$ whereas their difference is subtle. ∎

At this point, let us recall that a reduced order model resulting from truncation of a standard balanced stable realization is also stable. An issue naturally arises as to whether this property still holds in the case of L^2-sensitivity balanced realizations. Unfortunately, the answer is negative. To see this, we consider the following counterexample.

Example $H(z) = c(zI - A)^{-1}b$ where

$$A = \begin{bmatrix} 0.5895 & -0.0644 \\ 0.0644 & 0.9965 \end{bmatrix}, \quad b = \begin{bmatrix} 0.8062 \\ 0.0000 \end{bmatrix}, \quad c = \begin{bmatrix} 0.8062 & 0.0000 \end{bmatrix}$$

It is easily checked that the realization (A, b, c) is Lyapunov balanced with $||A||_2 = 0.9991 < 1$. Solving the relevant differential or difference equation, we find the solution of the L^2-sensitivity minimization problem to be

$$P_\infty = \begin{bmatrix} 1.0037 & -0.0862 \\ -0.0862 & 1.0037 \end{bmatrix}$$

from which a L^2-sensitivity-optimizing similarity transformation is constructed as

$$T_\infty = \begin{bmatrix} 0.0431 & -1.0009 \\ 1.0009 & -0.0431 \end{bmatrix}.$$

As a result, the L^2-sensitivity optimal realization $(\tilde{A}, \tilde{b}, \tilde{c})$ is

$$\tilde{A} = \begin{bmatrix} 1.0028 & -0.0822 \\ 0.0822 & 0.5832 \end{bmatrix}, \quad \tilde{b} = \begin{bmatrix} 0.0347 \\ 0.8070 \end{bmatrix}, \quad \tilde{c} = \begin{bmatrix} -0.0347 & 0.8070 \end{bmatrix}$$

with the L^2-sensitivity Gramians

$$\tilde{W}_c = \tilde{W}_o = \text{diag}(27.7640, 4.2585)$$

One observes that the spectral norm of $\tilde{A}$ is less than 1, but that the first order model resulting from the above described truncation procedure is unstable. It is also interesting to note that the total L^2-sensitivities of the original standard balanced realization and of the resulting sensitivity optimal one are 33.86 and 33.62, respectively, which shows little difference.

■

Main Points of Section Just as standard balancing leads to good model reduction techniques, so L^2-sensitivity balancing can achieve equally good model reduction. In the latter case, the reduced order model is, however, not L^2-sensitivity balanced, in general.

9.5 Sensitivity Minimization with Constraints

In the previous section, parameter sensitivity minimization of a linear system realization is achieved using an L^2-sensitivity norm. However, to minimize roundoff noise in digital filters due to the finite word-length implementations, it may be necessary to ensure that each of the states fluctuates over the same range. For such a situation, it is common to introduce an L^2 scaling constraint. One such constraint for controllable systems, studied in Roberts and Mullis (1987), is to restrict the controllability Gramian W_c, or rather

TW_cT' under a co-ordinate basis transformation T, so that its diagonal elements are identical. Here, without loss of generality we take these elements to be unity.

The approach taken in this section to cope with such scaling constraints, is to introduce gradient flows evolving on the general linear group of invertible $n \times n$ co-ordinate basis transformation matrices, T, but subject to the constraint diag $(TW_cT') = I_n$. Here Diag(X) denotes a diagonal matrix with the same diagonal elements as X, to distinguish from the notation diag$(X) = (x_{11}, \ldots, x_{nn})'$. Thus we restrict attention to the constraint set

$$M = \{T \in GL(n) \mid \text{diag}(TW_cT') = I_n\} \tag{5.1}$$

The simplest sensitivity function proposed in Roberts and Mullis (1987) for optimization is

$$\phi : M \to \mathbb{R}, \quad \phi(T) = \text{tr}\left((T')^{-1}W_oT^{-1}\right). \tag{5.2}$$

A natural approach in order to obtain the minimum of ϕ is to solve the relevant gradient flow of ϕ. In order to do this we first have to explore the underlying geometry of the constraint set M.

Lemma 5.1 *M is a smooth manifold of dimension $n(n-1)$ which is diffeomorphic to the product space $\mathbb{R}^{\frac{n(n-1)}{2}} \times O(n)$. Here $O(n)$ denotes the group of $n \times n$ orthogonal matrices. The tangent space $\text{Tan}_T(M)$ at $T \in M$ is*

$$\text{Tan}_T(M) = \{\xi \in \mathbb{R}^{n \times n} \mid \text{diag}(\xi W_c T') = 0\} \tag{5.3}$$

Proof Let $\gamma : GL(n) \to \mathbb{R}^n$ denote the smooth map defined by

$$\gamma(T) = \text{diag}(TW_cT')$$

The differential of γ at $T_0 \in GL(n)$ is the linear map $D\gamma|_{T_0} : \mathbb{R}^{n \times n} \to \mathbb{R}^n$ defined by

$$D\gamma|_{T_0}(\xi) = 2\,\text{diag}(\xi W_c T_0')$$

Suppose there exists a row vector $c' \in \mathbb{R}^n$ which annihilates the image of $D\gamma|_{T_0}$. Thus $0 = c' \text{diag}(\xi W_c T_0') = \text{tr}(D(c)\xi W_c T_0')$ for all $\xi \in \mathbb{R}^{n \times n}$, where $D(c) = \text{diag}(c_1, \cdots, c_n)$. This is equivalent to

$$W_c T_0' D(c) = 0 \Leftrightarrow D(c) = 0$$

since W_c and T_0 are invertible. Thus γ is a submersion and therefore for all $x \in \mathbb{R}^n$

$$\gamma^{-1}(x) = \{T \in GL(n) \mid \text{diag}(TW_cT') = x\}$$

9.5. SENSITIVITY MINIMIZATION WITH CONSTRAINTS

is a smooth submanifold of codimension n. In order to prove that M is diffeomorphic to $\mathbb{R}^{\frac{n(n-1)}{2}} \times O(n)$ we observe first that M is diffeomorphic to $\{S \in GL(n) \mid \text{diag}(SS') = I_n\}$ via the map

$$T \mapsto S := TW_c^{\frac{1}{2}}$$

Applying the Gram-Schmidt orthogonalization procedure to S yields

$$S = L \cdot U$$

where

$$L = \begin{bmatrix} * & & 0 \\ \vdots & \ddots & \\ * & \cdots & * \end{bmatrix}$$

is lower triangular with positive elements on the diagonal and $U \in O(n)$ is orthogonal. Hence L satisfies $\text{Diag}(LL') = I_n$ if and only if the row vectors of L have all norm one. The result follows. ∎

Lemma 5.2 *The function $\phi : M \to \mathbb{R}$ defined by (5.2) has compact sublevel sets.*

Proof M is a closed subset of $GL(n)$. By Lemma 8.2.1, the set

$$\{P = P' > 0 \mid \text{tr}(W_c P) = n, \text{tr}(W_o P^{-1}) \leq c\}$$

is a compact subset of $GL(n)$. Let $P = T'T$. Hence,

$$T \in \phi^{-1}((-\infty, c]) \Leftrightarrow \text{tr}((T')^{-1}W_o T^{-1}) \leq c \text{ and } \text{diag}(TW_c T') = I_n$$
$$\Rightarrow \text{tr}(W_o P^{-1}) \leq c \text{ and } \text{tr}(W_c P) = n$$

shows that $\phi^{-1}((-\infty, c])$ is a closed subset of a compact set and therefore is itself compact. ∎

Since the image of every continuous map $\phi : M \to \mathbb{R}$ with compact sublevel sets is closed in $\mathbb{R}$ and since $\phi(T) \geq 0$ for all $T \in M$, Lemma 5.2 immediately implies

Corollary 5.3 *A global minimum of $\phi : M \to \mathbb{R}$ exists.*

In order to define the gradient of ϕ we have to fix a Riemannian metric on M. In the sequel we will endow M with the Riemannian metric which is induced from the standard Euclidean structure of the ambient space $\mathbb{R}^{n \times n}$, i.e.

$$<\xi, \eta> := \text{tr}(\xi' \eta) \text{ for all } \xi, \eta \in \text{Tan}_T(M)$$

Let $\text{Tan}_T(M)^\perp$ denote the orthogonal complement of $\text{Tan}_T(M)$ with respect to the inner product in $\mathbb{R}^{n\times n}$. Thus

$$\eta \in \text{Tan}_T(M)^\perp \Leftrightarrow \text{tr}(\eta'\xi) = 0 \text{ for all } \xi \in \text{Tan}_T(M) \tag{5.4}$$

By Lemma 5.1., $\dim \text{Tan}_T(M)^\perp = n$. Given any diagonal matrix diag $(\eta_1,\cdots,\eta_n)$ let

$$\eta := \text{diag}(\eta_1,\cdots,\eta_n)TW_c \tag{5.5}$$

Since

$$\begin{aligned}\text{tr}(\eta'\xi) &= \text{tr}(W_cT'\text{diag}(\eta_1,\cdots,\eta_n)\xi)\\ &= \text{tr}(\text{diag}(\eta_1,\cdots,\eta_n)\text{Diag}(\xi W_c T'))\\ &= 0\end{aligned}$$

for all $\xi \in \text{Tan}_T(M)$. Thus any η of the form (5.5) is orthogonal to the tangent space $\text{Tan}_T(M)$ and hence is contained in $\text{Tan}_T(M)^\perp$. Since the vector space of such matrices η has the same dimension n as $\text{Tan}_T(M)^\perp$ we obtain.

Lemma 5.4 $\text{Tan}_T(M)^\perp = \{\eta \in \mathbb{R}^{n\times n} \mid \eta = \text{diag}(\eta_1,\cdots,\eta_n)TW_c \text{ for some } \eta_1,\cdots,\eta_n\}$.

We are now ready to determine the gradient of $\phi : M \to \mathbb{R}$. The gradient $\nabla\phi$ is the uniquely determined vector field on M which satisfies the condition

$$\begin{aligned}(i) \quad & \nabla\phi(T) \in \text{Tan}_T(M)\\ (ii) \quad & D\phi|_T(\xi) = -2\text{tr}((T')^{-1}W_o T^{-1}\xi T^{-1})\\ & = \langle\nabla\phi(T), \xi\rangle\\ & = \text{tr}(\nabla\phi(T)'\xi)\end{aligned} \tag{5.6}$$

for all $\xi \in \text{Tan}_T(M)$. Hence

$$\text{tr}((\nabla\phi(T)' + 2T^{-1}(T')^{-1}W_o T^{-1})\xi) = 0$$

for all $\xi \in \text{Tan}_T(M)$. Thus by Lemma 5.4 there exists a (uniquely determined) vector $\eta = (\eta_1,\cdots,\eta_n)$ such that

$$\nabla\phi(T)' + 2T^{-1}(T')^{-1}W_o T^{-1} = (\text{diag}(\eta)TW_c)'$$

or, equivalently,

$$\nabla\phi(T) = -2(T')^{-1}W_o T^{-1}(T')^{-1} + \text{diag}(\eta)TW_c \tag{5.7}$$

9.5. SENSITIVITY MINIMIZATION WITH CONSTRAINTS

Using (5.6) we obtain

$$\mathrm{diag}(\eta)\mathrm{Diag}(TW_c^2 T') = 2 \ \mathrm{Diag}((T')^{-1}W_o(T'T)^{-1}W_c T')$$

and therefore with W_c being positive definite

$$\mathrm{diag}(\eta) = 2 \ \mathrm{Diag}((T')^{-1}W_o(T'T)^{-1}W_c T') \ (\mathrm{Diag}(TW_c^2 T'))^{-1} \quad (5.8)$$

We therefore obtain the following result arising from Lemma 5.4.

Theorem 5.5 *The gradient flow $\dot{T} = -\nabla \phi(T)$ of the function $\phi : M \to \mathbb{R}$ defined by (5.2) is given by*

$$\boxed{\begin{aligned}\dot{T} &= 2(T')^{-1}W_o(T'T)^{-1} \\ &\quad -2\mathrm{Diag}((T')^{-1}W_o(T'T)^{-1}W_c T')(\mathrm{Diag}(TW_c^2 T'))^{-1}TW_c\end{aligned}} \quad (5.9)$$

For any initial condition $T_0 \in GL(n)$ the solution $T(t) \in GL(n)$ of (5.9) exists for all $t \geq 0$ and converges for $t \to +\infty$ to a connected component of the set of equilibria points $T_\infty \in GL(n)$.

The following lemma characterizes the equilibria points of (5.9). We use the following terminology.

Definition *A realization (A, B, C) is called essentially balanced if there exists a diagonal matrix $\Delta = \mathrm{diag}(\delta_1, \cdots, \delta_n)$ of real numbers such that for the following controllability and observability Gramians*

$$W_o = \Delta W_c \quad (5.10)$$

Note that for $\Delta = I_n$ this means that the Gramians coincide while for $\Delta = \mathrm{diag}(\delta_1, \cdots, \delta_n)$ with $\delta_i \neq \delta_j$ for $i \neq j$ the (5.10) implies that W_c and W_o are both diagonal.

Lemma 5.6 *A coordinate transformation $T \in GL(n)$ is an equilibrium point of the gradient flow (5.9) if and only if T is an essentially balancing transformation.*

Proof By (5.9), T is an equilibrium point if and only if

$$(T')^{-1}W_o T^{-1} = \mathrm{Diag}((T')^{-1}W_o T^{-1}(T')^{-1}W_c T')(\mathrm{Diag}(TW_c^2 T'))^{-1}TW_c T' \quad (5.11)$$

Set $W_c^T := TW_c T'$, $W_o^T = (T')^{-1}W_o T^{-1}$. Thus (5.11) is equivalent to

$$B = \mathrm{Diag}(B)(\mathrm{Diag}(A))^{-1}A \quad (5.12)$$

where
$$B = W_o^T(T')^{-1}W_cT', \quad A = TW_c^2T' \tag{5.13}$$

Now $B = \Delta A$ for a diagonal matrix Δ implies $(\text{Diag} B)(\text{Diag} A)^{-1} = \Delta$ and thus (5.12) is equivalent to

$$W_o^T = \Delta\, W_c^T \tag{5.14}$$

for a diagonal matrix $\Delta = \text{Diag}(W_o^T)(\text{Diag}(W_c^T))^{-1}$ ∎

Remark 1. By (5.12) and since $T \in M$ we have

$$\Delta = \text{Diag}(W_o^T) \tag{5.15}$$

so that the equilibrium points T_∞ are characterized by

$$W_o^T = \text{Diag}(W_o^T)W_c^T \tag{5.16}$$

2. If T is an *input balancing transformation*, i.e. if $W_c^T = I_n$, W_o^T diagonal, then T satisfies (5.16) and hence is an equilibrium point of the gradient flow. ∎

Trace Constraint The above analysis indicates that the task of finding the critical points $T \in M$ of the constrained sensitivity function (4.6) may be a difficult one. It appears a hard task to single out the sensitivity minimizing coordinate transformation $T \in M$. We now develop an approach which allows us to circumvent such difficulties. Here, let us restrict attention to the enlarged constraint set $M^* = \{T \in GL(n) \mid \text{tr}(TW_cT') = n\}$, but expressed in terms of $P = T'T$ as

$$N = \{P \in \mathcal{P}(n) \mid \text{tr}(W_cP) = n\}. \tag{5.17}$$

Here $\mathcal{P}(n)$ is the set of positive definite real symmetric $n \times n$ matrices. Of course, the diagonal constraint set M is a proper subset of M^*. Moreover, the extended sensitivity function on M^*, $\phi(T) = \text{tr}((T')^{-1}W_oT^{-1})$, expressed in terms of $P = T'T$ is now

$$\Phi: N \to \mathbb{R}, \quad \Phi(P) = \text{tr}(W_oP^{-1}). \tag{5.18}$$

Before launching into the global analysis of the constrained cost function (5.18), let us first clarify the relation between these optimization problems. For this, we need a lemma

Lemma 5.7 *Let $A \in \mathbb{R}^{n \times n}$ be symmetric with $\text{tr}A = n$. Then there exists an orthogonal coordinate transformation $\Theta \in O(n)$ such that $\text{Diag}(\Theta A \Theta') = I_n$.*

9.5. SENSITIVITY MINIMIZATION WITH CONSTRAINTS

Proof The proof is by induction on n. For $n = 1$ the lemma is certainly true. Thus let $n > 1$. Let $\lambda_1 \geq \ldots \geq \lambda_n$ denote the eigenvalues of Λ. Thus $\sum_{i=1}^{n} \lambda_i = n$, which implies $\lambda_1 \geq 1$ and $\lambda_n \leq 1$. Using the Courant-Fischer minmax Theorem 1.3.1, the mean value theorem from real analysis then implies the existence of a unit vector $x \in S^{n-1}$ with $x'Ax = 1$. Extend x to an orthogonal basis $\Theta' = (x_1, x_2, \ldots, x_n) \in O(n)$ of $\mathbb{R}^n$. Then

$$\Theta A \Theta' = \begin{bmatrix} 1 & \star \\ \star & A_1 \end{bmatrix}$$

with $A_1 \in \mathbb{R}^{(n-1) \times (n-1)}$ symmetric and $\mathrm{tr}\, A_1 = n-1$. Applying the induction hypothesis on A_1 thus completes the proof. ■

Effective numerical algorithms for finding orthogonal transformations $\Theta \in O(n)$ such as described in the above lemma are available. See Hwang (1977) for an algorithm based on Householder transformations. Thus we see that the tasks of minimizing the sensitivity function $\phi(T) = \mathrm{tr}((T')^{-1} W_o T^{-1})$ over M and M^* respectively are completely equivalent. Certainly the minimization of $\Phi: N \to \mathbb{R}$ over N is also equivalent to minimizing $\phi: M^* \to \mathbb{R}$ over M^*, using any square root factor T of $P = T'T$. Moreover the minimum value of $\phi(T)$ where $T \in M$ coincides with the minimum value of $\Phi(P)$ where $P \in N$. Thus these minimization tasks on M and N are equivalent.

Returning to the analysis of the sensitivity function $\Phi: N \to \mathbb{R}$ we begin with two preparatory results. The proofs are completely analogous to the corresponding ones from the first subsection and the details are than left as an easy exercise to the reader.

Lemma 5.8 $N = \{P \in \mathcal{P}(n) \mid \mathrm{tr}(W_c P) = n\}$ *is a smooth manifold of dimension* $d = \frac{1}{2}n(n+1) - 1$ *which is diffeomorphic to Euclidean space* $\mathbb{R}^d$. *The tangent space* $T_P N$ *at* $P \in N$ *is*

$$T_P N = \{\xi \in \mathbb{R}^{n \times n} \mid \xi' = \xi,\ \mathrm{tr}(W_c \xi) = 0\}. \tag{5.19}$$

Proof Follows the steps of Lemma 5.1. Thus let $\Gamma: \mathcal{P}(n) \to \mathbb{R}$ denoted the smooth map defined by

$$\Gamma(P) = \mathrm{tr}(W_c P).$$

The derivative of Γ at $P \in \mathcal{P}(n)$ is the surjective map $D\Gamma|_P: S(n) \to \mathbb{R}$ defined on the vector space $S(n)$ of symmetric matrices ξ by $D\Gamma|_P(\xi) = \mathrm{tr}(W_c \xi)$. Thus the first result follows from the fibre theorem from Appendix C. The remainder of the proof is virtually identical to the steps in the proof of Lemma 5.1. ■

Lemma 5.9 *The function $\Phi: N \to \mathbb{R}$ defined by (5.18) has compact sublevel sets. Moreover, a global minimum of $\Phi: N \to \mathbb{R}$ exists.*

Proof This is virtually identical to that of Lemma 5.2 and its corollary.

∎

In order to determine the gradient of Φ we have to specify a Riemannian metric. The following choice of a Riemannian metric on N is particularly convenient. For tangent vectors $\xi, \eta \in T_P N$ define

$$\langle\langle \xi, \eta \rangle\rangle := \operatorname{tr}(P^{-1}\xi P^{-1}\eta). \tag{5.20}$$

It is easily seen that this defines an inner product on $T_P N$ and does in fact define a Riemannian metric on N. We refer to this as the *intrinsic Riemannian metric* on N.

Theorem 5.10 *Let $\Phi: N \to \mathbb{R}$ be defined by (5.18).*

(a) The gradient flow $\dot{P} = -\nabla \Phi(P)$ of the function Φ with respect to the intrinsic Riemannian metric on N is given by

$$\boxed{\dot{P} = W_o - \lambda(P) P W_c P, \quad P(0) = P_0} \tag{5.21}$$

where

$$\lambda(P) = \frac{\operatorname{tr}(W_c W_o)}{\operatorname{tr}(W_c P)^2} \tag{5.22}$$

(b) There exists a unique equilibrium point $P_\infty \in N$. P_∞ satisfies the balancing condition

$$W_o = \lambda_\infty P_\infty W_c P_\infty$$

with $\lambda_\infty = \left(\frac{1}{n}\sum_{i=1}^n \sigma_i\right)^2$ and $\sigma_1, \ldots, \sigma_n$ are the singular values of the associated Hankel, i.e. $\sigma_i = \lambda_i^{\frac{1}{2}}(W_c W_o)$. P_∞ is the global minimum of $\Phi: N \to \mathbb{R}$.

(c) For every initial condition $P_0 \in N$ the solution $P(t)$ of (5.21) exists for all $t \geq 0$ and converges to P_∞ for $t \to +\infty$, with an exponential rate of convergence.

Proof Let $\nabla \Phi$ denote the gradient vector field. The derivative of $\Phi: N \to \mathbb{R}$ at an element $P \in N$ is the linear map on the tangent space $D\Phi|_P: T_P N \to \mathbb{R}$ given by

$$\begin{aligned} D\Phi|_P(\xi) &= -\operatorname{tr}(W_o P^{-1} \xi P^{-1}) \\ &= -\langle\langle W_o, \xi \rangle\rangle \end{aligned}$$

9.5. SENSITIVITY MINIMIZATION WITH CONSTRAINTS

for all $\xi \in T_P N$. It is easily verified that the orthogonal complement $(T_P N)^\perp$ of $T_P N$ in the vector space of symmetric matrices with respect to the inner product $\langle\langle\,,\,\rangle\rangle$ is

$$(T_P N)^\perp = \{\lambda P W_c P \mid \lambda \in \mathbb{R}\},$$

i.e. it coincides with the scalar multiplies of $PW_c P$. Thus

$$D\Phi|_P(\xi) = \langle\langle \nabla\Phi(P), \xi\rangle\rangle$$

for all $\xi \in T_P N$ if and only if

$$\nabla\Phi(P) = -W_o + \lambda P W_c P.$$

The Lagrange multiplier $\lambda(P)$ is readily computed such as to satisfy $(-W_o + \lambda P W_c P) \in T_P N$. Thus

$$\text{tr}(W_c(-W_o + \lambda P W_c P)) = 0$$

or equivalently,

$$\lambda(P) = \frac{\text{tr}(W_c W_o)}{\text{tr}(W_c P W_c P)}.$$

This shows (a).

To prove (b), consider an arbitrary equilibrium point P_∞ of (5.21). Thus

$$W_o = \lambda_\infty P_\infty W_c P_\infty.$$

for a real number $\lambda_\infty = \lambda(P_\infty) > 0$. By (8.2.3) we have

$$\lambda_\infty^{\frac{1}{2}} P_\infty = W_c^{-\frac{1}{2}} (W_c^{\frac{1}{2}} W_o W_c^{\frac{1}{2}})^{\frac{1}{2}} W_c^{-\frac{1}{2}}.$$

Therefore, using the constraint $\text{tr}(W_c P_\infty) = n$ we obtain

$$n\lambda_\infty^{\frac{1}{2}} = \text{tr}(W_c^{\frac{1}{2}} W_o W_c^{\frac{1}{2}})^{\frac{1}{2}} = \sigma_1 + \ldots + \sigma_n$$

and hence

$$\lambda_\infty = (\frac{1}{n}(\sigma_1 + \ldots + \sigma_n))^2.$$

Also P_∞ is uniquely determined and hence must be the global minimum of $\Phi: N \to \mathbb{R}$. This proves (b). A straightforward argument now shows that the linearization of the gradient flow (5.21) at P_∞ is exponentially stable. Global existence of solutions follows from the fact that $\Phi: N \to \mathbb{R}$ has compact sublevel sets. This completes the proof of the theorem. ∎

As a consequence of the above theorem we note that the constrained minimization problem for the function $\Phi: N \to \mathbb{R}$ is equivalent to the unconstrained minimization problem for the function $\Psi: \mathcal{P}(n) \to \mathbb{R}$ defined by

$$\Psi(P) = \text{tr}(W_o P^{-1}) + \lambda_\infty \text{tr}(W_c P) \tag{5.23}$$

where $\lambda_\infty = (\frac{1}{n}(\sigma_1 + \ldots + \sigma_n))^2$. This is of course reminiscent of the usual Lagrange multiplier approach, where λ_∞ would be referred to as a Lagrange multiplier.

So far we have considered the optimization task of minimizing the observability dependent measure $\operatorname{tr}(W_o P^{-1})$ over the constraint set N. A similar analysis can be performed for the minimization task of the constrained L^2-sensitivity cost function $\bar{S}: N \to \mathbb{R}$, defined for a matrix-valued transfer function $H(z) = C(zI - A)^{-1}B$. Here the constrained sensitivity index is defined as in Section 9.1. Thus in the single input, single output case

$$\bar{S}(P) = \frac{1}{2\pi i} \oint \operatorname{tr}(\mathcal{A}(z) P^{-1} \mathcal{A}(z)^* P) \frac{dz}{z} + \operatorname{tr}(W_c P) + \operatorname{tr}(W_o P^{-1}),$$

while for m input, p output systems (see (2.15))

$$\bar{S}(P) = \frac{1}{2\pi i} \oint \sum_{i=1}^{p} \sum_{j=1}^{m} \operatorname{tr}(\mathcal{A}_{ij}(z) P^{-1} \mathcal{A}_{ij}(z)^* P) \frac{dz}{z} + p \operatorname{tr}(W_c P) + m \operatorname{tr}(W_o P^{-1}) \qquad (5.24)$$

Here of course the constraint condition on P is $\operatorname{tr}(W_c P) = n$, so that this term does not influence the sensitivity index.

Following the notation of Section 9.3 define the L^2-sensitivity gramians

$$\tilde{W}_c(P) := \frac{1}{2\pi i} \oint \sum_{i=1}^{p} \sum_{j=1}^{m} (\mathcal{A}_{ij}(z) P^{-1} \mathcal{A}_{ij}(z)^* + \mathcal{B}_{ij}(z) \mathcal{B}_{ij}(z)^*) \frac{dz}{z}$$

$$\tilde{W}_o(P) := \frac{1}{2\pi i} \oint \sum_{i=1}^{p} \sum_{j=1}^{m} (\mathcal{A}_{ij}(z)^* P \mathcal{A}_{ij}(z) + \mathcal{C}_{ij}(z)^* \mathcal{C}_{ij}(z)) \frac{dz}{z}$$

A straightforward computation, analogous to that in the proof of Theorem 5.10, shows that the gradient flow $\dot{P} = -\nabla \bar{S}(P)$ with respect to the intrinsic Riemannian metric on N is

$$\boxed{\dot{P} = \tilde{W}_o(P) - P \tilde{W}_c(P) P - \lambda(P) P W_c P} \qquad (5.25)$$

where

$$\lambda(P) = \frac{\operatorname{tr}(W_c(\tilde{W}_o(P) + P \tilde{W}_c(P) P))}{\operatorname{tr}(W_c P)^2} \qquad (5.26)$$

Here $W_c = \sum_{k=0}^{\infty} A^k BB'(A')^k$ is the standard controllability Gramian while $\tilde{W}_o(P)$ and $\tilde{W}_c(P)$ denote the L^2-sensitivity Gramians.

Arguments virtually identical to those for Theorem 2.3 show

Theorem 5.11 *Consider any controllable and observable discrete-time asymptotically stable realization (A, B, C) of the matrix transfer function $H(z) = C(zI - A)^{-1}B$, $H(z) = (H_{ij}(z))$ with $H_{ij}(z) = c_i(zI - A)^{-1}b_j$. Then (a) There exist a unique positive definite matrix $P_{\min} \in N$ which minimizes the constrained L^2-sensitivity index $\bar{S}: N \to \mathbb{R}$ defined by (5.24) and $T_{\min} = P_{\min}^{\frac{1}{2}} \in M^*$ is a constrained L^2 sensitivity optimal coordinate transformation. Also, $P_{\min} \in N$ is the uniquely determined critical point of the L^2-sensitivity index $\bar{S}: N \to \mathbb{R}$. (b) The gradient flow $\dot{P} = -\nabla \bar{S}(P)$ of $S: N \to \mathbb{R}$ for the intrinsic Riemannian metric on N is given by (5.25), (5.26). Moreover, $P_{\min}$ is determined from the gradient flow (5.25), (5.26) as the limiting solution as $t \to +\infty$, for arbitrary initial conditions $P_o \in N$. The solution $P(t)$ exists for all $t \geq 0$, and converges exponentially fast to $P_{\min}$.*

Results similar to those developed in Section 9.3 hold for recursive Riccati-like algorithms for constrained L^2-sensitivity optimization, which by pass the necessity to evaluate the gradient flow (5.25) for constrained L^2-sensitivity minimization.

Main Points of Section In minimizing sensitivity functions of linear systems subject to scaling constraints, gradient flow techniques can be applied. Such flows achieve what are termed essentially balanced transformations.

Notes for Chapter 9

The books for Roberts and Mullis (1987), Middleton and Goodwin (1990) and Williamson (1991) give a background in digital control. An introductory textbook on digital signal processing is that of Oppenheim and Schafer (1989). A recent book on sensitivity optimization in digital control is Gevers and Li (1993). For important results on optimal finite-word-length representations of linear systems and sensitivity optimization in digital signal processing, see Mullis and Roberts (1976), Hwang (1977), Thiele (1986). The general mixed L^2/L^1 sensitivity optimization problem, with frequency weighted sensitivity cost functions, is solved in Li, Anderson and Gevers (1992).

Gradient flow techniques are used in Helmke and Moore (1993) to derive the existence and uniqueness properties of L^2-sensitivity optimal realizations. In Helmke (1992), corresponding results for L^p-sensitivity optimization are proved, based on the Azad-Loeb theorem. In recent work, Gevers and Li (1993) also consider the L^2-sensitivity problem. Their proof of parts (a) and (c) of Theorem 1.5 is similar in spirit to the second proof, but is derived independently of the more general balancing theory developed in Chapter 7, which is based on the Azad-Loeb theorem. Gevers and Li (1993) also give an interesting physical interpretation of the L^2-sensitivity index, showing that it is a more natural measure of sensitivity than the previously studied mixed L^2/L^1 indices. Moreover, a thorough study of L^2-sensitivity optimization with scaling constraints is given.

For further results on L^2-sensitivity optimal realizations including a preliminary study of L^2-sensitivity optimal model reduction we refer to Yan and Moore (1993). Efficient numerical methods for computing L^2-sensitivity optimal coordinate transformations are described in Yan, Moore and Helmke (1993). Also, sensitivity optimal controller design problems are treated by Madievski, Anderson and Gevers (1993).

Much more work remains to be done in order to develop a satisfactory theory of sensitivity optimization for digital control and signal processing. For example, we mention the apparently unsolved task of sensitivity optimal model reduction, which has been only briefly mentioned in Section 4. Also sensitivity optimal controller design such as developed in Madievski, Anderson and Gevers (1993) remains a challenge. Other topics of interest are sensitivity optimization with constraints and sensitivity optimal design for nonlinear systems.

The sensitivity optimization problems arising in signal processing are only in very exceptional cases amenable to explicit solution methods based on standard numerical linear algebra, such as the singular value decomposition. It is thus our conviction that the dynamical systems approach, as developed in this book, will prove to be a natural and powerful tool for tackling such complicated questions.

References

Gevers, M.R. and G. Li (1993) *Parametrizations in Control, Estimation and Filtering Problems: Accuracy Aspects*, Communications and Control Engineering Series, Springer Verlag, London

Helmke, U. (1992) A several complex variables approach to sensitivity analysis and standard singular values, *Journal of Mathematical Systems, Estimation and Control* 2, 1-13.

Helmke, U. and J.B. Moore (1993) L^2-sensitivity minimization of linear system representations via gradient flows, *J. of Mathematical Systems, Estimation and Control*, to appear.

Hitz, K.L. and B.D.O. Anderson (1972) Iterative method of computing the limiting solution of the matrix Riccati differential equation, *Proc. IEE*, 119, 9, 1402-1406.

Hwang, S.Y. (1977) Minimum uncorrelated unit noise in state space digital filtering, *IEEE Trans. on Acoustics Speech and Signal Processing*, ASSP25, 4, 273-281.

Li, G., Anderson, B.D.O. and M. Gevers (1992) Optimal FWL Design of State-Space Digital Systems with Weighted Sensitivity Minimization and Sparseness Consideration, *Proc. of MTNS conference, Kobe*, Japan.

Madievski, A.G., B.D.O. Anderson and M.R. Gevers (1993) Optimum FWL design of sampled data controllers, preprint.

Middleton, R.H. and Goodwin, G.C. (1990) *Digital Control and Estimation: A Unified Approach*, Prentice Hall, Englewood Cliffs, New Jersey.

Milnor, J. (1963) *Morse Theory*, Princeton University Press.

Oppenheim, A.V. and R.W. Shafer (1989) *Discrete-time signal processing*, Prentice Hall, Eaglewood Cliffs, N.J.

Roberts, R.A. and Mullis, C.T. (1987) *Digital Signal Processing*, Reeding, Mass: Addison-Wesley.

Thiele, L. (1986) On the Sensitivity of Linear State-Space Systems, *IEEE Trans. on Circuits and Systems*, CAS-33, 502-510.

Williamson, D. (1991) *Finite Word Length Digital Control Design and Implementation*, Prentice Hall, London.

Yan, W. and Moore, J.B. (1992) On L^2-sensitivity minimization of linear state-space systems, *IEEE Trans. on Circuits and Systems*, 39, 641 - 648.

Yan, W., Moore, J.B. and Helmke, U. (1993) Recursive algorithms for solving a class of nonlinear matrix equations with applications to certain sensitivity optimization problems, *SIAM J. of Control and Optimization*, to appear.

Appendix A

Linear Algebra

This appendix summarizes the key results of matrix theory and linear algebra results used in this text. For more complete treatments, see Barnett (1971), Bellman (1970).

A.1 Matrices and Vectors

Let $\mathbb{R}$ and $\mathbb{C}$ denote the fields of real numbers and complex numbers, respectively. The set of integers is denoted $\mathbb{N} = \{1, 2, \ldots\}$.

An $n \times m$ *matrix* is an array of n rows and m columns of elements x_{ij} for $i = 1, \cdots, n$, $j = 1, \cdots, m$ as

$$X = \begin{bmatrix} x_{11} & x_{12} & \cdots & x_{1m} \\ x_{12} & x_{22} & \cdots & x_{2m} \\ \vdots & & \vdots & \\ x_{n1} & x_{n2} & \cdots & x_{nm} \end{bmatrix} = (x_{ij}) \ , \quad x = \begin{bmatrix} x_1 \\ x_2 \\ \vdots \\ x_n \end{bmatrix} = (x_i).$$

The matrix is termed *square* when $n = m$. A column n-*vector* x is an $n \times 1$ matrix. The set of all n-vectors (row or column) with real arbitrary entries, denoted $\mathbb{R}^n$, is called n-space. With complex entries, the set is denoted $\mathbb{C}^n$ and is called complex n-space. The term *scalar* denotes the elements of $\mathbb{R}$, or $\mathbb{C}$. The set of real or complex $n \times m$ matrices is denoted by $\mathbb{R}^{n \times m}$ or $\mathbb{C}^{n \times m}$, respectively. The *transpose* of an $n \times m$ matrix X, denoted X', is the $m \times n$ matrix $X' = (x_{ji})$. When $X = X'$, the square matrix is termed *symmetric*. When $X = -X'$, then the matrix is *skew symmetric*. Let us denote the complex conjugate transpose of a matrix X as $X^* = \bar{X}'$. Then matrices X with $X = X^*$ are termed *Hermitian* and with $X = -X^*$ are termed *skew Hermitian*. The *direct sum*, of two square matrices X, Y, denoted $X \dotplus Y$, is

$$\begin{bmatrix} X & 0 \\ 0 & Y \end{bmatrix}$$ where 0 is a zero matrix consisting of zero elements.

A.2 Addition and Multiplication of Matrices

Consider matrices $X, Y \in \mathbb{R}^{n \times m}$ or $\mathbb{C}^{n \times m}$, and scalars $k, \ell \in \mathbb{R}$ or $\mathbb{C}$. Then $Z = kX + \ell Y$ is defined by $z_{ij} = kx_{ij} + \ell y_{ij}$. Thus $X + Y = Y + X$ and addition is commutative. Also, $Z = XY$ is defined for X an $n \times p$ matrix and Y an $p \times m$ matrix by $\xi_{ij} = \sum_{k=1}^{p} x_{ik} y_{kj}$, and is an $n \times m$ matrix. Thus $W = XYZ = (XY)Z = X(YZ)$ and multiplication is associative. Note that when $XY = YX$, which is not always the case, we say that X, Y are commuting matrices.

When $X'X = I$ and X is real then X is termed an *orthogonal* matrix and when $X^*X = I$ with X complex it is termed a *unitary* matrix. Note real vectors x, y are *orthogonal* if $x'y = 0$ and complex vectors are orthogonal if $x^*y = 0$. A *permutation matrix* π has exactly one unity element in each row and column and zeros elsewhere. Of course π is orthogonal. An $n \times n$ square matrix X with only diagonal elements and all other elements zero is termed a *diagonal* matrix and is written $X = \text{diag}(x_{11}, x_{22} \ldots, x_{nn})$. When $x_{ii} = 1$ for all i and $x_{ij} = 0$ for $i \neq j$, then X is termed an *identity* matrix, and is denoted I_n, or just I. Thus for an $n \times m$ matrix Y, then $YI_m = Y = I_nY$. A *sign* matrix S is a diagonal matrix with diagonal elements +1 or -1, and is orthogonal. For X, Y square matrices the *Lie bracket* of X and Y is

$$[X, Y] = XY - YX.$$

Of course, $[X, Y] = -[Y, X]$, and for symmetric X, Y then $[X, Y]' = [Y, X] = -[X, Y]$. Also, if X is diagonal, with distinct diagonal elements then $[X, Y] = 0$ implies that Y is diagonal. For $X, Y \in \mathbb{R}^{n \times m}$ or $\mathbb{C}^{n \times m}$, the *generalized Lie-bracket* is $\{X, Y\} = X'Y - Y'X$. It follows that $\{X, Y\}' = -\{X, Y\}$.

A.3 Determinant and Rank of a Matrix

A recursive definition of the determinant of a square $n \times n$ matrix X, denoted $\det(X)$, is

$$\det(X) = \sum_{j=1}^{n} (-1)^{i+j} x_{ij} \det(X_{ij})$$

where $\det(X_{ij})$ denotes the determinant of the submatrix of X constructed by detecting the i-th row and the j-th column deleted. The element $(-1)^{i+j} \det(X_{ij})$ is termed the *cofactor* of x_{ij}. The square matrix X is said to be a *singular* matrix if $\det(X) = 0$, and a *nonsingular* matrix otherwise. It can be proved that for square matrices $\det(XY) = \det(X)\det(Y)$, $\det(I_n + XY) = \det(I_n + YX)$. In the scalar case $\det(I + xy') = 1 + y'x$.

The *rank* of an $n \times m$ matrix X, denoted by $\text{rk}(X)$ or $\text{rank}(X)$, is the maximum positive integer r such that some $r \times r$ submatrix of X, obtained by deleting rows and columns is nonsingular. Equivalently, the rank is the maximum number of linearly independent rows and columns of X. If r is either m or n then X is *full rank*. It is readily seen that

$$\text{rank }(XY) \leq \min(\text{rank } X, \text{rank } Y).$$

A.4 Range Space, Kernel and Inverses

For an $n \times m$ matrix X, the *range space* or the *image space* $\mathcal{R}(X)$, also denoted by $\text{im}(X)$, is the set of vectors Xy where y ranges over the set of all m vectors. Its dimension is equal to the rank of X. The *kernel* $\ker(X)$ of X is the set of vectors z for which $Xz = 0$. It can be seen that for real matrices $\mathcal{R}(X')$ is orthogonal to $\ker(X)$, or equivalently, with $y_1 = X'y$ for some y and if $Xy_2 = 0$, then $y_1'y_2 = 0$.

For a square *nonsingular matrix* X, there exists a unique *inverse* of X, denoted X^{-1}, such that $X^{-1}X = XX^{-1} = I$. The ij-th element of X^{-1} is given from $\det(X)^{-1} \times$ cofactor of x_{ji}. Thus $(X^{-1})' = (X')^{-1}$ and $(XY)^{-1} = Y^{-1}X^{-1}$. More generally, a unique (Moore-Penrose) *pseudo-inverse* of X, denoted $X^{\#}$, is defined by the characterizing properties $X^{\#}Xy = y$ for all $y \in \mathcal{R}(X')$ and $X^{\#}y = 0$ for all $y \in \ker(X')$. Thus if $\det(X) \neq 0$ then $X^{\#} = X^{-1}$, if $X = 0, X^{\#} = 0, (X^{\#})^{\#} = X, X^{\#}XX^{\#} = X^{\#}$, $XX^{\#}X = X$

For a nonsingular $n \times n$ matrix X, a nonsingular $p \times p$ matrix A and an $n \times p$ matrix B, then provided inverses exist, the readily verified *Matrix Inversion Lemma* tells us that

$$\begin{aligned}(I + XBA^{-1}B')^{-1}X &= (X^{-1} + BA^{-1}B')^{-1} \\ &= X - XB(B'XB + A)^{-1}B'X\end{aligned}$$

and

$$\begin{aligned}(I + XBA^{-1}B')^{-1}XBA^{-1} &= (X^{-1} + BA^{-1}B')^{-1}BX^{-1} \\ &= XB(B'XB + A)^{-1}.\end{aligned}$$

A.5 Powers, Polynomials, Exponentials and Logarithms

The *p-th power* of a square matrix X, denoted X^p, is the p times product $XX\cdots X$ when p is integer. Thus $(X^p)^q = X^{pq}$, $X^p X^q = X^{p+q}$ for integers p, q. *Polynomials* of a matrix X are defined by $p(X) = \sum_{i=0}^{p} s_i X^i$, where s_i are scalars. Any two polynomials $p(X), q(X)$ commute. *Rational functions* of a matrix X, such as $p(X)q^{-1}(X)$, also pairwise commute.

The *exponential* of a square matrix X, denoted e^X, is given by the limit of the convergent series $e^X = \sum_{i=0}^{\infty} \frac{1}{i!} X^i$. It commutes with X. For X any square matrix, e^X is invertible with $(e^X)^{-1} = e^{-X}$. If X is skew-symmetric then e^X is orthogonal. If X and Y commute, then $e^{(X+Y)} = e^X e^Y = e^Y e^X$. For general matrices such simple expressions do not hold. One has always

$$e^X Y e^{-X} = Y + [X,Y] + \frac{1}{2!}[X,[X,Y]]$$
$$+ \frac{1}{3!}[X,[X,[X,Y]]] + \cdots.$$

When Y is nonsingular and $e^X = Y$, the logarithm is defined as $\log Y = X$.

A.6 Eigenvalues, Eigenvectors and Trace

For a square $n \times n$ matrix X, the *characteristic polynomial* of X is $\det(sI - X)$ and its *real or complex* zeros are the *eigenvalues* of X, denoted λ_i. The spectrum $\text{spec}(X)$ of X is the set of its eigenvalues. The *Cayley-Hamilton Theorem* tells us that X satisfies its own *characteristic equation* $\det(sI - X) = 0$. For eigenvalues λ_i, then $X v_i = \lambda_i v_i$ for some real or complex vector v_i, termed an *eigenvector*. The real or complex vector space of such vectors is termed the *eigenspace*. If λ_i is not a repeated eigenvalue, then v_i is unique to within a scalar factor. The eigenvectors are real or complex according to whether or not λ_i is real or complex. When X is diagonal then $X = \text{diag}(\lambda_1, \lambda_2, \ldots, \lambda_n)$. Also, $\det(X) = \Pi_{i=1}^{n} \lambda_i$ so that $\det(X) = 0$, if and only if at least one eigenvalue is zero. As $\det(sI - XY) = \det(sI - YX)$, XY has the same eigenvalues as YX.

A symmetric, or Hermitian, matrix has only real eigenvalues, a skew symmetric, or skew-Hermitian, matrix has only imaginary eigenvalues, and an orthogonal, or unitary, matrix has unity magnitude eigenvalues.

The *trace* of X, denoted $\text{tr}(X)$, is the sum $\sum_{i=1}^{n} x_{ii} = \sum_{i=1}^{n} \lambda_i$. Notice that $\text{tr}(X + Y) = \text{tr}(X) + \text{tr}(Y)$, and with XY square, then $\text{tr}(XY) =$

tr(YX). Also, tr($X'X$) = $\sum_{i=1}^{n}\sum_{j=1}^{n} x_{ij}^2$ and tr$^2(XY) \leq$ tr($X'X$) tr($Y'Y$). A rather useful identity is det(e^X) = $e^{\text{tr}(X)}$.

If X is square and nonsingular, then $\log \det(X) \leq \text{tr} X - n$ and with equality if and only if $X = I_n$.

The eigenvectors associated with distinct eigenvalues of a symmetric (or Hermitian) matrix are *orthogonal*. A *normal* matrix is one with an orthogonal set of eigenvectors. A matrix X is normal if and only if $XX' = X'X$.

If X, Y are symmetric matrices, the minimum eigenvalue of X is denoted by $\lambda_{\min}(X)$. Then

$$\lambda_{\min}(X+Y) \geq \lambda_{\min}(X) + \lambda_{\min}(Y).$$

Also, if X and Y are positive semidefinite, see (A.8), then tr(X)tr(Y) $\geq$ $[\Sigma_i \lambda_i^{\frac{1}{2}}(XY)]^2$, $\lambda_{\min}(XYX) \geq \lambda_{\min}(X^2)\lambda_{\min}(Y)$, and $\lambda_{\min}(XY) \geq \lambda_{\min}(X) \lambda_{\min}(Y)$. With $\lambda_{\max}(X)$ as the maximum eigenvalue, then for positive semidefinite matrices, see (A.8), tr(X) + tr(Y) $\geq 2\Sigma_i \lambda_i^{\frac{1}{2}}(XY)$, $\lambda_{\max}(X+Y) \leq \lambda_{\max}(X) + \lambda_{\max}(Y)$, $\lambda_{\max}(XYX) \leq \lambda_{\max}(X^2)\lambda_{\max}(Y)$, $\lambda_{\max}(XY) \leq \lambda_{\max}(X)\lambda_{\max}(Y)$.

If $\lambda_1(X) \geq \cdots \geq \lambda_n(X)$ are the ordered eigenvalues of a symmetric matrix X, then for $1 \leq k \leq n$ and symmetric matrices X, Y:

$$\sum_{i=1}^{k} \lambda_i(X+Y) \leq \sum_{i=1}^{k} \lambda_i(X) + \sum_{i=1}^{k} \lambda_i(Y).$$

$$\sum_{i=1}^{k} \lambda_{n-i+1}(X+Y) \geq \sum_{i=1}^{k} \lambda_{n-i+1}(X) + \sum_{i=1}^{k} \lambda_{n-i+1}(Y)$$

A.7 Similar Matrices

Two $n \times n$ matrices X, Y are called *similar* if there exists a nonsingular T such that $Y = T^{-1}XT$. Thus X is similar to X. Also, if X is similar to Y, then Y is similar to X. Moreover, if X is similar to Y and if Y is similar to Z, then X is similar to Z. Indeed, similarity of matrices is an *equivalence relation*, see Appendix (C.1). Similar matrices have the same eigenvalues.

If for a given X, there exists a *similarity transformation* T such that $\Lambda = T^{-1}XT$ is diagonal, then X is termed *diagonalizable* and $\Lambda = \text{diag}(\lambda_1, \lambda_2 \ldots \lambda_n)$ where λ_i are the eigenvalues of X. The columns of T are then the eigenvectors of X. All matrices with distinct eigenvalues are diagonalizable, as are orthogonal, symmetric, skew symmetric, unitary, Hermitian, and skew

Hermitian matrices. In fact, if X is symmetric, it can be diagonalized by a real orthogonal matrix and when unitary, Hermitian, or skew-Hermitian, it can be diagonalized by a unitary matrix. If X is Hermitian and T is any invertible transformation, then Sylvester's *inertia theorem* asserts that T^*XT has the same number P of positive eigenvalues and the same number N of negative eigenvalues as X. The difference $S = P - N$ is called the *signature* of X, denoted sig(X).

For square real X, there exists a unitary matrix U such that U^*XU is upper triangular the diagonal elements being the eigenvalues of X. This is called the *Schur form*. It is not unique. Likewise, there exists a unitary matrix U, such that U^*XU is a *Hessenberg* matrix with the form

$$\begin{bmatrix} x_{11} & \cdots & & x_{1n} \\ x_{21} & x_{22} & \cdots & x_{2n} \\ & \ddots & & \vdots \\ 0 & & x_{n,n-1} & x_{nn} \end{bmatrix}.$$

A *Jacobi* matrix is a (symmetric) Hessenberg matrix, so that in the Hessenberg form $x_{ij} = 0$ for $|i - j| \geq 2$. The Hessenberg form of any symmetric matrix is a Jacobi matrix.

A.8 Positive Definite Matrices and Matrix Decompositions

With $X = X'$ and real, then X is *positive definite (positive definite or nonnegative definite)* if and only if the scalar $x'Xx > 0$ ($x'Xx \geq 0$) for all nonzero vectors x. The notation $X > 0$ ($X \geq 0$) is used. In fact $X > 0$ ($X \geq 0$) if and only if all eigenvalues are positive (nonnegative). If $X = YY'$ then $X \geq 0$ and $YY' > 0$ if and only if Y is an $m \times n$ matrix with $m \leq n$ and rk$Y = m$. If $Y = Y'$, so that $X = Y^2$ then Y is unique and is termed the symmetric square root of X, denoted $X^{\frac{1}{2}}$. If $X \geq 0$, then $X^{\frac{1}{2}}$ exists.

If Y is lower triangular with positive diagonal entries, and $YY' = X$, then Y is termed a *Cholesky factor* of X. A successive row by row generation of the nonzero entries of Y is termed a *Cholesky decomposition*. A subsequent step is to form $Y\Lambda Y' = X$ where Λ is diagonal positive definite, and Y is lower triangular with 1's on the diagonal. The above decomposition also applies to Hermitian matrices with the obvious generalizations.

For X a real $n \times n$ matrix, then there exists a *polar decomposition* $X = \Theta P$ where P is positive semidefinite symmetric and Θ is orthogonal satisfying

$\Theta'\Theta = \Theta\Theta' = I_n$. While $P = (X'X)^{\frac{1}{2}}$ is uniquely determined, Θ is uniquely determined only if X is nonsingular.

The *singular values* of possibly complex rectangular matrices X, denoted $\sigma_i(X)$, are the positive square roots of the eigenvalues of X^*X. There exist unitary matrices U, V such that

$$V'XU = \begin{bmatrix} \sigma_1 & & 0 \\ & \ddots & \\ 0 & & \sigma_n \\ \cdots & \cdots & \cdots \\ 0 & & 0 \\ \vdots & & \vdots \\ 0 & & 0 \end{bmatrix} =: \Sigma \ .$$

If unitary U, V yield $V'XU$, a diagonal matrix with nonnegative entries then the diagonal entries are the singular values of X. Also, $X = V\Sigma U'$ is termed a *singular value decomposition* (SVD) of X.

Every real $m \times n$ matrix A of rank r has a factorization $A = XY$ by real $m \times r$ and $r \times n$ matrices X and Y with rk X = rk $Y = r$. With $X \in \mathbb{R}^{m \times r}$ and $Y \in \mathbb{R}^{r \times n}$, then the pair (X, Y) belong to the product space $\mathbb{R}^{m \times r} \times \mathbb{R}^{r \times n}$. If $(X, Y), (X_1, Y_1) \in \mathbb{R}^{m \times r} \times \mathbb{R}^{r \times n}$ are two full rank factorizations of A, i.e. $A = XY = X_1Y_1$, then there exists a unique invertible $r \times r$ matrix T with $(X, Y) = (X_1T^{-1}, TY_1)$.

For X a real $n \times n$ matrix, the *QR decomposition* is $X = \Theta R$ where Θ is orthogonal and R is upper triangular (zero elements below the diagonal) with nonnegative entries on the diagonal. If X is invertible then Θ, R are uniquely determined.

A.9 Norms of Vectors and Matrices

The norm of a vector x, written $||x||$, is any length measure satisfying $||x|| \geq 0$ for all x, with equality if and only if $x = 0$, $||sx|| = |s| \, ||x||$ for any scalar s, and $||x + y|| \leq ||x|| + ||y||$ for all x, y. The *Euclidean norm* or the 2-norm is $||x|| = (\sum_{i=1}^{n} x_i^2)^{\frac{1}{2}}$, and satisfies the *Schwartz inequality* $|x'y| \leq ||x|| \, ||y||$, with equality if and only if $y = sx$ for some scalar s.

The operator norm of a matrix X with respect to a given vector norm, is defined as $||X|| = \max_{||x||=1} ||Xx||$. Corresponding to the Euclidean norm is the 2-norm $||X||_2 = \lambda_{\max}^{\frac{1}{2}}(X'X)$, being the largest singular value of X. The *Frobenius norm* is $||X||_F = \text{tr}^{\frac{1}{2}}(X'X)$. The subscript F is deleted when it is clear that the Frobenius norm is intended. For all these norms $||Xx|| \leq ||X||\,||x||$. Also, $||X + Y|| \leq ||X|| + ||Y||$ and $||XY|| \leq ||X||\,||Y||$. Note that $\text{tr}(XY) \leq ||X||_F ||Y||_F$. The *condition number* of a nonsingular matrix X relative to a norm $||\cdot||$ is $||X||\,||X^{-1}||$.

A.10 Kronecker Product and Vec

With X an $n \times m$ matrix, the *Kronecker product* $X \otimes Y$ is the matrix,

$$\begin{bmatrix} x_{11}Y & \cdots & x_{1n}Y \\ \vdots & & \vdots \\ x_{n1}Y & \cdots & x_{nm}Y \end{bmatrix}.$$

With X, Y square the set of eigenvalues of $X \otimes Y$ is given by $\lambda_i(X)\lambda_j(Y)$ for all i, j and the set of eigenvalues of $(X \otimes I + I \otimes Y)$ is the set $\{\lambda_i(X) + \lambda_i(X) + \lambda_j(Y) + \lambda_j(Y)\}$ for all i, j. Also $(X \otimes Y)(V \otimes W) = XV \otimes YW$, and $(X \otimes Y)' = X' \otimes Y'$. If X, Y are invertible, then so is $X \otimes Y$ and $(X \otimes Y)^{-1} = X^{-1} \otimes Y^{-1}$.

Vec (X) is the column vector obtained by stacking the second column of X under the first, and then the third, and so on. In fact

$$\text{vec}(XY) = (I \otimes X)\,\text{vec}\,Y = (Y' \otimes I)\,\text{vec}(X)$$

$$\text{vec}(ABC) = (C' \otimes A)\text{vec}(B).$$

A.11 Differentiation and Integration

Suppose X is a matrix valued function of the scalar variable t. Then $X(t)$ is called differentiable if each entry $x_{ij}(t)$ is differentiable. Also,

$$\frac{dX}{dt} = \left(\frac{dx_{ij}}{dt}\right), \quad \frac{d}{dt}(XY) = \frac{dX}{dt}Y + X\frac{dY}{dt}, \quad \frac{d}{dt}e^{tX} = Xe^{tX} = e^{tX}X.$$

Also, $\int X dt = (\int x_{ij} dt)$. Now with ϕ a scalar function of a matrix X, then

$$\frac{\partial \phi}{\partial X} = \text{the matrix with } ij\text{-th entry } \frac{\partial \phi}{\partial x_{ij}}.$$

If Φ is a matrix function of a matrix X, then

$$\frac{\partial \Phi}{\partial X} = \text{a block matrix with } ij\text{-th block } \frac{\partial \phi_{ij}}{\partial X}.$$

The case when X, Φ are vectors is just a specialization of the above definitions. If X is square $(n \times n)$ and nonsingular $\frac{\partial}{\partial X}(\text{tr}(WX^{-1})) = -X^{-1}WX^{-1}$. Also $\log \det(X) \leq \text{tr} X - n$ and with equality if and only if $X = I_n$. Furthermore, if X is a function of time, then $\frac{d}{dt} X^{-1}(t) = -X^{-1} \frac{dX}{dt} X^{-1}$, which follows from differentiating $XX^{-1} = I$.

If $P = P'$, $\frac{\partial}{\partial x}(x'Px) = 2Px$.

A.12 Lemma of Lyapunov

If A, B, C are known $n \times n, m \times m$ and $n \times m$ matrices, then the linear equation $AX + XB + C = 0$ has a unique solution for an $n \times m$ matrix X if and only if $\lambda_i(A) + \lambda_j(B) \neq 0$ for any i and j. In fact $[I \otimes A + B' \otimes I]$ vec $(X) = -$vec (C) and the eigenvalues of $[I \otimes A + B' \otimes I]$ are precisely given by $\lambda_i(A) + \lambda_j(B)$. If $C > 0$ and $A = B'$, the *Lemma of Lyapunov* for $AX + XB + C = 0$ states that $X = X' > 0$ if and only if all eigenvalues of B have negative real parts.

The linear equation $X - AXB = C$, or equivalently, $[I_{n^2} - B' \otimes A]$ vec (X) = vec (C) has a unique solution if and only if $\lambda_i(A)\lambda_j(B) \neq 1$ for any i, j. If $A = B'$ and $|\lambda_i(A)| < 1$ for all i, then for $X - AXB = C$, the *Lemma of Lyapunov* states that $X = X' > 0$ for all $C = C' > 0$.

Actually, the condition $C > 0$ in the lemma can be relaxed to requiring for any D such that $DD' = C$ that (A, D) be completely controllable, or (A, D) be completely detectable, see definitions Appendix B.

A.13 Vector Spaces and Subspaces

Let us restrict to the real field $\mathbb{R}$ (or complex field $\mathbb{C}$), and recall the spaces $\mathbb{R}^n$ (or $\mathbb{C}^n$). These are in fact special cases of vector spaces over $\mathbb{R}$ (or $\mathbb{C}$) with the vector additions and scalar multiplications properties for its elements spelt out in (A.2). They are denoted *real (or complex) vector spaces*. Any space over an arbitrary field $\mathbb{K}$ which has the same properties is in fact a *vector* space V. For example, the set of all $m \times n$ matrices with entries in the field as $\mathbb{R}$ (or $\mathbb{C}$), is a vector space. This space is denoted by $\mathbb{R}^{m \times n}$ (or $\mathbb{C}^{m \times n}$).

A *subspace* W of V is a vector space which is a subset of the vector space V. The set of all linear combinations of vectors from a non empty subset S of V, denoted $L(S)$, is a subspace (the smallest such) of V containing S. The space $L(S)$ is termed the subspace *spanned* or *generated* by S. With the empty set denoted ϕ, then $L(\phi) = \{0\}$. The rows (columns) of a matrix X

viewed as row (column) vectors span what is termed the row (column) space of X denoted here by $[X]_r([X]_c)$. Of course, $\mathcal{R}(X') = [X]_r$ and $\mathcal{R}(X) = [X]_c$.

A.14 Basis and Dimension

A vector space V is n-*dimensional* (dim $V = n$), if there exists linearly independent vectors, the *basis vectors*, $\{e_1, e_2 \ldots e_n\}$ which span V. A basis for a vector space is non-unique, yet every basis of V has the same number n of elements. A subspace W of V has the property dim $W \leq n$, and if dim $W = n$, then $W = V$. The dimension of the row (column) space of a matrix X is the *row (column) rank* of X. The row and column ranks are equal and are in fact the rank of X. The co-ordinates of a vector x in V with respect to a basis are the (unique) tuple of coefficients of a linear combination of the basis vectors that generate x. Thus with $x = \sum_i a_i e_i$, then $a_1, a_2 \ldots a_n$ are the co-ordinates.

A.15 Mappings and Linear Mappings

For A, B arbitrary sets, suppose that for each $a \in A$ there is assigned a single element $f(a)$ of B. The collection f of such is called a *function, or map* and is denoted $f : A \to B$. The *domain* of the mapping is A, the *codomain* is B. For subsets A_s, B_s, of A, B then $f(A_s) = \{f(a) : a \in A_s\}$ is the *image* of A_s, and $f^{-1}(B_s) = \{a \in A : f(a) \in B_s\}$ is the *preimage or fiber* of B_s. If $B_s = \{b\}$ is a singleton set we also write $f^{-1}(b)$ instead of $f^{-1}(\{b\})$. Also, $f(A)$ is the *image* or *range* of f. The notation $x \mapsto f(x)$ is used to denote the image $f(x)$ of an arbitrary element $x \in A$.

The composition of mappings $f : A \to B$ and $g : B \to C$, denoted $g \circ f$, is an associative operation. The identity map $\mathrm{id}_A : A \to A$ is the map defined by $a \mapsto a$ for all $a \in A$.

A mapping $f : A \to B$ is *one-to-one* or *injective* if different elements of A have distinct images, i.e. if $a_1 \neq a_2 \Rightarrow f(a_1) \neq f(a_2)$. The mapping is *onto* or *surjective* if every $b \in B$ is the image of at least one $a \in A$. A *bijective* mapping is one-to-one and onto (surjective and injective). If $f : A \to B$ and $g : B \to A$ are maps with $g \circ f = \mathrm{id}_A$, then f is injective and g is surjective.

For vector spaces V, W over $\mathbb{R}$ or $\mathbb{C}$ (denoted $\mathbb{K}$) a mapping $F : V \to W$ is a *linear mapping* if $F(v + w) = F(v) + F(w)$ for any $v, w \in V$, and as $F(kv) = kF(v)$ for any $k \in \mathbb{K}$ and any $v \in V$. Of course $F(0) = 0$. A linear mapping is called an *isomorphism* if it is bijective. The vector spaces V, W are *isomorphic* if there is an isomorphism of V onto W. A linear mapping

$F : V \to U$ is called *singular* if it is not an isomorphism.

For $F : V \to U$, a linear mapping, the *image* of F is the set $Im(F) = \{u \in U \mid F(v) = u \text{ for some } v \in V\}$. The *kernel* of F, is $Ker(F) = \{u \in V \mid F(u) = 0\}$. In fact for finite dimensional spaces $\dim V = \dim Ker(F) + \dim Im(F)$.

Linear operators or transformations are linear mappings $T : V \to V$, i.e. from V to itself.

The *dual vector space* V^* of a $\mathbb{K}$-vector space V is defined as the $\mathbb{K}$-vector space of all $\mathbb{K}$-linear maps $\lambda : V \to \mathbb{K}$. It is denoted by $V^* = \mathrm{Hom}(V, \mathbb{K})$.

A.16 Inner Products

Let V be a real vector space. An *inner product* on V is a bilinear map $\beta : V \times V \to \mathbb{R}$, also denoted by $\langle \cdot , \cdot \rangle$, such that $\beta(u,v) = \langle u, v \rangle$ satisfies the conditions

$$\langle u, v \rangle = \langle v, u \rangle, \quad \langle u, sv + tw \rangle = s\langle u, v \rangle + t\langle u, w \rangle$$
$$\langle u, u \rangle > 0 \quad \text{for} \quad u \in V - \{0\}$$

for all $u, v, w \in V$, $s, t \in \mathbb{R}$.

An inner product defines a norm on V, by $||u|| := \langle u, u \rangle^{\frac{1}{2}}$ for $u \in V$, which satisfies the usual axioms for a norm. An *isometry* on V is a linear map $T : V \to V$ such that $\langle Tu, Tv \rangle = \langle u, v \rangle$ for all $u, v \in V$. All isometries are isomorphisms though the converse is not true. Every inner product on $V = \mathbb{R}^n$ is uniquely determined by a positive definite symmetric matrix $Q \in \mathbb{R}^{n \times n}$ such that $\langle u, v \rangle = u'Qv$ holds for all $u, v \in \mathbb{R}^n$. If $Q = I_n$, $\langle u, v \rangle = u'v$ is called the *standard Euclidean inner product* on $\mathbb{R}^n$. The induced norm $||x|| = \langle x, x \rangle^{\frac{1}{2}}$ is the Euclidean norm on $\mathbb{R}^n$. A linear map $A : V \to V$ is called *selfadjoint* if $\langle Au, v \rangle = \langle u, Av \rangle$ holds for all $u, v \in V$. A matrix $A : \mathbb{R}^n \to \mathbb{R}^n$ taken as a linear map is selfadjoint for the inner product $\langle u, v \rangle = u'Qv$ if and only if the matrix $QA = A'Q$ is symmetric. Every selfadjoint linear map $A : \mathbb{R}^n \to \mathbb{R}^n$ has only real eigenvalues.

Let V be a complex vector space. A *Hermitian inner product* on V is a map $\beta : V \times V \to \mathbb{C}$, also denoted by $\langle \cdot , \cdot \rangle$, such that $\beta(u,v) = \langle u, v \rangle$ satisfies

$$\langle u, v \rangle = \overline{\langle v, u \rangle}, \quad \langle u, sv + tw \rangle = s\langle u, v \rangle + t\langle u, w \rangle.$$
$$\langle u, u \rangle > 0 \quad \text{for} \quad u \in V - \{0\}$$

holds for all $u, v, w \in V$, $s, t \in \mathbb{C}$, where the superbar denotes the complex conjugate.

The induced norm on V is $||u|| = \langle u, u \rangle^{\frac{1}{2}}, u \in V$. One has the same notions of isometries and selfadjoint maps as in the real case. Every Hermitian inner product on $V = \mathbb{C}^n$ is uniquely defined by a positive definite Hermitian matrix $Q = Q^* \in \mathbb{C}^{n \times n}$ such that $\langle u, v \rangle = \bar{u}'Qv$ holds for all $u, v \in \mathbb{C}^n$. The *standard Hermitian inner product* on $\mathbb{C}^n$ is $\langle u, v \rangle = u^*v$ for $u, v \in \mathbb{C}^n$. A complex matrix $A : \mathbb{C}^n \to \mathbb{C}^n$ is selfadjoint for the Hermitian inner product $\langle u, v \rangle = u^*Qv$ if and only if the matrix $QA = A^*Q$ is Hermitian. Every selfadjoint linear map $A : \mathbb{C}^n \to \mathbb{C}^n$ has only real eigenvalues.

References

Barnett, S. (1971) *Matrices in Control Theory*, London: Van Nostrand Reinhold Company.

Bellman, R.E. (1970) *Introduction to Matrices*, 2nd Edition. New York: McGraw Hill Book Company.

Appendix B

Dynamical Systems

This appendix summarizes the key results of both linear and nonlinear dynamical systems theory required as a background in this text. For more complete treatments, see Irwin (1980), Isidori (1985), Kailath (1978) and Sontag (1990).

B.1 Linear Dynamical Systems

State equations Continuous-time, linear, finite-dimensional, dynamical systems initialized at time t_0 are described by

$$\dot{x} := \frac{dx}{dt} = Ax + Bu, \quad x(t_0) = x_0, \quad y = Cx + Du$$

where $x \in \mathbb{R}^n$, $u \in \mathbb{R}^m$, $y \in \mathbb{R}^p$ and A, B, C, D are matrices of appropriate dimension, possibly time varying.

In the case when $B = 0$, then the solution of $\dot{x} = A(t)x$, with initial state x_0 is $x(t) = \Phi(t, t_0)x_0$ where $\Phi(t, t_0)$ is the transition matrix which satisfies $\dot{\Phi}(t, t_0) = A(t)\Phi(t, t_0)$, $\Phi(t_0, t_0) = I$ and has the property $\Phi(t_2, t_1)\Phi(t_1, t_0) = \Phi(t_2, t_0)$. For any B, then

$$x(t) = \Phi(t, t_0)x_0 + \int_{t_0}^{t} \Phi(t, \tau)B(\tau)u(\tau)d\tau.$$

In the time-invariant case where $A(t) = A$, then $\Phi(t, t_0) = e^{(t-t_0)A}$.

In discrete-time, for $k = k_0, k_0 + 1, \cdots$ with initial state $x_{k_0} \in \mathbb{R}^n$

$$x_{k+1} = Ax_k + Bu_k, \quad y_k = Cx_k + Du_k.$$

The discrete-time solution is,

$$x_k = \Phi_{k,k_0} x_{k_0} + \sum_{i=k_o}^{k} \Phi_{k,i} B_i u_i$$

where $\Phi_{k,k_0} = A_k A_{k-1} \ldots A_{k_0}$. In the time invariant case $\Phi_{k,k_0} = A^{k-k_0}$.

A continuous-time, time-invariant system (A, B, C) is called *stable* if the eigenvalues of A have all real part strictly less than 0. Likewise a discrete-time, time-invariant system (A, B, C) is called stable, if the eigenvalues of A are all located in the open unit disc $\{z \in \mathbb{C} \mid |z| < 1\}$. Equivalently, $\lim_{t \to +\infty} e^{tA} = 0$ and $\lim_{k \to \infty} A^k = 0$, respectively.

Transfer functions The unilateral *Laplace transform* of a function $f(t)$ is the complex valued function

$$F(s) = \int_{0+}^{\infty} f(t) e^{-st} dt.$$

A sufficient condition for $F(s)$ to exist as a meromorphic function is that $|f(t)| \leq C e^{at}$ is exponentially bounded for some constants $C, a > 0$. In the time-invariant case, *transfer functions* for the continuous-time system are given in terms of the *Laplace transform* in the complex variable s as

$$H(s) = C(sI - A)^{-1} B + D.$$

Thus when $x_0 = 0$, then $Y(s) = (C(sI - A)^{-1} B + D) U(s)$ gives the Laplace transform $Y(s)$ of $y(\cdot)$, expressed as a linear function of the Laplace transform $U(s)$ of $u(\cdot)$. When $s = iw$ (where $i = (-1)^{\frac{1}{2}}$), then $H(iw)$ is the *frequency response* of the systems at a frequency w. The transfer function s^{-1} is an integrator. The *Z-transform* of a sequence $(h_k | k \in \mathbb{N}_0)$ is the formal power series in z^{-1}

$$H(z) = \sum_{k=0}^{\infty} h_k z^{-k}.$$

In discrete-time, the *Z-transform* yields the transfer function $H(z) = C(zI - A)^{-1} B + D$ in terms of the Z-transform variable z. For $x_0 = 0$, then $Y(z) = (C(zI - A)^{-1} B + D) U(z)$ expresses the relation between the Z-transforms of the sequences $\{u_k\}$ and $\{y_k\}$. The transfer function z^{-1} is a unit delay. The *frequency response* of a periodically sampled continuous-time signal with sampling period T is $H(z)$ with $z = e^{iwT}$.

B.2 Linear Dynamical System Matrix Equations

Consider the matrix differential equation

$$\dot{X} := \frac{dX(t)}{dt} = A(t) X(t) + X(t) B(t) + C(t), \quad X(t_0) = X_0.$$

B.3. CONTROLLABILITY AND STABILIZABILITY

Its solution when A, B are constant is

$$X(t) = e^{(t-t_0)A} X_0 e^{(t-t_0)B} + \int_{t_0}^{t} e^{(t-\tau)A} C(\tau) e^{(t-\tau)B} d\tau.$$

When A, B, and C are constant, and the eigenvalues of A, B have negative real parts, then $X(t) \to X_\infty$ as $t \to \infty$ where

$$X_\infty = \int_0^\infty e^{tA} C e^{tB} dt$$

which satisfies the linear equation $AX_\infty + X_\infty B + C = 0$. A special case is the Lyapunov equation $AX + XA' + C = 0, C = C' \geq 0$.

B.3 Controllability and Stabilizability

In the time-invariant, continuous-time, case, the pair (A, B) with $A \in \mathbb{R}^{n \times n}$ and $B \in \mathbb{R}^{n \times m}$ is termed *completely controllable* (or more simply *controllable*) if one of the following equivalent conditions holds:

- There exists a control u taking $\dot{x} = Ax + Bu$ from arbitrary state x_0 to another arbitrary state x_1, in finite time.

- Rank $(B, AB, \cdots, A^{n-1}B) = n$

- $(\lambda I - A, B)$ full rank for all (complex) λ

- $W_c(T) = \int_0^T e^{tA} BB' e^{tA'} dt > 0$ for all $T > 0$

- $AX = XA$ and $XB = 0$ implies $X = 0$

- $w' e^{tA} B = 0$ for all t implies $w = 0$

- $w' A^i B = 0$ for all i implies $w = 0$

- There exists a K of appropriate dimension such that $A + BK$ has arbitrary eigenvalues.

- There exists no co-ordinate basis change such that

$$A = \begin{bmatrix} A_{11} & A_{12} \\ 0 & A_{22} \end{bmatrix}, \quad B = \begin{bmatrix} B_1 \\ 0 \end{bmatrix}.$$

The symmetric matrix $W_c(T) = \int_0^T e^{tA} BB' e^{tA'} dt > 0$ is the *controllability Gramian*, associated with $\dot{x} = Ax + Bu$. It can be found as the solution at time T of $\dot{W}_c(t) = AW_c(t) + W_c(t)A' + BB'$ initialized by $W_c(t_0) = 0$. If

A has only eigenvalues with negative real part, in short $\text{Re}\lambda(A) < 0$, then $W_c(\infty) = \lim_{t \to \infty} W_c(t)$ exists.

In the time-varying case, only the first definition of controllability applies. It is equivalent to requiring that the Gramian

$$W_c(t, T) = \int_t^{t+T} \Phi(t+T, \tau) B(\tau) B'(\tau) \Phi'(t+T, \tau) d\tau$$

be positive definite for all t and some T. The concept of *uniform complete controllability* requires that $W_c(t, T)$ be uniformly bounded above and below from zero for all t and some $T > 0$. This condition ensures that a bounded energy control can take an arbitrary state vector x to zero in an interval $[t, t+T]$ for arbitrary t. A uniformly controllable system has the property that a bounded $K(t)$ exists such that $\dot{x} = (A + BK)x$ has an arbitrary degree of (exponential) stability.

The discrete-time controllability conditions, Gramians, etc are analogous to the continuous-time definitions and results. In particular, the N-controllability Gramian of a discrete-time system is defined by

$$W_c^{(N)} := \sum_{k=0}^N A^k BB'(A')^k$$

for $N \in \mathbb{N}$. The pair (A, B) is controllable if and only if $W_c^{(N)}$ is positive definite for all $N \geq n - 1$. If A has all its eigenvalues in the open unit disc $\{z \in \mathbb{C} \mid |z| < 1\}$, in short $|\lambda(A)| < 1$, then for $N = \infty$

$$W_c := \sum_{k=0}^\infty A^k BB'(A')^k$$

exists and is positive definite if and only if (A, B) is controllable.

B.4 Observability and Detectability

The pair (A, C) has observability/detectability properties according to the controllability/stabilizability properties of the pair (A', C'); likewise for the time-varying and uniform observability cases. The observability Gramians are known as duals of the controllability Gramians, e.g. in the continuous-time case

$$W_o(T) = \int_0^T e^{tA'} C' C e^{tA} dt, \quad W_o = \int_0^\infty e^{tA'} C' C e^{tA} dt$$

and in the discrete-time case,

$$W_o^{(N)} = \sum_{k=0}^N (A')^k C' C A^k, \quad W_o := \sum_{k=0}^\infty (A')^k C' C A^k.$$

B.5 Minimality

The state space systems of (B.1) denoted by the triple (A, B, C) are termed *minimal realizations*, in the time-invariant case, when (A, B) is completely controllable and (A, C) is completely observable. The *McMillan degree* of the transfer functions $H(s) = C(sI - A)^{-1}B$ or $H(z) = C(zI - A)^{-1}B$ is the state space dimension of a minimal realization. *Kalman's realization theorem* asserts that any $p \times m$ rational matrix function $H(s)$ with $H(\infty) = 0$ (that is $h(s)$ is *strictly proper*) has a minimal realization (A, B, C) such that $G(s) = C(sI - A)^{-1}B$ holds. Moreover, given two minimal realizations denoted (A_1, B_1, C_1) and (A_2, B_2, C_2). then there always exists a unique nonsingular transformation matrix T such that $TA_1T^{-1} = A_2, TB_1 = B_2, C_1T^{-1} = C_2$. All minimal realizations of a transfer function have the same dimension.

B.6 Markov Parameters and Hankel Matrix

With $H(s)$ strictly proper (that is for $H(\infty) = 0$), then $H(s)$ has the Laurent expansion at ∞

$$H(s) = \frac{M_0}{s} + \frac{M_1}{s^2} + \frac{M_2}{s^3} + \cdots$$

where the $p \times m$ matrices M_i are termed *Markov* parameters. The *Hankel* matrices of $H(s)$ of size $Np \times Nm$ are then

$$\mathcal{H}_N = \begin{bmatrix} M_0 & M_1 & \cdots & M_{N-1} \\ M_1 & M_2 & & M_N \\ \vdots & & \ddots & \vdots \\ M_{N-1} & M_N & \cdots & M_{2N-1} \end{bmatrix}, \quad N \in \mathbb{N}.$$

If the triple (A, B, C) is a realization of $H(s)$, then $M_i = CA^iB$ for all i. Moreover, if (A, B, C) is minimal and of dimension n, then rank $\mathcal{H}_N = n$ for all $N \geq n$. A Hankel matrix $\mathcal{H}_N$ of the transfer function $H(s) = C(sI - A)^{-1}B$ has the factorization $\mathcal{H}_N = O_N \cdot R_N$ where $R_N = (B, \ldots, A^{N-1}B)$ and $O_N = (C', \ldots, (A')^{N-1}C')'$ are the controllability and observability matrices of length N. Thus for singular values, $\sigma_i(\mathcal{H}_N) = \lambda_i(W_c^{(N)}W_o^{(N)})^{\frac{1}{2}}$.

B.7 Balanced Realizations

For a stable system (A, B, C), a realization in which the Gramians are equal and diagonal as

$$W_c = W_o = \text{diag}(\sigma_1, \cdots, \sigma_n)$$

is termed a *diagonally balanced* realization. For a minimal realization (A, B, C), the *singular values* σ_i are all positive. For a non minimal realization of McMillan degree $m < n$, then $\sigma_{m+i} = 0$ for $i > 0$. Corresponding definitions and results apply for Gramians defined on finite intervals T. Also, when the controllability and observability Gramians are equal but not necessarily diagonal the realizations are termed *balanced*. Such realizations are unique only to within orthogonal basis transformations.

Balanced truncation is where a system (A, B, C) with $A \in \mathbb{R}^{n \times n}$, $B \in \mathbb{R}^{n \times m}$, $C \in \mathbb{R}^{p \times n}$ is approximated by an rth order system with $r < n$ and $\sigma_r > \sigma_{r+1}$, the last $(n-r)$ rows of (A, B) and last $(n-r)$ columns of $\begin{bmatrix} A \\ C \end{bmatrix}$ of a balanced realization are set to zero to form a reduced rth order realization $(A_r, B_r, C_r) \in \mathbb{R}^{r \times r} \times \mathbb{R}^{r \times m} \times \mathbb{R}^{p \times r}$. A theorem of Pernebo and Silverman states that if (A, B, C) is balanced and minimal, and $\sigma_r > \sigma_{r+1}$, then the reduced r-th order realization (A_r, B_r, C_r) is also balanced and minimal.

B.8 Vector Fields and Flows

Let M be a smooth manifold and let TM denote its tangent bundle; see Appendix C. A *smooth vector field* on M is a smooth section $X : M \to TM$ of the tangent bundle. Thus $X(a) \in T_a M$ is a tangent vector for each $a \in M$. An *integral curve* of X is a smooth map $\alpha : I \to M$, defined on an open interval $I \subset \mathbb{R}$, such that $\alpha(t)$ is a solution of the differential equation on M.

$$\dot{\alpha}(t) = X(\alpha(t)), \quad t \in I. \qquad (*)$$

Here the left-hand side of (*) is the velocity vector $\dot{\alpha}(t) = D\alpha|_t(1)$ of the curve α at time t. In local coordinates on M, (*) is just an ordinary differential equation defined on an open subset $U \subset \mathbb{R}^n$, $n = \dim M$. It follows from the fundamental existence and uniqueness result for solutions of ordinary differential equations, that integral curves of smooth vector fields on manifolds always exist and are unique. The following theorem summarizes the basic properties of integral curves of smooth vector fields.

Theorem *Let $X : M \to TM$ be a smooth vector field on a manifold M.*

(1) For each $a \in M$ there exists an open interval $I_a \subset \mathbb{R}$ with $0 \in I_a$ and a smooth map $\alpha_a : I_a \to M$ such that the following conditions hold

(i) $\dot{\alpha}_a(t) = X(\alpha_a(t))$, $\alpha_a(0) = a$, $t \in I_a$.

(ii) I_a *is maximal with respect to (i). That is, if $\beta : J \to M$ is another integral curve of X satisfying (i) then $J \subset I_a$.*

B.8. VECTOR FIELDS AND FLOWS

(2) *The set* $\mathcal{D} = \{(t,a) \in \mathbb{R} \times M \mid t \in I_a\}$ *is an open subset of* $\mathbb{R} \times M$ *and*

$$\phi : \mathcal{D} \to M, \quad \phi(t,a) = \alpha_a(t),$$

is a smooth map.

(3) $\phi : \mathcal{D} \to M$ *satisfies*

 (i) $\phi(0,a) = a$ *for all* $a \in M$.

 (ii) $\phi(s, \phi(t,a)) = \phi(s+t, a)$ *whenever both sides are defined.*

 (iii) *The map*

$$\phi_t : \mathcal{D}_t \to M, \quad \phi_t(a) = \phi(t,a)$$

 on the open set $\mathcal{D}_t = \{a \in M \mid (t,a) \in \mathcal{D}\}$ *is a diffeomorphism of* $\mathcal{D}_t$ *onto its image. The inverse is given by*

$$(\phi_t)^{-1} = \phi_{-t}.$$

The smooth map $\phi : \mathcal{D} \to M$ or the associated one-parameter family of diffeomorphisms $\phi_t : \mathcal{D}_t \to M$ is called the *flow* of X. Thus, if (ϕ_t) is the flow of X, then $t \mapsto \phi_t(a)$ is the unique integral curve of (*) which passes through $a \in M$ at $t = 0$.

It is not necessarily true that a flow (ϕ_t) is defined for all $t \in \mathbb{R}$ (finite escape time phenomena) or equivalently that the domain of a flow is $\mathcal{D} = \mathbb{R} \times M$. We say that the vector field $X : M \to TM$ is *complete* if the flow (ϕ_t) is defined for all $t \in \mathbb{R}$. If M is compact then all smooth vector fields $X : M \to TM$ are complete. If a vector field X on M is complete then $(\phi_t : t \in \mathbb{R})$ is a *one-parameter group* of diffeomorphism $M \to M$ and satisfying

$$\phi_{t+s} = \phi_t \circ \phi_s, \quad \phi_0 = \mathrm{id}_M, \quad \phi_{-t} = (\phi_t)^{-1}$$

for all $s, t \in \mathbb{R}$.

In the case where the integral curves of a vector field are only considered for nonnegative time, we say that $(\phi_t \mid t \geq 0)$ (if it exists) is a one-parameter *semigroup* of diffeomorphisms.

An *equilibrium point* of a vector field X is a point $a \in M$ such that $X(a) = 0$ holds. Equilibria points of a vector field correspond uniquely to *fixed points* of the flow, i.e. $\phi_t(a) = a$ for all $t \in \mathbb{R}$. The *linearization* of a vector field $X : M \to TM$ at an equilibrium point is the linear map

$$T_a X = \dot{X}(a) : T_a M \to T_a M.$$

Note that the tangent map $T_a X : T_a M \to T_0(TM)$ maps $T_a M$ into the tangent space $T_0(TM)$ of TM at $X(a) = 0$. But $T_0(TM)$ is canonically isomorphic to $T_a M \oplus T_a M$, so that the above map is actually defined by composing $T_a X$ with the projection onto the second summand. The associated linear differential equation on the tangent space $T_a M$

$$\dot{\xi} = \dot{X}(a) \cdot \xi, \quad \xi \in T_a M \tag{**}$$

is then referred to as the *linearization* of (*). If $M = \mathbb{R}^n$ this corresponds to the usual concept of a linearization of a system of differential equations on $\mathbb{R}^n$.

B.9 Stability Concepts

Let $X : M \to TM$ be a smooth vector field on a manifold and let $\phi : \mathcal{D} \to M$ be the flow of X. Let $a \in M$ be an equilibrium point of X, so that $X(a) = 0$.

The equilibrium point $a \in M$ is called *stable* if for any neighborhood $U \subset M$ of a there exists a neighborhood $V \subset M$ of a such that

$$x \in V \longrightarrow \phi_t(x) \in U \text{ for all } t \geq 0.$$

Thus solutions which start in a neighborhood of a stay near to a for all time $t \geq 0$. In particular, if $a \in M$ is a stable equilibrium of X then all solutions of $\dot{x} = X(x)$ which start in a sufficiently small neighborhood of a are required to exist for all $t \geq 0$.

The equilibrium point $a \in M$ of X is *asymptotically stable* if it is stable and if a convergence condition holds: There exists a neighborhood $U \subset M$ of a such that for all $x \in U$

$$\lim_{t \to +\infty} \phi_t(x) = a.$$

Global asymptotic stability holds if a is a stable equilibrium point and if $\lim_{t \to +\infty} \phi_t(x) = a$ holds for all $x \in M$. Smooth vector fields generating such flows can exist only on manifolds which are homeomorphic to $\mathbb{R}^n$.

Let M be a Riemannian manifold and $X : M \to TM$ be a smooth vector field. An equilibrium point $a \in M$ is called *exponentially stable* if it is stable and there exists a neighborhood $U \subset M$ of a and constants $C > 0$, $\alpha > 0$ such that

$$\text{dist}(\phi_t(x), a) \leq C e^{-\alpha t} \text{dist}(x, a)$$

for all $x \in U, t \geq 0$. Here $\text{dist}(x,y)$ denotes the geodesic distance of x and y on M. In such cases, the solutions converging to a are said to converge exponentially.

The *exponential rate of convergence* $\rho > 0$ refers to the maximal possible α in the above definition.

Lemma *Given a vector field $X : M \to TM$ on a Riemannian manifold M with equilibrium point $a \in M$. Suppose that the linearization $\dot{X}(a) : T_a M \to T_a M$ has only eigenvalues with real part less than $-\alpha$, $\alpha > 0$. Then a is (locally) exponentially stable with an exponential rate of convergence $\rho \geq \alpha$.*

Let $X : M \to TM$ be a smooth vector field on a manifold M and let (ϕ_t) denote the flow of X. The ω-*limit set* $L_\omega(x)$ of a point x of the vector field X is the set of points of the form $\lim_{n \to \infty} \phi_{t_n}(x)$ with the $t_n \to +\infty$. Similarly, the α-*limit set* $L_\alpha(x)$ is defined by letting $t_n \to -\infty$ instead of converging to $+\infty$. A subset $A \subset M$ is called *positively invariant* or *negatively invariant*, respectively, if for each $x \in A$ the associated integral curve satisfies $\phi_t(a) \in A$ for all $t \geq 0$ or $\phi_t(a) \in A$ for $t \leq 0$, respectively. Also, A is called *invariant*, if it is positively and negatively invariant, and A is called *locally invariant*, if for each $a \in A$ there exists an open interval $]a, b[\subset \mathbb{R}$ with $a < 0 < b$ such that for all $t \in]a, b[$ the integral curve $\phi_t(a)$ of X satisfies $\phi_t(a) \in A$.

B.10 Lyapunov Stability

We summarize the basic results from Lyapunov theory for ordinary differential equations on $\mathbb{R}^n$.

Let

$$\dot{x} = f(x), \quad x \in \mathbb{R}^n \qquad (*)$$

be a smooth vector field on $\mathbb{R}^n$. We assume that $f(0) = 0$, so that $x = 0$ is an equilibrium point of $(*)$. Let $D \subset \mathbb{R}^n$ be a compact neighborhood of 0 in $\mathbb{R}^n$.

A *Lyapunov function* of $(*)$ on D is a smooth function $V : D \to \mathbb{R}$ having the properties

(i) $V(0) = 0$, $V(x) > 0$ for all $x \neq 0$ in D.

(ii) For any solution $x(t)$ of $(*)$ with $x(0) \in D$

$$\dot{V}(x(t)) = \tfrac{d}{dt} V(x(t)) \leq 0. \qquad (**)$$

Also, $V : D \to \mathbb{R}$ is called a *strict Lyapunov function* if the strict inequality holds

$$\dot{V}(x(t)) = \tfrac{d}{dt} V(x(t)) < 0 \text{ for } x(0) \in D - \{a\}. \qquad (***)$$

Theorem (Stability) *If there exists a Lyapunov function $V : D \to \mathbb{R}$*

defined on some compact neighborhood of $0 \in \mathbb{R}^n$, then $x = 0$ is a stable equilibrium point.

Theorem (Asymptotic Stability) *If there exists a strict Lyapunov function $V : D \to \mathbb{R}$ defined on some compact neighborhood of $0 \in \mathbb{R}^n$, then $x = 0$ is an asymptotically stable equilibrium point.*

Theorem (Global Asymptotic Stability) *If there exists a proper map $V : \mathbb{R}^n \to \mathbb{R}$ which is a strict Lyapunov function with $D = \mathbb{R}^n$, then $x = 0$ is globally asymptotically stable.*

Here properness of $V : \mathbb{R}^n \to \mathbb{R}$ is equivalent to $V(x) \to \infty$ for $\|x\| \to \infty$, see also Appendix (C.1).

Theorem (Exponential Asymptotic Stability) *If in Theorem (Asymptotic Stability) one has $\alpha_1 \|x\|^2 \leq V(x) \leq \alpha_2 \|x\|^2$ and $-\alpha_3 \|x\|^2 \leq \dot{V}(x) \leq -\alpha_4 \|x\|^2$ for some positive $\alpha_i, i = 1, \cdots, 4$, then $x = 0$ is exponentially asymptotically stable.*

Now consider the case where $X : M \to TM$ is a smooth vector field on a manifold M.

A smooth function $V : M \to \mathbb{R}$ is called a *(weak) Lyapunov function* if V has compact sublevel sets i.e. $\{x \in M | V(x) \leq c\}$ is compact for all $c \in \mathbb{R}$, and

$$\dot{V}(x(t)) = \frac{d}{dt} V(x(t)) \leq 0$$

for any solution $x(t)$ of $\dot{x}(t) = X(x(t))$.

Recall that a subset $\Omega \subset M$ is called *invariant* if every solution $x(t)$ of $\dot{x} = X(x)$ with $x(0) \in \Omega$ satisfies $x(t) \in \Omega$ for all $t \geq 0$.

La Salle's principle of invariance then asserts

Theorem (Principle of Invariance) *Let $X : M \to TM$ be a smooth vector field on a Riemannian manifold and let $V : M \to \mathbb{R}$ be a smooth weak Lyapunov function for X. Then every solution $x(t) \in M$ of $\dot{x} = X(x)$ exists for all $t \geq 0$. Moreover the ω-limit set $L_\omega(x)$ of any point $x \in M$ is a compact, connected and invariant subset of $\{x \in M \mid \langle \mathrm{grad} V(x), X(x) \rangle = 0\}$.*

References

Hirsch, M.W. and S. Smale, (1974) *Differential Equations, Dynamical Systems, and Linear Algebra*, Academic Press, New York.

Irwin, M.C., (1980) *Smooth Dynamical Systems*, Academic Press, 1980, London.

Isidori, A., (1985) *Nonlinear Control Systems: An Introduction*, Springer-Verlag, Berlin.

Kailath, T., (1978) *Linear Systems*, Englewood Cliffs, N.J., Prentice Hall, Inc 1978.

Sontag, E.D., (1990) *Mathematical Control Theory*, Springer-Verlag, New York.

Appendix C

Global Analysis

In this appendix we summarize some of the basic facts from general topology, manifold theory and differential geometry. As further references we mention Munkres (1975) and Hirsch (1976).

In preparing this appendix we have also profited from the appendices on differential geometry in Isidori (1985) as well as from unpublished lectures notes for a course on calculus on manifolds by Gamst (1975).

C.1 Point Set Topology

A *topological space* is a set X together with a collection of subsets of X, called *open sets*, satisfying the axioms

(1) The empty set ϕ and X are open sets.

(2) The union of any family of open sets is an open set.

(3) The intersection of finitely many open sets is an open set.

A collection of subsets of X satisfying (1)-(3) is called a *topology* on X. A *neighbourhood* of a point $a \in X$ is a subset which contains an open subset U of X with $a \in U$. A *basis* for a topology on X is a collection of open sets, called *basic open sets*, satisfying

(i) X is the union of basic open sets.

(ii) The intersection of two basic open sets is a union of basic open sets.

A subset A of a topological space X is called *closed* if the complement $X - A$ is open. The collection of closed subsets satisfies axioms dual to (1)-

(3). In particular, the intersection of any number of closed subsets is closed and the union of finitely many closed subsets is a closed subset.

The *interior* of a subset $A \subset X$, denoted by $\mathring{A}$, is the largest (possibly empty) open subset of X which is contained in A. A is open if and only if $A = \mathring{A}$. The *closure* of A in X, denoted by $\bar{A}$, is the smallest closed subset of X which contains A. A is closed if and only if $A = \bar{A}$.

A subset $A \subset X$ is called *dense* in X if its closure coincides with X. A subset $A \subset X$ is called *generic* if it contains a countable intersection of open and dense subsets of X. Generic subsets of manifolds are dense.

If X is a topological space and $A \subset X$ a subset, then A can be made into a topological space by defining the open subsets of A as the intersections $A \cap U$ by open subsets U of X. This topology on A is called the *subspace topology*.

Let X and Y be topological spaces. The cartesian product $X \times Y$ can be endowed with a topology consisting of the subsets $U \times V$ where U and V are open subsets of X and Y respectively. This topology is called the *product topology*.

A space X is called *Hausdorff*, if any two distinct points of X have disjoint neighborhoods. We say the points can be separated by open neighborhoods.

A topological space X is called *connected* if there do not exist disjoint open subsets U, V with $U \cup V = X$. Also, X is connected if and only if the empty set and X are the only open and closed subsets of X. Every topological space X is the union of disjoint connected subsets $X_i, i \in I$, such that every connected subset C of X intersects only one X_i. The subsets $X_i, i \in I$, are called the *connected components* of X.

Let A be a subset of X. A collection $\mathcal{C}$ of subsets of X is said to be a *covering* of A if A is contained in the union of the elements of $\mathcal{C}$. Also, $A \subset X$ is called *compact* if every covering of A by open subsets of X contains a finite subcollection covering A. Every closed subset of a compact space is compact. Every compact subset of a Hausdorff space is closed. A topological space X is said to be *locally compact* if and only if every point $a \in X$ has a compact neighborhood.

The standard Euclidean n-space is $\mathbb{R}^n$. A basis for a topology on the set of real numbers $\mathbb{R}$ is given by the collection of *open intervals* $]a,b[= \{x \in \mathbb{R} \mid a < x < b\}$, for $a < b$ arbitrary. We use the notations

$$[a,b] := \{x \in \mathbb{R} \mid a \leq x \leq b\}$$
$$[a,b[:= \{x \in \mathbb{R} \mid a \leq x < b\}$$
$$]a,b] := \{x \in \mathbb{R} \mid a < x \leq b\}$$

C.1. POINT SET TOPOLOGY

$$]a,b[\;:=\; \{x \in \mathbb{R} \mid a < x < b\}$$

Each of these sets is connected. For $a, b \in \mathbb{R}$, $[a, b]$ is compact.

A basis for a topology on $\mathbb{R}^n$ is given by the collection of open n-balls

$$B_r(a) = \{x \in \mathbb{R}^n \mid \sum_{i=1}^{n}(x_i - a_i)^2 < r\}$$

for $a \in \mathbb{R}^n$ and $r > 0$ arbitrary. The same topology is defined by taking the product topology via $\mathbb{R}^n = \mathbb{R} \times \cdots \times \mathbb{R}$ (n-fold product). A subset $A \subset \mathbb{R}^n$ is compact if and only if it is closed and bounded.

A map $f : X \to Y$ between topological spaces is called *continuous* if the pre-image $f^{-1}(Y) = \{p \in X \mid f(p) \in Y\}$ of any open subset of Y is an open subset of X. Also, f is called *open (closed)* if the image of any open (closed) subset of X is open (closed) in Y. The map $f : X \to Y$ is called *proper* if the pre-image of every compact subset of Y is a compact subset of X. The image $f(K)$ of a compact subset $K \subset X$ under a continuous map $f : X \to Y$ is compact. Let X and Y be locally compact Hausdorff spaces. Then every proper continuous map $f : X \to Y$ is closed. The image $f(A)$ of a connected subset $A \subset X$ of a continuous map $f : X \to Y$ is connected.

Let $f : X \to \mathbb{R}$ be a continuous function on a compact space X. The Weierstrass theorem asserts that f possesses a minimum and a maximum. A generalization of this result is: Let $f : X \to \mathbb{R}$ be a proper continuous function on a topological space X. If f is lower bounded, i.e. if there exists $a \in \mathbb{R}$ such that $f(X) \subset (a, \infty[$, then there exists a minimum $x_{\min} \in X$ for f with

$$f(x_{\min}) = \inf\{f(x) \mid x \in X\}$$

A map $f : X \to Y$ is called a *homeomorphism* if f is bijective and f and $f^{-1} : Y \to X$ are *continuous*. An *imbedding* is a continuous, injective map $f : X \to Y$ such that f maps X homeomorphically onto $f(X)$, where $f(X)$ is endowed with the subspace topology of Y. If $f : X \to Y$ is an imbedding, then we also write $f : X \hookrightarrow Y$. If X is compact and Y a Hausdorff space, then any bijective continuous map $f : X \to Y$ is a homeomorphism. If X is any topological space and Y a locally compact Hausdorff space, then any continuous, injective and proper map $f : X \to Y$ is an imbedding and $f(X)$ is closed in Y.

Let X and Y be topological spaces and let $p : X \to Y$ be a continuous surjective map. Then p is said to be a *quotient map* provided any subset $U \subset Y$ is open if and only if $p^{-1}(U)$ is open in X. Let X be a topological space and Y be a *set*. Then, given a surjective map $p : X \to Y$, there exists a unique topology on Y, called the *quotient topology*, such that p is a quotient map. It is the finest topology on Y which makes p a continuous map.

Let $\sim$ be a relation on X. Then $\sim$ is called an *equivalence relation*, provided it satisfies the three conditions

(i) $x \sim x$ for all $x \in X$

(ii) $x \sim y$ if and only if $y \sim x$ for $x, y \in X$.

(iii) For $x, y, z \in X$ then $x \sim y$ and $y \sim z$ implies $x \sim z$.

Equivalence classes are defined by $[x] := \{y \in X \mid x \sim y\}$. The *quotient space* is defined as the set $X/\sim := \{[x] \mid x \in X\}$ of all equivalence classes of $\sim$. It is a topological space, endowed with the quotient topology for the map $p : X \to X/\sim$, $p(x) = [x], x \in X$.

The *graph* Γ of an equivalence relation $\sim$ on X is the subset of $X \times X$ defined by $\Gamma := \{(x, y) \in X \times X \mid x \sim y\}$. The *closed graph theorem* states that the quotient space $X/\sim$ is Hausdorff if and only if the graph Γ is a closed subset of $X \times X$.

C.2 Advanced Calculus

Let U be an open subset of $\mathbb{R}^n$ and $f : U \to \mathbb{R}$ a function. $f : U \to \mathbb{R}$ is called a *smooth*, or C^∞, function if f is continuous and the mixed partial derivatives of any order exist and are continuous. A function $f : U \to \mathbb{R}$ is called *real analytic* if it is C^∞ and for each $x_0 \in U$ the Taylor series expansion of f at x_0 converges on a neighborhood of x_0 to $f(x)$. A smooth map $f : U \to \mathbb{R}^m$ is an m-tuple $(f_1, \cdots, f_m)$ of smooth functions $f_i : U \to \mathbb{R}$. Let $x_0 \in U$ and $f : U \to \mathbb{R}^m$ be a smooth map. The *Fréchet derivative* of f at x_0 is defined as the linear map $Df(x_0) : \mathbb{R}^n \to \mathbb{R}^m$ which has the following $\varepsilon - \delta$ characterization:

For any $\varepsilon > 0$ there exists $\delta > 0$ such that for $||x - x_0|| < \delta$

$$||f(x) - f(x_0) - Df(x_0)(x - x_0)|| < \varepsilon ||x - x_0||$$

holds. The derivative $Df(x_0)$ is uniquely determined by this condition. We also use the notation

$$Df|_{x_0}(\xi) = Df(x_0) \cdot \xi$$

to denote the Fréchet derivative $Df(x_0) : \mathbb{R}^n \to \mathbb{R}^m$, $\xi \mapsto Df(x_0)(\xi)$.

If $Df(x_0)$ is expressed with respect to the standard basis vectors of $\mathbb{R}^n$ and $\mathbb{R}^m$ then the associated $m \times n$-matrix of partial derivatives is the *Jacobi*

C.2. ADVANCED CALCULUS

matrix

$$J_f(x_0) = \begin{bmatrix} \frac{\partial f_1}{\partial x_1}(x_0) & \cdots & \frac{\partial f_1}{\partial x_n}(x_0) \\ \vdots & & \\ \frac{\partial f_m}{\partial x_1}(x_0) & \cdots & \frac{\partial f_m}{\partial x_n}(x_0) \end{bmatrix}.$$

The *second derivative* of $f : U \to \mathbb{R}^m$ at $x_0 \in U$ is a bilinear map $D^2 f(x_0) : \mathbb{R}^n \times \mathbb{R}^n \to \mathbb{R}^m$. It has the $\varepsilon - \delta$ characterization:

For all $\varepsilon > 0$ there exists $\delta > 0$ such that for $||x - x_0|| < \delta$

$$\begin{aligned} ||f(x) - f(x_0) - Df(x_0)(x - x_0) &- \frac{1}{2!} D^2 f(x_0)(x - x_0, x - x_0)|| \\ &< \varepsilon ||x - x_0||^2 \end{aligned}$$

holds. Similarly, for any integer $i \in \mathbb{N}$, the i-th derivative $D^i f(x_0) : \prod_{j=1}^i \mathbb{R}^n \to \mathbb{R}^m$ is defined as a multilinear map on the i-th fold products $\mathbb{R}^n \times \cdots \times \mathbb{R}^n$. The *Taylor series* is the formal power series

$$\sum_{i=0}^\infty \frac{D^i f(x_0)}{i!}(x - x_0, \cdots, x - x_0).$$

The matrix representation of the second derivative $D^2 f(x_0)$ of a function $f : U \to \mathbb{R}$ with respect to the standard basis vectors on $\mathbb{R}^n$ is the *Hessian*

$$H_f(x_0) = \begin{bmatrix} \frac{\partial^2 f}{\partial x_1^2}(x_0) & \cdots & \frac{\partial^2 f}{\partial x_1 \partial x_n}(x_0) \\ \vdots & & \\ \frac{\partial^2 f}{\partial x_n \partial x_1}(x_0) & \cdots & \frac{\partial^2 f}{\partial x_n^2}(x_0) \end{bmatrix}.$$

It is symmetric if f is C^∞.

The *chain rule* for the derivatives of smooth maps $f : \mathbb{R}^m \to \mathbb{R}^n$, $g : \mathbb{R}^n \to \mathbb{R}^k$ asserts that

$$D(g \circ f)(x_0) = Dg(f(x_0)) \circ Df(x_0)$$

and thus for the Jacobi matrices

$$J_{g \circ f}(x_0) = J_g(f(x_0)) \cdot J_f(x_0).$$

A *diffeomorphism* between open subsets U and V of $\mathbb{R}^n$ is a smooth bijective map $f : U \to V$ such that $f^{-1} : V \to U$ is also smooth. A map $f : U \to V$ is called a local diffeomorphism if for any $x_0 \in U$ there exists open neighborhoods $U(x_0) \subset U$ and $V(f(x_0)) \subset V$ of x_0 and $f(x_0)$, respectively, such that f maps $U(x_0)$ diffeomorphically onto $V(f(x_0))$.

Inverse Function Theorem Let $f : U \to V$ be a smooth map between open subsets of $\mathbb{R}^n$. Suppose the Jacobi matrix $J_f(x_0)$ is invertible at a point

$x_0 \in U$. Then f maps a neighborhood of x_0 in U diffeomorphically onto a neighborhood of $f(x_0)$ in V.

A *critical point* of a smooth map $f : U \to \mathbb{R}^n$ is a point $x_0 \in U$ where rank $J_f(x_0) < \min(m, n)$. Also, $f(x_0)$ is called a *critical value*. *Regular values* $y \in \mathbb{R}^n$ are those such that $f^{-1}(y)$ contains no critical point. If $f : U \to \mathbb{R}$ is a function, critical points $x_0 \in U$ are those where $Df(x_0) = 0$. A local *minimum (local maximum)* of f is a point $x_0 \in U$ such that $f(x) \geq f(x_0)$ ($f(x) \leq f(x_0)$) for all points $x \in U_0$ in a neighborhood $U_0 \subset U$ of x_0. Moreover, $x_0 \in U$ is called a global minimum (*global maximum*) if $f(x) \geq f(x_0)$ ($f(x) \leq f(x_0)$) holds for all $x \in U$. Local minima or maxima are critical points. All other critical points $x_0 \in U$ are called *saddle points*.

A critical point $x_0 \in U$ of $f : U \to \mathbb{R}$ is called a *nondegenerate critical point* if the Hessian $H_f(x_0)$ is invertible. A function $f : U \to \mathbb{R}$ is called a *Morse function* if all its critical points $x_0 \in U$ are nondegenerate. Nondegenerate critical points are always *isolated*.

Let $x_0 \in U$ be a critical point of the function f. If the Hessian $H_f(x_0)$ is positive definite, i.e. $H_f(x_0) > 0$, then x_0 is a local minimum. If $H_f(x_0) < 0$, then x_0 is a local maximum. The Morse lemma explains the behaviour of a function around a saddle point.

Morse Lemma *Let $f : U \to \mathbb{R}$ be a smooth function and let $x_0 \in U$ be a nondegenerate critical point of f. Let k be the number of positive eigenvalues of $H_f(x_0)$. Then there exists a local diffeomorphism ϕ of a neighborhood $U(x_0)$ of x_0 onto a neighborhood $V \subset \mathbb{R}^n$ of 0 such that*

$$(f \circ \phi^{-1})(x_1, \cdots, x_n) = \sum_{j=1}^{k} x_j^2 - \sum_{j=k+1}^{n} x_j^2.$$

In constrained optimization, conditions for a point to be a local minimum subject to constraints on the variables are often derived using *Lagrange multipliers*.

Let $f : \mathbb{R}^m \to \mathbb{R}$ be a smooth function. Let $g : \mathbb{R}^m \to \mathbb{R}^n$, $g = (g_1, \cdots, g_n)$, $n < m$, be a smooth map such that constraints are defined by $g(x) = 0$. Assume that 0 is a regular value for g, so that $\mathrm{rk} Dg(x) = n$ for all $x \in \mathbb{R}^m$ satisfying $g(x) = 0$.

First-Order Necessary Condition *Let $x_0 \in \mathbb{R}^m$, $g(x_0) = 0$, be a local minimum of the restriction of f to the constraint set $\{x \in \mathbb{R}^m \mid g(x) = 0\}$. Assume that 0 is a regular value of g. Then there exists real numbers $\lambda_1, \cdots, \lambda_n$ such that x_0 is a critical point of $f(x) + \sum_{j=1}^{n} \lambda_j g_j(x)$, i.e.*

$$D(f + \sum_{j=1}^{n} \lambda_j g_j)(x_0) = 0.$$

The parameters $\lambda_1, \cdots, \lambda_n$ appearing in the above necessary condition are called the *Lagrange multipliers*.

Second-Order Sufficient Condition Let $f: \mathbb{R}^n \to \mathbb{R}$ be a smooth function and let 0 be a regular value of $g: \mathbb{R}^n \to \mathbb{R}^m$. Also let $x_0 \in \mathbb{R}^m, g(x_0) = 0$, satisfy
$$D(f + \lambda_j g_j)(x_0) = 0$$
for real numbers $\lambda_1, \cdots, \lambda_n$. Suppose that the Hessian
$$H_f(x_0) + \sum_{j=1}^{n} \lambda_j H_{g_j}(x_0)$$
of the function $f + \sum \lambda_j g_j$ at x_0 is positive definite on the linear subspace $\{\xi \in \mathbb{R}^n \mid Dg(x_0) \cdot \xi = 0\}$ of $\mathbb{R}^n$. Then x_0 is a local minimum for the restriction of f on $\{x \in \mathbb{R}^n \mid g(x) = 0\}$.

C.3 Smooth Manifolds

Let M be a topological space. A *chart* of M is a triple (U, ϕ, n) consisting of an open subset $U \subset M$ and a homeomorphism $\phi : U \to \mathbb{R}^n$ of U onto an open subset $\phi(U)$ of $\mathbb{R}^n$. The integer n is called the dimension of the chart. We use the notation $\dim_x M = n$ to denote the dimension of M at any point x in U. $\phi = (\phi_1, \cdots, \phi_n)$ is said to be a *local coordinate system* on U.

Two charts (U, ϕ, n) and (V, ψ, m) of M are called C^∞ *compatible charts* if either $U \cap V = \emptyset$, the empty set, or if

(i) $\phi(U \cap V)$ and $\psi(U \cap V)$ are open subsets of $\mathbb{R}^n$ and $\mathbb{R}^m$.

(ii) The *transition functions*
$$\psi \circ \phi^{-1} : \phi(U \cap V) \to \psi(U \cap V)$$
and
$$\phi \circ \psi^{-1} : \psi(U \cap V) \to \phi(U \cap V)$$
are C^∞ maps, see Figure C.C.3.1. It follows that, if $U \cap V \neq \emptyset$ then the dimensions of the charts m and n are the same.

A C^∞ *atlas* of M is a set $\mathcal{A} = \{(U_i, \phi_i, n_i) \mid i \in I\}$ of C^∞ compatible charts such that $M = \cup_{i \in I} U_i$. An atlas $\mathcal{A}$ is called *maximal* if every chart (U, ϕ, n) of M which is C^∞ compatible with every chart from $\mathcal{A}$ also belongs to $\mathcal{A}$.

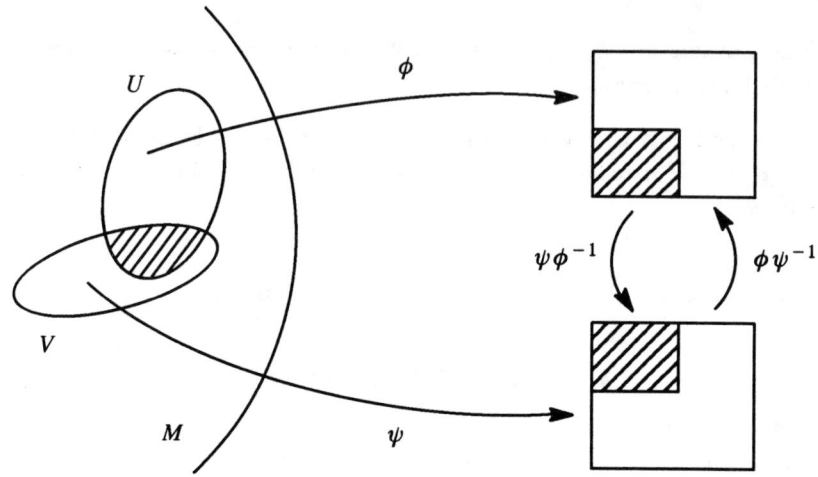

Figure C.3.1: Coordinate charts

Definition A *smooth* or C^∞, *manifold* M is a topological Hausdorff space, having a countable basis, that is equipped with a maximal C^∞ atlas. If all coordinate charts of M have the same dimension n, then M is called an *n-dimensional manifold*.

If M is connected, then all charts of M must have the same dimension and therefore the dimension of M coincides with the dimension of any coordinate chart.

Let $U \subset M$ be an open subset of a smooth manifold. Then U is a smooth manifold. The charts of M which are defined on open subsets of U form a C^∞ atlas of U. Also, U is called an *open submanifold* of M.

Let M and N be smooth manifolds. Any two coordinate charts (U, ϕ, m) and (V, ψ, n) of M and N define a coordinate chart $(U \times V, \phi \times \psi, m+n)$ of $M \times N$. In this way, $M \times N$ becomes a smooth manifold. It is called the *product manifold* of M and N. If M and N have dimensions m and n, respectively, then $\dim(M \times N) = m + n$.

Let M and N be smooth manifolds. A continuous map $f : M \to N$ is called *smooth* if for any charts (U, ϕ, m) of M and (V, ψ, n) of N with $f(U) \subset V$ the map $\psi \circ f \circ \phi^{-1} : \phi(U) \to \mathbb{R}^n$ is C^∞. The map $\psi f \phi^{-1}$ is called the *expression of f in local coordinates*.

A continuous map $f : M \to \mathbb{R}^n$ is smooth if and only if each component function $f_i : M \to \mathbb{R}, f = (f_1, \cdots, f_n)$, is smooth. The identity map $\mathrm{id}_M : M \to M, x \mapsto x$, is smooth. If $f : M \to N$ and $g : N \to P$ are smooth maps, then also the composition $g \circ f : M \to P$. If $f, g : M \to \mathbb{R}^n$ are smooth maps, so is every linear combination. If $f, g : M \to \mathbb{R}$ are smooth, so is the sum $f + g : M \to \mathbb{R}$ and the product $f \cdot g : M \to \mathbb{R}$.

A map $f : M \to N$ between smooth manifolds M and N is called a *diffeomorphism* if $f : M \to N$ is a smooth bijective map such that $f^{-1} : N \to M$ is also smooth. M and N are called *diffeomorphic* if there exists a diffeomorphism $f : M \to N$.

Let M be a smooth manifold and (U, ϕ, n) a coordinate chart of M. Then $\phi : U \to \mathbb{R}^n$ defines a diffeomorphism $\phi : U \to \phi(U)$ of U onto $\phi(U) \subset \mathbb{R}^n$. An example of a smooth homeomorphism which is not a diffeomorphism is $f : \mathbb{R} \to \mathbb{R}$, $f(x) = x^3$. Here $f^{-1} : \mathbb{R} \to \mathbb{R}, f^{-1}(x) = x^{\frac{1}{3}}$ is not smooth at the origin.

A smooth map $f : M \to N$ is said to be a *local diffeomorphism at* $x \in M$ if there exists open neighborhoods $U \subset M, V \subset N$ of x and $f(x)$ respectively with $f(U) \subset V$, such that the restriction $f : U \to V$ is a diffeomorphism.

C.4 Spheres, Projective Spaces and Grassmannians

The *n-sphere* in $\mathbb{R}^{n+1}$ is defined by $S^n = \{x \in \mathbb{R}^{n+1} \mid x_1^2 + \cdots + x_{n+1}^2 = 1\}$. It is a smooth, compact manifold of dimension n. Coordinate charts of S^n can be defined by *stereographic projection* from the north pole $x_N = (0, \cdots, 0, 1)$ and south pole $x_S = (0, \cdots, 0, -1)$.

Let $U_N := S^n - \{x_N\}$ and define $\psi_N : U_N \to \mathbb{R}^n$ by

$$\psi_N(x_1, \cdots, x_{n+1}) = \frac{x - x_{n+1} x_N}{1 - x_{n+1}} \quad \text{for } x = (x_1, \cdots, x_{n+1}) \in U_N.$$

Then $\psi_N^{-1} : \mathbb{R}^n \to U_N$ is given by

$$\psi_N^{-1}(y) = \frac{(\|y\|^2 - 1) x_N + 2(y, 0)}{\|y\|^2 + 1}, \quad y \in \mathbb{R}^n.$$

where $\|y\|^2 = y_1^2 + \cdots + y_n^2$. Similarly for $U_S = S^n - \{x_S\}$, the stereographic projection from the south pole is defined by $\psi_S : U_S \to \mathbb{R}^n$,

$$\psi_S(x) = \frac{x + x_{n+1} x_S}{1 + x_{n+1}} \quad \text{for } x = (x_1, \cdots, x_{n+1}) \in U_S$$

and

$$\psi_S^{-1}(y) = \frac{1 - \|y\|^2}{1 + \|y\|^2} x_S + \frac{2}{1 + \|y\|^2}(y, 0), \quad y \in \mathbb{R}^n.$$

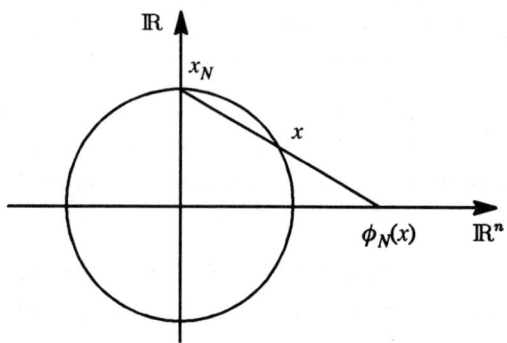

Figure C.4.1: Stereographic projection

By inspection, the transition functions $\psi_S \circ \psi_N^{-1}, \psi_N \circ \psi_S^{-1} : \mathbb{R}^n - \{0\} \to \mathbb{R}^n - \{0\}$ are seen to be C^∞ maps and thus $\mathcal{A} = \{(U_N, \psi_N, n), (U_S, \psi_S, n)\}$ is a C^∞ atlas for S^n.

The *real projective n-space* $\mathbb{R}\mathbb{P}^n$ is defined as the set of all lines through the origin in $\mathbb{R}^{n+1}$. Thus $\mathbb{R}\mathbb{P}^n$ can be identified with the set of equivalence classes $[x]$ of nonzero vectors $x \in \mathbb{R}^{n+1} - \{0\}$ for the equivalence relation on $\mathbb{R}^{n+1} - \{0\}$ defined by

$$x \sim y \iff x = \lambda y \text{ for } \lambda \in \mathbb{R} - \{0\}.$$

Similarly, the *complex projective n-space* $\mathbb{C}\mathbb{P}^n$ is the defined as the set of *complex* one-dimensional subspaces of $\mathbb{C}^{n+1}$. Thus $\mathbb{C}\mathbb{P}^n$ can be identified with the set of equivalence classes $[x]$ of nonzero complex vectors $x \in \mathbb{C}^{n+1} - \{0\}$ for the equivalence relation

$$x \sim y \iff x = \lambda y \text{ for } \lambda \in \mathbb{C} - \{0\}$$

defined on $\mathbb{C}^{n+1} - \{0\}$. It is easily verified that the graph of this equivalence relation is closed and thus $\mathbb{R}\mathbb{P}^n$ and $\mathbb{C}\mathbb{P}^n$ are Hausdorff spaces.

If $x = (x_0, \cdots, x_n) \in \mathbb{R}^{n+1} - \{0\}$, the one-dimensional subspace of $\mathbb{R}^{n+1}$ generated by x is denoted by

$$[x] = [x_0 : \cdots : x_n] \in \mathbb{R}\mathbb{P}^n.$$

The coordinates $x_0, \cdots, x_n$ are called the *homogeneous coordinates* of $[x]$. They are uniquely defined up to a common scalar multiple by a nonzero real number.

Coordinate charts for $\mathbb{R}\mathbb{P}^n$ are defined by

$$U_i := \{[x] \in \mathbb{R}\mathbb{P}^n \mid x_i \neq 0\}$$

and $\phi_i : U_i \to \mathbb{R}^n$ with

$$\phi_i([x]) = \left(\frac{x_0}{x_i}, \cdots, \frac{x_{i-1}}{x_i}, \frac{x_{i+1}}{x_i}, \cdots, \frac{x_n}{x_i}\right) \in \mathbb{R}^n$$

for $i = 0, \cdots, n$. It is easily seen that these charts are C^∞ compatible and thus define a C^∞ atlas of $\mathbb{R}\mathrm{P}^n$. Similarly for $\mathbb{C}\mathrm{P}^n$.

Since every one dimensional real linear subspace of $\mathbb{R}^{n+1}$ is generated by a unit vector $x \in S^n$ one has a surjective continuous map

$$\pi : S^n \to \mathbb{R}\mathrm{P}^n, \quad \pi(x) = [x],$$

with $\pi(x) = \pi(y) \iff y = \pm x$. One says that $\mathbb{R}\mathrm{P}^n$ is defined by identifying antipodal points on S^n. We obtain that $\mathbb{R}\mathrm{P}^n$ is a smooth compact and connected manifold of dimension n. Similarly $\mathbb{C}\mathrm{P}^n$ is seen as a smooth compact and connected manifold of dimension $2n$.

Both $\mathbb{R}\mathrm{P}^1$ and $\mathbb{C}\mathrm{P}^1$ are homeomorphic to the circle S^1 and the Riemann sphere S^2 respectively. Also, $\mathbb{R}\mathrm{P}^2$ is a non orientable surface which cannot be visualized as a smooth surface imbedded in $\mathbb{R}^3$.

For $1 \leq k \leq n$ the *real Grassmann manifold* $\mathrm{Grass}_{\mathbb{R}}(k, n)$ is defined as the set of k-dimensional real linear subspaces of $\mathbb{R}^n$. Similarly the *complex Grassmann manifold* $\mathrm{Grass}_{\mathbb{C}}(k, n)$ is defined as the set of k-dimensional complex linear subspaces of $\mathbb{C}^n$, where k means the complex dimension of the subspace.

$\mathrm{Grass}_{\mathbb{R}}(k, n)$ is a smooth compact connected manifold of dimension $k(n-k)$. Similarly $\mathrm{Grass}_{\mathbb{C}}(k, n)$ is a smooth compact connected manifold of dimension $2k(n-k)$. For $k = 1$, $\mathrm{Grass}_{\mathbb{R}}(1, n)$ and $\mathrm{Grass}_{\mathbb{C}}(1, n)$ coincide with the real and complex projective spaces $\mathbb{R}\mathrm{P}^{n-1}$ and $\mathbb{C}\mathrm{P}^{n-1}$ respectively.

The points of the Grassmann manifold $\mathrm{Grass}_{\mathbb{R}}(k, n)$ can be identified with equivalence classes of an equivalence relation on rectangular matrices. Let

$$ST(k, n) = \{X \in \mathbb{R}^{n \times k} \mid \mathrm{rk}\, X = k\}$$

denote the *noncompact Stiefel manifold*, i.e. the set of full rank $n \times k$ matrices. Any such matrices are called k-*frames* Let $[X]$ denote the equivalence class of the equivalence relation defined on $ST(k, n)$ by

$$X \sim Y \iff X = YT \text{ for } T \in GL(k, \mathbb{R}).$$

Thus $X, Y \in ST(k, n)$ are equivalent if and only if their respective column-vectors generate the same vector space. This defines a surjection

$$\pi : ST(k, n) \to \mathrm{Grass}_{\mathbb{R}}(k, n), \pi(X) = [X],$$

and $\text{Grass}_{\mathbb{R}}(k,n)$ can be identified with the quotient space $ST(k,n)/\sim$. It is easily seen that the graph of the equivalence relation is a closed subset of $ST(k,n) \times ST(k,n)$. Thus $\text{Grass}_{\mathbb{R}}(k,n)$, endowed with the quotient topology, is a Hausdorff space. Similarly $\text{Grass}_{\mathbb{C}}(k,n)$ is seen to be Hausdorff.

To define coordinate charts on $\text{Grass}_{\mathbb{R}}(k,n)$, let $I = \{i_1 < \cdots < i_k\} \subset \{1, \cdots, n\}$ denote a subset having k elements. Let $I' = \{1, \cdots, n\} - I$ denote the complement of I. For $X = (x_1', \ldots, x_n')' \in \mathbb{R}^{n \times k}$, let $X_I = (x_{i_1}', \ldots, x_{i_k}')' \in \mathbb{R}^{k \times k}$ denote the submatrix of X of those row vectors x_i of X with $i \in I$. Similarly let $X_{I'}$ be the $(n-k) \times k$ submatrix formed by those row vectors x_i with $i \notin I$.

Let $U_I := \{[X] \in \text{Grass}_{\mathbb{R}}(k,n) \mid \det(X_I) \neq 0\}$ with $\psi_I : U_I \to \mathbb{R}^{k(n-k)}$ defined by

$$\phi_I([X]) = X_{I'} \cdot (X_I)^{-1} \in \mathbb{R}^{(n-k) \times k}$$

It is easily seen that the system of $\binom{n}{k}$ coordinate charts

$$\{(U_I, \psi_I, (n-k)k) \mid I \subset \{1, \cdots, n\} \text{ and } |I| = k\}$$

is a C^∞ atlas for $\text{Grass}_{\mathbb{R}}(k,n)$. Similarly coordinate charts and a C^∞ atlas for $\text{Grass}_{\mathbb{C}}(k,n)$ are defined.

As the Grassmannian is the image of the continuous surjective map $\pi : St(k,n) \to \text{Grass}_{\mathbb{R}}(k,n)$ of the compact Stiefel manifold $St(k,n) = \{X \in \mathbb{R}^{n \times k} \mid X'X = I_k\}$ it is compact. Two orthogonal k-frames X and $Y \in St(k,n)$ are mapped to the same point in the Grassmann manifold if and only if they are orthogonally equivalent. That is, $\pi(X) = \pi(Y)$ if and only if $X = YT$ for $T \in O(k)$ orthogonal. Similarly for $\text{Grass}_{\mathbb{C}}(k,n)$.

C.5 Tangent Spaces and Tangent Maps

Let M be a smooth manifold. There are various equivalent definitions of the tangent space of M at a point $x \in M$. The following definition is particularly appealing from a geometric viewpoint.

A *curve* through $x \in M$ is a smooth map $\alpha : I \to M$ defined on an open interval $I \subset \mathbb{R}$ with $0 \in I$, $\alpha(0) = x$. Let (U, ϕ, n) be a chart of M with $x \in U$. Then $\phi \circ \alpha : I \to \phi(U) \subset \mathbb{R}^n$ is a smooth map and its derivative at 0 is called the *velocity vector* $(\phi\alpha)'(0)$ of α at x with respect to the chart (U, ϕ, n). If (V, ψ, n) is another chart with $x \in V$, then the velocity vectors of α with respect to the two charts are related by

$$(\psi\alpha)'(0) = D(\psi\phi^{-1})|_{\phi(x)} \cdot (\phi\alpha)'(0) \tag{$*$}$$

C.5. TANGENT SPACES AND TANGENT MAPS

Two curves α and β through x are said to be *equivalent*, denoted by $\alpha \sim_x \beta$, if $(\phi\alpha)'(0) = (\phi\beta)'(0)$ holds for some (and hence any) coordinate chart (U, ϕ, n) around x. Using (*), this is an equivalence relation on the set of curves through x. The equivalence class of a curve α is denoted by $[\alpha]_x$.

Definition A *tangent vector* of M at x is an equivalence class $\xi = [\alpha]_x$ of a curve α through $x \in M$. The *tangent space* $T_x M$ of M at x is the set of all such tangent vectors.

Working in a coordinate chart (U, ϕ, n) one has a bijection

$$\tau_x^\phi : T_x M \to \mathbb{R}^n, \quad \tau_x^\phi([\alpha]_x) = (\phi\alpha)'(0)$$

Using the vector space structure on $\mathbb{R}^n$ we can define a vector space structure on the tangent space $T_x M$ such that τ_x^ϕ becomes an isomorphism of vector spaces. The vector space structure on $T_x M$ is easily verified to be independent of the choice of the coordinate chart (U, ϕ, n) around x. Thus if M is a manifold of dimension n, the tangent spaces $T_x M$ are n-dimensional vector spaces.

Let $f : M \to N$ be a smooth map between manifolds. If $\alpha : I \to M$ is a curve through $x \in M$ then $f \circ \alpha : I \to N$ is a curve through $f(x) \in N$. If two curves $\alpha, \beta : I \to M$ through x are equivalent, so are the curves $f \circ \alpha, f \circ \beta : I \to N$ through $f(x)$. Thus we can consider the map $[\alpha]_x \mapsto [f \circ \alpha]_{f(x)}$ on the tangent spaces $T_x M$ and $T_{f(x)} N$.

Definition Let $f : M \to N$ be a smooth map, $x \in M$. The *tangent map* $T_x f : T_x M \to T_{f(x)} N$ is the linear map on tangent spaces defined by $T_x f([\alpha]_x) = [f \circ \alpha]_{f(x)}$ for all tangent vectors $[\alpha]_x \in T_x M$.

Expressed in local coordinates (U, ϕ, m) and (V, ψ, n) of M and N at x and $f(x)$, the tangent map coincides with the usual derivative of f (expressed in local coordinates). Thus

$$\tau_{f(x)}^\psi \circ T_x f \circ (\tau_x^\phi)^{-1} = D(\psi f \phi^{-1})|_{\phi(x)}.$$

One has a chain rule for the tangent map. Let $\mathrm{id}_M : M \to M$ be the identity map and let $f : M \to N$ and $g : N \to P$ be smooth. Then

$$T_x(\mathrm{id}_M) = \mathrm{id}_{T_x M}, \quad T_x(g \circ f) = T_{f(x)} g \circ T_x f.$$

The inverse function theorem for open subsets of $\mathbb{R}^n$ implies

Inverse Function Theorem Let $f : M \to M$ be smooth and $x \in M$. Then $T_x f : T_x M \to T_{f(x)} N$ is an isomorphism if and only if f is a local diffeomorphism at x.

Let M and N be smooth manifolds, and $x \in M, y \in N$. The tangent space of the product manifold $M \times N$ at a point (x,y) is then canonically isomorphic to

$$T_{(x,y)}(M \times N) = (T_x M) \times (T_y N).$$

The tangent space of $\mathbb{R}^n$ (or of any open subset $U \subset \mathbb{R}^n$) at a point x is canonically isomorphic to $\mathbb{R}^n$, i.e.

$$T_x(\mathbb{R}^n) = \mathbb{R}^n.$$

If $f : M \to \mathbb{R}^n$ is a smooth map then we use the above isomorphism of $T_{f(x)}(\mathbb{R}^n)$ with $\mathbb{R}^n$ to identify the tangent map $T_x f : T_x M \to T_{f(x)}(\mathbb{R}^n)$ with the linear map

$$Df|_x : T_x M \to \mathbb{R}^n.$$

Let $f : M \to \mathbb{R}$ be a smooth function on M. A *critical point* of f is a point $x \in M$ where the derivative $Df|_x : T_x M \to \mathbb{R}$ is the zero map, i.e. $Df|_x(\xi) = 0$ for all tangent vectors $\xi \in T_x M$.

A point $x_0 \in M$ is a *local minimum* (or *local maximum*) if there exists an open neighborhood U of x in M such that $f(x) \geq f(x_0)$ (or $f(x) \leq f(x_0)$) for all $x \in U$. Local minima maxima are critical points of f.

The *Hessian* $H_f(x_0)$ of $f : M \to \mathbb{R}$ at a critical point $x_0 \in M$ is the symmetric bilinear form

$$H_f(x_0) : T_{x_0} M \times T_{x_0} M \to \mathbb{R}$$
$$H_f(x_0)(\xi, \xi) = (f\alpha)''(0)$$

for any tangent vector $\xi = [\alpha]_{x_0} \in T_{x_0} M$. Thus for tangent vectors $\xi, \eta, \in T_{x_0} M$

$$\begin{aligned} H_f(x_0)(\xi, \eta) &= \frac{1}{2}(H_f(x_0)(\xi + \eta, \xi + \eta) \\ &\quad - H_f(x_0)(\xi, \xi) - H_f(x_0)(\eta, \eta)). \end{aligned}$$

Note that the Hessian of f is only well defined at a critical point of f.

A critical point $x_0 \in M$ where the Hessian $H_f(x_0)$ is positive definite on $T_x M$ is a local minimum. Similarly, a critical point x_0 with $H_f(x_0)(\xi, \xi) < 0$ for all $\xi \in T_{x_0} M, \xi \neq 0$, is a local maximum.

A critical point $x_0 \in M$ is called *nondegenerate* if the Hessian $H_f(x_0)$ is a nondegenerate bilinear form; i.e. if

$$H_f(x_0)(\xi, \eta) = 0, \text{ for all } \eta \in T_{x_0} M \Longrightarrow \xi = 0$$

Nondegenerate critical points are always isolated and there are at most countably many of them. A smooth function $f : M \to \mathbb{R}$ with only nondegenerate critical points is called a *Morse function*.

The *Morse index* $\text{ind}_f(x_0)$ of a critical point $x_0 \in M$ is defined as the *signature* of the Hessian:

$$\text{ind}_f(x_0) = \text{sig } H_f(x_0).$$

Thus $\text{ind}_f(x_0) = \dim T_{x_0} M$ if and only if $H_f(x_0) > 0$, i.e. if and only if x_0 is a local minimum of f.

If x_0 is a nondegenerate critical point then the index

$$\text{ind}_f(x_0) = \dim M - 2n^-,$$

where n^- is the dimension of the negative eigenspace of $H_f(x_0)$. Often the number n^- of negative eigenvalues of the Hessian is known as the Morse index of f at x_0.

C.6 Submanifolds

Let M be a smooth manifold of dimension m. A subset $A \subset M$ is called a smooth *submanifold* of M if every point $a \in A$ possesses a coordinate chart (V, ψ, m) around $a \in V$ such that $\psi(A \cap V) = \psi(V) \cap (\mathbb{R}^k \times \{0\})$ for some $0 \leq k \leq n$.

In particular, submanifolds $A \subset M$ are also manifolds. If $A \subset M$ is a submanifold, then at each point $x \in A$, the tangent space $T_x A \subset T_x M$ is considered as a sub vector space of $T_x M$. Any open subset $U \subset M$ of a manifold is a submanifold of the same dimension as M. There are subsets A of a manifold M, such that A is a manifold but not a submanifold of M. The main difference is topological: In order for a subset $A \subset M$ to qualify as a submanifold, the topology on A must be the subspace topology.

Let $f : M \to N$ be a smooth map of manifolds.

(i) f is an *immersion* if the tangent map $T_x f : T_x M \to T_{f(x)} N$ is injective at each point $x \in M$.

(ii) f is a *submersion* if the tangent map $T_x f : T_x M \to T_{f(x)} N$ is surjective at each point $x \in M$.

(iii) f is a *subimmersion* if the tangent map $T_x f : T_x M \to T_{f(x)} N$ has constant rank on each connected component of M.

(iv) f is an *imbedding* if f is an injective immersion such that f induces a homeomorphism of M onto $f(M)$, where $f(M)$ is endowed with the subspace topology of N.

Submersions are open mappings. If $f : M \to N$ is an imbedding, then $f(M)$ is a submanifold of N and $f : M \to f(M)$ is a diffeomorphism. Conversely, if M is a submanifold of N than the inclusion map inc $: M \to N$, $\text{inc}(x) = x$ for $x \in M$, is an imbedding.

The image $f(M)$ of an injective immersion $f : M \to N$ is called an *immersed submanifold* of N, although it is in general not a submanifold of N. A simple example of an immersed submanifold which is not a submanifold are the so-called "irrational lines on a torus". Here $N = S^1 \times S^1$ is the two-dimensional torus and $f : \mathbb{R} \to S^1 \times S^1$, $f(t) = (e^{it}, e^{2\pi it})$, is an injective immersion with dense image $f(\mathbb{R})$ in $S^1 \times S^1$. Thus $f(\mathbb{R})$ cannot be a submanifold of $S^1 \times S^1$.

A simple condition for an injective immersion to be an imbedding is stated as a proposition.

Proposition C.6.1 *Let $f : M \to N$ be an injective immersion. If M is compact then f is an imbedding.*

An important way to produce submanifolds is as fibers of smooth maps.

Fiber Theorem *Let $f : M \to N$ be a subimmersion. For $a \in M$ let $A = f^{-1}(f(a))$ be the fiber of f through a. Then A is a smooth submanifold of M with*

$$T_x A = \text{Ker} T_x f \quad \text{for all } x \in A.$$

In particular, for $x \in A$

$$\dim_x A = \dim_x M - \text{rk} T_x f$$

Important special cases are

(a) Let $q(x) = x'Ax = \sum_{i,j=1}^n a_{ij} x_i x_j$ be a quadratic form on $\mathbb{R}^n$ with $A = (a_{ij})$ a symmetric invertible $n \times n$ matrix. Then $M = \{x \in \mathbb{R}^n \mid q(x) = 1\}$ is a submanifold of $\mathbb{R}^n$ with tangent spaces

$$T_x M = \{\xi \in \mathbb{R}^n \mid x'A\xi = 0\}, \quad x \in M.$$

(b) Let $f : \mathbb{R}^n \to \mathbb{R}$ be smooth with

$$\nabla f(x) = \left(\frac{\partial f}{\partial x_1}(x), \cdots, \frac{\partial f}{\partial x_n}(x) \right)' \neq 0$$

for all $x \in \mathbb{R}^n$ satisfying $f(x) = 0$. Then $M = \{x \in \mathbb{R}^n \mid f(x) = 0\}$ is a smooth submanifold with tangent spaces

$$T_x M = \{\xi \in \mathbb{R}^n \mid \nabla f(x)'\xi = 0\}, \quad x \in M.$$

If $M \neq \emptyset$, then M has dimension $n - 1$.

(c) Let $f : \mathbb{R}^n \to \mathbb{R}^k$ be smooth and $M = \{x \in \mathbb{R}^n \mid f(x) = 0\}$. If the Jacobi matrix

$$J_f(x) = \left(\frac{\partial f_i}{\partial x_j}(x) \right) \in \mathbb{R}^{k \times n}$$

has constant rank $r \leq \min(n, k)$ for all $x \in M$ then M is a smooth submanifold of $\mathbb{R}^n$. The tangent space of M at $x \in M$ is given by

$$T_x M = \{\xi \in \mathbb{R}^n \mid J_f(x)\xi = 0\}.$$

If $M \neq \phi$, it has dimension $n - r$.

(d) Let $f : M \to N$ be a submersion. Then $A = f^{-1}(f(a))$ is a submanifold of M with $\dim A = \dim M - \dim N$, for every $a \in M$.

C.7 Groups, Lie Groups and Lie Algebras

A *group* is a set G together with an operation $\circ$ which assigns to every ordered pair (a, b) of elements of G a unique element $a \circ b$ such that the following axioms are satisfied

(i) $(a \circ b) \circ c = a \circ (b \circ c)$

(ii) there exists some element $e \in G$ with $e \circ a = a$ for all $a \in G$.

(iii) for any $a \in G$ there exists $a^{-1} \in G$ with $a^{-1} \circ a = e$

A *subgroup* $H \subset G$ is a subset of G which is also a group with respect to the group operation $\circ$ of G.

A *Lie group* is a group G, which is also a smooth manifold, such that the map

$$G \times G \to G, \quad (x, y) \mapsto xy^{-1}$$

is smooth. A *Lie subgroup* $H \subset G$ is a subgroup of G which is also a submanifold. By *Ado's theorem*, closed subgroups of a Lie group are Lie subgroups. Compact Lie groups are Lie groups which are compact as topological spaces. A compact subgroup $H \subset G$ is called *maximal compact* if there is no compact subgroup H' with $G \subsetneq H' \subset G$. If H is a compact Lie group, then the only maximal compact Lie subgroup is G itself.

Similar definitions exist for complex Lie groups, or algebraic groups, etc. Thus a *complex Lie group* G is a group G which is also a complex manifold, such that the map $G \times G \to G$, $(x,y) \mapsto xy^{-1}$, is holomorphic

The tangent space $\mathcal{G} = T_e G$ of a Lie group G at the identity element $e \in G$ has in a natural way the structure of a Lie algebra. A *Lie algebra* is a real vector space which is endowed with a product structure, $[\ ,\ ] : V \times V \to V$, called the *Lie bracket*, satisfying

(i) $[x,y] = -[y,x]$ for all $x,y \in V$.

(ii) For $\alpha, \beta \in \mathbb{R}$ and $x,y,z \in V$

$$[\alpha x + \beta y, z] = \alpha [x,z] + \beta [y,z].$$

(iii) The *Jacobi identity* is satisfied

$$[x,[y,z]] + [z,[x,y]] + [y,[z,x]] = 0$$

for all $x,y,z \in V$.

Examples 1. The *general linear group* $GL(n, \mathbb{K}) = \{T \in \mathbb{K}^{n \times n} \mid \det(T) \neq 0\}$ for $\mathbb{K} = \mathbb{R}$, or $\mathbb{C}$ is a Lie group of dimension n^2, or $2n^2$, under matrix multiplication. A maximal compact subgroup of $GL(n, \mathbb{K})$ is the *orthogonal group*

$$O(n) = O(n, \mathbb{R}) = \{T \in \mathbb{R}^{n \times n} \mid TT' = I_n\}$$

for $\mathbb{K} = \mathbb{R}$ and the *unitary group*

$$U(n) = U(n, \mathbb{C}) = \{T \in \mathbb{C}^{n \times n} \mid TT^* = I_n\}$$

for $\mathbb{K} = \mathbb{C}$. Here T^* denotes the Hermitian transpose of T. Also $GL(n, \mathbb{R})$ and $O(n, \mathbb{R})$ each have two connected components,

$$\begin{aligned} GL^+(n, \mathbb{R}) &= \{T \in GL(n, \mathbb{R}) \mid \det(T) > 0\}, \\ GL^-(n, \mathbb{R}) &= \{T \in GL(n, \mathbb{R}) \mid \det(T) < 0\}. \\ O^+(n, \mathbb{R}) &= \{T \in O(n, \mathbb{R}) \mid \det(T) = 1\}, \\ O^-(n, \mathbb{R}) &= \{T \in O(n, \mathbb{R}) \mid \det(T) = -1\}. \end{aligned}$$

while $GL(n, \mathbb{C})$ and $U(n, \mathbb{C})$ are connected.

The Lie algebra of $GL(n, \mathbb{K})$ is

$$gl(n, \mathbb{K}) = \{X \in \mathbb{K}^{n \times n}\}$$

endowed with the Lie bracket product structure

$$[X,Y] = XY - YX$$

C.7. GROUPS, LIE GROUPS AND LIE ALGEBRAS

for $n \times n$ matrices X, Y. The Lie algebra of $O(n, \mathbb{R})$ is the vector space of skew-symmetric matrices

$$\text{skew}(n, \mathbb{R}) = \{X \in \mathbb{R}^{n \times n} \mid X' = -X\}$$

and the Lie algebra of $U(n, \mathbb{C})$ is the vector space of skew-Hermitian matrices

$$\text{skew}(n, \mathbb{C}) = \{X \in \mathbb{C}^{n \times n} \mid X^* = -X\}.$$

In both cases the Lie algebras are endowed with the Lie bracket product $[X, Y] = XY - YX$.

2. The *special linear group* $SL(n, \mathbb{K}) = \{T \in \mathbb{K}^{n \times n} \mid \det(T) = 1\}$. A maximal compact subgroup of $SL(n, \mathbb{K})$ is the *special orthogonal group*

$$SO(n) = SO(n, \mathbb{R}) = \{T \in \mathbb{R}^{n \times n} \mid TT' = I_n, \det(T) = 1\}.$$

for $\mathbb{K} = \mathbb{R}$ and the *special unitary* group

$$SU(n) = SU(n, \mathbb{C}) = \{T \in \mathbb{C}^{n \times n} \mid TT^* = I_n, \det(T) = 1\}$$

for $\mathbb{K} = \mathbb{C}$. The groups $SO(n, \mathbb{R})$ and $SU(n, \mathbb{C})$ are connected. Also, $SO(2, \mathbb{R})$ is homeomorphic to the circle S^1 and $SO(3, \mathbb{R})$ is homeomorphic to real projective 3-space $\mathbb{RP}^3$, and $SU(2, \mathbb{C})$ is homeomorphic to the 3-sphere S^3.

The Lie algebra of $SL(n, \mathbb{K})$ is

$$\text{sl}(n, \mathbb{K}) = \{X \in \mathbb{K}^{n \times n} \mid \text{tr}(X) = 0\},$$

endowed with the matrix Lie bracket product structure $[X, Y] = XY - YX$. The Lie algebra of $SO(n, \mathbb{R})$ coincides with the Lie algebra $\text{skew}(n, \mathbb{R})$ of $O(n, \mathbb{R})$. The Lie algebra of $SU(n, \mathbb{C})$ is the subspace of the Lie algebra of $U(n, \mathbb{C})$ defined by $\{X \in \mathbb{C}^{n \times n} \mid X^* = -X, \text{tr}(X) = 0\}$.

3. The *real symplectic group* $Sp(n, \mathbb{R}) = \{T \in GL(2n, \mathbb{R}) \mid T'JT = J\}$ where

$$J = \begin{bmatrix} 0 & -I_n \\ I_n & 0 \end{bmatrix}.$$

A maximal compact subgroup of $Sp(n, \mathbb{R})$ is

$$Sp(n) = Sp(n, \mathbb{R}) \cap O(2n, \mathbb{R}).$$

The Lie algebra of $Sp(n, \mathbb{R})$ is the vector space of $2n \times 2n$ *Hamiltonian* matrices

$$\text{Ham}(n) = \{X \in \mathbb{R}^{2n \times 2n} \mid JX + X'J = 0\}.$$

The *Lie bracket* operation is $[X,Y] = XY - YX$.

4. The *Lie group* $O(p,q) = \{T \in GL(n, \mathbb{R}) \mid T'I_{pq}T = I_{pq}\}$ where

$$I_{pq} = \begin{bmatrix} I_p & 0 \\ 0 & -I_q \end{bmatrix}$$

and $p, q \geq 0$, $p+q = n$. A maximal compact subgroup of $O(p,q)$ is the direct product $O(p) \times O(q)$ of orthogonal groups $O(p)$ and $O(q)$. The Lie algebra of $O(p,q)$ is the vector space of *signature skew-symmetric matrices*

$$O(p,q) = \{X \in \mathbb{R}^{n \times n} \mid (XI_{pq})' = -XI_{pq}\}.$$

∎

The *exponential map* is a local diffeomorphism $\exp : \mathcal{G} \to G$, which maps the Lie algebra of G into the connected component of $e \in G$. In the above cases of matrix Lie groups the exponential map is given by the classical matrix exponential

$$\exp(X) = e^X = \sum_{i=0}^{\infty} \frac{1}{i!} X^i.$$

Thus if X is a skew-symmetric, skew-Hermitian, or Hamiltonian matrix, then e^X is an orthogonal, unitary, or symplectic matrix.

For $A \in \mathbb{R}^{n \times n}$ let $\mathrm{ad}_A : \mathbb{R}^{n \times n} \to \mathbb{R}^{n \times n}$ be defined by $\mathrm{ad}_A(X) = AX - XA$. The map ad_A is called the *adjoint transformation*. Let $\mathrm{ad}_A^i(B) = \mathrm{ad}_A(\mathrm{ad}_A^{i-1}B)$, $\mathrm{ad}_A^0 B = B$ for $i \in \mathbb{N}$. Then the *Baker-Campbell-Hausdorff* formula holds

$$\begin{aligned} e^{tA}Be^{-tA} &= B + t[A,B] + \frac{t^2}{2!}[A,[A,B]] + \cdots \\ &= \sum_{k=0}^{\infty} \frac{t^k}{k!} \mathrm{ad}_A^k B \end{aligned}$$

for $t \in \mathbb{R}$ and $A, B \in \mathbb{R}^{n \times n}$ arbitrary.

C.8 Homogeneous Spaces

A *Lie group action* of a Lie group G on a smooth manifold M is a smooth map

$$\delta : G \times M \to M, \quad (g, x) \mapsto g \cdot x$$

satisfying for all $g, h \in G$ and $x \in M$

$$g \cdot (h \cdot x) = (gh) \cdot x, \quad e \cdot x = x.$$

C.8. HOMOGENEOUS SPACES

A group action $\delta : G \times M \to M$ is called *transitive* if there exists an element $x \in M$ such that every $y \in M$ satisfies $y = g \cdot x$ for some $g \in G$. A space M is called *homogeneous* if there exists a transitive G-action on M.

The *orbit* of $x \in M$ is defined by

$$\mathcal{O}(x) = \{g \cdot x \mid g \in G\}.$$

Thus the homogeneous spaces are the orbits of a group action.

Any Lie group action $\delta : G \times M \to M$ induces an equivalence relation $\sim$ on M defined by

$$x \sim y \iff \text{There exists } g \in G \text{ with } y = g \cdot x$$

for $x, y \in M$. Thus the equivalence classes of $\sim$ are the orbits of $\delta : G \times M \to M$. The quotient space of this equivalence relation is called the *orbit space* and is denoted by M/G or by $M/\sim$. Thus

$$M/G = \{\mathcal{O}(x) \mid x \in M\}.$$

M/G is endowed with the quotient topology, i.e. with the finest topology on M/G such that the quotient map

$$\pi : M \to M/G, \quad \pi(x) = \mathcal{O}(x)$$

is continuous. Also, M/G is Hausdorff if and only if the *graph* of the G-action, defined by

$$\Gamma = \{(x, g \cdot x) \in M \times M \mid x \in M, g \in G\}$$

is a closed subset of $M \times M$. Moreover, M/G is a smooth manifold if and only if Γ is a closed submanifold of $M \times M$.

Given a Lie group action $\delta : G \times M \to M$ and a point $x \in M$, the *stabilizer subgroup* of x is defined by

$$\text{Stab}(x) = \{g \in G \mid g \cdot x = x\}.$$

$\text{Stab}(x)$ is a closed Lie subgroup of G.

For any Lie subgroup $H \subset G$ the orbit space of the H-action $\alpha : H \times G \to G$, $(h, g) \mapsto gh^{-1}$, is the set of coset classes

$$G/H = \{g \cdot H \mid g \in G\}.$$

G/H is an homogeneous space for the G-action $\alpha : G \times G/H \to G/H$, $(f, g \cdot H) \to fg \cdot H$. If H is a closed Lie subgroup of G then G/H is a smooth manifold. In particular, $G/\mathrm{Stab}(x)$, $x \in M$, is a smooth manifold for any Lie group action $\delta : G \times M \to M$.

A group action $\delta : G \times M \to M$ of a complex Lie group on a complex manifold M is called *holomorphic*, if $\delta : G \times M \to M$ is a holomorphic map. If $\delta : G \times M \to M$ is a holomorphic group action then the homogeneous spaces $G/\mathrm{Stab}(x)$, $x \in M$, are complex manifolds.

Let G be a complex algebraic group acting algebraically on a complex variety M. The *closed orbit lemma* states that every orbit is then a smooth locally closed subset of M whose boundary is a union of orbits of strictly lower dimension. This is wrong for arbitrary holomorphic Lie group actions. However, it is true for real algebraic and even semialgebraic group actions (see below).

For a point $x \in M$ consider the smooth map

$$\delta_x : G \to M, \quad g \mapsto g \cdot x.$$

Thus the image

$$\delta_x(G) = \mathcal{O}(x)$$

coincides with the G-orbit of x and δ_x induces a smooth injective immersion

$$\bar{\delta}_x : G/\mathrm{Stab}(x) \to M$$

of $G/\mathrm{Stab}(x)$ onto $\mathcal{O}(x)$. Thus $\mathcal{O}(x)$ is always an immersed submanifold of M. It is an *imbedded submanifold* in any of the following two cases

(i) G is compact.

(ii) $G \subset GL(n, \mathbb{R})$ is a *semialgebraic* Lie subgroup and $\delta : G \times \mathbb{R}^N \to \mathbb{R}^N$ is a *smooth semialgebraic action*. This means that the graph $\{(x, g \cdot x) \in \mathbb{R}^N \times \mathbb{R}^N \mid g \in G, x \in \mathbb{R}^N\}$ is a semialgebraic subset of $\mathbb{R}^N \times \mathbb{R}^N$. This condition is, for example, satisfied for real algebraic subgroups $G \subset GL(n, \mathbb{R})$ and $\delta : G \times \mathbb{R}^N \to \mathbb{R}^N$ a real algebraic action. Examples of such actions are the similarity action

$$(S, A) \mapsto SAS^{-1}$$

of $GL(n, \mathbb{R})$ on $\mathbb{R}^{n \times n}$ and more generally, the similarity action

$$(S, (A, B, C)) \mapsto (SAS^{-1}, SB, CS^{-1})$$

of $GL(n, \mathbb{R})$ on matrix triples

$$(A, B, C) \in \mathbb{R}^{n \times n} \times \mathbb{R}^{n \times m} \times \mathbb{R}^{p \times n}.$$

In these two cases (i) or (ii) the map $\bar{\delta}_x$ is a diffeomorphism of $G/\mathrm{Stab}(x)$ onto $\mathcal{O}(x)$. For a proof of the following proposition we refer to Gibson (1979).

Proposition *Let $G \subset GL(n, \mathbb{R})$ be a semialgebraic Lie subgroup and let $\delta: G \times \mathbb{R}^N \to \mathbb{R}^N$ be a smooth semialgebraic action. Then each orbit $\mathcal{O}(x), x \in \mathbb{R}^N$, is a smooth submanifold of $\mathbb{R}^N$ and*

$$\bar{\delta}_x: G/\text{Stab}(x) \to \mathcal{O}(x)$$

is a diffeomorphism.

C.9 Tangent Bundle

Let M be a smooth manifold. A vector bundle E on M is a family of real vector spaces E_x, $x \in M$, such that E_x varies continuously with $x \in M$. In more precise terms, a *smooth vector bundle* of rank k on M is a smooth manifold E together with a smooth surjective map $\pi: E \to M$ satisfying

(i) For any $x \in M$ the fiber $E_x = \pi^{-1}(x)$ is a real vector space of dimension k.

(ii) *Local Triviality*: Every point $x \in M$ possesses an open neighborhood $U \subset M$ such that there exists a diffeomorphism

$$\theta: \pi^{-1}(U) \to U \times \mathbb{R}^k$$

which maps E_y linearly isomorphically onto $\{y\} \times \mathbb{R}^k$ for all $y \in U$.

The smooth map $\pi: E \to M$ is called the *bundle projection*. As a set, a vector bundle $E = \bigcup_{x \in M} E_x$ is a disjoint union of vector spaces E_x, $x \in M$.

A *section* of a vector bundle E is a smooth map $s: M \to E$ such that $\pi \circ s = \text{id}_M$. Thus a section is just a smooth right inverse of $\pi: E \to M$. A smooth vector bundle E of rank k on M is called *trivial* if there exists a diffeomorphism

$$\theta: E \to M \times \mathbb{R}^k$$

which maps E_x linearly isomorphically onto $\{x\} \times \mathbb{R}^k$ for all $x \in M$. A rank k vector bundle E is trivial if and only if there exist k (smooth) sections $s_1, \cdots, s_k: M \to E$ such that $s_1(x), \cdots, s_k(x) \in E_x$ are linearly independent for all $x \in M$.

The *tangent bundle* TM of a manifold M is defined as

$$TM = \bigcup_{x \in M} T_x M.$$

Thus TM is the disjoint union of all tangent spaces $T_x M$ of M. It is a smooth vector bundle on M with bundle projection

$$\pi: TM \to M, \pi(\xi) = x \text{ for } \xi \in T_x M.$$

Also, TM is a smooth manifold of dimension $\dim TM = 2 \dim M$.

Let $f : M \to N$ be a smooth map between manifolds. The *tangent map* of f is the map
$$Tf : TM \to TN, \quad \xi \mapsto T_{\pi(\xi)}f(\xi).$$
Thus $Tf : TM \to TN$ maps a fiber $T_x M = \pi_M^{-1}(x)$ into the fiber $T_{f(x)}N = \pi_N^{-1}(f(x))$ via the tangent map $T_x f : T_x M \to T_{f(x)}N$. If $f : M \to N$ is smooth, so is the tangent map $Tf : TM \to TN$.

One has a chain rule for the tangent map
$$T(g \circ f) = Tg \circ Tf, \quad T(\mathrm{id}_M) = \mathrm{id}_{TM}$$
for arbitrary smooth maps $f : M \to N, g : N \to P$.

A smooth *vector field* on M is a (smooth) section $X : M \to TM$ of the tangent bundle. The tangent bundle TS^1 of the circle is trivial, as is the tangent bundle of any Lie group. The tangent bundle $T(\mathbb{R}^n) = \mathbb{R}^n \times \mathbb{R}^n$ is trivial. The tangent bundle of the 2-sphere S^2 is not trivial.

The *cotangent bundle* T^*M of M is defined as
$$T^*M = \bigcup_{x \in M} T_x^*M$$
where $T_x^*M = \mathrm{Hom}(T_x M, \mathbb{R})$ denotes the dual (cotangent) vector space of $T_x M$. Thus T^*M is the disjoint union of all cotangent spaces T_x^*M of M. Also, T^*M is a smooth vector bundle on M with bundle projection
$$\pi : T^*M \to M, \quad \pi(\lambda) = x \text{ for } \lambda \in T_x^*M$$
and T^*M is a smooth manifold of dimension $\dim T^*M = 2 \dim M$. While the tangent bundle provides the right setting for studying differential equations on manifolds, the cotangent bundle is the proper setting for the development of Hamiltonian mechanics.

A smooth section $\alpha : M \to T^*M$ of the cotangent bundle is called a *1-form*. Examples of 1-forms are the derivatives $D\Phi : M \to T^*M$ of smooth functions $\Phi : M \to \mathbb{R}$. However, not every 1-form $\alpha : M \to T^*M$ is necessarily of this form (i.e. α need not be *exact*).

The *bilinear bundle* $\mathrm{Bil}(M)$ of M is defined as the disjoint union
$$\mathrm{Bil}(M) = \bigcup_{x \in M} \mathrm{Bil}(T_x M),$$
where $\mathrm{Bil}(T_x M)$ denotes the real vector space consisting of all symmetric bilinear maps $\beta : T_x M \times T_x M \to \mathbb{R}$. Also, $\mathrm{Bil}(M)$ is a smooth vector bundle on M with bundle projection
$$\pi : \mathrm{Bil}(M) \to M, \quad \pi(\beta) = x \text{ for } \beta \in \mathrm{Bil}(T_x M)$$

and Bil(M) is a smooth manifold of dimension $\dim \text{Bil}(M) = n + \frac{1}{2}n(n+1)$, where $n = \dim M$.

A *Riemannian metric* on M is a smooth section $s : M \to \text{Bil}(M)$ of the bilinear bundle Bil(M) such that $s(x)$ is a positive definite inner product on T_xM for all $x \in M$. Riemannian metrics are usually denoted by $\langle \ , \ \rangle_x$ where x indicates the dependence of the inner product of $x \in M$. A *Riemannian manifold* is a smooth manifold, endowed with a Riemannian metric. If M is a Riemannian manifold, then the tangent bundle TM and the cotangent bundle T^*M are isomorphic.

Let M be a Riemannian manifold with Riemannian metric $\langle \ , \ \rangle$ and let $A \subset M$ be a submanifold. The *normal bundle* of A in M is defined by

$$T^\perp A = \bigcup_{x \in A} (T_xA)^\perp$$

where $(T_xA)^\perp \subset T_xM$ denotes the orthogonal complement of T_xA in T_xM with respect to the positive definite inner product $\langle \ , \ \rangle_x : T_xM \times T_xM \to \mathbb{R}$ on T_xM. Thus

$$(T_xA)^\perp = \{\xi \in T_xM \mid \langle \xi, \eta \rangle_x = 0 \text{ for all } \eta \in T_xA\}$$

and $T^\perp A$ is a smooth vector bundle on A with bundle projection $\pi : T^\perp A \to A$, $\pi(\xi) = x$ for $\xi \in (T_xA)^\perp$. Also, $T^\perp A$ is a smooth manifold of dimension $\dim T^\perp A = \dim M$.

C.10 Riemannian Metrics and Gradient Flows

Let M be a smooth manifold and let TM and T^*M denote its tangent bundle and cotangent bundle, respectively, see Appendix C.9.

A *Riemannian metric* on M is a family of nondegenerate inner products $\langle \ , \ \rangle_x$, defined on each tangent space T_xM, such that $\langle \ , \ \rangle_x$ depends smoothly on $x \in M$. Thus a Riemannian metric is a smooth section in the bundle of bilinear forms defined on TM, such that the value at each point $x \in M$ is a positive definite inner product $\langle \ , \ \rangle_x$ on T_xM. Riemannian metrics exist on every smooth manifold. Once a Riemannian metric is specified, M is called a *Riemannian manifold*. A Riemannian metric on $\mathbb{R}^n$ is a smooth map $Q : \mathbb{R}^n \to \mathbb{R}^{n \times n}$ such that for each $x \in \mathbb{R}^n$, $Q(x)$ is a real symmetric positive definite $n \times n$ matrix. Every positive definite inner product on $\mathbb{R}^n$ defines a (constant) Riemannian metric.

If $f : M \to N$ is an immersion, any Riemannian metric on N pulls back to a Riemannian metric on M by defining

$$\langle \xi, \eta \rangle_x := \langle T_xf(\xi), T_xf(\eta) \rangle_{f(x)}$$

for all $\xi, \eta \in T_x M$. In particular, if $f : M \to N$ is the inclusion map of a submanifold M of N, any Riemannian metric on N defines by restriction of TN to TM a Riemannian metric on M. We refer to this as the *induced Riemannian metric* on M.

Let $\Phi : M \to \mathbb{R}$ be a smooth function defined on a manifold M and let $D\Phi : M \to T^* M$ denote the differential, i.e. the section of the cotangent bundle $T^* M$ defined by

$$D\Phi(x) : T_x M \to \mathbb{R}, \quad \xi \mapsto D\Phi(x) \cdot \xi,$$

where $D\Phi(x)$ is the derivative of Φ at x. We also use the notation

$$D\Phi|_x(\xi) = D\Phi(x) \cdot \xi$$

to denote the derivative of Φ at x. To define the gradient vector field of Φ we have to specify a Riemannian metric $\langle \, , \, \rangle$ on M. The gradient vector field $\operatorname{grad} \Phi$ of Φ with respect to this choice of a Riemannian metric on M is then uniquely characterized by the following two properties.

Tangency Condition

(i) $\operatorname{grad} \Phi(x) \in T_x M$ for all $x \in M$.

Compatibility condition

(ii) $D\Phi(x) \cdot \xi = \langle \operatorname{grad} \Phi(x), \xi \rangle$ for all $\xi \in T_x M$.

There exists a uniquely determined smooth vector field $\operatorname{grad} \Phi : M \to TM$ on M such that (i) and (ii) hold. It is called the *gradient vector field* of Φ.

If $M = \mathbb{R}^n$ is endowed with its standard constant Riemannian metric defined by

$$\langle \xi, \eta \rangle = \xi' \eta \quad \text{for } \xi, \eta \in \mathbb{R}^n,$$

the associated gradient vector field is the column vector

$$\nabla \Phi(x) = \left(\frac{\partial \Phi}{\partial x_1}(x), \cdots, \frac{\partial \Phi}{\partial x_n}(x) \right)'$$

If $Q : \mathbb{R}^n \to \mathbb{R}^{n \times n}$ is a smooth map with $Q(x) = Q(x)' > 0$ for all $x \in \mathbb{R}^n$, the gradient vector field with respect to the general Riemannian metric on $\mathbb{R}^n$ defined by

$$\langle \xi, \eta \rangle_x = \xi' Q(x) \eta, \quad \xi, \eta \in T_x(\mathbb{R}^n) = \mathbb{R}^n$$

is

$$\operatorname{grad} \Phi(x) = Q(x)^{-1} \nabla \Phi(x).$$

The linearization of the gradient flow

$$\dot{x}(t) = \text{grad}\Phi(x(t))$$

at an equilibrium point $x_0 \in \mathbb{R}^n$, $\nabla\Phi(x_0) = 0$, is the linear differential equation

$$\dot{\xi} = A\xi$$

with $A = Q(x_0)^{-1}H_\Phi(x_0)$ where $H_\Phi(x_0)$ is the Hessian at x_0. Thus A is similar to $Q(x_0)^{-\frac{1}{2}}H_\Phi(x_0)Q(x_0)^{-\frac{1}{2}}$ and has only real eigenvalues. The numbers of positive and negative eigenvalues of A coincides with the numbers of positive and negative eigenvalues of the Hessian $H_\Phi(x_0)$.

Let $\Phi : M \to \mathbb{R}$ be a smooth function on a Riemannian manifold and let $V \subset M$ be a submanifold, endowed with the Riemannian metric induced from M. If $x \in V$, then the gradient $\text{grad}(\Phi|V)(x)$ of the restriction $\Phi : V \to \mathbb{R}$ is the image of $\text{grad}\Phi(x) \in T_xM$ under the orthogonal projection map $T_xM \to T_xV$.

Let $M = \mathbb{R}^n$ be endowed with the standard, constant Riemannian metric and let $V \subset \mathbb{R}^n$ be a submanifold endowed with the induced Riemannian metric. Let $\Phi : \mathbb{R}^n \to \mathbb{R}$ be a smooth function. The gradient $\text{grad}(\Phi|V)(x)$ of the restriction $\Phi : V \to \mathbb{R}$ is thus the image of the orthogonal projection of $\nabla\Phi(x) \in \mathbb{R}^n$ onto T_xV under $\mathbb{R}^n \to T_xV$.

C.11 Stable Manifolds

Let $a \in M$ be an equilibrium point of a smooth vector field $X : M \to TM$.

(a) The *stable manifold* of a (or the *inset* of a) is

$$W^s(a) = \{x \in M \mid L_\omega(x) = \{a\}\}.$$

(b) The *unstable manifold* of a (or the *outset* of a) is

$$W^u(a) = \{x \in M \mid L_\alpha(x) = \{a\}\}.$$

The stable and unstable manifolds of a are invariant subsets of M. In spite of their names, they are in general *not* submanifolds of M.

Let $a \in M$ be an equilibrium point of a smooth vector field $X : M \to TM$. Let $\dot{X}(a) : T_aM \to T_aM$ be the linearization of the vector field at a. The equilibrium point $a \in M$ is called *hyperbolic* if $\dot{X}(a)$ has only nonzero eigenvalues.

Let $E^+ \subset T_a M$ and $E^- \subset T_a M$ denote the direct sums of the generalized eigenspaces of $\dot{X}(a)$ corresponding to eigenvalues of $\dot{X}(a)$ having positive or negative real part respectively. Let

$$n^+ = \dim E^+, \quad n^- = \dim E^-.$$

If $a \in M$ is hyperbolic, then $E^+ \oplus E^- = T_a M$ and $n^+ + n^- = \dim M$.

A *local stable manifold* is a locally invariant smooth submanifold $W^s_{\text{loc}}(a)$ of M with $a \in W^s_{\text{loc}}(a)$ such that $T_a W^s_{\text{loc}}(a) = E^-$. Similarly a *local unstable manifold* is a locally invariant smooth submanifold $W^u_{\text{loc}}(a)$ of M with $a \in W^u_{\text{loc}}(a)$ such that $T_a W^u_{\text{loc}}(a) = E^+$.

Stable/Unstable Manifold Theorem *Let $a \in M$ be a hyperbolic equilibrium point of a smooth vector field $X : M \to TM$, with integers n^+ and n^- as defined above.*

(a) There exists local stable and local unstable manifolds $W^s_{\text{loc}}(a)$ and $W^u_{\text{loc}}(a)$. They satisfy $W^s_{\text{loc}}(a) \subset W^s(a)$ and $W^u_{\text{loc}}(a) \subset W^u(a)$. Every solution starting in $W^s_{\text{loc}}(a)$ converges exponentially fast to a.

(b) There are smooth injective immersions

$$\varphi^+ : \mathbb{R}^{n^+} \to M, \varphi^+(0) = a,$$
$$\varphi^- : \mathbb{R}^{n^-} \to M, \varphi^+(0) = a,$$

such that $W^s(a) = \varphi^-(\mathbb{R}^{n^-})$ and $W^u(a) = \varphi^+(\mathbb{R}^{n^+})$. The derivatives

$$T_0 \varphi^+ : T_0(\mathbb{R}^{n^+}) \to T_a M, \quad T_0(\varphi^-) : T_0(\mathbb{R}^{n^-}) \to T_a M$$

map both $T_0(\mathbb{R}^{n^+}) \cong \mathbb{R}^{n^+}$ and $T_0(\mathbb{R}^{n^-}) \cong \mathbb{R}^{n^-}$ isomorphically onto the eigenspaces E^+ and E^-.

Thus, in the case of a hyperbolic equilibrium point a, the stable and unstable manifolds are immersed submanifolds, tangent to the generalized eigenspaces E^- and E^+ of the linearization $\dot{X}(a) : T_a M \to T_a M$.

Let $f : M \to \mathbb{R}$ be a smooth Morse function on a compact manifold M. Suppose M has a C^∞ Riemannian metric which is induced by *Morse charts*, defined by the Morse lemma, around the critical point. Consider the gradient flow of $\text{grad} f$ for this metric. Then the stable and unstable manifolds $W^s(a)$ and $W^u(a)$ are connected C^∞ submanifolds of dimensions n^- and n^+ respectively, where n^- and n^+ are the numbers of negative and positive eigenvalues of the Hessian of f at a. Moreover $W^s(a)$ is diffeomorphic to $\mathbb{R}^{n^-}$ and $W^u(a)$ is diffeomorphic to $\mathbb{R}^{n^+}$.

The following concept becomes important in understanding the behaviour of a flow around a continuum of equilibrium points; see Hirsch, Pugh and

C.11. STABLE MANIFOLDS

Shub (1977). Let $f: M \to M$ be a diffeomorphism of a smooth Riemannian manifold M. A compact invariant submanifold $V \subset M$, i.e. $f(V) = V$, is called *normally hyperbolic* if the tangent bundle of M restricted to V, denoted as $T_V M$ splits into three continuous subbundles

$$T_V M = N^u \oplus TV \oplus N^s,$$

invariant by the tangent map Tf of f, such that

(a) Tf expands N^u more sharply than TV.

(b) Tf contracts N^s more sharply than TV.

Thus the normal behaviour (to V) of Tf is hyperbolic and dominate the tangent behaviour.

A vector field $X: M \to TM$ is said to be *normally hyperbolic* with respect to a compact invariant submanifold $V \subset M$ if V is normally hyperbolic for the time-one map $\phi_1: M \to M$ of the flow of X.

The stable manifold theorem for hyperbolic equilibrium points has the following extension to normally hyperbolic invariant submanifolds.

Fundamental Theorem of Normally Hyperbolic Invariant Manifolds *Let $X: M \to TM$ be a complete smooth vector field on a Riemannian manifold and let V be a compact invariant submanifold, which is normally hyperbolic with respect to X.*

Through V pass stable and unstable C^1 manifolds, invariant under the flow and tangent at V to $TV \oplus N^s$, $N^u \oplus TV$. The stable manifold is invariantly fibered by C^1 submanifolds tangent at V to the subspaces N^s. Similarly, for the unstable manifold and N^u.

Let $a \in M$ be an equilibrium point of the smooth vector field $X : M \to TM$. Let E^0 denote the direct sum of the generalised eigenspaces of the linearization $\dot{X}(a) : T_a M \to T_a M$ corresponding to eigenvalues of $\dot{X}(a)$ with real part zero. A submanifold $W^c \subset M$ is said to be a *center manifold* if $a \in W^c, T_a W^c = E^o$ and W^c is locally invariant under the flow of X.

Center Manifold Theorem *Let $a \in M$ be an equilibrium point of a smooth vector field $X : M \to TM$. Then for any $k \in \mathbb{N}$ there exists a C^k-center manifold for X at a.*

Note that the above theorem asserts only the existence of center manifolds which are C^k submanifolds of M for any finite $k \in \mathbb{N}$. Smooth, i.e. C^∞ center manifolds do not exist, in general.

A center manifold $W^c(a)$ captures the main recurrence behaviour of a vector field near an equilibrium point.

Reduction Principle *Let $a \in M$ be an equilibrium point of a smooth vector field $X : M \to TM$ and let n°, n^+ and n^- denote the numbers of eigenvalues λ of the linearization $\dot{X}(a) : T_a M \to T_a M$ with $\mathrm{Re}\,\lambda = 0$, $\mathrm{Re}\,\lambda > 0$ and $\mathrm{Re}\,\lambda < 0$, respectively (counted with multiplicities). Thus $n^\circ + n^+ + n^- = n$ the dimension of M. Then there exists a homeomorphism $\varphi : U \to \mathbb{R}^n$ from a neighborhood $U \subset M$ of a onto a neighborhood of $0 \in \mathbb{R}^n$ such that φ maps integral curves of X to integral curves of*

$$\begin{aligned} \dot{x}_1 &= X_1(x_1), \quad x_1 \in \mathbb{R}^{n^\circ} \\ \dot{y} &= y, \quad y \in \mathbb{R}^{n^+} \\ \dot{z} &= -z, \quad z \in \mathbb{R}^{n^-}. \end{aligned}$$

Here the flow of $\dot{x}_1 = X_1(x_1)$ is equivalent to the flow of X on a center manifold $W^c(a)$.

C.12 Convergence of Gradient Flows

Let M be a Riemannian manifold and let $\Phi : M \to \mathbb{R}$ be smooth function. Let $\mathrm{grad}\,\Phi$ denote the gradient vector field with respect to the Riemannian metric on M. The critical points of $\Phi : M \to \mathbb{R}$ coincide with the equilibria of the gradient flow on M.

$$\dot{x}(t) = -\mathrm{grad}\,\Phi(x(t)). \tag{*}$$

For any solution $x(t)$ of the gradient flow

$$\begin{aligned} \frac{d}{dt}\Phi(x(t)) &= \langle \mathrm{grad}\,\Phi(x(t)), \dot{x}(t) \rangle \\ &= -\|\mathrm{grad}\,\Phi(x(t))\|^2 \leq 0 \end{aligned}$$

and therefore $\Phi(x(t))$ is a monotonically decreasing function of t.

Proposition C.12.1 *Let $\Phi : M \to \mathbb{R}$ be a smooth function on a Riemannian manifold with compact sublevel sets, i.e. for all $c \in \mathbb{R}$ the sublevel set $\{x \in M \mid \Phi(x) \leq c\}$ is a (possibly empty) compact subset of M. Then every solution $x(t) \in M$ of the gradient flow (*) on M exists for all $t \geq 0$. Furthermore, $x(t)$ converges to a connected component of the set of critical points of Φ as $t \to +\infty$.*

Note that the condition of the proposition is automatically satisfied if M is compact. Solutions of a gradient flow (*) have a particularly simple convergence behaviour. There are no periodic solutions or strange attractors, and there is no chaotic behaviour. Every solution converges to a connected component of the set of equilibria points. This does not necessarily mean

C.12. CONVERGENCE OF GRADIENT FLOWS

that every solution actually converges to an equilibrium point rather than to a whole subset of equilibria.

Let $C(\Phi) \subset M$ denote the set of critical points of $\Phi : M \to \mathbb{R}$. Recall that the ω-limit set $L_\omega(x)$ of a point $x \in M$ for a vector field X on M is the set of points of the form $\lim_{n\to\infty} \phi_{t_n}(x)$, where (ϕ_t) is the flow of X and $t_n \to +\infty$.

Proposition C.12.2 *Let $\Phi : M \to \mathbb{R}$ be a smooth function on a Riemannian manifold with compact sublevel sets. Then*

(a) The ω-limit set $L_\omega(x), x \in M$, of the gradient flow (*) is a nonempty, compact and connected subset of the set $C(\Phi)$ of critical points of $\Phi : M \to \mathbb{R}$. Moreover, for any $x \in M$ there exists $c \in \mathbb{R}$ such that $L_\omega(x)$ is a nonempty compact and connected subset of $C(\Phi) \cap \{x \in M \mid \Phi(x) = c\}$.

(b) Suppose $\Phi : M \to \mathbb{R}$ has isolated critical points in any level set $\{x \in M \mid \Phi(x) = c\}$, $c \in \mathbb{R}$. Then $L_\omega(x), x \in M$, consists of a single critical point. Therefore every solution of the gradient flow (*) converges for $t \to +\infty$ to a critical point of Φ.

In particular, the convergence of a gradient flow to a set of equilibria rather than to single equilibrium points occurs only in nongeneric situations.

Let $\Phi : M \to \mathbb{R}$ be a smooth function on a manifold M and let $C(\Phi) \subset M$ denote the set of critical points of Φ. We say Φ is a *Morse-Bott* function, provided the following three conditions are satisfied.

(i) $\Phi : M \to \mathbb{R}$ has compact sublevel sets.

(ii) $C(\Phi) = \bigcup_{j=1}^k N_j$ with N_j disjoint, closed and connected submanifolds of M such that Φ is constant on $N_j, j = 1, \cdots, k$.

(iii) $\mathrm{Ker}(H_\Phi(x)) = T_x N_j$ for all $x \in N_j$, $j = 1, \cdots, k$.

The original definition of a Morse-Bott function also includes a global topological condition on the negative eigenspace bundle defined by the Hessian. This further condition is omitted here as it is not relevant to the subsequent result. Condition (ii) implies that the tangent space $T_x N_j$ is always contained in the kernel of the Hessian $H_\Phi(x)$ at each $x \in N_j$. Condition (iii) then asserts that the Hessian of Φ is full rank in the directions normal to N_j at x.

Proposition C.12.3 *Let $\Phi : M \to \mathbb{R}$ be a Morse-Bott function on a Riemannian manifold M. Then the ω-limit set $L_\omega(x), x \in M$, for the gradient flow (*) is a single critical point of Φ. Every solution of (*) converges as $t \to +\infty$ to an equilibrium point.*

References

Gamst, J., (1975) *Calculus on Manifolds*, unpublished lecture notes, Department of Mathematics, University of Bremen.

Gibson, C.G., (1979) *Singular points of smooth mappings*, Research Notes in Mathematics, Pitman, London.

Hirsch, M.W., (1976) *Differential Topology*, Graduate Text in Math. 33, Springer Verlag, New York.

Hirsch, M.W. and S. Smale, (1974) *Differential Equations, Dynamical Systems, and Linear Algebra*, Academic Press, New York.

Hirsch, M.W., Pugh, C.C. and M. Shub (1977) *Invariant Manifolds*, Lecture Notes in Mathematics 583, Springer Verlag, Berlin.

Kailath, T., (1978) *Linear Systems*, Englewood Cliffs, N.J., Prentice Hall, Inc.

Munkres, J.R., (1975) *Topology – A First Course*, Prentice-Hall, Inc., New Jersey.

Isidori, A., (1985) *Nonlinear Control Systems*, Springer Verlag, New York.

Lang, S., (1962) *Introduction to Differentiable Manifolds*, Interscience Publ., John Wiley & sons, Inc., New York.

Spivak, M. (1965) *Calculus on Manifolds*, W.A. Benjamin, Ind., New York.

Bibliography

Adamyan, Arov and Krein (1971) Analytic properties of Schmidt pairs for a Hankel operator and the generalized Schur-Takagi problem, *Math. USSR Sbornik* 15, 31-73.

Aiyer, S.V.B., Niranjan, M. and F. Fallside (1990) A theoretical investigation into the performance of the Hopfield model, *IEEE Transactions on Neural Networks* 1, 204-215.

Akin, E. (1979) *The geometry of population genetics*, Lecture Notes in Biomathematics 31, Springer Verlag, Berlin.

Ammar, G.S. and C.F. Martin (1986) The geometry of matrix eigenvalue methods, *Acta Applicandae Math.* 5, 239-278.

Anderson, B.D.O. and J.B. Moore (1990) *Optimal Control: Linear Quadratic Methods*, Prentice Hall, New York

Anderson, B.D.O. and R.R. Bitmead (1977) The matrix Cauchy index: Properties and applications, *SIAM J. Appl. Math.* 33, 655-672.

Anderson, T.W. and I. Olkin (1978) An extremal problem for positive definite matrices, *Linear and Multilinear Algebra*, 6, pp. 257-262.

Andronov, A.A. and L. Pontryagin (1937) *Systèmes grossiers*, C.R. (Dokl.) Acad. URSS 14, 247-251

Andronov, A.A., Vitt, A.A. and S.E. Khaikin (1987) *Theory of Oscillators*, Dover Publ., Inc., New York.

Arnold, V.I. (1989) *Mathematical Methods of Classical Mechanics*, Second Edition, Graduate Texts in Mathematics 60, Springer, New York.

Arnold, V.I. (1990) Dynamics of intersections, *Analysis et cetera* (P. Rabinowich and E. Zehnder, Eds.) Academic Press, New York.

Atiyah, M.F. (1982) Convexity and commuting Hamiltonians, *Bull. London Math Soc.*, 16, 1-15.

Atiyah, M.F. and R. Bott (1982) On the Yang-Mills equations over Riemann surfaces, *Phil. Trans. Roy. Soc. London A*, 308, 523 - 615.

Atiyah, M.F. (1983) Angular momentum, convex polyhedra and algebraic geometry, *Proceedings of the Edinburgh Math. Soc.*, 26, 121-138.

Auchmuty, G. (1991) Globally and rapidly convergent algorithms for symmetric eigen problems, *SIAM Journal of Matrix Analysis and Applications*, 12, 690-706.

Azad, H. and J.J. Loeb (1990) On a theorem of Kempf and Ness, *Indiana Univ. J.*, 39, 61-65.

Baldi P. and K. Hornik (1989) Neural networks and principal component analysis: learning from examples without local minima, *Neural Networks* 2, 53-58.

Barnett, S. (1971) *Matrices in Control Theory*, London: Van Nostrand Reinhold Company.

Batterson, S. and J. Smilie (1989) The dynamics of Rayleigh quotient iteration, *SIAM Journal of Numerical Analysis*, 26, 624-636.

Bayer, D.A. and J.C.Lagarias (1989) The nonlinear geometry of linear programming I, II, *Trans. Amer. Math. Soc.*, 314, 499-580.

Beattie, C. and D.W. Fox (1989) Localization criteria and containment for Rayleigh quotient iteration, *SIAM Journal of Matrix Analysis and Applications*, 10, 80-93.

Bellman, R.E. (1970) *Introduction to Matrices*, 2nd Edition. New York: McGraw Hill Book Company.

Bhatia, R. (1987) *Perturbation bounds for matrix eigenvalues*, Pitman Research Notes in Mathematics Series 162, Longman Scientific & Technical, New York.

Bloch, A.M. (1985a) A completely integrable Hamiltonian system associated with line fitting in complex vector spaces, *Bull. Amer. Math. Soc.*, 12, 250 - 254.

Bloch, A.M. (1985b) Estimation, principal components and Hamiltonian systems, *Systems and Control Letters*, 6, 1-15.

Bloch, A.M. and C.I. Byrnes (1986) An infinite-dimensional variational problem arising in estimation theory, *Algebraic and Geometric Methods in Nonlinear Control Theory*, (M. Fliess and H. Hazewinkel, Eds.), Reidel Publ.

Bloch, A.M. (1986) An infinite-dimensional variational problem arising in estimation theory, *Algebraic and Geometric Methods in Nonlinear Control Theory* (M. Fliess and M. Hazewinkel, Eds.), Reidel Publ.

Bloch, A.M. (1987) An infinite-dimensional Hamiltonian system on projective Hilbert space, *Transactions of the American Mathematical Society* 302, 787-796.

Bloch, A.M., R.W. Brockett, Y. Kodama and R. Ratiu (1989) Spectral equations for the long wave limit of the Toda lattice equations, *Hamiltonian systems, Transformation groups and Spectral Transform Methods* (J. Harnad and J.E. Marsden, Eds.).

Bloch, A.M., (1990) Steepest descent, linear programming and Hamiltonian flows, *Contemporary Math.* 114, 77-88.

Bloch, A.M. (1990) The Kähler structure of the total least squares problem, Brockett's steepest descent equations, and constrained flows, *Realization and Modelling in Systems Theory* (M. Kaashoek, J.H. van Schuppen and A.C. Ran, Eds.), Birkhäuser Publ., Boston.

Bloch, A.M., R.W. Brockett and T.Ratiu (1990) A new formulation of the generalized Toda lattice equations and their fixed point analysis via the moment map, *Bull. Amer. Math. Soc.*,23, 447-456..

Bloch, A.M., H. Flaschka and T. Ratiu (1990) A convexity theorem for isospectral sets of Jacobi matrices in a compact Lie algebra, *Duke Math J.* 61, 41-65.

Bloch, A.M. (1991) The Kähler structure of the total least squares problem, Brockett's steepest descent equations and constrained flows, *Realization and Modelling in System Theory* (M. Kaashoek, J.W. van Schuppen and A.C.M. Ran, Eds.), Birkhäuser, Boston.

Bloch, A.M., R.W. Brockett and T.Ratiu (1992) Completely integrable gradient flows, *Communications in Mathematical Physics*, 23, 47-456.

Bott, R. (1954) Nondegenerate critical manifolds, *Ann. of Mathematics*, 60, 248-261.

Bott, R. and H. Samelson (1958) Application of the theory of Morse to symmetric spaces, *Amer. J. Math.* 80, 964 - 1029

Bott, R. (1982) Lectures on Morse theory, old and new, *Bulletin of the American Mathematical Society*, 7, 331-358.

Bröcker, Th. and T. tom Dieck (1985) *Representations of Compact Lie Groups*, Springer-Verlag, New York.

Brockett, R.W. and R.A. Skoog (1971) A new perturbation theory for the synthesis of nonlinear networks, SIAM-AMS Proc. Vol. III, *Mathematical Aspects of Electrical Network Analysis* Providence.R.I. 17-33.

Brockett, R.W. and P.S. Krishnaprasad (1980) A scaling theory for linear systems, *IEEE Transactions on Automatic Control* AC-25, 197-207.

Brockett, R.W. and C.I. Byrnes (1981) Multivariable Nyquist criteria, root loci and pole placement: A geometric viewpoint, *IEEE Transactions on Automatic Control* AC-26, 271-284.

Brockett, R.W. (1988) Dynamical systems that sort lists and solve linear programming problems, *Proc. 27th IEEE Conference on Decision and Control, Austin, TX*, 779-803. See also *Linear Algebra Appl.* 146 (1991)., 79-91.

Brockett, R.W. (1989) Least squares matching problems, *Linear Algebra Appl.*, 122-127, 701-777.

Brockett, R.W. (1989) Smooth dynamical systems which realize arithmetical and logical operations, *Three Decades of Mathematical System Theory*, 135, Lecture Notes in Control and Information Sciences, Springer-Verlag, 19-30.

Brockett, R.W. and A.M.Bloch (1989) Sorting with the dispersionless limit of the Toda Lattice, in *Hamiltonian Systems, Transformation Groups and Spectral Transform Methods*, CRM, (J. Harnad and J.E. Marsden, Eds.) Université de Montréal, Montréal, Canada, 103-112.

Brockett, R.W. (1991a) Dynamical systems that learn subspaces, *Mathematical System Theory - The Influence of Kalman*, (A.C. Antoulas, Ed.), 579-592.

Brockett, R.W. (1991b) Dynamical systems that sort lists, diagonalize matrices and solve linear programming problems, *Linear Algebra Appl.*, 146, 79-91.

Brockett, R.W. and L.E. Faybusovich (1991) Toda flows, inverse spectral transform and realization theory, *Systems and Control Letters*, 16, 779-803.

Brockett, R.W. and W.S.Wong (1991) A gradient flow for the assignment problem, *New Trends in System Theory*, Progress in System and Control Theory, (G. Conte, A.M. Perdon and B. Wyman, Eds.), Birkhäuser, 170-177.

Brockett, R.W. (1993) Differential geometry and the design of gradient algorithms, *Proceedings of Symposia in Pure Mathematics*, 54, 69-91.

Bucy, R.S. (1975) Structural stability for the Riccati equation, *SIAM J. on Optimization and Control* 13, 749-753.

Byrnes, C.I. (1978) On certain families of rational functions arising in dynamics, *Proc. IEEE*, 1002-1006.

Byrnes, C.I. and J.C. Willems (1986) Least-squares estimation, linear programming and momentum: A geometric parametrization of local minima, *IMA J. of Math. Control and Inf.*, 3, 103-118.

Byrnes, C.I. (1989) Pole assignment by output feedback, *Lecture Notes in Control and Information Sciences* 135, Springer-Verlag, Berlin, 31-78.

Byrnes, C.I. and T.W. Duncan (1989) On certain topological invariants arising in system theory, in *New Directions in Applied Mathematics*, (P.J. Hilton and G.S. Young, eds.), Springer Verlag, pp. 29-72.

Chu, M.T. (1984a) On the global convergence of the Toda Lattice for real normal matrices and its application to the eigenvalue problems, *SIAM J. Math. Anal*, 15, 98-104

Chu, M.T. (1984b) The generalized Toda flow, the QR algorithm and the center manifold theory, *SIAM J. Disc. Meth.*, 5, 187-201.

Chu, M.T. (1986) A differential equation approach to the singular value decomposition of bidiagonal matrices, *Linear Algebra and its Appl.*, 80, 71-80.

Chu, M.T. (1988) On the continuous realization of iterative processes, *SIAM Review*, 30, 375-387.

Chu, M.T. and K.R.Driessel (1990) The projected gradient method for least squares matrix approximations with spectral constraints, *SIAM J. Numer. Anal.* 27, 1050-1060.

Chu, M.T. (1992a) Matrix differential equations: A continuos realization process for linear algebra problems, *Nonlinear Analysis, TMA* 18, 1125-1146.

Chu, M.T. (1992b) Numerical methods for inverse singular value problems, *SIAM J. Numer. Anal.* 29, 885-903.

Chua, L.O. and G.N.Lin (1984) Nonlinear programming without computation, *IEEE Trans. Circuits and Systems*, 31, 182-188.

Colonius, F. and W. Kliemann (1990) Linear control semigroup acting on projective space, Report No. 224, Universität Augsburg.

Courant, R. (1922) Zur Theorie der kleinen Schwingungen, *Zts. f. angew. Math. und Mech.*, 2, 278-285.

Crouch, P.E. and R. Grossman (1991) Numerical integration of ordinary differential equations on manifolds, *J. of Nonlinear Science*, to appear.

Crouch, P.E., R. Grossmann and Y. Yan (1992a) On the numerical integration of the dynamic attitude equations, *Proc. of IEEE Conference on Decision and Control*, Tuscon, Arizona, 1497-1501.

Crouch, P.E., R. Grossman and Y. Yan (1992b) A third order Runge-Kutta algorithm on a manifold, preprint

Dantzig, G.B. (1963) *Linear Programming and Extensions*, Princeton University Press, Princeton, N.J.

Davis, M.W. (1987) Some aspherical manifolds, *Duke Math. Journal*, 5, 105-139.

Deift, P., T. Nanda and C. Tomei (1983) Ordinary differential equations for the symmetric eigenvalue problem, *SIAM J. Numer. Anal.*, 20, 1-22.

Deift, P., L.C. Li and C. Tomei (1985) Toda flows with infinitely many variables, *J. of Functional Analysis* 64, 358-402.

de la Rocque Palis, G. (1978) Linearly induced vector fields and $\mathbb{R}^2$ actions on spheres, *Journal of Differential Geometry*, 13, 163-191.

de Mari, F. and M.A. Shayman (1988) Generalized Eulerian numbers and the topology of the Hessenberg variety of a matrix, *Acta Applicandae Math.* 12, 213-235.

de Mari, F., Procesi, C. and M.A. Shayman (1992) Hessenberg varieties, *Transactions of the American Mathematical Society* 332, 529-534.

Demmel, J.W. (1987) On condition numbers and the distance to the nearest ill-posed problem, *Numer. Math.* 51, 251-289.

de Moor, B.L.R., and G.H. Golub (1989) The restricted singular value decomposition: Properties and applications, *Technical Report* NA-89-03, Department of Computer Science, Stanford University, Stanford.

de Moor, B. and J. David (1992) Total linear least squares and the algebraic Riccati equation, *Systems and Control Letters* 5, 329-337.

Dempster, A.P. (1969) *Elements of Continuous Multivariable Analysis*, Reading, MA: Addison-Wesley.

Dieudonné, J. and Carrell, J.B. (1971) *Invariant theory, old and new*, Academic Press, New York.

Dikin, I.I. (1967) Iterative solutions of problems of linear and quadratic programming, *Soviet Math. Dokl.*, 8, 674-675.

Doolin, B.F. and C.F. Martin (1990) *Introduction to Differential Geometry for Engineers*, Marcel Dekker, Inc. New York

Driessel, K.R. (1986) On isospectral gradient flow-solving matrix eigen- problems using differential equations, *Inverse Problems* (J.R. Cannon and U. Hornung, Eds.) ISNM 77, Birkhäuser Publ. 69 - 91.

Driessel, K.R. (1987 a) On isospectral surfaces in the space of symmetric tridiagonal matrices, *Technical Report* # 544, Department of Mathematical Sciences, Clemson University.

Driessel, K.R. (1987 b) On finding the eigenvalues and eigenvectors of a matrix by means of an isospectral gradient flow, *Technical Report* # 541, Department of Mathematical Sciences, Clemson University.

Duistermaat, J.J., J.A.C. Kolk and V.S. Varadarajan (1983) Functions, flow and oscillatory integrals on flag manifolds and conjugacy classes in real semisimple Lie groups, *Compositio Math.* 49, 309-398.

Eckart, G. and Young, G. (1936) The approximation of one matrix by another of lower rank, *Psychometrika*, 1. 221-218.

Fan, K. (1949) On a theorem of Weyl concerning eigenvalues of linear transformations 1, *Proc. Nat. Acad. Sci.*, U.S.A.,35, 652-655.

Faybusovich, L.E. (1989) QR-type factorizations, the Yang-Baxter equations, and an eigenvalue problem of control theory, *Linear Algebra and its Appl.* 122/123/124, 943 - 971

Faybusovich, L.E. (1991a) Dynamical systems which solve optimization problems with linear constraints, *IMA J. Math. Control and Info.*, 8, 135-149.

Faybusovich, L.E. (1991b) Hamiltonian structure of dynamical systems which solve linear programming problems, *Physica*, D 53, 217-232.

Faybusovich, L.E. (1991c) Interior print methods and entropy, *Proc of IEEE Conf. on Decision and Control*, Brighton, UK, 2094-2095.

Faybusovich, L.E. (1992) Dynamical systems that solve linear programming problems, *Proc. of IEEE Conference on Decision and Control*, Tuscon, Arizona, 1626-1631.

Faybusovich, L. (1992) Toda flows and isospectral manifolds, *Proc. American Mathematical Society* 115, 837-847.

Fernando, K.V.M. and H. Nicholson (1980) Karhunen-Loeve expansion with reference to singular value decomposition and separation of variables, *Proc. IEE*, part D, vol. 127, 204-206.

Fischer, E. (1905) Über quadratische Formen mit reelen Koeffizienten, *Monatsh. f. Math. und Phys.*, 16, 234-249.

Flanders, H. (1975) An extremal problem on the space of positive definite matrices, *Linear and Multilinear Algebra*, 3, pp. 33-39.

Flaschka, H. (1974) The Toda lattice, I, *Phys. Rev. B* 9, 1924-1925.

Flaschka, H. (1975) Discrete and periodic illustrations of some aspects of the inverse methods, in dynamical system, theory and applications, J. Moser, ed., *Lecture Notes in Physics*, 38 Springer-Verlag, Berlin.

Fletcher, R. (1985) Semi-definite matrix constraints in optimization, *SIAM J. Control Optim.* 23, 493-513.

Francis, J.G.F. (1961) The QR transformation, a unitary analogue to the LR transformation, *Comput. J.*, 4, 265-280.

Frankel, T. (1963) Critical submanifolds of the classical groups and Steifel manifolds, *Differential and Combinatorial Topology*, (S.S. Cairns, Ed.,) Princeton University Press, Princeton, N.J.

Frankel, T. (1965) Critical submanifolds of the classical groups and Stiefel manifolds, *Differential and Combinatorial Topology*, (S.S Cairns, Ed.,) Princeton University Press, Princeton, N.J.

Fried, D. (1986) The cohomology of an isospectral flow, *Proc. Amer. Math. Soc.*, 98, 363-368.

Gamst, J., (1975) *Calculus on Manifolds*, unpublished lecture notes, Department of Mathematics, University of Bremen.

Gelfand, I.M. and V.V. Serganova (1987) Combinatorial geometries and torus strata on homogeneous compact manifolds, *Russian Math. Surveys* 42, 133-168.

Ge-Zhong and J.E. Marsden (1988) Lie-Poisson Hamilton-Jacobi theory and Lie-Poisson integration, *Phys. Lett.*, A-133, 134 - 139.

Ghosh, B.K. (1988) An approach to simultaneous system design. Part II: Nonswitching gain and dynamic feedback compensation by algebraic geometric methods, *SIAM J. Control and Optimization* 26, 919-963.

Gibson, C.G., (1979) *Singular points of smooth mappings*, Research Notes in Mathematics, Pitman, London.

Glover, K. (1984) All optimal Hankel-norm approximations of linear multivariable systems and their L^∞-error bounds, *International Journal of Control* 39, 1115-1193.

Gohberg, I.C. and M.G. Krein (1969) *Introduction to the Theory of Linear Nonselfadjoint operators*, Transl. Math. Monographs, Vol. 18, Amer. Math. Soc., Providence, R.I.

Golub, G.H. and W. Kahan (1965) Calculating the singular values and pseudo-inverse of a matrix, *SIAM J. Numerical Anal.*, Series B2, 205-224.

Golub, G.H. and C. Reinsch (1970) Singular value decomposition and least squares solutions, *Numer. Math.* 14, 403-420.

Golub, G.H. and C. Van Loan (1980) An analysis of the total least squares problem, *SIAM J. Numer. Anal.*, 17, 883-843.

Golub, G.H. and C.F. Van Loan (1989) *Matrix Computations*. Second Edition. The John Hopkins University Press, Baltimore.

Grötschel, M., L. Lovász and A. Schrijver (1988) *Geometric algorithms and combinatorial optimization*, Springer-Verlag, Berlin.

Godbout, L.F. and D. Jordan (1980) Gradient matrices for output feedback systems, *International Journal of Control* 32, 411-433.

Gray, W.S. and E.I. Verriest (1989) On the sensitivity of generalized state-space systems, *Proc. of the 28th Conference on Decision and Control*, Tampa, 1337-1342.

Gray, W.S. and E.I. Verriest (1987) Optimality properties of balanced realizations: minimum sensitivity, *Proc. of the 26th Conference on Decision and Control*, Los Angeles, 124-128.

Guillemin, V. and S. Sternberg (1984) *Symplectic Techniques in Physics*, Cambridge University Press, Cambridge.

Halmos, P. (1972) Positive approximants of operators, *Indiana Univ. Math. U.*, 21, 951-960.

Hangan, T. (1968) A Morse function on Grassmann manifolds, *Journal Diff. Geometry* 2, 363 - 367.

Hansen, P.C. (1990) Truncated singular value decomposition solutions to discrete ill-posed problems with ill-determined numerical rank, *SIAM J. Sci. Stat. Comp.* 11, 503-518.

Hardy, G.H., Littlewood, J.E. and G. Polya (1952) *Inequalities*, 2nd edition. Cambridge University Press, Cambridge, England.

Hayden, T.L. and Well, J. (1988) Approximation by matrices positive semi-definite on a subspace, *Linear Algebra and its Appl.*, 109, 115-130.

Helmke, U. (1991) Isospectral flows on symmetric matrices and the Riccati equation, *Systems and Control Letters*, 16, 159-166.

Helmke, U. (1992) A several complex variables approach to sensitivity analysis and structured singular values, *Journal of Mathematical Systems, Estimation, and Control*, 2, 1-13.

Helmke, U. and J.B. Moore (1992) Singular value decomposition via gradient and self equivalent flows, *Linear Algebra and its Appl.* 69, 223-248.

Helmke, U. (1993) Balanced realizations for linear systems: a variational approach, *SIAM J. Control and Opt.* 31, 1-15.

Helmke, U. (1993) Isospectral flows and linear programming, *Journal of Australian Mathematical Soc.*, Series B, 34, 495-510.

Helmke, U., J.B. Moore and J.E. Perkins (1993) Dynamical systems that compute balanced realizations and the singular value decomposition, to appear in *SIAM J. Matrix Analysis*.

Helmke, U., Prechtel, M. and M.A. Shayman (1993) Riccati-like flows and matrix approximations, submitted to *Kybernetika*.

Helmke, U. and M.A. Shayman (1993) Critical points of matrix least squares distance functions, *Linear Algebra and its Applications*, to appear.

Henniart, G. (1983) Les inegalities de Morse, *Seminaire Bourbaki*, 617, 1-17.

Hermann, R. (1962) Geometric aspects of potential theory in the bounded symmetric domains, *Math. Annalen*, 148, 349 - 366

Hermann, R. (1963) Geometric aspects of potential theory in the symmetric bounded domains II, *Math. Ann.*, 151, 143 - 149

Hermann, R. (1964) Geometric aspects of potential theory in symmetric spaces III, *Math. Annalen*, 153, 384 - 394.

Hermann, R. and C.F. Martin (1977) Applications of algebraic geometry to system theory, Part I, *IEEE Transactions on Automatic Control* AC-22, 19-25.

Hermann, R. (1979) *Cartanian geometry, nonlinear waves, and control theory, Part A*, Interdisciplinary Mathematics, Vol. 20, Math. Sci. Press, Brookline MA 02146.

Hermann, R. and C.F. Martin (1982) Lie and Morse theory of periodic orbits of vector fields and matrix Riccati equations I: General Lie-theoretic methods, *Math. Systems Theory* 15, 277-284.

Herzel, S., M.C.Recchioni and F.Zirilli (1991) A quadratically convergent method for linear programming, *Linear Algebra Appl.*, 151, 255-290.

Higham, N.J. (1986) Computing the polar decomposition - with applications, *SIAM J. Sci. Stast. Comput.* 7, 1160-1174.

Higham, N.J. (1988) Computing a nearest symmetric positive semidefinite matrix, *Linear Algebra and its Appl.*, 103, 103-118.

Hirsch, M.W. and S. Smale (1974) *Differential Equations, Dynamical Systems, and Linear Algebra*, Academic Press, New York

Hirsch, M.W., (1976) *Differential Topology*, Graduate Text i Math. 33, Springer Verlag, New York.

Hirsch, M.W., Pugh, C.C. and M. Shub (1977) *Invariant Manifolds*, Lecture Notes in Mathematics 583, Springer Verlag, Berlin.

Hitz, K.L. and B.D.O. Anderson (1972) Iterative method of computing the limiting solution of the matrix Riccati differential equation, *Proc. IEE* 119, 1402-1406.

Hopfield, J.J. (1982) Neural networks and physical systems with emergent collective computational abilities, *Proc. Nat. Acad. Sci. USA* 79, 2554.

Hopfield, J.J. (1984) Neurons with graded response have collective computational properties like those of two-state neurons, *Proc. National Academy of Sciences USA* 81, 3088-3092.

Hopfield, J.J. and D.W. Tank (1985) Neural computation of decisions in optimization problems, *Biolog. Cybern.* 52, 1-25.

Horn, R. (1953) Doubly stochastic matrices and the diagonal of a rotation matrix, *American Journal of Mathematics*, 76, 620-630.

Hotelling, H. (1933) Analysis of a complex of statistical variables into principal components, *J. Educ. Psych.* 24, 417-441, 498-520.

Hotelling, H. (1935) Simplified calculation of principal components, *Psychometrika* 1, 27-35.

Humphreys, J.E. (1972) *Introduction to Lie Algebras and Representation Theory*, Springer-Verlag, New York.

Hung, Y.S. and A.G.J. MacFarlane (1982) *Multivariable Feedback: A Quasi-Classical Approach*, Lecture Notes in Control and Information Sciences, Vol. 40, Springer, Berlin.

Hwang, S.Y. (1977) Minimum uncorrelated unit noise in state space digital filtering, *IEEE Trans. Acoust. Speech Signal Process.*, ASSP-25, 273-281.

Irwin, M.C., (1980) *Smooth Dynamical Systems*, Academic Press, 1980, London.

Isidori, A., (1985) *Nonlinear Control Systems: An Introduction*, Springer-Verlag, Berlin.

Jonckheere, E. and L.M. Silverman (1983) A new set of invariants for linear systems: Applications to reduced order compensator design, *IEEE Transactions on Automatic Control* 28, 953-964.

Kailath, T. (1980) *Linear Systems*, Prentice Hall, Englewood Cliffs.

Karmarkar, N. (1984) A new polynomial time algorithm for linear programming, *Combinatorica*, 4, 373-395.

Karmarkar, N. (1990) Riemannian geometry underlying interior point methods for linear programming, *Contemporary Mathematics*, 114, AMS, 51-76.

Kempf, G. and L. Ness (1979) The length of vectors in representation spaces, *Algebraic Geometry*, (K. Lonsted, ed.), *Lecture Notes in Math.*, 732, Springer Verlag, 233-244.

Khachian, L.G. (1979) A polynomial algorithm in linear programming, *Doklady Akademit Nauk SSSR* 244S, 1093-1096, translated in *Soviet Mathematics Doklady* (1979) 201, 191-194.

Kimura, H. (1975) Pole assignment by gain output feedback, *IEEE Transactions on Automatic Control* AC-20, 509-516.

Klema, V.C. and A.J. Laub (1980) The singular value decomposition: Its computation and some applications, *IEEE Transactions on Automatic Control* AC-25, 164-176.

Knuth, D.E. (1973) *The art of computer programming, Vol. 3: Sorting and Searching*, Addison-Wesley, Reading.

Kostant, (1973) On convexity, the Weyl Group and the Iwasawa decomposition, *Ann. Sci. Ecole Norm. Sup.*, 6, 413-455.

Kostant, B. (1979) The Solution to a generalized Toda lattice and representation theory, *Advances in Mathematics* 34, 195 - 338.

Kraft, H. (1984) *Geometrische Methoden in der Invariantentheorie*, Aspects of Mathematics, D1, Vieweg, Braunschweig.

Krantz, S.G. (1982) *Function Theory of Several Complex Variables*, John Wiley, New York.

Krishnaprasad, P.S. (1979) Symplectic mechanics and rational functions, *Richerche Automat.* 10, 107-135.

Kung, S.Y. and D.W. Lin (1981) Optimal Hankel-norm model reductions: multivariable systems, *Transactions on Automatic Control* AC-26, 832-852.

Lagarias, J. and M.J. Todd (Eds.) (1990) *Mathematical Development Arising from Linear Programming*, Providence, R.I., Contemporary Math. 114, AMS.

Lagarias, J.U. (1991) Monotonicity properties of the Toda flow, the QR flow and subspace iteration, *SIAM Journal of Matrix Analysis and Applications*, 12, 449-462.

Lang, S., (1962) *Introduction to Differentiable Manifolds*, Interscience Publ., John Wiley & sons, Inc., New York.

Laub, A.J., M.T. Heath, C.C. Paige and R.C. Ward (1987) Computation of system balancing transformations and other applicaitons of simultaneous diagonalization algorithms, *IEEE Trans. Automatic Control* AC-32, 115-121.

Lawson, C.L. and R.J. Hanson (1974) *Solving Least Squares Problems*, Prentice-Hall, Englewood Cliffs, N.J.

Li, G., Anderson, B.D.O. and M. Gevers (1992) Optimal FWL Design of State-Space Digital Systems with Weighted Sensitivity Minimization and Sparseness Consideration, *Proc. of MTNS conference, Kobe*, Japan.

Luenberger, D.G. (1969) *Optimization by Vector Space Methods*, John Wiley, New York.

Luenberger, D.G. (1973) *Introduction to linear and nonlinear programming*, Addison-Wesley Publ. Co., Reading.

Madievski, A.G., B.D.O. Anderson and M.R. Gevers (1993) Optimum FWL design of sampled data controllers, preprint.

Mahony, R.E. and U. Helmke (1993) System assignment and pole placement for symmetric realizations, preprint.

Mahony, R.E. and J.B.Moore (1992) Recursive interior-point linear programming based on Lie-Brockett flows, *Proc. of Int. Conf. on Optimization Techniques and Applications*, Singapore.

Martin, C.F. (1981) Finite escape time for Riccati differential equations, *Systems and Control Letters* 1, 127-131.

Marshall, A.W. and I. Olkin (1979) *Inequalities: Theory of Majorization and its Applications*, Academic Press, New York.

Meggido, N. and M. Shub (1989) Boundary behaviour of interior-point algorithms in linear programming, *Math. Oper. Research* 14, 97-146.

Middleton, R.H. and Goodwin, G.C. (1990) *Digital Control and Estimation: A Unified Approach*, Prentice Hall, Englewood Cliffs, New Jersey.

Milnor, J. (1963) *Morse Theory*, Ann. of Math. Studies, 51, Princeton University Press, Princeton, N.J.

Mirsky, L. (1960) Symmetric gauge functions and unitarily invariant norms, *Quart. J. Math. Oxford*, 11, 50-59.

Moonen, M., van Dooren, P. and J. Vandewalle (1990) SVD updating for tracking slowly time-varying systems. A parallel implementation, *Signal Processing, Scattering and Operator Theory, and Numerical Methods* (M.A. Kaashoek, J.H. van Schuppen and A.C.M. Ran, Eds.), Birkhäuser, Boston, 487-494.

Moore, B.C. (1981) Principal component analysis in linear systems: controllability, observability, and model reduction, *IEEE Transactions on Automatic Control* AC-26, 17-32.

Moore, J.B., R.E. Mahony and U. Helmke (1993) Numerical gradient algorithms for eigenvalue and singular value decomposition, *SIAM J. of Matrix Analysis*, to appear.

Moser, J. (1975) Finitely many mass points on the line under the influence of an exponential potential - An integrable system, in J. Moser, Ed., *Dynamic Systems Theory and Applications* Springer-Verlag, Berlin - New York. 467-497.

Moser, J. and A.P. Veselov (1991) Discrete versions of some classical integrable systems and factorization of matrix polynomials, *Commun. Math. Phys.* 139, 217-243.

Mullis, C.T. and R.A. Roberts (1976) Synthesis of minimum roundoff noise fixed point digital filters, *IEEE Transactions on Circuits and Systems*, CAS-23, 551-562.

Mumford, D. and J. Fogarty (1982) *Geometry Invariant Theory*, Ergebnisse der Mathematik und ihrer Grenzgebiete 124, Springer, Berlin.

Munkres, J.R., (1975) *Topology - A First Course*, Prentice-Hall, Inc., New Jersey.

Nakamura, Y. (1988) Fractional transformation group induced by QR factorization and linear prediction problems, *Systems and Control Letters*, 10, 181 - 184

Nanda, T. (1985) Differential equations and the QR algorithm, *SIAM J. Numer. Anal.* 22, 310-321.

Ober, R. (1987) Balanced realizations: Canonical form, parametrization, model reduction, *International Journal of Control* 46, 643-670.

Oja, E. (1982) A simplified neuron model as a principal component analyzer, *J. of Math. Biology.* 15, 267-273.

Oja, E. (1990) Neural networks, principal components and subspaces, *Int. Journal of Neural Systems*, Vol. 1, No. 1, 61-68.

Oppenheim, A.V. and R.W. Shafer (1989) *Discrete-time signal processing*, Prentice Hall, Eaglewood Cliffs, N.J.

Palais, R.S. (1962) Morse theory on Hilbert manifolds, *Topology*, 2, 299-340.

Palis, J. and W. de Melo (1982) *Geometric Theory of Dynamical Systems*, Springer-Verlag, New York.

Parlett, B.N. and W.G. Poole (1973) A geometric theory for the QR, LU and Power iterations, *SIAM J. Numer. Anal.* 10, 389-412.

Parlett, B.N. (1978) Progress in numerical analysis, *SIAM Rev.* 20, 443-456.

Paul, St., K. Hueper and J.A. Nossek (1992) A class of nonlinear lossless dynamical systems, *Archiv für Elektronik und Übertragungstechnik* 46, 219-227.

Pearson, K. (1901) On lines and planes of closest fit to points in space, *Phil. Mag.*, 559-572.

Peixoto, M.M. (1962) Structural stability on two-dimensional manifolds, *Topology* 1, 101-120.

Pernebo, L. and L.M. Silverman (1982) Model reduction via balanced state space representations, *IEEE Transactions on Automatic Control* 27, 282-287.

Peterson C. and B. Soederberg (1989) A new method for mapping optimization problems onto neural networks, *International Journal of Neural Systems* 1, 3-22.

Pressley, A.N. (1982) The energy flow on the loop space of a compact Lie group, *J. London Math. Soc.*,26,557-566.

Pressley, A.N. and G. Segal (1986) *Loop Groups*, Oxford Mathematical Monographs, Oxford.

Pyne, I.B. (1956) Linear programming on an analogue computer, *Trans. AIEE*, Part I, 75, 139-143.

Reid, W.T. (1972) *Riccati Differential Equations*, Academic Press, New York.

Riddell, R.C. (1984) Minimax problems on Grassmann manifolds. Sums of eigenvalues, *Advances in Mathematics*, 54, 107-109.

Rosenthal, J. (1992) New results in pole assignment by real output feedback, *SIAM J. Control and Optimization* 30, 203-211.

Rutishauser, H. (1954) Ein Infinitesimales Analogon zum Quotienten-Differenzen-Algorithmus, *Arch. Math.*, (Basel), 5, 132-137.

Rutishauser, H. (1958) Solution of eigenvalue problems with the LR-transformation, *National Bureau of Standards Applied Mathematics Series* 49, 47-81.

Safonov, M.G. and R.Y. Chiang (1989) A Schur method for balanced-truncation model reduction, *IEEE Transactions on Automatic Control*, July, 729-733.

Schneider, C.R. (1973) Global aspects of the matrix Riccati equation, *Math. System Theory* 1, 281-286.

Schur, I (1923) Über eine Klasse von Mittelbildungen mit Anwendungen auf die Determinanten Theorie, *Sitzungsber. der Berliner Math. Gesellschaft.* 22, 9-20.

Schuster, P., K.Sigmund and R. Wolff (1978) Dynamical systems under constant organisation I: Topological analysis of a family of nonlinear differential equations, *Bull. Math. Biophys.*, 40, 743-769.

Shafer, D. (1980) Gradient vectorfields near degenerate singularities, *Global Theory of Dynamical Systems* (Z. Nitecki and C. Robinson, Eds.), Lecture Notes in Mathematics 819, Springer-Verlag, Berlin, 410-417.

Shayman, M.A. (1982) Morse functions for the classical groups and Grassmann manifolds, unpublished manuscript.

Shayman, M.A. (1986) Phase portrait of the matrix Riccati equation, *SIAM J. Control and Optimization*, 24, 1-65.

Shub, M. and A.T. Vasquez (1987) Some linearly induced Morse-Smale systems, the QR algorithm and the Toda lattice, in: *The Legacy of Sonya Kovaleskaya* (L. Keen, ed.) Contemporary Mathematics 64, A.M.S., Providence, 181-194.

Sirat, J.A. (1991) A fast neural algorithm for principal component analysis and singular value decompositions, *International Journal of Neural Systems* 2, 147-155.

Slodowy, P. (1989) Zur Geometrie der Bahnen reell reduktiver Gruppen, *Algebraic Transformation Groups and Invariant Theory* (H. Kraft, P. Slodowy and T. Springer, eds.), Birkhäuser, Boston, pp. 133-144.

Smale, S. (1960) Morse inequalities for a dynamical system, *Bull. Amer. Math. Soc.* 66, 43-49.

Smale, S. (1961) On gradient dynamical systems, *Annals of Mathematics* 74, 199-206.

Smale, S. (1976) On the differential equations of species in competition, *Journal of Mathematical Biology* 3, 5-7.

Smith, S.T., (1991) Dynamical systems that perform the singular value decomposition, *Systems and Control Letters* 16, 319-328.

Smith, S.T. (1993) *Geometric optimization methods for adaptive filtering*, PhD Thesis, Harvard University.

Sontag, E.D. (1990) *Mathematical Control Theory*, Springer-Verlag, New York.

Sonnevend, G. and J. Stoer (1990) Global ellipsoidal approximations and homotopy methods for solving convex programs, *Applied Mathematics and Optimization* 21, 139-165.

Sonnevend, G., Stoer, J. and G. Zhao (1990) On the complexity of following the central path of linear programs by linear extrapolation, *Methods of Operations Research* 62, 19-31.

Sonnevend, G., Stoer, J. and G. Zhao (1991) On the complexity of following the central path of linear programs by linear extrapolation II, *Mathematical Programming* (Series B).

Sontag, E.D. (1990) *Mathematical Control Systems: Deterministic Finite Dimensional Systems*, Texts in Applied Mathematics, Springer Verlag, New York.

Spivak, M. (1965) *Calculus on Manifolds*, W.A. Benjamin, Ind., New York.

Stone, R.E. and C.A.Tovey (1991) The simplex and projective scaling algorithms as iteratively reweighted least squares methods, *SIAM Review* 33, 220-237.

Symes, W.W. (1980) Systems of the Toda type, inverse spectral problems, and representation theory, *Inventiones Mathematicae* 59, 13-51.

Symes, W.W. (1980) Hamiltonian Group Actions and Integrable Systems, *Physica* 1D, 339-374.

Symes, W.W. (1982) The QR algorithm and scattering for the finite nonperiodic Toda lattice, *Physica.* 4D, 275-280.

Takeuchi, M. (1965) Cell decompositions and Morse equalities on certain symmetric spaces, *J. Fac. of Sci. Univ. of Tokyo*, 12, 81 - 192.

Tank, D.W. and J.J.Hopfield (1985) Simple 'neural' optimization networks: an A/D converter, signal decision circuit and a linear programming circuit, *IEEE Transactions, Circuits and Systems*, CAS-33, 533-541, preprint.

Thiele, L. (1986) On the Sensitivity of Linear State-Space Systems, *IEEE Trans. on Circuits and Systems*,CAS-33, 502-510.

Thom, R. (1949) Sur une partition en cellules associee anne fonction sur une varite, *C.R. Acad. Sci. Paris*, Series A-B, 228, 973-975.

Tomei, C. (1984) The topology of isospectral manifolds of tri-diagonal matrices, *Duke Math. Journal* 51, 981-996.

van Loan, C.F. (1985) How near is a stable matrix to an unstable matrix? *Linear Algebra and its Role in Systems Theory* (B.N. Datta, Ed.), Contemporary Math. 47, Amer. Math. Soc., 465-478.

Verriest, E.I. (1983) On generalized balanced realizations, *IEEE Transaction on Automatic Control* AC-28, 833-844.

Verriest, E.I. (1986) Model reduction via balancing, and connections with other methods, *Modelling and Approximation of Stochastic Processes* (U.B. Desai, Ed.), Kluwer Academic Publ., 123-154.

Verriest, E.I. (1988) Minimum sensitivity implementation for multi-mode systems, *Proc. IEEE Conf. on Decision and Control*, Austin, TX, 2165-2170.

Verriest, E.I. and W.S. Gray (1988) Robust design problems: A geometric approach, in *Linear Circuits, Systems and Signal Processing: Theory and Applications* (Byrnes, Martin and Saeks, eds.), North-Holland, Amsterdam, 321-328.

Veselov, A.P. (1992) Growth and integrability in the dynamics of mappings, *Commun. Math. Phys.* 145, 181-193.

Vladimirov, V.S. (1966) *Methods of the Theory of Functions of Many Complex Variables*, Cambridge, MA: MIT Press.

von Neumann, J. (1937) Some matrix-inequalities and metrization of matric-spaces, *Tomsk Univ. Rev.*, 1, 286-300. See also *John von Neumann Collected Works*, Vol. IV, (A.H. Taub Ed.) Pergamon Press, New York, 1962, 205-218.

Wang, X. (1991) On output feedback via Grassmannians, *SIAM J. Control and Optimization* 29, 926-935.

Watkins, D.S. (1984) Isospectral Flows, *SIAM Rev.*, 26, 379-391.

Watkins, D. (1982) Understanding the QR algorithm, *SIAM Rev.* 24, 379-391.

Watkins, D.S. and Elsner, L., (1988) Self-similar flows, *Linear Algebra and its Appl.* 110, 213-242.

Watkins, D. and L. Elsner (1989) Self-equivalent flows associated with the singular value decomposition, *SIAM J. Matrix Anal. Appl..* 10, 244-258.

Werner, J. (1992) *Numerische Mathematik 2*, Vieweg Verlag, Braunschweig.

Weyl, H. (1946) *The Classical Groups*, Princeton University Press, N.J.

Wilf, H.S. (1981) An algorithm-inspired proof of the spectral theorem in E^n, *Amer. Math. Monthly* 88, 49-50.

Willems, J.C. and W.H. Hesselink (1978) Generic properties of the pole placement problem, *Proc. IFAC World Congress*, Helsinki.

Williamson, D. (1986) A property of internally balanced and low noise structures, *IEEE Transactions on Automatic Control*, AC-31, 633-634.

Williamson, D. and R.E. Skelton (1989) Optimal q-Markov COVER for finite wordlength implementation, *Math. Systems Theory* 22, 253-273.

Wimmer, K. (1988) Extremal problems for Hölder norms of matrices and realizations of linear systems, *SIAM J. Matrix Anal. Appl.* 9, 314-322.

Witten, E. (1982) Supersymmetry and Morse theory, *Journal of Diff. Geometry*, 17, 661-692.

Wonham, W.M. (1967) On pole assignment in multi-input controllable linear systems, *IEEE Transactions on Automatic Control* AC-12, 660-665.

Wonham, W.M. (1985) *Linear Multivariable Control*, Third Edition, Springer-Verlag, New York

Wu, W.T. (1965) On critical sections of convex bodies, *Sci. Sinica* 14, 1721 - 1728

Yan, W.Y, U. Helmke and J.B. Moore (1993) Global analysis of Oja's flow for neural networks, *IEEE Trans. on Neural Networks*, to appear.

Yan, W., Moore, J.B. and U. Helmke (1993) Recursive algorithms for solving a class of nonlinear matrix equations with applications to certain sensitivity optimization problems, *SIAM J. of Control and Optimization*, to appear.

Youla, D.C. and P. Tissi (1966) N-port synthesis via reactance extraction - Part I, *IEEE Intern. Convention Record*, 183-205.

Yuille, A.L. (1990) Generalized deformable models, statistical physics and matching problems, *Neural Computation* 2, 1-24.

Zha, H. (1989a) A numerical algorithm for computing the restricted singular value decomposition of matrix triplets, *Scientific Report* 89-1, Konrad-Zuse Zentrum für Informationstechnik, Berlin.

Zha, H. (1989b) Restricted singular value decomposition of matrix triplets, *Scientific Report* 89-2, Konrad-Zuse Zentrum für Informationstechnik, Berlin.

Zeeman, E.C. (1980) Population dynamics from game theory, in *Global Theory of Dynamical Systems*, (Z. Nitecki and C. Robinson, Eds.), Lecture Notes in Mathematics 819, Springer Verlag Berlin, 471-497.

Author Index

Adamyan, 164
Aiyer, 128
Akin, 128
Ammar, 7, 40
Anderson, 39, 67, 202, 225, 264, 269, 287, 303, 304
Andronov, 39
Arnold, 79, 202
Arov, 164
Atiyah, 78, 127
Auchmuty, 19
Azad, 172, 208

Baldi, 164, 201
Barnett, 307
Batterson, 18
Bayer, 2, 108, 127
Beattie, 18
Bellman, 307
Bhatia, 40, 102
Bitmead, 225
Bloch, 2, 50, 60, 78, 79, 108, 127, 163
Bott, 39, 78
Brockett, 1, 2, 45, 50, 54, 67, 69, 78, 79, 109, 127, 151, 164, 257, 264
Bröcker, 102
Bucy, 40
Byrnes, 78, 79, 127, 163, 164, 225

Carrell, 171, 201
Chiang, 231
Chu, 1, 2, 30–32, 35, 36, 45, 69, 78, 79, 85, 103
Chua, 108, 128
Colonius, 78
Courant, 1
Crouch, 79

Dantzig, 127
David, 163

Davis, 78
de la Rocque Palis, 24
de Mari, 78
de Melo, 39
de Moor, 103, 163
Deift, 2, 30, 79
Demmel, 164
Dempster, 102
Dieudonné, 171, 201
Doolin, 39
Driessel, 45, 78, 79, 85, 103
Duistermaat, 53, 54, 62
Duncan, 225

Eckart, 142, 163
Elsner, 30, 35, 36

Faybusovich, 2, 79, 108, 117, 119, 127, 264
Fernando, 102
Fischer, 1
Flanders, 178, 202
Flaschka, 2, 60, 61, 78, 79
Fletcher, 164
Fogarty, 201
Fox, 19
Frankel, 77

Gamst, 331
Gardner, 201
Gauss, 1
Ge-Zhong, 79
Gelfand, 78
Gevers, 269, 303, 304
Ghosh, 164
Gibson, 352
Glover, 102
Godbout, 165
Gohberg, 102
Golub, 6, 18, 33, 35, 102, 103, 163
Goodwin, 268, 303

Gray, 221, 228
Grossmann, 79
Grötschel, 127
Guillemin, 202

Halmos, 164
Hamiltonian flow, 163
Hangan, 77
Hansen, 101, 102
Hardy, 58
Heath, 231
Helmke, 27, 69, 74, 79, 85, 103, 122,
    127, 134, 161, 163, 165, 202,
    205, 228, 257, 264, 267, 278,
    287, 303, 304
Henniart, 39
Hermann, 1, 39, 78, 164, 264
Herzel, 119, 127
Hesselink, 164
Higham, 139, 164
Hirsch, 20, 128, 328, 331, 358
Hitz, 264, 287
Hoffman, 163
Hopfield, 108, 128
Horn, 127, 141
Hornik, 164, 201
Hotelling, 102
Hueper, 79
Humphreys, 102
Hung, 102
Hwang, 207, 228, 299, 303

Irwin, 319
Isidori, 319, 331

Johnson, 141
Jonkeere, 228
Jordan, 165

Kahan, 35
Kailath, 164, 201, 228, 319, 362
Karmarkar, 2, 114, 127
Kempf, 1, 171, 208
Khachian, 2, 113, 127
Khaikin, 39
Kimura, 164
Klema, 102
Kliemann, 78
Knuth, 127
Kodama, 79

Kolk, 53, 54, 62
Kostant, 2, 60, 78, 127
Kraft, 171, 201
Krantz, 208
Krein, 102, 164
Krishnaprasad, 79, 264
Kung, 164
Ky Fan, 26

Lagarias, 2, 79, 108, 127
Lang, 362
Laub, 102, 231
Lawson, 101
Li, 79, 269, 303
Lin, 108, 128, 164
Littlewood, 58
Loeb, 172, 208
Lovász, 127
Luenberger, 127

MacFarlane, 102
Madievski, 304
Mahony, 69, 74, 79, 103, 122, 161, 165
Marsden, 79
Marshall, 40
Martin, 7, 39, 40, 164
Meggido, 127
Middleton, 268, 303
Milnor, 20, 39, 68, 77, 260
Mirsky, 163
Moore, 27, 39, 67, 69, 74, 79, 85, 102,
    103, 122, 207, 264, 267, 278,
    287, 303, 304
Moser, 60, 78, 79
Mullis, 102, 207, 228, 268, 294, 303
Mumford, 201
Munkres, 331

Nanda, 30, 79
Ness, 1, 171, 208
Nicholson, 102
Nossek, 79

Ober, 228
Oja, 22, 27, 102
Olkin, 40, 202
Oppenheim, 303

Paige, 231
Palais, 39
Palis, 39

AUTHOR INDEX

Parlett, 6, 7, 40, 164
Paul, 79
Pearson, 163
Peixoto, 39
Pernebo, 228, 264
Peterson, 128
Polya, 58
Pontryagin, 39
Poole, 6, 7, 40
Prechtel, 134, 163
Pressley, 78
Procesi, 78
Pyne, 108, 128

Ratiu, 2, 50, 78, 79
Recchioni, 119
Reid, 39
Reinsch, 102
Roberts, 102, 207, 228, 268, 294, 303
Rosenthal, 164
Rutishauser, 1, 30, 40

Safanov, 231
Schafer, 303
Schneider, 40
Schrijver, 127
Schur, 127
Schuster, 113, 128
Segal, 78
Serganova, 78
Shafer, 39
Shayman, 40, 77, 78, 134, 151, 163
Shub, 79, 127
Sigmund, 113, 128
Silverman, 228, 264
Sirat, 102
Skelton, 228
Skoog, 67
Slodowy, 172, 202
Smale, 39, 128, 328
Smilie, 18
Smith, 85, 103
Soederberg, 128
Sonnevend, 127
Sontag, 164, 228, 319
Spivak, 362

Sternberg, 202
Stewart, 163
Stoer, 127
Symes, 2, 30, 79

Takeuchi, 78
Tank, 108, 128
Thiele, 268, 281, 303
Thom, 39
Tissi, 67
Todd, 127
tom Dieck, 102
Tomei, 2, 30, 61, 78, 79

Van Loan, 6, 18, 33, 102, 163, 164
Varadarajan, 53, 54, 62
Vasquez, 79
Verriest, 221, 228, 257
Veselov, 79
Vitt, 39
Vladimirov, 208
von Neumann, 1, 102

Wang, 164
Ward, 231
Watkins, 30, 35, 36
Weyl, 201
Wilf, 79
Willems, 78, 127, 163, 164
Williamson, 228, 268, 303
Wimmer, 202
Witten, 39
Wolff, 113, 128
Wong, 45, 127
Wonham, 164
Wu, 77

Yan, 27, 79, 264, 267, 278, 287, 304
Youla, 67
Young, 142, 163
Yuille, 128

Zeeman, 113, 114, 128
Zha, 103
Zhao, 127
Zirilli, 119

Subject Index

C^∞-atlas, 337
L^2-sensitivity Gramians, 276
L^2-sensitivity balanced, 276
L^2-sensitivity balanced realization, 292
L^2-sensitivity measure, 268
L^2-sensitivity model reduction, 291
L^2-sensitivity optimal co-ordinate transformation, 273
N-Gramian norm, 221
N-balanced, 220
Z-transform, 320
α-limit set, 19, 327
ω-limit set, 19, 327, 361
n-dimensional manifold, 338
1-form, 354
2-norm, 313

adjoint orbits, 78
adjoint transformation, 350
Ado's theorem, 347
affine constraint, 37
algebraic group action, 171
approximation and control, 133
approximations by symmetric matrices, 134
artificial neural networks, 2, 108
asymptotically stable, 326
Azad-Loeb theorem, 210

Baker-Campbell-Hausdorff formula, 350
balanced matrix factorizations, 169
balanced realizations, 232, 323
balanced truncation, 207, 324
balancing transformation, 177, 184
balancing via gradient flows, 231
basic open sets, 331
basins of attraction, 74
basis, 331
Betti numbers, 260

bijective, 316
bilinear bundle, 354
bilinear systems, 227
binomial coefficient, 136
Birkhoff minimax theorem, 260
bisection method, 164

Cartan decomposition, 102
Cauchy index, 67
Cauchy-Maslov index, 225
Cayley-Hamilton theorem, 310
cell decomposition, 39
center manifold, 359
center manifold theorem, 359
chain rule, 335
characteristic polynomial, 310
chart, 337
Cholesky decomposition, 312
Cholesky factor, 312
closed, 331
closed graph theorem, 334
closed loop system, 153
closed orbit lemma, 148, 217, 352
closed orbits, 202
closure, 332
combinatorial assignment problems, 127
compact, 332
compact Lie groups, 46, 347
compact Stiefel manifold, 342
compact sublevel sets, 19, 178
compatibility condition, 16, 356
completely integrable gradient flows, 2
completely integrable Hamiltonian systems, 127
complex Grassmann manifold, 8, 341
complex homogeneous space, 211
complex Lie group, 348
complex projective n-space, 340

complex projective space, 7
complexity, 79
computational considerations, 76
computer vision, 78
condition number, 210, 314
congruence group action, 135
connected components, 135, 332
constant Riemannian metric, 151
constant step-size selection, 74
constrained L^2-sensitivity cost function, 302
constrained least squares, 37
continuation methods, 281
continuous map, 333
continuous-time, 319
contour integration, 271
controllability, 67, 321
controllability Gramian, 206, 321
convergence of gradient flows, 19, 360
convex constraint set, 117
convex functions, 208
coordinate charts, 8
cotangent bundle, 15, 354
Courant-Fischer minimax theorem, 14
covering, 332
critical point, 19, 336
critical value, 15, 336

dense, 332
detectability, 322
determinant, 308
diagonal balanced factorization, 169
diagonal balanced realization, 232
diagonal balancing flows, 199
diagonal balancing transformations, 186, 248
diagonal matrix, 308
diagonalizable, 311
diagonalization, 45
diagonally N-balanced, 221
diagonally balanced realization, 207, 324
diffeomorphism, 48, 335, 339
differential geometry, 1
digital filter design, 4
discrete integrable systems, 79
discrete Jacobi method, 79
discrete-time, 319
discrete-time balancing flows, 255

discrete-time dynamical system, 5
discretization, 196
dominant eigenspace, 9
dominant eigenvalue, 5
dominant eigenvector, 5
double bracket equation, 4, 45, 50, 85, 107
double bracket flow, 123, 158
dual flows, 240
dual vector space, 317
dynamic Riccati equation, 67
dynamical systems, 319

Eckart-Young-Mirsky theorem, 133, 163
eigenspace, 310
eigenstructure, 153
eigenvalue assignment, 4, 153
eigenvalue inequalities, 40
eigenvalues, 310
eigenvectors, 310
Eising distance, 258
equilibrium point, 19, 325
equivalence classes, 334
equivalence relation, 47, 334
essentially balanced, 297
Euclidean diagonal norm balanced, 258
Euclidean inner product, 16
Euclidean norm, 5, 258, 313
Euclidean norm balanced, 258
Euclidean norm balanced realizations, 273
Euclidean norm balancing, 223
Euclidean norm controllability Gramian, 258
Euclidean norm observability Gramian, 258
Euclidean norm optimal realizations, 257
Euclidean topology, 202
Euler characteristic, 61
Euler iteration, 77
exponential, 310
exponential convergence, 37, 73, 192
exponential map, 350
exponential rate of convergence, 17, 327
exponentially stable, 326

SUBJECT INDEX

Faybusovich flow, 118
feedback control, 4
feedback gain, 153
feedback preserving flow, 161
fiber, 316
fiber theorem, 25, 346
finite escape time, 13, 40, 68, 159
finite Gramians, 206
finite-word-length constraints, 268
first fundamental theorem of invariant theory, 201
first-order necessary condition, 336
fixed points, 72
flag, 62
flag manifolds, 62
flows on orthogonal matrices, 54
flows on the factors, 190
frequency response, 320
frequency shaped sensitivity minimization, 281
Frobenius norm, 46, 313
Fréchet derivative, 14, 334
function balanced realization, 214
function minimizing, 214

general linear group, 46
generalized Lie-bracket, 308
generalized Rayleigh quotient, 24
generic, 332
genericity assumption, 73
geometric invariant theory, 215, 228
global analysis, 331
global asymptotic stability, 326
global maximum, 336
global minimum, 137, 336
gradient flows, 15, 355
gradient vector field, 16, 19, 356
gradient-like behaviour, 149
gradient-like flow, 152
Gram-Schmidt orthogonalization, 29
graph, 351
Grassmann manifold, 7, 163
Grassmannians, 339
group, 347
group actions, 171

Hamiltonian matrix, 67, 349
Hamiltonian mechanics, 78
Hamiltonian realization, 224
Hamiltonian system, 2, 61

Hankel matrix, 164, 170, 201, 323
Hankel operators, 164
Hausdorff, 47, 332
Hebbian learning, 27
Hermitian, 307
Hermitian inner product, 317
Hermitian projection, 8
Hessenberg form, 59, 276
Hessenberg matrix, 59, 312
Hessenberg varieties, 78
Hessian, 19, 50, 53, 335, 344
high gain output feedback, 157
Hilbert-Mumford criterion, 202
holomorphic group action, 211, 352
homeomorphism, 48, 333
homogeneous coordinates, 8, 340
homogeneous space, 46, 49, 350
Householder transformations, 299
hyperbolic, 357

identity matrix, 308
image space, 309
imbedded submanifold, 352
imbedding, 333, 346
immersed submanifold, 346
immersion, 345
induced Riemannian metric, 15, 54, 55, 179, 190, 252, 272, 356
inequality constraints, 119
infinite dimensional spaces, 39
injective, 316
inner product, 52, 317
input, 153
input balancing transformation, 298
integrable Hamiltonian system, 30
integrable systems, 79
integral curve, 324
interior, 332
interior point algorithms, 113
interior point flows, 113
intrinsic Riemannian metric, 189, 300
invariant submanifold, 100
invariant theory, 171, 205
inverse eigenvalue problems, 45, 133
inverse function theorem, 335
isodynamical flows, 249, 279
isomorphism, 316
isospectral differential equation, 30
isospectral flow, 33, 45, 50, 85
isospectral manifold, 93

Iwasawa decomposition, 164

Jacobi identity, 348
Jacobi matrix, 59, 312, 335
Jordan algebra, 142

Kalman's realization theorem, 154, 207, 323
Karhunen-Loeve expansion, 102
Karmarkar algorithm, 113
Karmarkar's algorithms, 107
Kempf-Ness theorem, 176, 260
kernel, 309
Khachian's algorithms, 107
Killing form, 55, 97, 102
Kronecker product, 180, 314
Krylov-sequence, 5
Kähler structure, 163

La Salle's principle of invariance, 148, 328
Lagrange multiplier, 15, 302, 336
Landau-Lifshitz equation, 78
Laplace transform, 320
least squares estimation on a sphere, 37
least squares matching, 78, 127, 151
least squares optimization, 1
Lemma of Lyapunov, 315
Levi form, 209
Levi's problem, 208
Lie algebra, 46, 348
Lie bracket, 31, 45, 47, 308, 348
Lie group, 46, 347
Lie group action, 47, 350
Lie subgroup, 347
Lie-bracket recursion, 69
linear algebra, 307
linear algebraic group action, 171
linear cost function, 114
linear dynamical systems, 152, 319
linear index gradient flow, 179, 233
linear induced flow, 68
linear neural networks, 133
linear operators, 317
linear optimal control, 67
linear programming, 45, 107, 108
linearization, 17, 325
local coordinate system, 337
local diffeomorphism, 335, 339

local maximum, 336
local minimum, 137, 336
local stable manifold, 358
local unstable manifold, 358
locally compact, 332
locally invariant, 327
locally stable attractor, 19
lossless electrical network, 79
Lyapunov balanced realization, 291
Lyapunov equation, 196, 277
Lyapunov function, 148, 327
Lyapunov stability, 327

Markov parameters, 323
matching problems, 257
matrix, 307
matrix eigenvalue methods, 1
matrix exponential, 10
matrix inversion lemma, 122, 309
matrix least squares approximation, 45
matrix least squares index, 46
matrix logarithm, 11
matrix nearness problems, 164
matrix Riccati equation, 65
maximal compact, 347
McMillan degree, 67, 154, 323
micromagnetics, 78
minimal realizations, 323
mixed L^2/L^1 sensitivity, 269
mixed equality-inequality constraints, 117
model reduction, 102, 164, 207
moment map, 78, 202
Morse charts, 358
Morse function, 39, 68, 260, 336
Morse function on Grassmann manifolds, 77
Morse index, 78, 95, 345
Morse inequalities, 261, 275
Morse lemma, 20, 336
Morse theory, 39, 77, 260
Morse-Bott function, 18, 20, 26, 54, 98, 361
multi-mode systems, 228
multivariable sensitivity norm, 284

negatively invariant, 327
neighbourhood, 331
neural network, 22, 27, 102, 201

SUBJECT INDEX

Newton method, 3, 76
Newton-Raphson gradient flows, 244
noncompact Stiefel manifold, 27, 341
nondegenerate critical point, 336
nonlinear artificial neurons, 128
nonlinear constraint, 37
nonlinear programming, 127
norm balanced realization, 214
norm minimal, 214
normal bundle, 355
normal matrix, 173, 311
normal Riemannian metric, 52, 54, 94, 143, 147, 156, 190, 252, 280
normal space, 25
normally hyperbolic subset, 186
numerical analysis, 40
numerical integration, 79
numerical matrix eigenvalue methods, 40

observability, 322
observability Gramian, 206, 322
observable realization, 67
Oja flow, 22
one-parameter group, 325
open intervals, 332
open sets, 331
open submanifold, 338
optimal eigenvalue assignment, 160
optimal vertex, 114
optimally clustered, 220
orbit, 47, 171, 351
orbit closure lemma, 157
orbit space, 47, 351
orthogonal group, 46, 348
orthogonal Lie-bracket algorithm, 74
orthogonal matrix, 308
orthogonal projection, 24
orthogonally invariant, 172
output, 153
output feedback control, 152
output feedback equivalent, 153
output feedback group, 154
output feedback optimization, 153
output feedback orbit, 154

Padé approximation, 76, 124, 201
parallel processing, 2
pattern analysis, 78

perfect Morse-Bott function, 173
permutation matrix, 46, 56, 308
Pernebo and Silverman theorem, 324
phase portrait analysis, 79
plurisubharmonic functions, 208, 209
polar decomposition, 139, 150, 211, 312
pole and zero sensitivities, 281
pole placement, 153
polynomial invariants, 202
polytope, 116
population dynamics, 113, 128
positive definite approximant, 150
positive definite matrices, 178, 312
positive semidefinite matrices, 122, 149
positively invariant, 327
power method, 5, 9
principal component analysis, 27
product manifold, 338
product topology, 332
projected gradient flow, 58
projective spaces, 7, 339
proper function, 148
proper map, 333
pseudo-inverse, 37, 309

QR algorithm, 4, 29
QR-decomposition, 29, 313
QR-factorization, 29
quadratic convergence, 127
quadratic cost function, 114
quadratic index gradient flow, 181, 234
quantization, 45
quotient map, 333
quotient space, 334
quotient topology, 47, 333

rank, 309
rank preserving flow, 146, 147
rates of exponential convergence, 183
Rayleigh quotient, 14, 54, 110
Rayleigh quotient method, 18
real algebraic Lie group action, 155
real algebraic subgroups, 352
real analytic, 334
real Grassmann manifold, 9, 341
real projective n-space, 340
real projective space, 7

real symplectic group, 349
recursive L^2-sensitivity balancing, 287
recursive balancing matrix factorizations, 196
recursive flows on orthogonal matrices, 74
recursive linear programming, 122
reduction principle, 360
reductive groups, 202
regular values, 336
research problems, 152, 264
Riccati equation, 3, 10, 27, 59, 62, 147, 189, 236, 288
Riccati-like algorithms, 303
Riemann sphere, 7
Riemann surface, 62
Riemannian manifold, 15, 355
Riemannian metric, 117, 120, 355
roundoff noise, 293
row rank, 316
row space, 316
Runge-Kutta, 33, 287

saddle point, 21, 336
scaling constraints, 294
Schur complement formula, 141
Schur form, 312
Schur-Horn theorem, 127
second-order sufficient condition, 337
section, 353
self equivalent differential equation, 35
self-equivalent flow, 35, 87, 91, 103
selfadjoint, 317
semialgebraic Lie subgroup, 352
sensitivity minimization with constraints, 293
sensitivity of Markov parameters, 283
sensitivity optimal controller design, 304
sensitivity optimal singular systems, 228
sensitivity optimization, 267
Shahshahani metric, 128
shift strategies, 4
sign matrix, 57, 308
signal processing, 4
signature, 312
signature symmetric realization, 224

signature-symmetric balanced realization, 227
signatures, 11
similar matrices, 311
similarity action, 171, 251
similarity orbit, 270
similarity transformation, 311
simplex, 109
simulations, 75, 125, 235, 242
simultaneous eigenvalue assignment, 153
singular value decomposition, 4, 34, 85, 88, 169, 313
singular values, 34, 313
singular vectors, 34
skew Hermitian, 307
skew symmetric, 307
smooth, 334
smooth manifolds, 337
smooth semialgebraic action, 352
sorting, 45, 107
sorting algorithms, 127
sorting of function values, 79
special linear group, 46, 349
special orthogonal group, 349
special unitary group, 349
spectral theorem, 8
spheres, 339
stability, 326
stability theory, 73
stabilizability, 321
stabilizer, 171
stabilizer subgroup, 48, 351
stable, 320, 326
stable manifold, 186, 357
stable manifold theorem, 358
stable points, 215
standard Euclidean inner product, 317
standard Hermitian inner product, 318
standard inner product, 143
standard Riemannian metric, 16, 97
standard simplex, 108
state vector, 153
steepest descent, 88
step-size selection, 70, 197
Stiefel manifold, 24, 54, 90
strict Lyapunov function, 327

SUBJECT INDEX

strictly plush, 209
strictly proper, 323
structural stability, 24, 40, 79
subgroup, 347
subharmonic, 208
subimmersion, 345
submanifold, 345
submersion, 345
subspace, 315
subspace learning, 78
subspace topology, 332
surjective, 316
symmetric, 307
symmetric approximant, 138
symmetric projection operators, 63
symmetric realization, 67, 157, 224
symplectic geometry, 2, 78, 127, 202
system balancing, 205
systems with symmetries, 153

tangency condition, 16, 356
tangent bundle, 353
tangent map, 343, 354
tangent space, 25, 49, 343
tangent vector, 343
Taylor series, 335
Taylor's theorem, 70
Toda flow, 2, 30, 59
Toda lattice equation, 61
Toeplitz matrix, 164
topological closure, 174
topological space, 331
topology, 331
torus actions, 2
torus varieties, 78
total L^2-sensitivity function, 271
total least squares, 163
total least squares estimation, 133

total linear least squares, 133
trace, 310
trace constraint, 298
transfer function, 154, 320
transition function, 337
transitive, 47
transitive action, 351
transpose, 307
travelling salesman problem, 45, 128
trivial vector bundle, 353
truncated singular value decomposition, 133, 163

uniform complete controllability, 322
unit sphere, 14
unitarily invariant, 211
unitarily invariant matrix norms, 102
unitary group, 46, 348
unitary matrix, 308
universal covering space, 78
unstable manifold, 39, 53, 357
upper triangular, 31

variable step size selection, 74
vec, 314
vec operation, 180
vector bundle, 353
vector field, 324, 354
vector spaces, 315
VLSI, 79
Volterra-Lotka equation, 113, 128

weak Lyapunov function, 328
Weierstrass theorem, 333
Wielandt's minmax principle, 40
Wielandt-Hoffman inequality, 54, 78, 103

Zariski topology, 202